CRYSTALLOGRAPHY AND CRYSTAL CHEMISTRY OF MATERIALS WITH LAYERED STRUCTURES

PHYSICS AND CHEMISTRY OF MATERIALS
WITH LAYERED STRUCTURES

VOLUME 2

CRYSTALLOGRAPHY
AND CRYSTAL CHEMISTRY
OF MATERIALS
WITH LAYERED STRUCTURES

Edited by

F. LÉVY

Cavendish Laboratory, University of Cambridge, U.K.

D. REIDEL PUBLISHING COMPANY

DORDRECHT-HOLLAND / BOSTON-U.S.A.

Library of Congress Cataloging in Publication Data

Main entry under title:

Crystallography and crystal chemistry of materials with
    layered structures.

    (Physics and chemistry of materials with layered
strucures ; v. 2)
    Includes bibliographical references and indexes.
    1. Layer structure (Solids)—Addresses, essays,
lectures. 2. Solid state chemistry—Addresses, essays,
lectures. I. Lévy, Francis Alain, 1940–   II. Series.
QD478.P47   530.4'ls 75–43882
ISBN–13:978-94-010-1435-9   e-ISBN–13:978-94-010-1433-5
DOI: 10.1007/978-94-010-1433-5

Published by D. Reidel Publishing Company
P.O. Box 17, Dordrecht, Holland

Sold and distributed in the U.S.A., Canada, and Mexico
by D. Reidel Publishing Company, Inc.
Lincoln Building, 160 Old Derby Street, Hingham, Mass. 02043, U.S.A.

TABLE OF CONTENTS

PREFACE   VII

A. A. BALCHIN / Growth and the Crystal Characteristics of Dichalcogenides
Having Layer Structures   1

P. M. WILLIAMS / Phase Transitions and Charge Density Waves in the Layered
Transition Metal Dichalcogenides   51

W. BRONGER / The Layered Structures of Ternary Chalcogenides with Alkali
and Transition Metals   93

R. J. D. TILLEY / Structural Aspects of Non-Stoichiometry in Materials with
Layered Structures   127

S. CAILLÈRE and S. HÉNIN / Physical and Chemical Properties of
Phyllosilicates   185

G. C. TRIGUNAYAT and AJIT RAM VERMA / Polytypism and Stacking Faults
in Crystals with Layer Structure   269

S. S. MAJOR, JR. / Thermal Behavior of Stacking Faults   341

INDEX OF NAMES   359

INDEX OF SUBJECTS   363

INDEX OF FORMULAS   366

ABBREVIATIONS   370

PREFACE

In the last ten years, the chemistry and physics of materials with layered structures became an intensively investigated field in the study of the solid state. Research into physical properties of these crystals and especially investigations of their physical anisotropy related to the structural anisotropy has led to remarkable and perplexing results. Most of the layered materials exist in several polytypic modifications and can include stacking faults. The crystal structures are therefore complex and it became apparent that there was a great need for a review of the crystallographic data of materials approximating two-dimensional solids.

This second volume in the series 'Physics and Chemistry of Materials with Layered Structures' has been written by specialists of different classes of layered materials. Structural data are reviewed and the most important relations between the structure and the chemical and physical properties are emphasized.

The first three contributions are devoted to the transition metal dichalcogenides whose physical properties have been investigated in detail. The crystallographic data and crystal growth conditions are presented in the first paper. The second paper constitutes an incisive review of the phase transformations and charge density waves which have been observed in the metallic dichalcogenides.

In two contributions the layered structures of newer ternary compounds are described and the connection between structure and non-stoichiometry is discussed.

The article on the structural chemistry of phyllosilicates represents a very valuable introduction into a vast field of poorly known materials. It may well serve as stimulation for new interests in research.

The last contribution reviews the current knowledge of polytypism and stacking faults in layered materials.

Because of its contents, the present volume is a reference book rather than a textbook and is intended for scientists actively engaged in research. Although it was not entirely compiled by crystallographers it is of interest to them and encourages precise determination of structures which are necessary to arrive at a comprehensive interpretation of the physical properties of layered materials.

The systematic structural chemistry of layered materials has outgrown the present volume, for which it was originally planned and, therefore will constitute a new volume by itself.

Lausanne, 1976   F. LÉVY

GROWTH AND THE CRYSTAL CHARACTERISTICS OF DICHALCOGENIDES HAVING LAYER STRUCTURES

A. A. BALCHIN

Crystallography Laboratory, Dept. of Applied Physics, Brighton Polytechnic, Moulsecoomb, Brighton, BN2 4GJ, England

1. Basic Structures

Of the 101 compounds which are listed in Landolt-Bornstein [1] or Wyckoff [2] as having layer like structures related to those of cadmium iodide, cadmium chloride or molybdenum disulphide, thirty are dichalcogenides, i.e. metallic disulphides, diselenides or ditellurides. They comprise the compounds listed in Table I. The remainder essentially are the di-halides and di-hydroxides of the same metals.

Tubbs [3] has reviewed the optical properties and chemical decomposition of halides with layer structures, and the structural, optical and electronic properties of the transition metal dichalcogenides are extensively reviewed by Wilson and Yoffe [4]. A description of the preparation and properties of some of the Group $IV-VI_2$ chalcogenides having the cadmium iodide layer structure has been given by Greenaway and Nitsche [5].

Wyckoff distinguishes between the cadmium iodide structure and the cadmium chloride structure; the cadmium iodide structure has an hexagonal unit cell containing a single molecule, with cations at special positions $(0, 0, 0)$ and anions at positions $(\frac{1}{3}, \frac{2}{3}, u)$ and $(\frac{2}{3}, \frac{1}{3}, -u)$ of space-group $P\bar{3}m1$, where u is approximately $\frac{1}{4}$; the cadmium chloride structure is described as having a rhombohedral unit cell containing a single molecule, with the cations at special positions $(0, 0, 0)$ and the anions at the points (u, u, u) and $(-u, -u, -u)$ of space-group $R\bar{3}m$.

The cadmium iodide lattice (Figure 1) has an almost perfect hexagonal close pack-

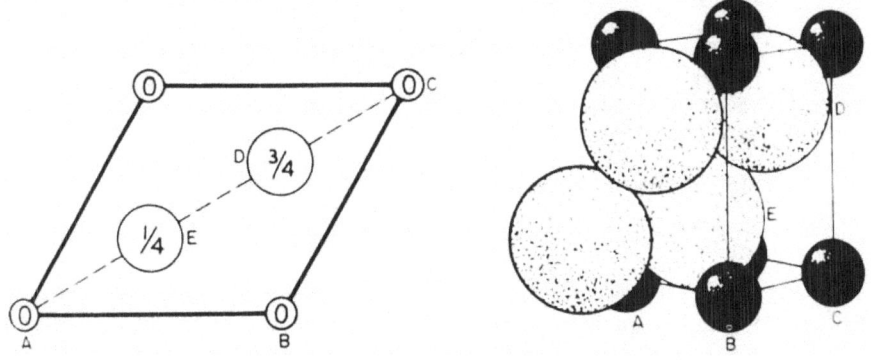

Fig. 1. The basal plan and the stacking arrangement in the hexagonal unit cell of cadmium iodide. (Reproduced, with permission, from Wyckoff, R. G., *Crystal Structures*, Interscience Publishers Ltd.).

F. Lévy (ed.), Crystallography and Crystal Chemistry of Materials with Layered Structures. 1–50. All Rights Reserved.
Copyright © 1976 by D. Reidel Publishing Company, Dordrecht-Holland.

A. A. BALCHIN

TABLE I

	$a_0(\text{Å})$	$c_0(\text{Å})$
(a) Chalcogenides having the cadmium iodide layer structure		
$CoTe_2$	3.784	5.403
HfS_2	3.635	5.837
$HfSe_2$	3.748	6.159
$IrTe_2$	3.93	5.393
$NiTe_2$	3.861	5.297
$PdTe_2$	4.0365	5.1262
PtS_2	3.537	5.019
$PtSe_2$	3.724	5.062
$PtTe_2$	4.010	5.201
$RhTe_2$	3.92	5.41
$SiTe_2$	4.28	6.71
SnS_2	3.639	5.868
$SnSSe$	3.716	6.050
$SnSe_2$	3.811	6.137
$\alpha\text{-}TaS_2$	3.35	5.86
TiS_2	3.412	5.695
$TiSe_2$	3.541	5.986
$TiTe_2$	3.757	6.513
Tl_2S		
VSe_2 ($VSe_{1.62}$–$VSe_{1.97}$)		
ZrS_2	3.662	5.813
$ZrSe_2$	3.771	6.138–6.149
$ZrTe_2$	3.950	6.630
(b) Chalcogenides having the cadmium chloride layer structure		
$\gamma\text{-}TaS_2$	3.32	18.29
$\beta\text{-}TaSe_2$	3.428	17.100
$NbTe_2$	10.904	19.88
$MoTe_2$	10.904	20.075
(c) Chalcogenides having the molybdenum disulphide layer structure		
MoS_2	3.1604	12.295
WS_2	3.18	12.5
WSe_2	3.280	12.950
$MoTe_2$	3.5182	13.9736
$\alpha\text{-}NbSe_2$	3.449	12.998
$\beta\text{-}NbSe_2$	3.439	25.188
$\alpha\text{-}TaSe_2$	3.431	12.737
$MoSe_2$	3.288	12.900
MoS_2 (rhombohedral form)		
	3.16	18.45
NbS_2 (rhombohedral form)		
	3.33	17.91

ing of anions with the smaller cations nestled in octahedral interstices between alternate layers of the anions. Only half of the octahedral interstices are filled. This produces strongly bonded sandwiches of anions which are held together only by long range van der Waal's forces. The result is that the atomic lattice is highly anisotropic, with easy cleavage and extended growth perpendicular to the unique hexagonal crystallographic axis. The sandwiches themselves are strongly, partially covalently bonded. The intralayer forces are at least one hundred times greater than the inter-layer forces [6]

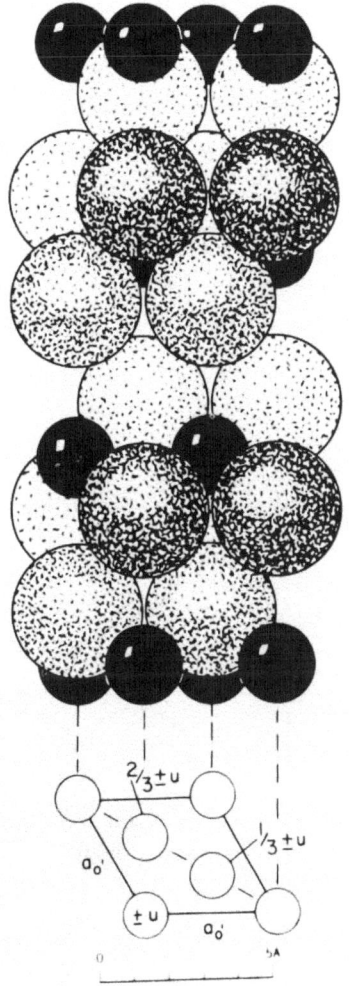

Fig. 2. The stacking of the cadmium chloride atomic arrangement. (Reproduced, with permission, from Wyckoff, R. G., *Crystal Structures*, Interscience Publishers Ltd.).

The octahedral co-ordination of cation by anion is retained in the structure of cadmium chloride (Figure 2), but here the close packing of the anions is cubic rather than hexagonal, and the octahedral cation co-ordination polyhedra, which in both this and cadmium iodide are joined in sheets perpendicular to c by sharing anions in common, are in differing relative orientations. In the iodide lattice the sheets are stacked one above the other, with the octahedra aligned parallel to a single direction.

4   A. A. BALCHIN

In the chloride the stacking is staggered and the octahedra are in two alternating alignments. These two modifications of the basic layer lattice lead to a large number of possible stacking arrangements, and these compounds have many polymorphs and polytypes.

A few compounds, notably molybdenum disulphide and niobium disulphide adopt a third layer lattice structure in which the cation is in trigonal prismatic coordination between anion sheets. The normal structure of molybdenum disulphide (Figure 3a) is hexagonal, with two molecules per unit cell. Atoms are in special positions of space group $P6_3/mmc$ with cations at $\pm(\frac{1}{3}, \frac{2}{3}, \frac{1}{4})$ and anions at $\pm(\frac{1}{3}, \frac{2}{3}, u)$, $\pm(\frac{2}{3}, \frac{1}{3}, \frac{1}{2}+u)$ where u for molybdenum disulphide is 0.629. A second form of molybdenum disulphide (Figure 3b) is known to be rhombohedral in space-group $R3m$ with atoms at sites $(0, 0, 0)$ – molybdenum – and $(0, 0, \frac{1}{12})$, $(0, 0, \frac{5}{12})$ – sulphurs. Niobium disulphide has assigned to it this structure.

Since all of these atomic arrangements are based upon nets of close-packed anions, many variations of stacking order may arise. The close-packing of anions is preserved,

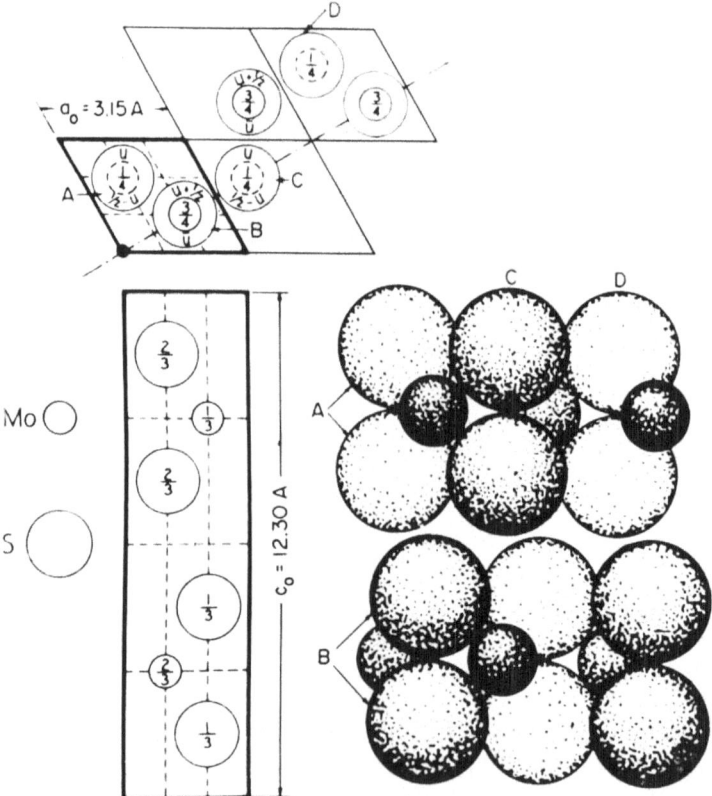

Fig. 3a–b.   (a) Stacking diagram of the normal molybdenum disulphide structure on to the sides of the unit cell, and the projections of the structure. (Reproduced, with permission, from Wyckoff, R. G., *Crystal Structures*, Interscience Publishers Ltd.).

but cation coordination may vary from octahedral (as in cadmium iodide and chloride) to trigonal prismatic (as in molybdenum disulphide). Layers having different cation coordination may be interleaved in many different ways, forming unit cells many layers high. In polytypic cadmium iodide itself the number of layers forming a repeat unit, with anion stacking varying from hexagonal to cubic and back again, may be as high as 120. In the dichalcogenides polytypism is not so common, and changes of stacking are usually associated with polymorphic phase changes. The largest number

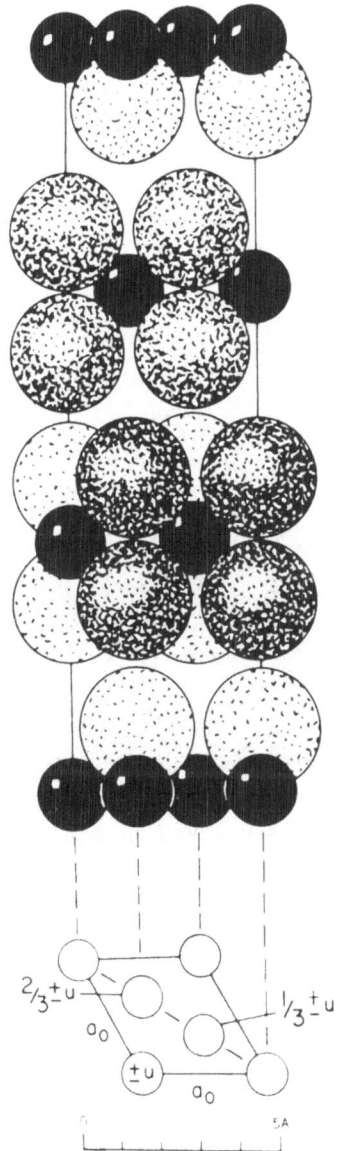

Fig. 3a–b. (b) Stacking diagram of the rhombohedral form of molybdenum disulphide. (Reproduced, with permission, from Wyckoff, R. G., *Crystal Structures*, Interscience Publishers, Ltd.).

of layers forming a repeat unit is six. Figure 4 shows the stacking of a number of polymorphs of various dichalcogenide materials, in which variations of cation coordination from octahedral to trigonal prismatic may be recognised from sections taken of the unit cells through the (11$\bar{2}$0) plane.

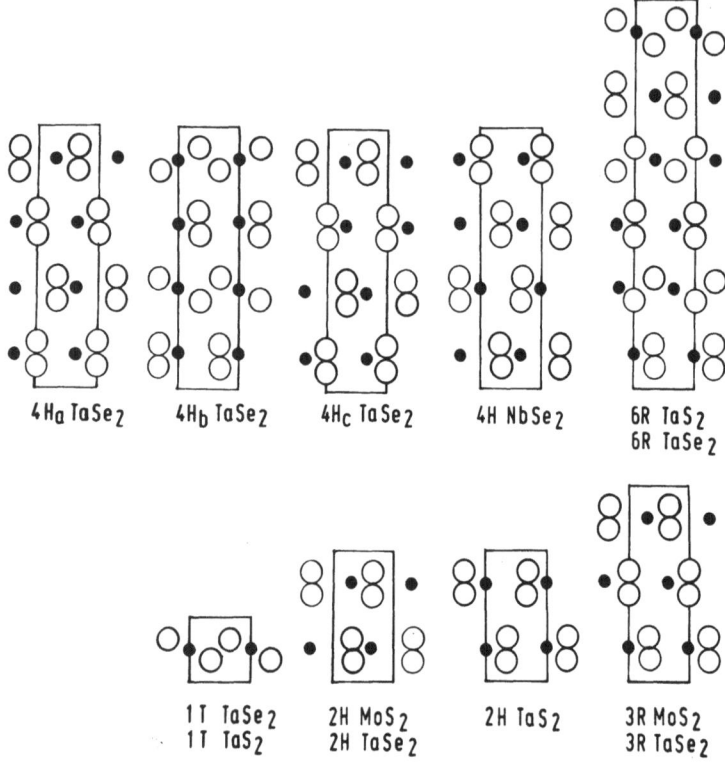

Fig. 4.   Sections through (11$\bar{2}$0) planes of polymorphic or polytypic phases, showing variations of octahedral and trigonal prismatic stacking.

A good example of polymorphism is offered by tantalum diselenide. The four-layer polymorph of tantalum diselenide $4s(c)$–$TaSe_2$, (Huisman and Jellinek [7]) has the trigonal prismatic coordination also found in $2s(a)$–$TaSe_2$, $3s$–$TaSe_2$, and $4s(a)$–$TaSe_2$. In $1s$–$TaSe_2$ the metal coordination is octahedral and in $4s(b)$–$TaSe_2$ and $6s$–$TaSe_2$ the two types of coordination coexist. Octahedrally coordinated $TaSe_2$ is stable above 880 °C, trigonal coordination is stable below 800 °C and at intermediate temperatures the two coordinations can co-exist as a single phase. Transitions between $1s$, $6s$, and $2s$, and between $1s$, $4s(b)$ and $4s(a)$ are reversible, although high temperature forms can be retained at room temperature by quenching. In $TaSe_2$ the $3s$ form is metastable, but can be obtained by heating quenched $1s$ or $6s$–$TaSe_2$.

2. Bonding in Layer Structures

Evans [8] discusses the occurrence of layered lattices such as cadmium iodide and cadmium chloride as a transition between ionic and molecular compounds brought about by the increasing polarisation of the anions by the cations. The influence of the octahedral crystalline field upon the d-orbitals is described by Hulliger [9], for the transition metal dichalcogenides. He points out that compounds with d^0 and d^6 cations would be expected to be non-metallic, whereas in fact most transition metal dichalcogenides, having $d^0, d^1, d^5, d^6, d^7, d^8$, cations, are metallic or semiconducting, sometimes (as in the case of α-TaS$_2$ or PdTe$_2$) superconducting. Trigonal prismatic co-ordination, involving d^4sp cationic bonds, is commonly met if the cation carries only two or less d-electrons. In compounds with this co-ordination, d^1 compounds are expected to be metallic and Pauli-paramagnetic, or diamagnetic, whereas d^2 cations would give rise to a diamagnetic semi-conductor. Hulliger predicts [10, 11] that transition metal layer compounds will include insulators, semi-conductors, semi-metals or even superconductors.

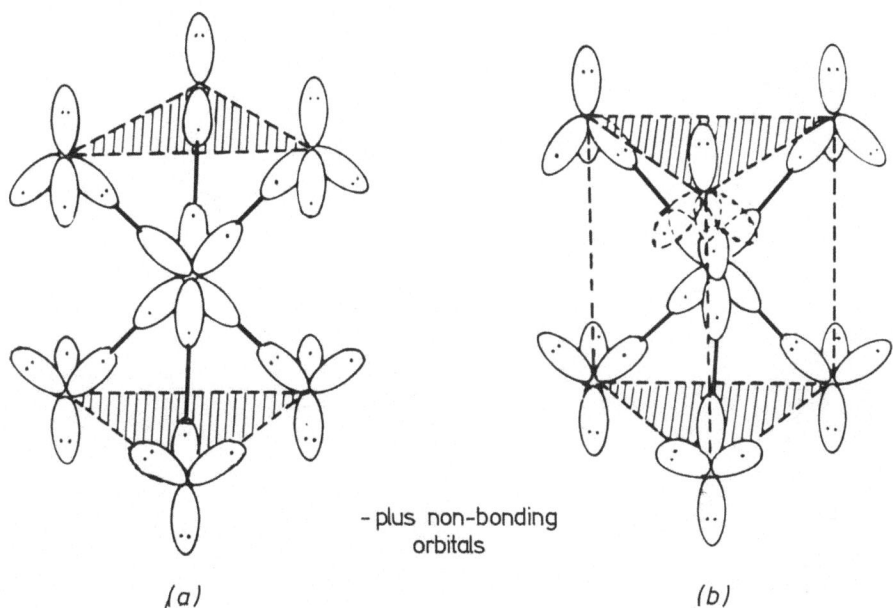

- plus non-bonding
orbitals

(a)   (b)

Fig. 5.   Occupation of bonding orbitals on the simple valency model. (a) Octahedral co-ordination.
(b) Trigonal prismatic co-ordination. (After Wilson and Yoffe).

The band structure of semiconducting transition metal and rare earth compounds depends on the radial extension of the wavefunctions, and their symmetry upon the distance and configurations of the neighbouring atoms, and upon the degree of covalent overlap of the anion lattice. Wilson and Yoffe [4] discuss a valency bond arrangement which gives a banding scheme consistent with the observed electrical, magnetic, and structural properties of the whole TX$_2$ dichalcogenide family. The valency bond picture requires both hybridisation and resonance. Figure 5 represents diagrammatically the arrangement of electrons in the bonded directions for the two types of layer dichalcogenide sandwich. Each chalcogenide atom puts a lone pair of electrons into the van der Waal's region and each metal atom then needs to supply four electrons for the bonding states to be completely full. (in ZrS_2 these would be s^2d^2 and in SnS_2, s^2p^2). Any further electrons (MoS_2, has s^2d^4) must enter the non bonding orbitals, which may accommodate up to six electrons, as in PtS_2, s^2d^8. The semi-conductivity of octahedral ZrS_2 and PtS_2 then follows, while that of the trigonal MoS_2 family is seen to be a consequence of the adoption of the trigonal-prismatic structure. It is pointed out, however, that the hybridisation of p and d states involved in the bonding scheme is proportionally much less than would be implied from the use of six equivalent metal hybrid bonding functions of the octahedral formulation sp^3d^2 or for the trigonal prismatic d^4sp hybridisation. In $ZrSe_2$, for example, the zirconium

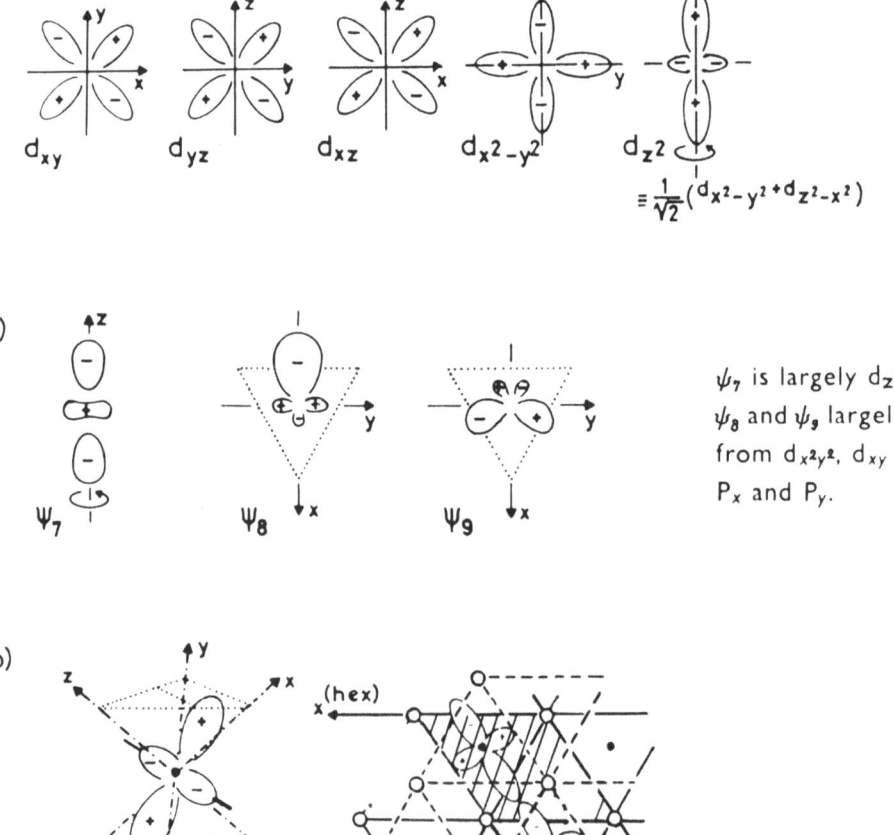

Fig. 6. Shape and orientation of non-bonding orbitals in the TX_2 layer structures. (a) Trigonal prismatic co-ordination. (b) Octahedral co-ordination. (After Wilson and Yoffe).

$5p$ states are almost completely confined to the conduction band, while in $SnSe_2$ the tin $5d$ states show virtually no mixing as occupied states into the valence band.

The non-bonding orbitals required to accommodate electrons not involved in bonding arise largely from hybridisation of metal d and p states. These are summarised in Figure 6. The five d-orbitals and the three p-orbitals combine to give wavefunctions ψ_7, made up largely of d_{z^2}, ψ_8 made up largely from $d_{(x^2-y^2)}$ and p_x, and ψ_9 made up largely from d_{xy}, p_x, and p_y. In octahedral co-ordination ψ_7, ψ_8, ψ_9 are degenerate, and the lobes bisect the bond-angles, giving good metal-metal overlap in the layers. For the trigonal prism, ψ_7 yields poor metal-metal overlap and ψ_8 and ψ_9 are degenerate.

For group IVA metals in octahedral co-ordination a complete electronic band structure has been calculated by Murray *et al.* [12]. The results are in good agreement with data derived from optical reflectivity and absorption data.

3. Electrical Properties

The electronic energy band structures of many of the dichalcogenide layer compounds are now known in detail, having been derived either by calculation or by measurement of optical properties of the materials. Many are potentially useful semi-conductors or super-conductors. Reflection spectra and energy band structures have been measured [13] for molybdenum disulphide itself, for molybdenum diselenide and molybdenum ditelluride, and for tungsten diselenide and rhenium diselenide, and although complex, have been correlated with published semi-quantitative calculated band structures. Theoretical band structures have been calculated for molybdenum disulphide [14, 15]. Spectra of molybdenum disulphide intercalated with alkali metals in the van der Waals gap between the sulphur layers have also been interpreted in terms of the band structure of MoS_2 [16], and sodium and potassium intercalates of MoS_2 are reported to be superconducting [17], with transition temperatures of the order of 1.3 K for the sodium intercalated material, and approximately 4.5 K for the potassium intercalate. Optical and structural studies of molybdenum diselenide made by Evans and Hazelwood [18] permit comparison of the measured activation energy of intrinsic electrical conduction with the direct optical band gap deduced from exciton spectra. Brixner [19] from an investigation of electrical conductivity, thermal conductivity and thermoelectric properties of crystals grown by vapour transport reactions, classifies molybdenum diselenide, molybdenum ditelluride and tungsten diselenide as semi-conductors, whereas niobium diselenide, tantalum diselenide, niobium ditelluride and tantalum ditelluride are metallic conductors.

The system niobium-sulphur [20] contains a number of layered phases, viz. NbS_3 (a layered diamagnetic semi-conductor), $2s-NbS_2$, $3s-NbS_2$, and $2s-Nb_{(1+x)}S_2$. These latter are metallic with nearly temperature independent paramagnetism.

In the niobium-selenium system [21] four polymorphs of $NbSe_2$ exist, viz $1s-NbSe_2$ stable above 980 °C, $4s(d)-NbSe_2$ stable between 980 °C and 910 °C and the $2s(a)$ and the $4s(a)$ forms stable at lower temperatures. All polymorphs have hexagonal or trigonal layer structures. In $1s-NbSe_2$ the metal has octahedral co-ordination, in the $2s(a)$ and the $4s(a)$ forms the stacking is trigonal prismatic, while in $4s(d)-NbSe_2$ probably layers with the two types of coordination co-exist (see Figure 7).

Kalikhman and Duksina [22] have synthesised a non-stoichiometric phase $Nb_{(1+x)}Se_2$ by a method based on the removal of Se from $NbSe_2$. A like phase, studied by Huisman, *et al.* [23] (in conjunction with the similar phase $Ta_{(1+x)}Se_2$ in the tantalum-selenium system) has been shown to have a structure built up from the

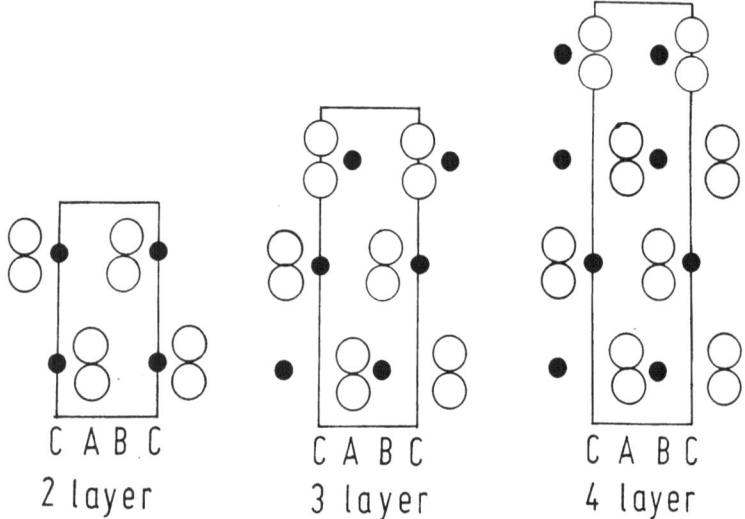

Fig. 7.   Three different layer stackings for niobium diselenide, for which niobium is in trigonal prismatic co-ordination. Sections through (11$\bar{2}$0) planes. (After Antonova *et al.*).

normal MSe_2 layers with the metal in trigonal prismatic co-ordination. The x additional metal atoms lie in octahedral holes between the layers. Both niobium and tantalum phases $M_{(1+x)}Se_2$ are metallic with nearly temperature independent paramagnetism.

Superconductivity in the double layer modification of niobium diselenide has been studied by Antonova *et al.* [24, 27]. The composition interval between 31.7 and 34.3 atomic per cent niobium is non-stoichiometric. The magnetic properties and critical currents of niobium diselenide are related to the anisotropy of the crystal lattice, and its regular disordering due to non-stoichiometry. Strong anisotropy of magnetic properties and critical currents is observed, depending on the direction of the magnetic field relative to the crystal planes. There is a most favourable direction for superconduction in the material, passing through the niobium planes. The degree of occupancy of the niobium and selenium sub-lattices is closely related to the critical superconducting transition temperature, and in niobium diselenide, superconductivity is predominantly planar in origin. Superconducting $NbSe_2$ has already found a use in the construction of weak link loop devices (SQUID) [28].

The optical properties of $\varepsilon - NbSe_2$, $\zeta - NbSe_2$, $\varepsilon - Nb_{1.04}Se_2$. $\zeta - Nb_{1.04}Se_2$ have been studied by Myers and Montet [25]. The most notable feature of their study is the similarity of the gross features of the visible and near infra-red absorption spectra for all four compounds, indicating that the absorption is governed by a mechanism operating largely within the $Se - Nb - Se$ layers, and is relatively independent of the structure or the presence of interstitial excess atoms.

Anisotropy in the resistivity of niobium diselenide has been observed by Edwards and Frindt [26]. Resistivity was measured by a four-probe technique across the layers parallel to c. The temperature-dependent part of the resistivity, $\varrho'\|c$, varies linearly with temperature from 300 K to 80 K and has a T^6 dependence below 60 K. The ratios of the resistivities $\varrho'\|c / \varrho'\perp c$ are constant at 31:1 from 300 K to 80 K but then fall linearly with temperature.

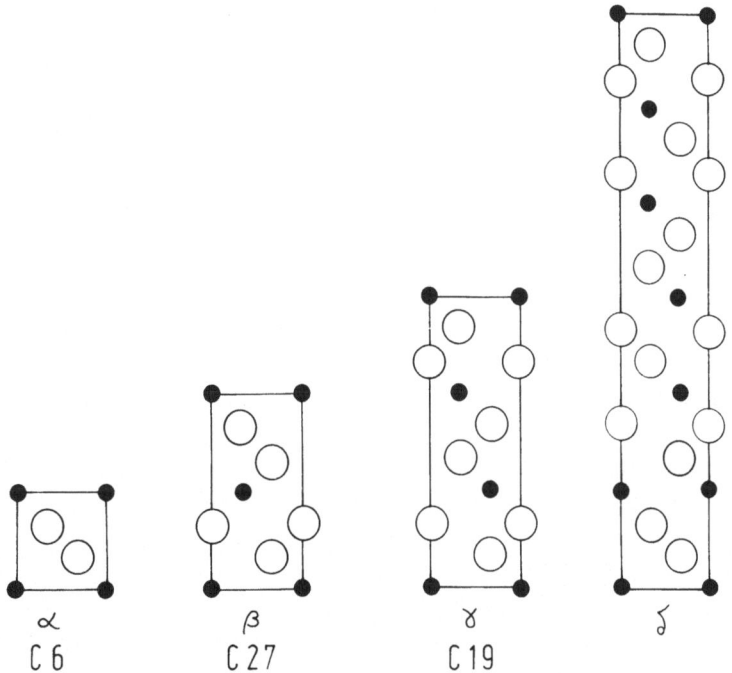

Fig. 8.   Sections through the (11$\bar{2}$0) planes of tantalum disulphide. (After Hahn and Schonberg).

Tantalum disulphide exists in the four crystallographic modifications shown in Figure 8. (Hahn and Schonberg [29]). In its 1T modification it is a material which is metallic at room temperatures, but reverts to semi-conduction at low temperatures. The transition between the metallic state and the semi-conducting state has been studied as a function of temperature by Thompson *et al.* [36]. The electrical properties of $1T - TaS_2$ indicate two-phase transformations at temperatures of 190 K and 348 K. As the temperature increases between these limits the electrical resistivity decreases by an order of magnitude. The higher temperature transition is a semi-conductor – metallic transition through which the resistance decreases by a factor of

two. A third transition exists at 315 K. Chu and Huang [31] find that the semi-con-
ductor-metal transition is suppressed linearly by compression up to a pressure of
about 15 k bars with $\Delta T_0 / \Delta p = (3.0 \pm 0.2)$ K/k bar. A two band model with a small but
temperature and pressure dependent overlap is required to explain these observations.

At liquid helium temperatures intercalated tantalum disulphide is a super-conductor,
and diamagnetic anisotropy associated with the transition to the superconducting
state is seen above 3.5 K [32]. The anisotropy grows inversely with temperature obey-
ing roughly a diamagnetic Curie Law. It seems plausible to ascribe it to electron correlations which at lower temperature are responsible for superconduction. Measurement
of the superconducting heat capacity anomaly [33] as a function of layer spacing in
intercalated TaS_2 shows it to be independent of the layer spacing if this lies within
the range 3 to 30 Å.

Magnetic and crystallographic studies, made by van Laar et al. [34] on the com-
pounds Me_xNbS_2 and Me_xTaSe_2 where $x = \frac{1}{3}$ or $\frac{1}{4}$ and Me is a transition metal, can
best be described in terms of an orderly distribution of transition metal atoms over
the available octahedral holes between adjacent sulphur layers of a $2s - NbS_2$ structure (see Figure 9). Where Me = Mn they found a ferromagnetic ordering of magnetic
moments over a range of temperature, whereas for $Fe_{1/3}NbS_2$ the ordering was antiferromagnetic.

Low temperature magnetic studies of niobium diselenide containing first row transition metals have been made [35] from liquid helium to room temperature. All transition metals studied (V, Cr, Mn, Fe, Co, Ni) showed a localised magnetic moment,
except for Co, which showed a temperature independent magnetic susceptibility.

Ehrenfreund et al. [36] query the antiferromagnetic ordering inferred for niobium
diselenide and tantalum diselenide. They point out that these are deduced from
magnetic susceptibility maxima, Hall coefficient sign reversals and from a kink in the
resistivity/temperature curve (at 110 K for $NbSe_2$ and at 70 K for $TaSe_2$). From their
own study of nuclear magnetic resonance of ^{93}Nb in niobium diselenide and of ^{77}Se
in tantalum diselenide above and below these temperatures, and from the observation
of resonances between 4.2 K and 300 K, they infer that the materials are not magneti-
cally ordered. In niobium diselenide an axially symmetric ^{93}Nb quadrupole powder
pattern is observed at 77 K and 300 K, but at 4.2 K this is found to be further split,
implying that the transitions are instead a low temperature structural distortion. This
distortion could be responsible for the previously observed anomalies and for the
absence of magnetic order.

Among the range of cadmium iodide type compounds, Greenaway and Nitsche [5]
report measurements of optical absorption showing that tin disulphide, zirconium
disulphide, hafnium disulphide and hafnium diselenide are indirect gap semi-conductors, having energy gaps of respectively 2.21 eV, 1.68 eV, 1.96 eV and 1.13 eV. Titanium disulphide, titanium diselenide and titanium ditelluride have all been reported [37] as degenerate semiconductors. Vilanova [38], from a study of energy loss
spectra, shows that it is possible to excite the oscillation of some of the valence electrons, and that the interband transitions in these compounds are anisotropic. Switch-

ing behaviour, similar to that found in tin disulphide and zirconium disulphide, is found also in hafnium disulphide [39] and has now been established for zirconium diselenide. Both zirconium disulphide and zirconium diselenide are semiconductors, having energy gaps of respectively 1.70 eV and 1.20 eV. [40]

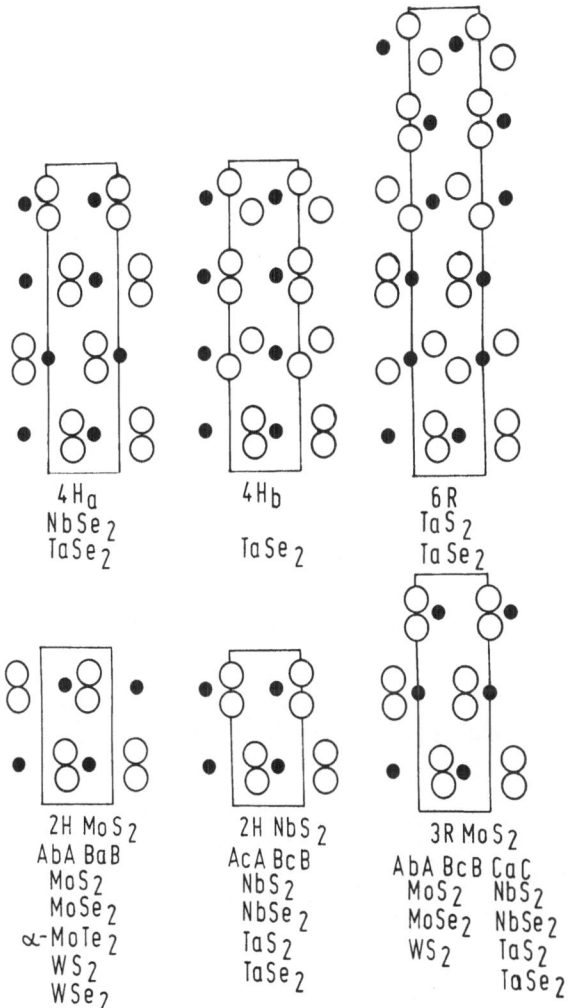

Fig. 9. Stacking polytypes found in Group V and Group VI TX₂ layer dichalcogenides. (11$\bar{2}$0) sections. (After Wilson and Yoffe).

Tin disulphide and zirconium disulphide exhibit [41] negative resistance and bistable switching when an electric field is applied parallel to the unique c-axis. Electrical and photoconductive properties of tin disulphide have been measured by Patil and Tredgold [42]. The crystals, as in the work of Lee and Said, were prepared by vapour transport techniques. Using a four-probe technique a value of 0.18 eV was obtained for the difference in the thermal activation energies associated with electrical conduction parallel and perpendicular to the layers. From the spectral response of photoconductivity a band gap of 2.34 eV at 300 K was measured, with a temperature coefficient of -1.02×10^{-3} eV K^{-1}.

The electrical conduction mechanism in tin disulphide has been elucidated by Said and Lee [43]. From the measurement of conductivity in a direction parallel to the trigonal c-axis, prior to forming, space charge injection is shown to be the predominant mechanism, while subsequently, for the formed crystal the relation $I \propto V^n$ applies where n may be as high as 11. Their results show marked similarity to those obtained for amorphous chalcogenide glasses. Patil [44] investigates a.c. conductivity in tin disulphide over the temperature range $-100°C$ to 120°C, and from the frequency dependence of a.c. conductivity in the low temperature region suggests that perpendicular to c-electron transport is by a hopping mechanism. This is in agreement with earlier results by Gowers and Lee [45], who from measurements of electron mobility parallel to the c-axis, associate this with a hopping mechanism, whereas the mobility parallel to the layers shows normal lattice scattering. Band calculations for tin disulphide have been made [46] which allow a satisfactory interpretation of the optical transitions contributing to the absorption edge. Similar pseudopotential calculations have been made for the series of solid solutions SnS_xSe_{2-x} [47].

Other models of conduction mechanisms in chalcogenide materials in which switching and negative resistance effects occur have been suggested by Coward [48], who links switching with the formation of high conductivity filaments during the forming process, by Suntola [49], and by Iida and Hameda [50] who associate switching with electron tunnelling through an insulating layer.

Measurements of refractive index and optical absorption in tin diselenide crystals grown by the iodine vapour transport technique [51] show that this material is an n-type semiconductor with an energy gap of 1.00 ± 0.03 eV, due to forbidden indirect transitions, having an electron mobility of 33 cm^2 V^{-1} s^{-1} and an effective electron mass $m = (0.4 \pm 0.2) m_0$. The optical and electrical properties of tin diselenide have been the subject of investigation by Evans and Hazelwood [52], who find it to be n-type. The intrinsic absorption edge, at 77 K, arises from forbidden indirect transitions having energy gaps not greater than 1.03 eV and 1.30 eV, and allowed direct transitions across an energy gap of 1.97 eV. The band structure of $SnSe_2$ is derived by Sobolev and Donetskikh [53] from measured reflection spectra of cleaved single crystals at 77 K and 293 K.

Tin diselenide grown by sublimation and the Bridgman technique is p-type [54] with an energy gap of about 1 eV. Popov and Zemlyanov [55] have studied photoelectron emission, quantum yield and optical absorption in tin diselenide and have constructed a zone scheme on the basis of their results. A complete review of the electrical properties of the series of solid solutions SnS_xSe_{2-x}, where x varies between 0 and 2, is given by Said [56].

The electronic structure and optical properties of tin diselenide have been calculated by Young and Cohen [57] using an empirical pseudo-potential approach. The optical constants calculated are comparable with experimental data. Fong [58] has used a similar method to calculate the electronic band structures of tin disulphide and tin diselenide. The imaginary part $\varepsilon_2(w)$ of the dielectric function is calculated for SnS_2 and some comparison is made between theory and existing experimental data.

Albers and Verbekt [59] describe the $SnSe - SnSe_2$ eutectic as a p–n multilayer junction exhibiting 10^3 to 10^4 heterojunctions per cm, and they describe the crystallographic relationship between the SnSe and the $SnSe_2$ lamellae.

4. The Growth of Layer Compounds

The group IV transition metals can react with sulphur, selenium and tellurium in several stoichiometric ratios. In addition non-stoichiometric phases of considerable width are known to exist. A wide range of study has been made of phase equilibrium in these systems. Much of the early work was restricted to the preparation and crystallographic characterisation of polycrystalline samples of the chalcogenide phases. In the titanium-sulphur system, for example, there exists monoclinic TiS_3, hexagonal TiS_2, pseudo-hexagonal Ti_2S_3, hexagonal Ti_3S_4, rhombohedral $Ti_{(1+x)}S$ $(x=0.1)$, three hexagonal modifications of TiS, and hexagonal Ti_6S. [60–65]

Studies of the titanium-selenium and the titanium-tellurium systems [66–70], the zirconium-sulphur system [71, 72], the zirconium-selenium [73] and the zirconium-tellurium [74] systems and the hafnium-sulphur system [75] show a similar complexity. Although it was established that most of the MX_2 compounds had the cadmium iodide layer structure, only fortuitously were single crystals of these compounds prepared and no systematic attempt to grow single crystals was possible. Al-Hilli and Evans [76] have used fired samples of polycrystalline material to show that the system $W_{1-x}Ta_xSe_2$ formed a continuous range of solid solutions of the molybdenum-disulphide structure type.

With the advent of the halogen vapour transport technique large single crystals of many of these compounds were prepared with relative ease. It is this factor more than any other which has allowed systematic studies of the electrical and other physical properties of the compounds.

In the preparation of 'mussivgold' (SnS_2) by the old Wohler synthesis $(Sn+S+Hg+NH_4Cl)$ hydrogen chloride formed at high temperatures was responsible for the formation of large single crystals. This synthesis is the precursor of the vapour transport technique, used by Nitsche [77, 78] for the preparation of many binary, ternary and mixed chalcogenides. Niobium diselenide and tantalum diselenide may be grown using iodine as transporter [79], and Brixner [80] has used the vapour transport technique in the growth of $NbSe_2$, $NbTe_2$, $TaSe_2$, $TaTe_2$, WSe_2, WTe_2, $MoSe_2$, $MoTe_2$, using iodine or bromine as transporting agent. Nitsche has prepared TiS_2, $TiSe_2$, $TiTe_2$, ZrS_2, $ZrSe_2$, HfS_2, $HfSe_2$, VS_2, VSe_2, NbS_2, TaS_2, MoS_2, WS_2, $(TiS_2:TiSe_2)$, $(TiS_2:TiTe_2)$, $(ZrS_2:ZrSe_2)$, $(TiS_2:VS_2)$ by similar methods. Iodine was also used as the vapour transporter in a series of crystal preparations by Bro [82] of the ditellurides and tri-tellurides of lanthanum, cerium, praseodymium and neodymium.

Nitsche found it impracticable to prepare $ZrTe_2$ or $HfTe_2$, although both of these compounds have been reported to have a cadmium iodide structure. Bartram and Smeggil [81] have since succeeded in growing a compound $HfTe_{2-x}$ $(x=0.061)$ by vapour transport. It is described as crystallising in space-group $P\bar{3}m1$ with $a=3.9492\pm 4$ Å and $c=6.6514\pm7$ Å. The compound is characterised by non-stoichiometry in the tellurium sub-lattice, and a random, but unequal, distribution of Hf over two possible sites. Nitsche noted that all the compounds he had succeeded in growing possessed the cadmium iodide structure, and that c-axis rotation photographs gave sharp reflections and showed no signs of disordering, although these materials can exist in non-stoichiometric phases of considerable width.

The vapour transport technique is discussed in detail by Schafer [83]. In its essentials it consists first of the formation of a polycrystalline charge of the dichalcogenide by heating the stoichiometric mixture of the elements in vacuo. This charge is then volatilised in the form of a chemically intermediate phase at a high temperature T_1 (the reaction temperature). The intermediate phase diffuses or convects to a region of lower temperature T_2 (the growth temperature). Here the chemical intermediate decomposes and deposits the dichalcogenide, now in a single crystal form. At correctly chosen values of T_1 and T_2, and with good temperature stabilisation, thin crystals of a high degree of perfection and several square cm in superficial area can be grown.

In the growth of dichalcogenides by iodine vapour transport it is believed that the chemically intermediate phase is an iodide vapour, and that the equilibrium

$$2MeX_2 + 2I_2 \rightleftharpoons 2MeI_2 + 2X_2$$

is responsible for transport. At the growth temperature dissociation of the di-iodide to the tetra-iodide occurs, depositing the dichalcogenide

$$2MeI_2 + 2X_2 \rightleftharpoons MeI_4 + MeX_2 + X_2$$

The tetra-iodide cycles to the reaction end of the system, where it reacts with more dichalcogenide, the process continuing until all the charge is transported to the growth region.

Revelli et al. [84] have used iodine vapour transport for the preparation of a range of solid solutions in the system TaS_xSe_{2-x} and obtain a complete range of isomorphous solid solutions. Rimmington, Whitehouse and Balchin have perfected the use of iodine as the transporter in closed ampoules for the growth of SnS_2, $SnSe_2$, TiS_2, $TiSe_2$, $TiTe_2$, ZrS_2, $ZrSe_2$, HfS_2, $HfSe_2$, and in addition grew complete ranges of isomorphous solid solutions of these materials [85–88]. The solid solutions grown comprise the systems SnS_xSe_{2-x} $(0\leqslant x\leqslant 2)$, TiS_xSe_{2-x}, $TiSe_xTe_{2-x}$, TiS_xTe_{2-x}, ZrS_xSe_{2-x}, HfS_xSe_{2-x}. The optimum growth conditions for solid solutions of each series have been determined. Alamy and Balchin [89] extend this work to the growth of solid solutions of composition $Sn_xZr_{1-x}S_2$ $(0\leqslant x\leqslant 1)$.

The growth conditions quoted by these workers for the end members of each series are given in Table II. The growth conditions given for HfS_2 and $HfSe_2$ are not, how-

TABLE II

Optimum conditions for the growth of some layer compounds
by the iodine vapour transport technique

Compound	T_1 °C	T_2 °C	Growth Time (h)	Crystal size mm $\times$ mm $\times$ mm
SnS_2	680	640	8	$20 \times 10 \times 0.1$
$SnSe_2$	590	570	8	$20 \times 8 \times 0.2$
TiS_2	800	720	400	$15 \times 15 \times 0.2$
$TiSe_2$	780	740	300	$10 \times 10 \times 0.6$
$TiTe_2$	750	690	300	$15 \times 10 \times 0.3$
ZrS_2	900	820	250	$15 \times 15 \times 0.3$
$ZrSe_2$	850	800	350	$10 \times 5 \times 1.8$
HfS_2	1010	1000	1000	$5 \times 5 \times 0.1$
$HfSe_2$	900	850	700	$5 \times 5 \times 0.1$

ever, the optimum, being near the upper operating limit of the furnaces available. Initial results by Whitehouse suggest that a high yield of larger crystals of hafnium compounds may be obtained if the temperature of the growth zone is programmed through a cooling cycle while growth takes place.

Balchin et al. find that once the optimum reaction and growth temperatures for the two end members of a series are determined experimentally, conditions for growing solid solutions of these materials, of a given composition parameter, x, are obtained by a linear interpolation between these optimum conditions. Only slight adjustments of temperature are then required to establish optimum conditions for growth of solid solutions. Occasionally small deviations of temperature away from the optimum result in the formation of tri-chalcogenides.

The method of crystal preparation used by these authors is found to yield crystals of very high crystallographic quality. Examination by X-ray topographic methods [90] indicate that the dislocation density may fall as low as 100 cm^{-2} for the growth products.

In the growth method they describe, the metal fractions, of high purity, are placed in cleaned fused quartz ampoules some 20 cm long $\times$ 18 mm bore, having a volume of some 20 ml. Each ampoule has attached to it a pumping stem joined by a constriction of bore 2mm for sealing. The ampoule is evacuated to a pressure of 10^{-3} Torr and then heated to white heat, with its metal contents, to drive off any volatile contamination. The ampoule is allowed to cool and the chalcogenide fraction, in the correct molar proportions to form the dichalcogenide or solid solution required, is added to the ampoule to make up a charge of some 2 g.

The transporter is then added to the charge to provide a transporter concentration normally in the range 1–10 g. per litre of ampoule volume. Different choices of transporter have been tried (see, for example, Table III) and in different proportions (see, for example, Figure 10), but commercial '3N' iodine in the form of 99.5 weight per cent purity flakes proves to be one of the most effective. It is, however, stressed that

TABLE III

Illustrating the variation of growth rate of dichalcogenide
(SnS$_2$) with choice of transporter

Transporter	Concentration mg ml^{-1}	Growth rate of single crystals mg h^{-1}
HCl	4.0	15
Bromine	6.0	25
Pure water	5.0	20
Redistilled iodine	5.0	18
HIO$_3$	7.0	65
Pure water + redistilled iodine	4.0	140
Commercial iodine	4.0	145

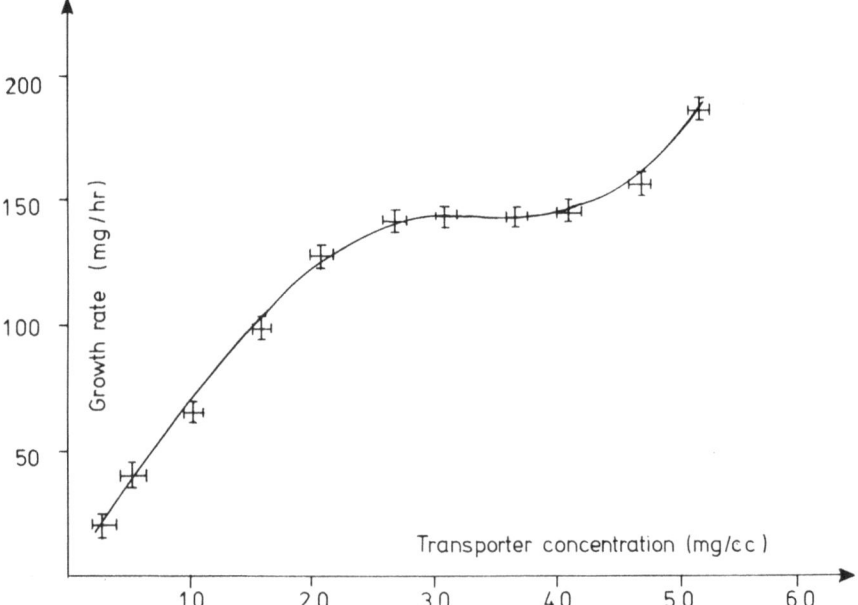

Fig. 10.   Growth rate of tin disulphide single crystals as a function of transporter concentration (iodine)

the effectiveness of iodine as a transporter depends largely on the presence of residual water in the transporting material, and that when this is eliminated the transport rate drops sharply to about one-tenth of its value when water vapour is present.

After adding the transporter the ampoule is re-evacuated to a pressure of 10^{-4} Torr with the lower part of the ampoule, containing the charge, supported in a bath of liquid nitrogen. This prevents evaporation or sublimation of the contents of the ampoule while it is sealed. To degas the ampoule, whilst retaining the volatile materials, especially water vapour, the top part of the ampoule is heated to dull red heat with its lower end cooled. The ampoule is then sealed, at a final pressure measured inside the ampoule of 10^{-5} Torr.

The charged sealed ampoule is transferred to a crystal growing furnace comprising a horizontal tube with a centre-tapped split electrical heater winding. The temperature of each half of the furnace may be adjusted separately. The charged end of the ampoule is heated to the reaction temperature, measured by chromel-alumel thermocouples and stabilised to $\pm 1\,°C$ by Eurotherm controllers. The empty end of the ampoule is maintained at a lower temperature similarly measured and stabilised. The furnace presents an almost linear temperature gradient along the length of the ampoule. Reaction and growth temperatures are maintained for times varying from eight hours to several weeks, during which time large crystals of typical flaky habit accumulate at the growth end of the ampoule. When crystal growth has gone to completion, usually leaving some 10% of the untransported charge at the reaction end of the ampoule, the ampoule is removed from the furnace and cooled under a water jet to condense the transporter away from the crystals grown. These are then extracted by cutting the ampoule with a dry diamond wheel, and are washed in alcohol.

Nitsche *et al.* [78] give the following rules for the successful growth of crystals by vapour transport techniques: –

(a) the rate of transport must not exceed the rate of growth of the seeds,

(b) the optimum crystallisation temperature must be evaluated empirically for each system grown, taking into account the possibility of polymorphism,

(c) the crystallisation chamber should be large in order to prevent intergrowth between adjacent seeds. Asymmetric heating is sometimes useful.

(d) the temperature distribution in the crystallisation chamber should be as uniform as possible to avoid partial re-evaporation of already grown crystals,

(e) well developed crystals form more easily in large diameter tubes, where transporter convection determines the rate of transport,

(f) the temperature difference between the reaction and the growth chambers can be made smaller when wider tubes are used (thus facilitating an even distribution of growth products along the crystallisation chamber) since the gas flow is here the rate determining parameter.

The sizes of crystals grown by vapour transport may be further increased by subjecting them to a process known as 'mineralisation' or the 'Pendelofen process'. Mineralisation is also a method whereby poorly crystalline material, or material with a distorted crystal lattice may be converted into a good crystalline state. Usually mineralisers form a melt that dissolves a part of the solid substance to be recrystallised, and this is then allowed to re-precipitate from the melt in a more stable form. Instead of a melt, a gas phase can be used as a solvent for the solid substance, and consequently will act as a mineraliser, provided a transport reaction exists for the solid substance. Mineralisation via the gas phase also frequently offers the advantage that the amount of mineraliser used may be infinitesimally small relative to the amount of solid material regrown. Mineralisation, in the case of halogen vapour transport, is thus a special case of the transport process where the transported crystals are grown directly on the charge, or at a point at most only a few microns from it.

Rimmington [91] has recrystallised small crystals of tin disulphide, zirconium disulphide and hafnium disulphide from an atmosphere of iodine in a single zone horizontal furnace. To achieve mineralisation a periodic temperature fluctuation about the growth temperature was used. Having established, within $\pm 10\,°C$, the optimum growth temperature for the compound, small crystals of the material were subjected, in an atmosphere of iodine, to periodic temperature fluctuations of up to $20\,°C$. This periodic fluctuation in the growth temperature was achieved by feeding a square wave of some 1.5 mV amplitude in series with the thermocouple output, and presenting the combined waveform to the temperature controller. The period and the amplitude of the mineralisation waveform could be varied separately. The results obtained are summarised in Table IV. Using the technique very large single crystals of layer compounds

TABLE IV

Crystallisation by the mineralisation process
(Rimmington (91))

(a) Compounds grown by mineralisation

Compound	Growth temperature (°C)	Time (hr)	Crystal size mm × mm × mm
SnS_2	640	600	$45 \times 20 \times 0.1$
$SnS_{1.8}Se_{0.2}$	630	600	$50 \times 35 \times 0.1$
ZrS_2	810	920	$40 \times 25 \times 0.3$
HfS_2	970	1000	$35 \times 20 \times 0.1$

(b) Growth rate of SnS_2 as a function of temperature amplitude

Amplitude (°C)	Growth rate mg h^{-1}	Growth time (hr)
0	145	8.25
5	0.00124	650
10	0.00115	720
15	0.00112	750
20	0.00100	750

(c) Growth rate of SnS_2 as function of fluctuation frequency

Cycle time (min.)	Growth rate mg h^{-1}	Growth time (hr)
∞	145	7.25
5	0.00140	700
10	0.00125	715
15	0.00120	730
20	0.00112	730
25	0.00107	750

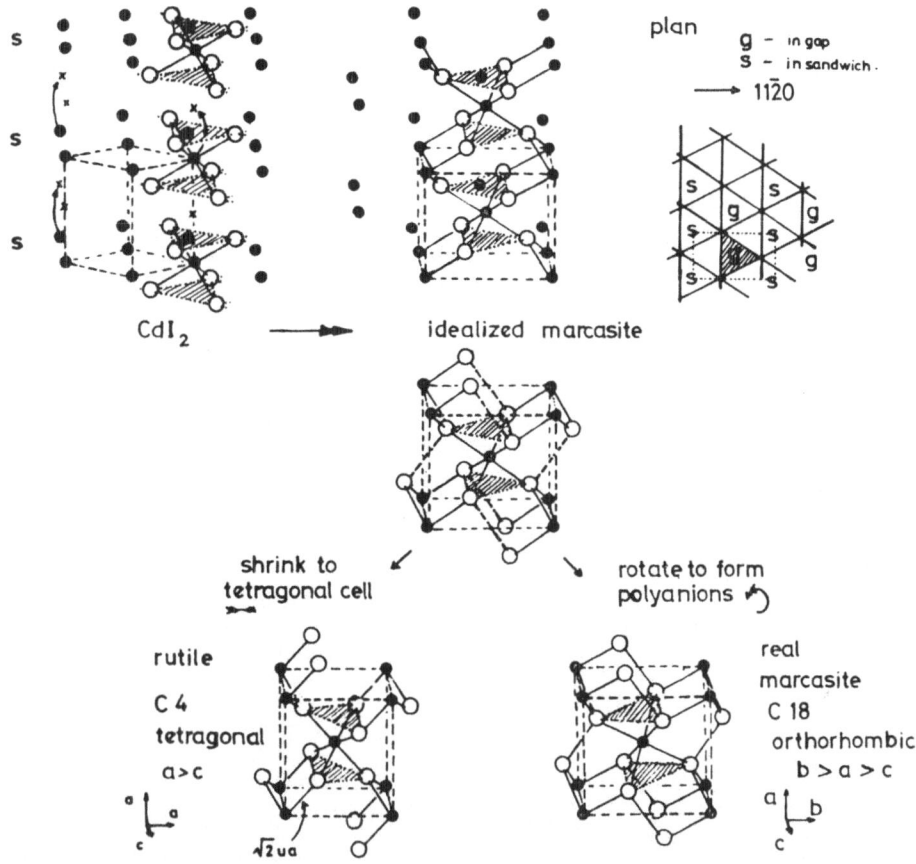

Fig. 11.   Structural relationships between the cadmium iodide layer structure, the marcasite structure and the rutile structure. (After Wilson and Yoffe).

are obtained, although at the expense of greatly extended growth times. The amplitude of the temperature fluctuation is found to be the dominant factor influencing the growth rate, and for the highest growth rates this should be kept as small as possible.

5. Structural Relationships with Other Compounds

Wilson and Yoffe [4] examine a suggestion put forward in its original form by Hulliger and Mooser [92] that, as the metal in the transition metal dichalcogenides acquires successively greater numbers of d electrons, there is a trend away from the octahedrally coordinated layer compounds towards the octahedrally co-ordinated pyrite and marcasite structures, the compounds TX_2 reverting to their layered structure at Group VIII. The evidence for this is summarised in Table V and the structural relationships involved are shown in Figure 11. Hulliger and Mooser also point out the relationship between the cadmium iodide layer structure and the nickel arsenide structure.

TABLE V

Structural types found in the TX_2 dichalcogenides

M / X_2		$-S_2$	$-Se_2$	$-Te_2$	Electrical character
IV	Ti	L octa	L octa	L octa	Diamagnetic semiconductors.
d^2	Zr	L octa	L octa	L octa	$\varrho \geqslant 1$ ohm cm, $E_g \sim 0.2$–2.0 eV.
	Hf	L octa	L octa	(L octa)	Ti, as with all first series metals, yields non-stoichiometric products.
V	V	(L octa)	L octa	L distd. octa	Undistorted compounds are narrow band metals,
d^3	Nb	L trig pr	L trig pr	L distd. octa	$\varrho \sim 10^{-4}$ ohm cm. Pauli
	Ta	L $\{$ trig pr $\}$ (dist) octa	L $\{$ trig pr $\}$ (dist.) octa	L distd. octa	paramagnetic $\rightarrow$ band antiferro-magnetic. Superconducting. Free-carrier absorption in I.R. Others diamagnetic semi-metals. Octa TaS_2/Se_2 perhaps semi-conducting.
VI	Cr	–	–	–	(i) The undistorted compounds are diamagnetic semiconductors.
d^4	Mo	L trig pr	L trig pr	L $\{$ trig pr $\}$ distd. octa	$\varrho \geqslant 1$ ohm cm. $E_g^{opt} \geqslant 1.5$ eV, $E_g^{th} \sim 0.15$ eV.
	W	L trig pr	L trig pr	L distd. octa	(ii) The distorted octahedral structures are semi-metals. $\varrho \sim 10^{-3}$ ohm cm. Diamagnetic, low Seebeck coefficients.
VII	Mn		Pyrites		(i) Mn compounds quasi-ionic antiferromagnetic semiconductors.
d^5	Tc	L dist octa	L dist octa	L distd.	(ii) TcS_2 possibly also a pyrite form.
	Re	L dist octa	L dist octa	L distd.	(iii) ReS_2 and $ReSe_2$ are small gap semiconductors, diamagnetic, and with no free-carrier absorption.
VIIIa	Fe	Pyrite	Marcasite		'Non-magnetic' semiconductors
d^6	Ru				
	Os		Pyrites		
VIIIb	Co	Pyrite	$\{$ Pyrite $\}$ Marcasite	$\{$ Marcasite $\}$ L octa	Pyrite, marcasite and layer types metallic. CoS_2 ferromagnetic.
d^7	Rh	(Pyrite)	$\{$ $IrSe_2$ $\}$ Pyrite	$\{$ Pyrite $\}$ L octa	β-$RhSe_2$ and $RhTe_2$ super-conductors.
	Ir	$IrSe_2$	$IrSe_2$	L octa	$IrSe_2$ types semiconducting.
VIIIc	Ni		Pyrites	L octa	(i) Tellurides, metallic Pauli paragmagnetic $\varrho \sim 10^{-5}$ ohm cm. $PdTe_2$ superconducting.
d^8	Pd	Layer type showing X-X links	L octa	L octa	
	Pt	L octa	L octa	L octa	(ii) PdS_2/Se_2 $\{$ semiconducting. PtS_2/Se_2 $\{$ $E_g \approx 0.4$ eV. $\varrho \approx 1$ ohm cm. Diamagnetic, large Seebeck coefficients.

L octa: layer structure – octahedral coordination. L trig pr: layer structure – trigonal prismatic coordination. L distd.: layer structure showing distorted coordination.

The cadmium iodide structure and the nickel arsenide structure are closely related, and one may be converted into the other. In both structures the chalcogenide arrangement is the same, but an extra layer of metal atoms per unit sandwich, inserted into the layer structure, will convert it into the nickel arsenide arrangement. The filling of the van der Waals' region by additional atoms destroys, however, the layered nature of the lattice, as the van der Waals' gap no longer exists as a bond free region. The nickel arsenide structure has tightly bonded atoms in this region. In the cadmium iodide structure, non-bonded intercalated materials may also be introduced into the van der Waals' gap, with little effect on the chemical reactivity of the intercalated

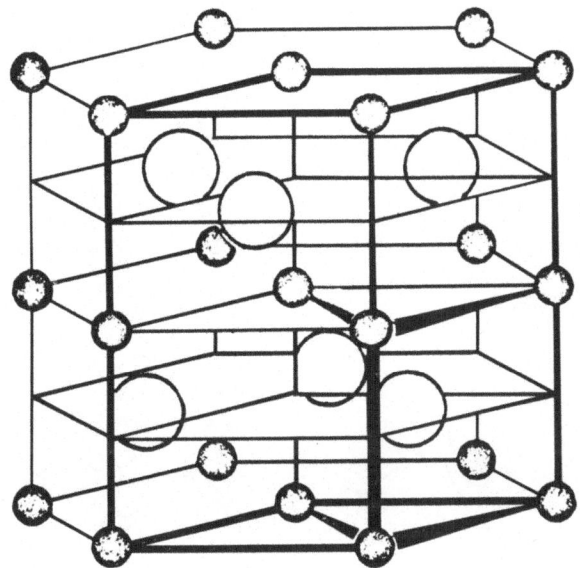

Fig. 12.   The structure of nickel arsenide.

materials, but giving the possibility of altering the physical properties of the materials.

A complete range of structural transitions, from the cadmium iodide type structure to the nickel arsenide type structure, may be brought about as the octahedral interstices in the van der Waals' region of the cadmium iodide structure are filled with metal atoms. The structure of the nickel arsenide lattice is shown in Figure 12, and the filling of the octahedral interstices in a cadmium iodide lattice to form the nickel arsenide structure is shown in Figure 13. The transition from $CoTe_2$ to $CoTe$ (Tengner [94]) is sometimes quoted as an example of this transition.

Jellinek [93] points out that stacking disorder is frequent in layer compounds, but that this may be greatly reduced by the insertion of additional interstitial metal atoms in octahedral interstices between the layers X–M–X. This results in phases of composition $M_{1+p}X_2$ $(0 < p < 1)$. The distribution of metal atoms and vacancies within the interstitial layers may be statistical or ordered, depending for a given M and X on the magnitude of p and the temperature. Jellinek suggests ordered arrangements of atoms in the van der Waals' region which would allow a series of structural transitions between cadmium iodide type and nickel arsenide type structures, and which would explain the stoichiometric ratios of the intermediate phases. These arrangements are illustrated in Figure 14. A further possibility, relevant to transition metal dichalco-

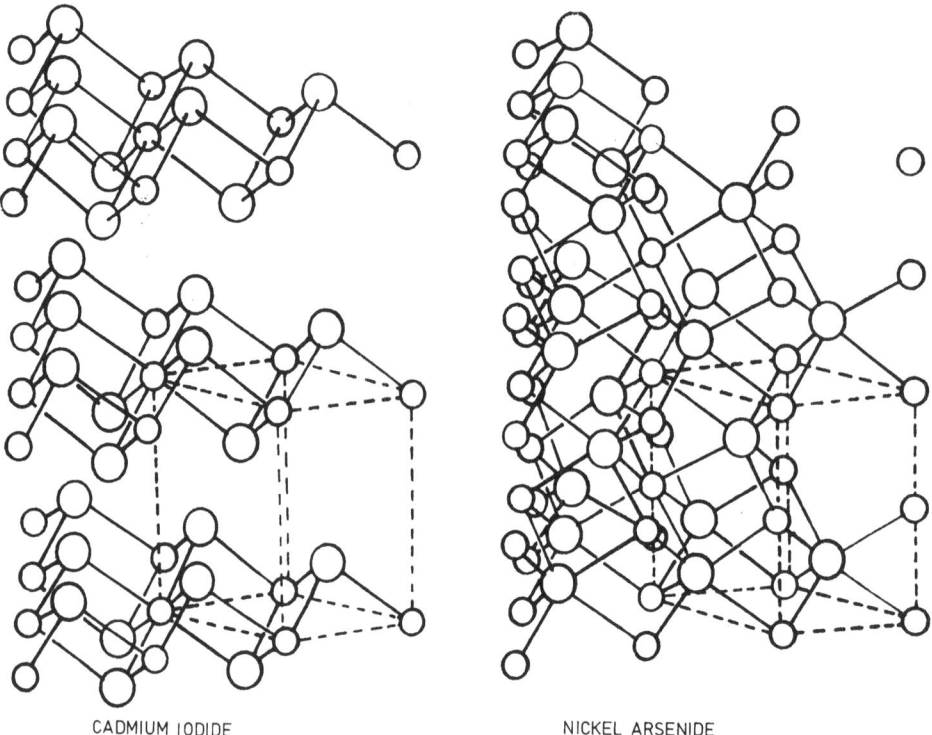

CADMIUM IODIDE   NICKEL ARSENIDE

Figs. 13.   The transition from the cadmium iodide to the nickel arsenide structure.

genides such as V, Cr, Mn, Fe, Co, Ni, is that the placement of metal atoms in tetrahedral interstices would allow of a transition from the nickel arsenide to the spinel structure.

In view of these possibilities a number of observations have been made of structural transitions occurring in systems containing layer compounds, notably in the system titanium-sulphur, in which TiS is known to crystallise in the nickel arsenide arrangement, TiS_2 in the cadmium iodide arrangement, and in which it is known that **a** wide range of non-stoichiometric phases exist. Blitz, *et al.* [60] made a comprehensive study of phases existing in the Ti−S system, and were the first to speculate on the existence of a sesquisulphide phase, Ti_2S_3, recognising this from powder photographs of a two-phase mixture. The existence of this phase was finally established by Hahn and Harder [61] who defined a series of phases of differing composition in the Ti−S system, and were the first to point out the relationship between the structure of TiS and NiAs.

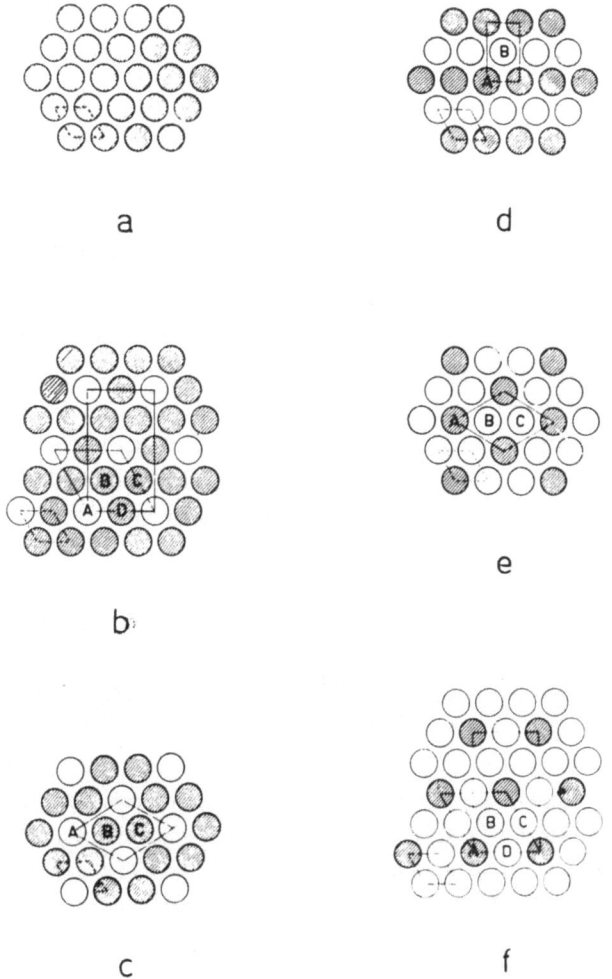

Fig. 14.   Occupation of interstitial sites in ordered arrays within the van der Waals' gap, leading to structural transitions between MX and MX_2. (a) Fully occupied M_2X_2; (b) M_7X_8; (c) M_5X_6; (d) M_3X_4; (e) M_2X_3; (f) M_5X_8. (After Jellinek)

The transition from titanium disulphide to titanium sesquisulphide and from titanium sesquisulphide to titanium monosulphide has been examined in detail by Benard and Jeannin [95], who found that the disulphide, in the absence of sufficient sulphur in the reacting materials, took the form of a material with a unit cell of the same size as stoichiometric TiS_2 but with an $S:Ti$ ratio less than 2.00. The deviation in composition they ascribed to the addition of extra interstitial titanium atoms between layers. Benard and Jeannin found it impossible to isolate stoichiometric TiS_2 in samples prepared at $1000\,^{\circ}C$ or $800\,^{\circ}C$, since it always decomposed to give sulphur vapour and a non-stoichiometric sulphide. To determine the lattice parameter of TiS_2 from their preparations, measurements were extrapolated to an ideal stoichiometric composition of 2.00 in the ratio $S:Ti$. On the sulphur-low side, the phase limit of the TiS_2 phase was placed at an $S:Ti$ ratio of 1.82, since it was at this composition that extra lines on their powder photographs suggested the formation of a second phase. Earlier, Ehrlich [96] had found a discontinuity in the 'c'-parameter at $S:Ti = 1.7$, which he thought indicated the phase limit of TiS_2.

For the TiS_2 phase the variation of lattice parameter with the $S:Ti$ ratio is linear. However, as $S:Ti$ is decreased below 1.82, superlattice lines corresponding to integral multiples of the unit cell height, c, appear. These occupy the range of composition between $S:Ti = 1.8$ and $S:Ti = 1.6$ and are explicable in terms of variations in the stacking order of titanium and sulphur. As the $S:Ti$ ratio is decreased to less than 1.59, the single phase pattern of the sesquisulphide, Ti_2S_3, is obtained. This phase occupies a homogeneity range from $S:Ti = 1.59$ to 1.38, after which line splitting denotes a superlattice of lower symmetry. Benard and Jeannin did not extend their examination as far as the composition TiS. From measurements of densities of the individual phases they decided that sulphur vacancies must be created concurrently with the insertion of titanium.

The structure of the non-stoichiometric phase $Ti_{2+x}S_4$ has been derived from single crystal data by Norrby and Franzen [97]. They confirm the crystal structure proposed earlier by Wadsley [98]. The crystal structure they derive is shown in Figure 15. A value of 0.455 ± 0.001 is assigned to their compositional parameter, x, and their structure contains two crystallographically independent sites for the titanium atoms, one of which is filled, but the other is only 23% occupied in a random fashion.

In the system titanium-selenium, Ehrlich [66] describes a complete series of solid solutions from $TiSe$ to $TiSe_2$, with a hexagonal structure and continuously changing lattice constants. The structure is described as nickel-arsenide-like at the composition $TiSe$ and cadmium-iodide-like at the composition $TiSe_2$. The structure of $TiSe_2$ was determined by Oftedal [99]. There was, however, some discrepancy between calculated and observed X-ray intensities.

A study of the titanium-selenium system undertaken by Gronvold and Langmyhr [70] established the existence of a deformed nickel arsenide structure in the region $TiSe_{1.2}$ to $TiSe_{1.4}$. Additional work by Hahn and Ness [62] showed that the $TiSe$ phase extended with hexagonal symmetry to an $Se:Ti$ ratio as low as 0.75. Below this composition, at an $Se:Ti$ ratio between 0.70 and 0.50 a new phase, Ti_3Se_2 was described as nickel-arsenide-like with excess titanium occupying the vacant selenium positions. At $Se:Ti = 0.95$ a distorted orthorhombic NiAs-type phase was observed. Around the composition $Se:Ti = 1.05$ an hexagonal nickel arsenide structure was found, merging at $Se:Ti = 1.10$ into a two phase region. A single phase was again observed at $TiSe_{1.20}$, its cell dimensions gradually changing with composition. Its upper composition limit is close to $TiSe_{2.00}$, in agreement with Ehrlich's results. However, structural distortions were observed within its range of compositions. Its structure starts off as hexagonal between Se:Ti = 1.20 to 1.30, reverting to monoclinic between S:Ti = 1.30 to 1.40 and back to hexagonal between 1.50 and 2.00. Some low symmetry line splitting was observed also near Se:Ti = 1.60.

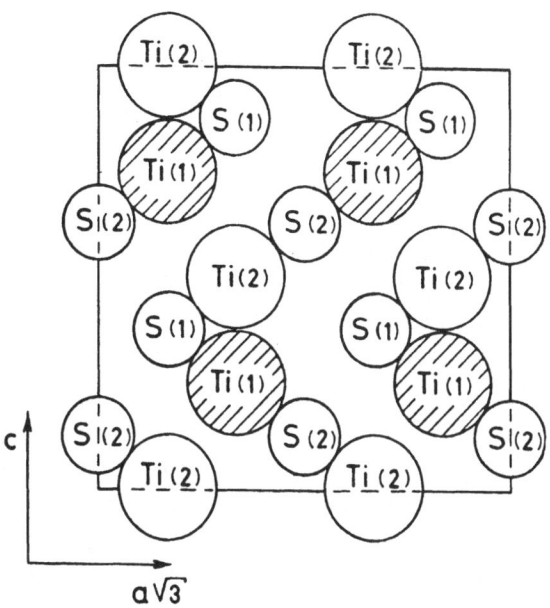

Fig. 15.   The crystal structure of $Ti_{2+x}S_4$ represented by a section through the hexagonal (11$\bar{2}$0) plane. The completely filled Ti (1) sites are shaded. The Ti (2) sites are only 23% occupied. (After Norrby and Franzen)

Bernusset and Jeannin [67] dispute the width of this cadmium iodide like phase $TiSe_2$, concluding from their results that in the titanium-selenium system the nonstoichiometric $TiSe_2$ phase exists between Se:Ti = 1.42 and 1.96. Chevreton and Bertaut [68] discussed the occurrence of additional phases Ti_3Se_4 (and in the titanium-tellurium system Ti_3Te_4), structurally based on nickel arsenide, but slightly deformed monoclinically. Brunix and Chevreton [100] similarly assigned a nickel arsenide like structure to Ti_5Te_8.

In the titanium-tellurium system the existence of two phases Ti_5Te_4 and $Ti_{2-x}Te_2$ was established by Raam, et al. [69]. The phase Ti_5Te_4 had no appreciable homogeneity range, having a tetragonal unit cell of stoichiometric composition. The phase $Ti_{2-x}Te_2$ has a homogeneity range over the limits 55.4 to 59.2 atomic % Te, the structure being monoclinic and related to the nickel arsenide structure. Over the composition range 60.0 to 66.7 atomic % Te a hexagonal phase with a nickel arsenide/cadmium iodide type structure appeared, showing a continuous variation of lattice parameters towards the cadmium iodide like phase of $TiTe_2$. The occurrence of the solid solution was attributed to the substitution of Ti atoms by Te as the Ti content decreased from $Ti_{1.61}Te_2$ to $TiTe_2$.

Sulphides, selenides and tellurides of titanium, zirconium and hafnium have been prepared by McTaggart and Wadsley [75], although they have not attempted to define composition limits. They do, however, obtain a compound $ZrSe_{1.98}$ by thermal decomposition of $ZrSe_3$. Hahn and Ness [73] detect a phase which they denote as $Zr_{1+x}Se_2$ with a homogeneity range $ZrSe_2$ to $Zr_{1.2}Se_2$. For the compound $ZrSe_2$, stoichiometry has been examined by Gleizes and Jeannin [101]. They find it impossible to prepare stoichiometric $ZrSe_2$ at 800 °C. The highest $Se:Zr$ ratio they obtain is 1.945, probably containing selenium vacancies at an average concentration of 0.055 per unit cell. At the selenium-low limit they relate the $Se:Zr$ ratio to substitution of selenium atoms by zirconium, the unit cell mass continually increasing as zirconium atoms enter the lattice, occupying empty selenium sites. The mechanism suggested leads to a unit cell formula $Zr_1(Zr, Se)_xSe_{1.945}Se_{0.055-x}$. This mechanism predicts a low selenium composition limit of 1.844 for the cadmium iodide-like phase, in good agreement with the experimental value of 1.850. This value is obtained when all the selenium vacancies are occupied by zirconium.

An examination of the zirconium-sulphur system by Hahn et al. [72] established the formation of six different crystallographic phases. viz. ZrS_3, ZrS_2, Zr_4S_3, Zr_3S_4, Zr_3S_2, and ZrS. Zr_3S_4 forms as a single phase between $S:Zr$ ratios of 1.5 to 0.9 and Zr_3S_2 between $S:Zr$ ratios of 0.8 to 0.5. ZrS was found to have a sodium chloride type lattice, Zr_4S_3 was tetragonal, Zr_3S_4 was cubic, of the sodium chloride type but with superlattice effects, Zr_3S_2 was hexagonal, WC type. ZrS_2 was of the CdI_2 structure type. They found no evidence of the existence of a nickel-arsenide like structure near ZrS, as found in the titanium-sulphur system.

In the zirconium-tellurium system, however, the same workers [74] find a complete structural transition between the nickel arsenide like $ZrTe$ and the cadmium iodide like $ZrTe_2$. In all five phases were detected; monoclinic $ZrTe_3$, probably isostructural with $ZrSe_3$, tetragonal Zr_4Te_3, hexagonal Zr_3Te_2 having a WC-type structure over a homogeneity range $ZrTe_{0.5}$ to $ZrTe_{0.75}$, $ZrTe_2$ a cadmium iodide-like layer structure and $ZrTe$ having a nickel arsenide like structure. These workers found a continuous transition between $ZrTe_{0.8}$ (NiAs-like) to $ZrTe_{2.2}$ (CdI_2-like), a range which includes both $ZrTe$ and $ZrTe_2$.

Jeannin [102] has examined the results of a number of phase equilibrium determinations in the systems of elements containing layer compounds, pointing out the diversity of results obtained by different workers, and tabulating the occurrence of different phases between the compounds TiS and TiS_2. (see Figure 16). He considers from these results that a direct transition from an NiAs structure to a CdI_2 structure is comparatively rare, and usually this transition is made through a series of intermediate phases. These phases may be made up from the nickel arsenide structure by considering either: –

(a) vacancies placed at random in possible metal positions,

(b) vacancies localised in determined planes – the repetition of filled and unfilled planes could be variable.

(c) the vacancies could be ordered in certain planes as in chromium sulphides, [103].

This would be in accordance with the scheme of vacancy sites put forward by Jellinek [93]. Figures 17 and 18 show the occurrence and phase widths of the CdI_2 structure and the NiAs structure in different metal-chalcogenide systems.

In materials grown by iodine vapour transport the range of homogeneity for the

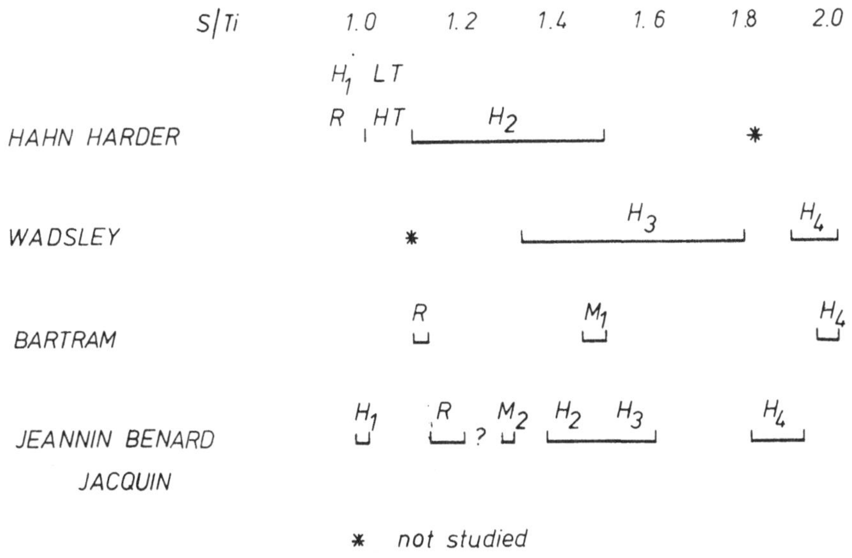

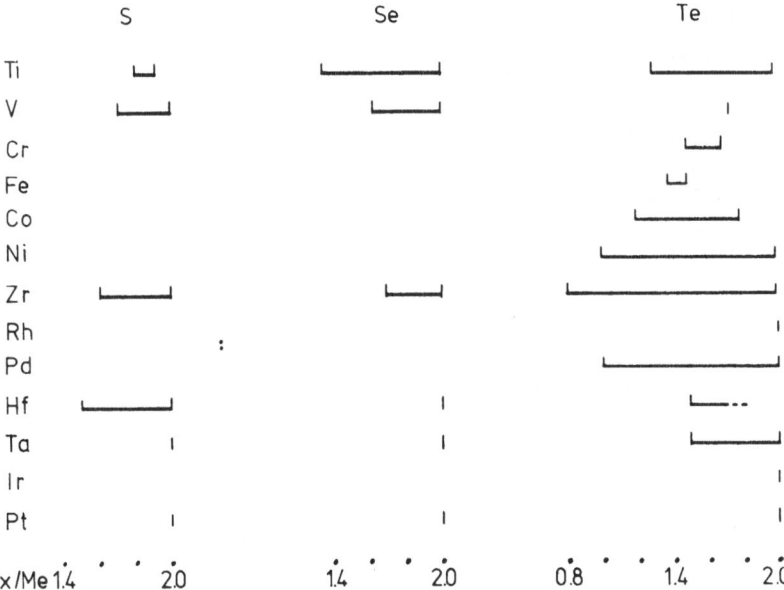

Fig. 16.   Results obtained by different authors who have studied the system Ti−S. H_1, H_2, H_3, H_4 are four different hexagonal phases. R is rhombohedral. Hahn and Harder indicate H_1 to be a low temperature form, and R a high temperature form, of TiS. M_1 and M_2 are two monoclinic phases. The widths of the phases are indicated. (After Jeannin)

Fig. 17.   Widths of homogenous regions of non-stoichiometry of which the crystal type is CdI_2. (After Jeannin)

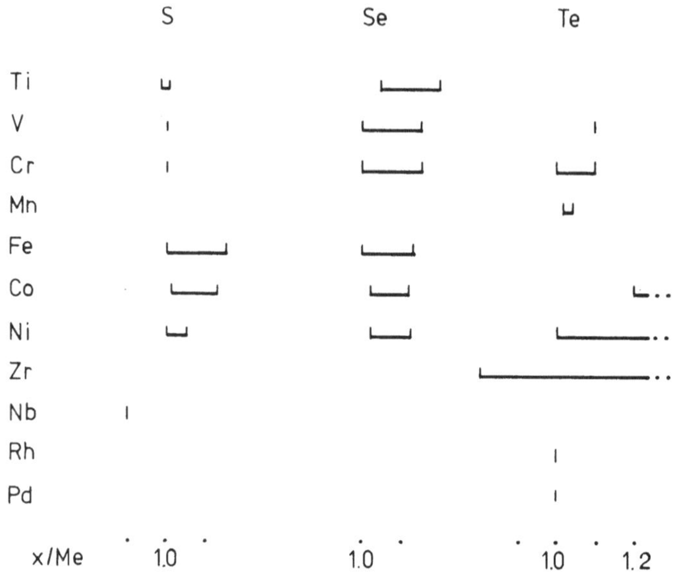

Fig. 18.   Widths of homogenous regions of non-stoichiometry of which the crystal type is NiAs.
(After Jeannin)

MX_2 compounds appears to be much lower than those indicated in Figure 17. Thus Rimmington quotes the following homogeneity ranges for vapour transport grown material

MX_2	X/M high	X/M low
SnS_2	1.98	
$SnSe_2$	1.99	
TiS_2	1.96	1.89
$TiSe_2$	1.94	1.83
$TiTe_2$	1.96	1.89
ZrS_2	1.96	1.88
$ZrSe_2$	1.95	1.88
HfS_2	1.98	1.91
$HfSe_2$	1.98	1.92

and he ascribes the variation of S:Sn away from 2.00 in SnS_2, from density and electrical properties, as due to the replacement of sulphur by tin.

6. Intercalation Complexes

The main interest in intercalated materials lies in the fact that many of their crystallographic and electrical properties make them promising systems in which to investigate mechanisms of superconductivity which may rely upon two-dimensional constraints. The effect of intercalation is to introduce into the van der Waals gap alkali metal atoms or molecules of organic complexes which may modify their structural and electrical characteristics. Almost all layer compounds have the property of intercalating alkalis or organic materials, usually with the result that the modular sandwiches of the lattice are pushed further apart. Rimmington [91] for example, has taken single crystals of several layer compounds and reacted them for varying times with pyridine in an attempt to determine the maximum amount of intercalating material that the materials could take up. The single crystals eventually shattered into small fragments. For pyridine intercalated materials he found that the lattice expansions listed in Table VI represented the maximum possible amount of pyridine take-up. The pyridine was introduced into the lattice by gentle heating below 100 °C, but could be removed from the lattice, the crystal parameters reverting to their former values, by heating at temperatures in excess of 200 °C.

TABLE VI

Maximum lattice expansion caused by pyridine intercalation in layer compounds (Rimmington [91])

Compound	Normal length of c-axis (Å)	c-axis after maximum intercalation (Å)	Maximum lattice expansion (%)
CdI_2	5.8744	7.2038	23
SnS_2	5.8853	6.5915	12
SnS Se	6.0483	6.5225	8
$SnSe_2$	6.1255	6.8536	12
TiS_2	5.6883	6.1404	8
$TiSe_2$	6.0036	6.5376	9
$TiTe_2$	6.5935	6.9296	5
ZrS_2	5.8193	6.1175	5
$ZrSe_2$	6.1324	6.6186	8
HfS_2	5.8456	6.0086	3
$HfSe_2$	6.1487	6.4507	5

The first account of intercalation in layer compounds was by Rüdorff [104] during a study of the take-up of alkali metals from liquid ammonia solution. Some of these compounds have since been shown to be superconducting. Intercalation tends to produce a material of composition $A_x MX_2$ where x is fixed for a given MX_2, but is usually less than one. For example WS_2, MoS_2 and NbS_2 will intercalate potassium with $x = 0.5, 0.6, 0.8$ respectively. Both the a-axis and the c-axis, but more particularly the c-axis, is increased in length by intercalation. The relative c-expansion is smaller as x increases, and suggests an increasing tendency towards ionic behaviour. The metal-insulator transition in these compounds is such that metallic conduction does not appear until the alkali metal content is quite large.

Studies have been made of superconducting layered compounds intercalated with organic materials, e.g. pyridine, forming for example $NbS_2(Py)_{1/2}$ [105] or $TaS_2(Py)_{1/2}$

[106]. Studies of the temperature dependence of electrical resistivity, Hall effect, and the magnetic susceptibility indicate that the crystallographic distortion apparent in metallic layer dichalcogenides at low temperatures is absent after intercalation. The crystal distortion referred to has been the object of study in $NbSe_2$ down to 15 K by Marezio *et al.* [107]. At 40 K the 2H polytype of $NbSe_2$ undergoes a transition which is accompanied by anomalies in the electrical resistivity and Hall effect. Magnetic susceptibility does not show a pronounced anomaly at the transition. The crystal structure of 2H-$NbSe_2$ has been determined from single crystal data above and below the transition. At room temperature $2H-NbSe_2$ has the normal layered hexagonal structure with the niobium atoms at the centres of trigonal selenium prisms, i.e. of the molybdenum disulphide type. At the transition the unit cell doubles along the '*a*'-axis. The hexagonal symmetry is conserved and $\frac{2}{3}$rds of the niobium atoms are displaced from their $\bar{6}m2$ symmetry. The coupling of these niobium atoms seems to be the driving mechanism for the crystallographic distortion.

The structure of MoS_2 and $NbSe_2$ has been studied by transmission electron microscopy [108] in thin (0001) crystals after intercalation with sodium in ammonia solution. Three distinct superlattice structures have been identified, one $(a\sqrt{3} \times 2)$ with orthorhombic symmetry associated with the presence of sodium and the other two, with hexagonal symmetry, associated with ammonia. In $NbSe_2$ and $TaSe_2$ intercalated with pyridine (and with aniline) superlattices also are formed [109]. Diffraction contrast studies have shown the presence of intercalation domains which are ordered along crystallographic directions. The predominant feature is a rhombohedral 2×2 unit cell consistent with the stoichiometry of $NbSe_2(Py)\frac{1}{2}$. At low temperatures 3×3 and 4×4 rhombohedral unit cells are also identified. These may be interpreted as due either to the dilution of an intercalate within a layer or to ordering of the orientations of the benzene-like rings in the pyridine or the aniline molecules.

Expansion of the lattice of TaS_2 intercalated with pyridine may be followed directly by high resolution electron microscopy [110]. The interlayer spacing increases from 0.6 to 1.0 nm on intercalation.

High voltage electron microscopy at liquid helium temperatures has been used [111, 113] to study the crystal structures of TaS_2, NbS_2 and $NbSe_2$ intercalated with pyridine, aniline and $N-N-$dimethylaniline. Significant detail of the atomic lattice, resolved at temperatures between 1.8 K and 4.2 K showed characteristic changes in the fine structure of these complexes.

At room temperature, resolutions of 2 Å to 3 Å were obtained in powders and crystals of $2H-TaS_2$, $2H-TaS_2(C_5H_5N)_{1/2}$ and $2H-TaS_2(C_5H_5N)_{1/4}$, under optimum conditions for direct imaging of the atomic lattice and related structural details. With the incident beam normal to the *a*-plane of thin TaS_2 crystals regular hexagonal moire patterns were seen as transmission images of overlapping layers of TaS_2 planes; this represents an indirect resolution of the atomic array at 1 Å spacing.

Electron microscopy of the ultra-thin crystalline lamellae of the $2H-TaS_2$ $(C_5H_5N)_{1/2}$ compound revealed a regular arrangement of dense layers separated by less dense layers which generally featured a faint intermediate line. The (0001) reflection corresponded to the layer spacing of 12.001 Å and the calculated a-axis parameter was 3.335 ± 0.005 Å. These values and the slightly lower lattice spacings of 11.847 Å found in other specimens are in agreement with the corresponding X-ray data [112].

Specimens of fairly pure second stage complex $TaS_2 (C_5H_5N)_{1/4}$, in which two layers of TaS_2 are separated by intercalated pyridine revealed a distinctive pattern of two dense lines separated by a uniform light layer to give a repeating period of 18 Å. This corresponded to an average repeat period of 17.639 Å in selected area diffraction patterns, and is in agreement with corresponding data derived from X-ray diffraction studies.

In both $TaS_2 (C_5H_5N)_{1/2}$ and $TaS_2 (C_5H_5N)_{1/4}$ compounds the uniform light layer was approximately 8–10 Å thick, representing a major component of the periodic layer structure. These dimensions and the intermediate line found in the central region of the light layer suggest that they represent the relatively electron transparent intercalated pyridine molecules which separate the electron dense TaS_2 layers. This direct demonstration of atomically thin metallic layers separated by an organic barrier is of interest in the general context of superconductivity. The lattice spacings are well within the range of tunnelling phenomena and these materials could exhibit characteristics related to Josephson tunnelling.

More recent work, centred on $2H - NbS_2$ and related compounds examined by high resolution electron microscopy and diffraction with specimen cooling at 4.2 K to 1.8 K, shows that when thin layers, about 200 Å to 600 Å thick, of either $2H - NbS_2$ or $2H - NbSe_2$ are imaged with a 200 kV electron beam normal to the layers, characteristic doughnut shaped structures 100 Å to 200 Å in diameter are repeatedly imaged. These have a typical fine structure, exhibiting an electron dense ring 20 Å to 40 Å in diameter with a lighter core. In many areas the doughnut shaped rings seem to be arranged in regular patterns. In view of the reproduceability of the images, recorded at specimen cooling of 4.2 K at magnetic fields of 6000 to 8000 G, it is believed that these images may represent electron microscope images of fluxoids in the form of quantised vortices described by C. M. Verma (1972).

Among the layer compounds intercalated with alkali metals, MoS_2 and $NbSe_2$ are known [27] to be superconducting at low temperatures. When MoS_2 is intercalated with sodium or potassium a superconducting transition temperature of approx. 1.3 K is obtained [17] for the sodium complex, and approximately 4.5 K for the potassium intercalated complex.

The metal-semiconductor transition in alkali-metal intercalated compounds is discussed by Wilson and Yoffe [4, 16]. They compare these intercalation complexes to the tungsten bronzes. These are based on a cubic unit cell, body centred by tungsten, face-centred by oxygen and accommodating any added alkali metals at cubic corner sites. An empty non-bonding band accepts electrons from the alkali metals, giving a situation which finds a counterpart in the metallic compound ReO_3. This sequence is very much like the sequence $TiSe_2$, intercalated $TiSe_2$ and VSe_2. Alkali metal intercalation of TiS_2 and MoS_2 both produce metallic compounds.

In an examination of superconductivity in the sodium, potassium and rubidium intercalates of MoS_2 [14] very wide (several degrees) superconducting transitions have been observed. An X-ray analysis of these systems indicated partial intercalation of some of the samples, with expansion of the unit cell dimensions upon intercalation, and the existence of two differing expansions for the potassium and sodium intercalates respectively. In a further study of intercalated natural and vapour grown MoS_2 by the alkali metals lithium, sodium, potassium, rubidium and caesium [115] stoichiometric and X-ray data were determined, and a complete indexing of the X-ray patterns of the materials $K_{0.4}MoS_2$ and $Rb_{0.3}MoS_2$ was achieved. The alkali metals enter the octahedral interstices in the van der Waals' region. The X-ray results showed insignificant changes in the a_0 lattice parameter (see Table VII) but considerable expansion of the c-axis. All intercalated crystals were superconducting, believed to be due to an electron transfer from the alkali metal to an empty band of MoS_2, resulting in an increase of electron density as well as an increase in the density of states at the Fermi surface.

TABLE VII

X-ray and superconductivity data for intercalates of MoS_2 with alkali metals [115]

Compound	c_0 (Å)	a_0 (Å)	Δc_0 (Å)	Δd_0 (Å)	T_{onset} (K)
MoS_2 (2H)	12.2943	3.1603			
MoS_2 (3R)	18.3670	3.1620			
$Li_x MoS_2$ (2H) $0.4 \leqslant x \leqslant 1.0$	19.039	not refined	6.745	3.372	3.7 ± 0.1
$Na_x MoS_2$ (2H) $0.3 \leqslant x \leqslant 0.6$	14.998		2.704	1.352	4.15
$K_{0.4} MoS_2$ (2H)	16.5804	3.2036	4.286	2.143	6.1
$K_{0.4} MoS_2$ (3R)	24.7915	3.2072	6.422	2.141	5.5
$Rb_{0.3} MoS_2$ (2H)	17.1937	3.2039	4.899	2.450	6.25
$Cs_{0.3} MoS_2$ (2H)	19.606		7.312	3.656	6.30

The X-ray data for the lithium and sodium intercalates differed from the others. It has been reported by Rüdorff [104] that ammonia intercalates into MoS_2 along with lithium, and the possible presence of lithium amide or lithium metal solvated by NH_3 may explain the differences in X-ray data. In the sodium compounds low angle lines indicated a possible tetragonal or orthorhombic superlattice of sodium atoms. The sodium and lithium intercalates could, however, be considered as disordered intercalation compounds in the sense that many of their X-ray lines are accounted for by assigning them to the group $(000l)$ based on an hexagonal unit cell, the remaining lines being unaccounted for. Upon intercalation the crystal expands in such a way that there exists good vertical correlation between the layers, but without horizontal correlation or registration. This could be the result of extensive horizontal layer displacement during intercalation, and the sodium and lithium intercalates could then be regarded as disordered intercalation crystals. The ability to index some of the

X-ray lines of Na_xMoS_2 based on a tetragonal unit cell suggests that these compounds may be similar to the $Li_xTi_{1.1}S_2$ compounds observed by Barz et al. [116]. The formation of an intercalated bridge compound is also possible, since non-stoichiometric MoS_2 i.e. $Mo_{1-x}S_2$ is known [117]. However, non-stoichiometric MoS_2 has never been found as a result of reactions with stoichiometric MoS_2, and its presence, and the formation of a bridge compound, is unlikely.

In intercalated TaS_2 the spacing between the layers can vary from 3 Å to 30 Å, with little effect on the entropy associated with the superconducting specific heat anomaly [33] which must therefore arise mainly from two-dimensional correlations. Tracey et al. [118] have prepared intercalates between TaS_2 and $NbSe_2$ and a number of metals (e.g.Hg, Ga, In, Cd, Sn, Pb,) and alloys. X-ray data for these is listed in Table VIII.

TABLE VIII

Intercalated compounds between TaS_2, $NbSe_2$ and differing metals (Tracey et al. [118])

Compound	a_0 Å	c_0 Å	Reaction time (days)/Temp °C
TaS_2	3.315	12.08	
TaS_2 (Hg) red	3.325	9.058	4/200
TaS_2 (Hg) black	3.325	8.923	5/200
TaS_2 $(In)_{0.5}$	3.320	7.97	7/450
TaS_2 $(Cd)_{0.9}$ pure phase	3.311	17.44	4/550
TaS_2 $(Cd)_{0.9}$ two phase mixture	3.311	13.58 ⎫ 17.44 ⎬	5/550
TaS_2 $(Sn:Cd)_x$	~3.32	$6.7 \times n$	2/500
$TaS_2(Sn:Cd:Bi:Pb)_x$	3.3	$6.68 \times n$ ⎫ $7.08 \times n$ ⎬	5/450
$NbSe_2$	3.45	12.54	
$NbSe_2$ $(Hg)_{0.8}$	3.445	18.40	14/250
$NbSe_2$ $(In)_{0.75}$	3.470	16.50	7/450
$NbSe_2$ $(Sn:Cd)_x$	3.50	13.64	8/450

7. Polytypism

Polytypism may be defined as the ability of a substance to crystallise in a number of different modifications, in all of which two dimensions of the unit cell are the same, while the third is a variable integral multiple of a common unit. The different polytypic modifications can be regarded as layers of structure stacked parallel to each other at constant intervals along the variable dimension. The two unit cell dimensions parallel to these layers are the same for all modifications. The length of the third axis depends upon the stacking sequence, but is always an integral multiple of the layer spacing. Different manners of stacking result in structures which not only have different morphologies, but may even have different lattice types and space-groups.

Polytypism may be regarded as a special case of polymorphism involving one-dimensional rearrangement. It is, however, distinguishable from polymorphism in that polymorphs usually exhibit a stability governed by the phase rule and the criterion of minimum Gibbs' free energy, having a distinct temperature range over which they exist. Different polytypes form under the same conditions of temperature and pressure and in some cases may exhibit the phenomenon of 'syntactic coalescence' in which different regions of a single crystal may consist of different polytypes in consecutive layers along a prominent crystallographic direction. Most polytypes have the same first and second nearest neighbour co-ordination, and differences of internal energy involve only long-range interactions between widely separated atoms. However, the distinction between polytypism and polymorphism is very indistinct, since many high order polytypes can be converted to those of a lower order by heating.

One of the prototype structures of the dichalcogenides under discussion, cadmium iodide, is a material characterised by a high degree of polytypism. Over 160 different crystal modifications, involving rearrangements of octahedral stacking have been reported [119–121] in cadmium iodide, and over 20 in the like compound, lead iodide. The major interest in these studies is the bearing which they have on current theories of crystal growth. The occurrence of growth spirals, which is also a feature of dichalcogenide crystals grown by the chemical vapour transport technique, has been linked with Frank's screw dislocation mechanism of crystal growth. This mechanism, however, proves to be insufficient to explain the growth of all polytypes observed, despite a large amount of experimental evidence in its support, and other theories of crystal growth e.g. those of Jagodzinski, Vand and Hanoka, Mardix and Steinberger, etc. are currently under study.

In view of their similarity to cadmium iodide, it is surprising that the layer dichalcogenides do not appear to form a wide range of polytype structures, although polymorphism is common. Only two, or possibly three, true polytypes are known for the materials under discussion.

The report by Hahn and Harder of unindexed lines in the powder pattern of TiS_2 suggests that it may be possible to observe polytypism in this substance. Tin disulphide exists in two polytypic forms – the common 2H type and the rarer hexagonal modification 4H, C27 type, with space group $C6_3mc$. In the 2H type, structure $C6$, the stacking is $A\gamma B\ A\gamma B\ A\gamma B\ A\gamma B$ whereas in the $C27$ structure type the stacking becomes $A\gamma B\ C\alpha B\ A\gamma B\ C\alpha B$, and the c-axis is doubled from 5.901 Å to 11.802 Å. The crystal structure of a similar 4H polytype is under study by C. R. Whitehouse. These crystals are all prepared by iodine vapour transport. Molybdenum disulphide can exist in 2H and 3R polytypes, as can $MoSe_2$, WS_2 and WSe_2.

Two polytypic forms of tantalum disulphide have been prepared by chemical transport [123]. In $1s - TaS_2$ the co-ordination is octahedral whereas in $2s - TaS_2$ it is trigonal prismatic. The former is a diamagnetic semi-conductor of low resistivity, whereas the latter is metallic. The electrical properties of the metallic $2s$ polytype may be correlated with energy band models that have been found applicable to other layer compounds, but this is not possible for the $1s$ polytype.

Polymorphism is of common occurrence in the niobium and tantalum dichalcogenides. NbS_2, for example, will crystallise [20] in 2H and 3R polytypes forming $2s-NbS_2$ and $3s-NbS_2$, as well as a non-stoichiometric phase $2s-Nb_{1+x}S_2$. In the system Nb−Se [21] four polymorphic forms exist at differing temperatures, co-ordination in the $1s$ form being octahedral, in $2s(a)$ and $4s(a)$ trigonal prismatic, and in $4s(d)$ a mixture of the two. Structural distortions of the 2H form occur at low temperatures ($\sim 40\,K$) [107]. Six polymorphic forms have been found of $TaSe_2$ [7] viz. 1T, 2H, 3R, $4H_a$, $4H_b$, and 6H.

8. Dislocations and Stacking Faults

In Frank's model of crystal growth, the occurrence of polytypes in cadmium iodide requires the intersection of stacking faults and screw dislocations in such a way that single stacking faults are carried into the structure of a bulk polytype by their repetition around the screw axis. Little use has, however, been made of direct methods of imaging stacking faults or dislocations in cadmium iodide itself. Suitable methods for doing this are transmission electron microscopy and X-ray topography. The first of these is unsuitable for the examination of cadmium iodide, which decomposes rapidly on exposure to the electron beam [124]. Some micrographs of dislocations in cadmium iodide have, however, been obtained [125] showing dissociated dislocations and partials, three- and four-fold ribbons, from which it is shown [126, 127] that dissociation is a prominent feature of dislocations observed in CdI_2. Partials, compounded together with axial screw dislocations, could possibly lead to the formation of polytypes by means of a mechanism hypothesised originally by Daniels [128] and elaborated by Hanoka and Vand [129], in which an nH polytype, together with a Schottky partial dislocation $a\sqrt{3}(0\bar{1}10)$ could result in the formation of a 3nR polytype. Alternatively the movement of partial dislocations could lead to the conversion of one polytype to another by the transposition of stacking faults, as required in Jagodzinski's model of crystal formation.

The application of the electron microscope to chalcogenide layer compounds is found in the work of Amelincx [130, 131] who describes many beautiful effects seen in tin disulphide crystals, although even here beam irradiation can cause damage to the crystals. Tin disulphide is converted [132] by local beam heating to tetragonal tin grown epitaxially on inner lattice planes of the SnS_2 lattice, with $[100]_{Sn}$ parallel to $[100]_{SnS_2}$ and $[001]_{Sn}$ parallel to $[001]_{SnS_2}$.

X-ray topographic methods do not have the disadvantage of causing radiation damage. Lang X-ray topography was first applied to a series of layer compounds, viz. SnS_2, $SnSe_2$, TiS_2, ZrS_2, HfS_2, grown from the vapour phase [90], and an extended study of a related compound, $\gamma-In_2S_3$ has been made [133]. The Lang method requires setting the single crystal at the Bragg reflecting position relative to a narrow, highly collimated, beam of X-rays and traversing the crystal in synchronism with a photographic plate recording the diffracted beam. As the X-ray beam scans the crystal an X-ray image of the dislocations and stacking faults is recorded on the plate, with a one-to-one correspondence between points in the image and points in the crystal projection. Local variations of intensity on the plate record images of such features as line and screw dislocations, stacking faults, and strain fields surrounding precipitates or agglomerated vacancies.

Many indirect studies of stacking faults, dislocation distribution and one-dimensional disorder have, however, been made on cadmium iodide crystals. Foremost among these is the work of Agrawal and Trigunayat [134] who explain the phenomenon of 'arcing' seen on X-ray Laue photographs of cadmium iodide crystals in terms of the accumulation of line dislocations in vertical alignment to form tilt boundaries. 'Arcing' affects over 42% of cadmium iodide crystals examined, and 5% of the lead iodide crystals. Measurement of arc length gives a convenient method for estimation of dislocation densities in these crystals, and for solution grown cadmium iodide a dislocation density of 10^5 to 10^6 cm^{-2} is quoted; beyond the limit of resolution of X-ray topographic methods. The tilt boundaries result from a vertical alignment of triple nodes of dislocations created by simultaneous slip along close-packed directions. In most cases the slip is confined to the basal plane and occurs at regular intervals.

The arcing exhibited by these crystals is affected by temperature. In most cases when the crystals are heated the arcing is reduced as the crystal is converted into the most stable polytype 4H. This is attributed to the migration during heating of unit edge and partial dislocations formed in the lattice during growth away from the tilt boundaries. In a few cases [135, 136] an increase of arcing is observed upon heating, which may be explained by the migration of dislocations, previously pinned against some obstacle, into the tilt boundary. The streaking of X-ray reflections, caused by random one-dimensional disorder, progressively decreases during successive heating runs, showing that the random stacking faults resulting from partial dislocations are gradually eliminated, and that no fresh dislocations are created upon heating. Changes in streaking and arcing are also found at room temperature [137] and are attributed to small fluctuations in ambient temperature.

The effects of stacking faults and dislocations in cadmium iodide are reviewed and interpreted in the light of the differing models of crystal growth by Trigunayat [138].

The mechanism whereby stacking faults and line dislocations arise in the cadmium iodide and molybdenum disulphide structures may be explained in terms of the easy glide arising from the weak bonding between anion-anion layers, producing low energy stacking faults, or in terms of easy glide between close-packed cation-anion layers, giving stacking faults of higher energy. For hexagonal close-packed structures it is convenient to use Thompson's notation [139] for the Burger's vectors of dislocations or the fault vectors of stacking faults.

In Thompson's notation an hexagonal close packed lattice (Figure 19) is inscribed with a tetrahedron *ABCS* joining the atomic positions, as shown. The centres of the faces of this tetrahedron are denoted by small letters a, b, c, σ opposite the vertices *ABCS* of the tetrahedron. These faces represent the possible glide planes in the lattice, while the Burger's vectors of line dislocations are represented by the edges of the tetrahedron. The Burger's vectors of Shockley partials are denoted by *Aσ*, *Ab*, etc.

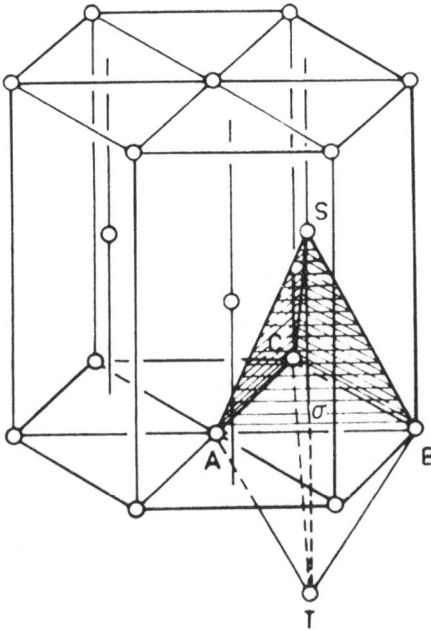

Fig. 19.   Thompson's nomenclature for dislocations in hexagonal close packed structures.

The cadmium iodide structure can be described by the stacking symbol
-----$A\gamma B\ A\gamma B\ A\gamma B$----- the capital letters denoting the relative sequence of the anion
layers, and the greek letters representing the cations. The close-packed layers of cations are sandwiched between two close-packed layers of anions, occupying the
octahedral interstices, but the binding between the two successive anion layers is much
weaker than that between an anion-cation layer. This results in easy cleavage and glide
between two close-packed anion layers, and as a result partials with Burger's vectors
of the type $A\sigma$, $B\sigma$ or σA, σB will be common. They have indeed been found by
Amelincx [131] in cadmium iodide, tin disulphide, tin diselenide and tin sulpho-
selenide. In tin disulphide the ribbons of dissociated partials are particularly wide.
However, the situation is slightly complicated by the presence of two possible (0001)
glide planes, viz that between anion-anion layers, and that between anion-cation
layers;
   (a) If the glide plane lies between layers of anions and falls within the van der
Waals' gap, on gliding along this plane a stacking fault of the type

$$A\gamma B\ A\gamma B\ |\ C\beta A\ C\beta A$$

develops. This can be considered as two sandwich layers stacked in the cadmium
chloride arrangement, for which the stacking symbol is $A\gamma BC\beta AB\alpha C$. The sandwiched
layers of cation and anion remain unsheared in this process and the glide vectors are
of the type $A\sigma$, $B\sigma$, $C\sigma$. Stacking faults of this type, extending in width up to a mil-

limetre, have been detected in vapour transported tin disulphide [90] (see Figure 20). These have a fault vector $\frac{1}{3}a \langle 1\bar{1}00 \rangle$ and are bounded by partials of Burger's vector $\frac{1}{3}a \langle 1\bar{1}00 \rangle$. The faults occur from glide of close packed sulphur layers over each other, and due to the weak bonding between layers are of low energy.

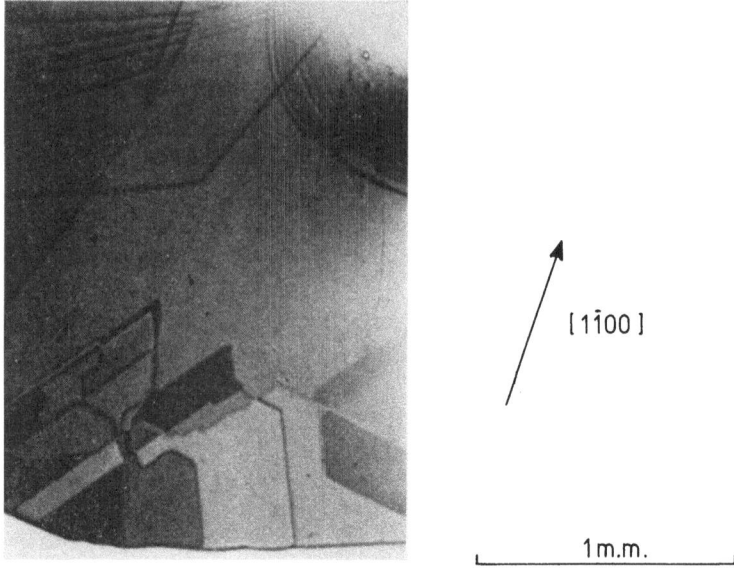

[$1\bar{1}00$]

1 m.m.

Fig. 20.   Stacking faults and line dislocations in tin disulphide, imaged by Lang X-ray topography.

(b) If the glide is between anion and cation within the modular sandwich, glide of two kinds can occur, depending upon whether the cation maintains its original oc-tahedral co-ordination, or whether it reverts to trigonal-prismatic co-ordination. These two types of glide are illustrated in Figure 21. The first (Figure 21a) involves the simultaneous movement of a layer of cations and a layer of anions relative to a fixed layer of anions. If the cations are to retain their octahedral coordination, rather than to move into tetrahedral interstices, they move in a direction different from that of the anions. The term 'synchroshear' has been used to describe this movement. It re-sults in a stacking of the form

$$-----A\gamma BA \mid \beta CB\alpha CB\alpha C-----$$

giving once again two lamellae stacked relative to each other in the cadmium chloride stacking. The direction of movement of the cations is at 60° to the direction of move-ment of the anions and of the remaining part of the crystal.

In the alternative possibility, when glide is considered between anion and cation without synchroshear, the formation results of a fault

$$-----A\gamma BA \mid \beta AC\beta AC-----$$

(Figure 21b). This may be considered as one lamella of the molybdenum disulphide type, for which the stacking is

$$----- A\beta A\beta A\beta A\beta A -----$$

This is less probable, however, as it would require an increase in the anion-anion separation within a lamella, increasing the thickness of the sandwiches and pushing the anions into the van der Waals' gap. A stacking fault of high energy would result.

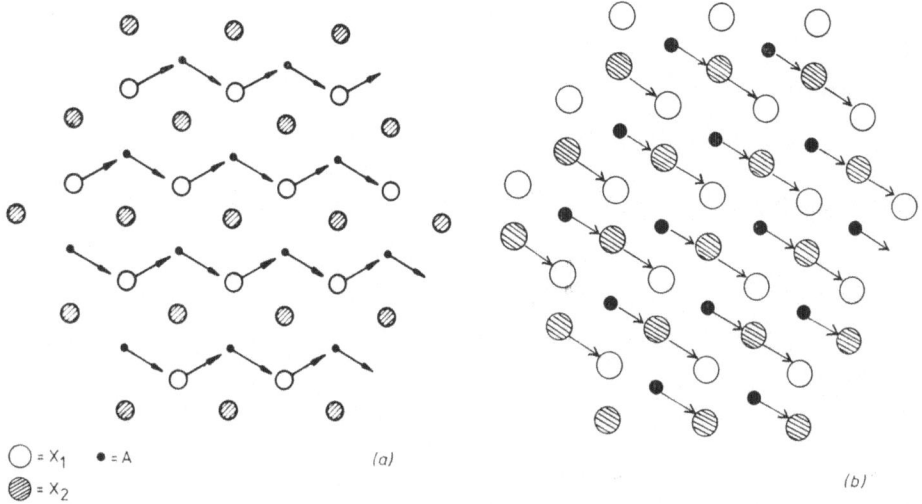

$\bigcirc = X_1$ $\bullet = A$ (a)

$\oslash = X_2$ (b)

Fig. 21.   Glide on anion-cation layers in the CdI₂ lattice. (a) Synchroshear of anions and cations preserves the octahedral co-ordination of the cation. (b) Glide of anions and cations places the cation in trigonal prismatic co-ordination.

Studies of stacking fault energy in deformed $MoSe_2$ [140] from X-ray line profile analysis shows glide along the selenium-selenium layers with a high stacking fault probability of 10%. Similar studies of deformed MoS_2 samples may be accounted for [141] on the basis of sulphur-sulphur glide with a lower stacking fault probability of 3%. In deformed samples a dislocation density of the order of 2.6×10^{10} cm^{-2} is estimated.

Glide processes between cation-anion layers are again thought to operate in molybdenum disulphide like structures. This can, however, distort the coordination of the metal ion from trigonal prismatic to octahedral, and the resulting bond distortion again produces a fault of high energy. In molybdenum disulphide the complete Burger's vector is of the type AB (Figure 22). No matter whether the glide plane is between anion and anion or between anion and cation, glide takes place between close-packed layers and dissociation into partials is possible. In practice, three types of stacking fault and three types of dislocation having different Burger's vectors are found. However, the structure of MoS_2 only allows of two kinds of stacking fault – a low energy stacking fault resulting from S−S glide and a high energy fault involving

Mo−S glide and requiring Mo to take up an octahedral co-ordination. The third type of stacking fault found is probably associated with the occurrence of polytypism in this compound.

The directions and magnitudes of possible Burger's vectors for molybdenum disulphide are shown in Figure 22. If Mo−S glide over a distance $A\sigma$ occurs, the resultant stacking becomes

$$-----A\beta AB\alpha BA \mid \gamma BC\beta C-----$$

in which an octahedral environment for Mo is observed in one layer. This would represent the high energy stacking fault. The low energy stacking fault would arise from S−S glide, giving rise to a stacking

$$-----A\beta AB\alpha BA\beta A \mid C\beta CB\gamma B-----$$

whereby the trigonal prismatic co-ordination of the molybdenum atom is conserved.

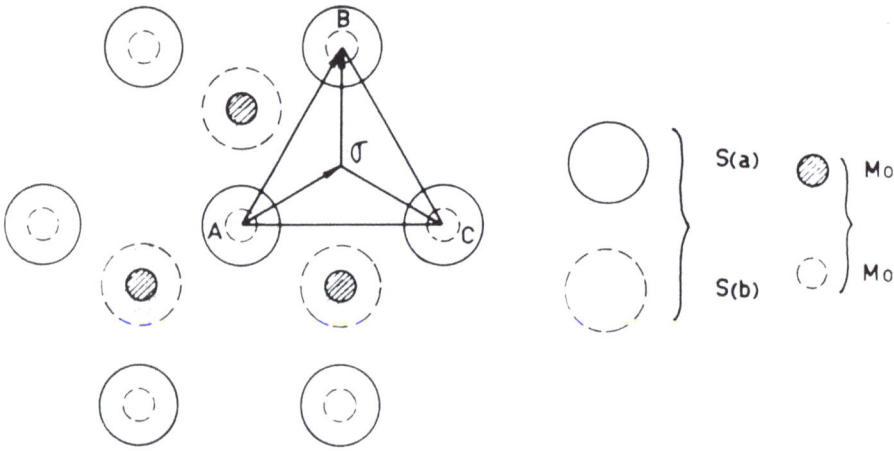

Fig. 22.   Possible Burger's vectors in the MoS₂ lattice.

In tin disulphide, studies of dislocation and stacking fault ribbon formation in the transmission electron microscope show ribbons consisting of three partial dislocations with the same Burger's vectors resulting from the fusion of two single ribbons. This indicates that the two single ribbons are present with vectors $A\sigma + \sigma B$ and $\sigma C + A\sigma$ and provide direct evidence that the two types of glide plane operate in this substance.

If two dislocations with Burger's vectors $\mathbf{b}_1$ and $\mathbf{b}_2$ meet they may form a segment of dislocation with a Burger's vector $\mathbf{b}_3 = \mathbf{b}_1 + \mathbf{b}_2$, provided that elastic energy is not increased in the process. Since the energy of a dislocation is proportional to the square of its Burger's vector, this condition may be written as

$$b_3^2 < b_1^2 + b_2^2.$$

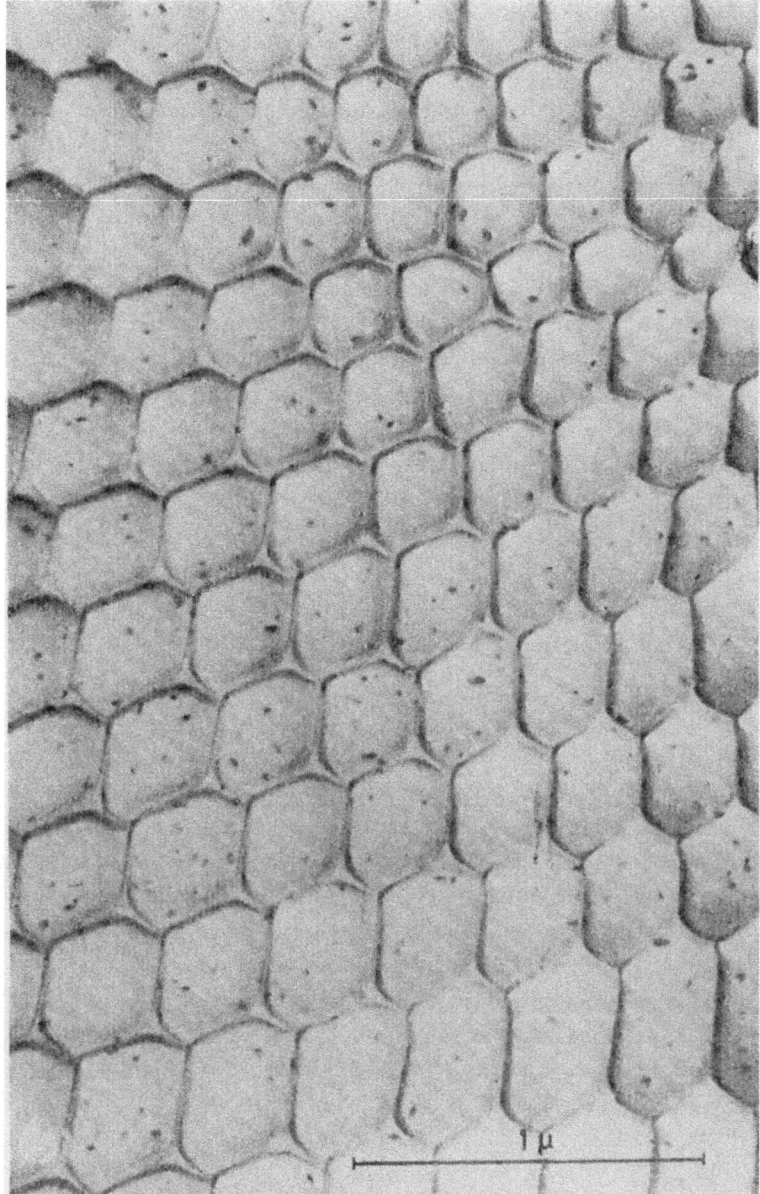

Fig. 23. The formation of hexagonal networks of partial dislocations when two sets of dissimilar dislocations meet in a hexagonal lattice. (SnS_2) (After Amelincx)

When extended ribbons of partial dislocations meet, as for example the three-fold ribbons observed in SnS_2, extensive hexagonal arrays of dislocations are formed (Figure 23). These may be seen in the electron microscope, where they are separated by areas of stacking fault contrast. For example, when two families of dislocations having differing Burger's vectors meet at an angle of 120° (or 60°) an hexagonal grid of screw dislocations may be formed. In tin disulphide all such networks are confined to the c-plane, since this is the main glide plane. Ribbons of partials having Burger's vectors $B\sigma + \sigma A$ and $C\sigma + \sigma A$ are pushed against each other by an applied shear stress. σA condenses with $C\sigma$ to form a perfect dislocation CA having a Burger's vector perpendicular to $B\sigma$. The two dislocations cross under the action of the shear stress and dissociate once again into two partials $C\sigma + \sigma A$.

Hexagonal networks of dislocations formed by this mechanism have been observed in SnS_2 by Amelincx [131]. Extended jogs are observed in the same material and arise from dislocations crossing each other and coalescing with cross glide.

The structures of niobium ditelluride and tantalum ditelluride are rather strongly distorted lattices based upon that of cadmium iodide [142]. The symmetry distortion is monoclinic and the compounds are isomorphous, with $a = 19.39$ Å $b = 3.642$ Å, $c = 9.375$ Å, $\beta = 134°35$ for niobium ditelluride, and $a = 19.31$ Å, $b = 3.651$ Å, $c = 9.377$ Å, $\beta = 134°13$ for tantalum ditelluride. As a result of the monoclinic distortion, ions which are in equivalent positions in the hexagonal cadmium iodide lattice, are no longer so in the deformed structure. The main features of the structure of niobium ditelluride are that the tellurium co-ordination octahedra lie in well-defined chains parallel to b, and that these chains form corrugated sheets parallel to the ab plane. The chains arise from a displacement of the metal cations away from the centres of the octahedra; these displacements can arise in a number of ways, and give rise to domain structures and anti-phase boundaries. These effects have been observed by van Landuyt et al. [143, 144] in an electron microscope examination of cleaved flakes of niobium ditelluride prepared by iodine vapour transport. Similar domain structures are to be expected for $TaTe_2$, WTe_2, $MoTe_2$ and $ReSe_2$, or indeed for any structure of low symmetry which is formed by small distortions of a structure of higher symmetry. The low symmetry structure can be envisaged as having been formed by the cooling of a high symmetry perfect cadmium-iodide-like lattice, the distortions taking effect in one of three equivalent directions.

In the domain structure of $NbTe_2$ domain walls arise from the junction of octahedron chains along the boundaries $(31\bar{1})$ or $(91\bar{3})$ in a herringbone formation, and by the overlapping of chains parallel to (010) – see Figure 24. Antiphase boundaries having displacement vectors

$$R_1 = \tfrac{1}{6}(130)$$

or

$$R_2 = 2R_1 = \tfrac{1}{3}(130)$$

are also observed, where chains running parallel to b are broken, and their ends offset relative to each other by $\frac{1}{6}$th or $\frac{1}{3}$rd of a unit cell respectively. Dislocations in the lattice may be analysed in terms of the perfect cadmium iodide lattice, but show characteristic fine structure consisting of multiribbons of partials, usually three fold but also possibly two-fold or six-fold. These give rise to a series of differing stacking configurations.

In passing from one domain to another the dislocation multiribbons can interact with the domain walls in a number of ways, either by changing in width, by refraction, or by piling up against the boundary. Changes in width arise from differences in the relative orientation of the Burger's vector of the dislocation and the domain lattice on either side of the boundary. This may cause, for example, a multiribbon of three pseudo-perfect dislocations coupled by faults to become three independent dislocations. The repulsion between three such dislocations is then no longer counteracted

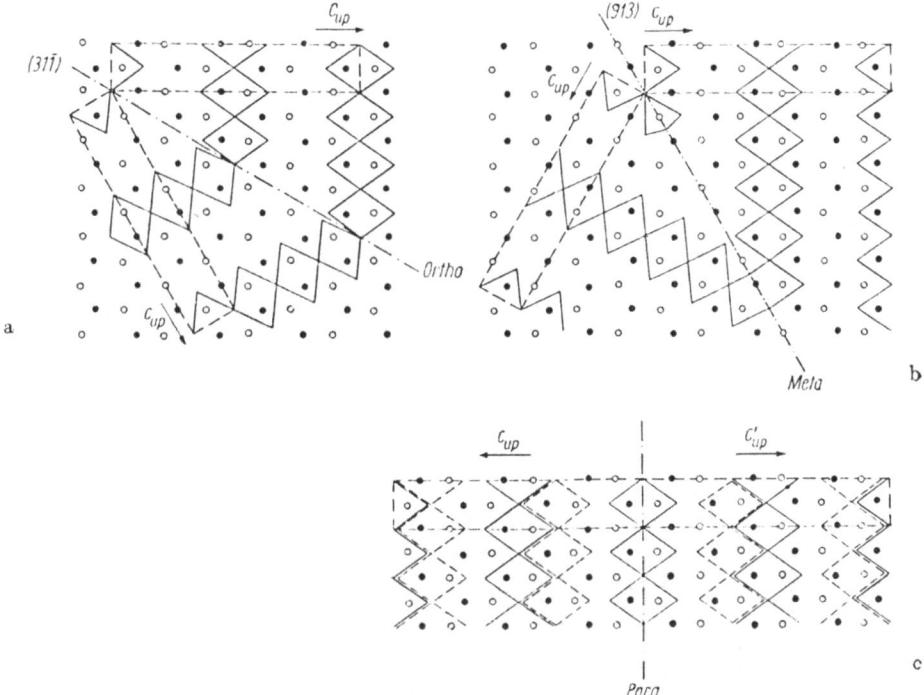

Fig. 24.   Structural models projected on to the c-plane for the three types of domain boundary found in NbTe₂. The symmetry relationships are shown by the chains of cations marked by full lines. Open and filled small circles represent anions above and below the plane of the drawing respectively.
(After van Landuyt, Remaut, and Amelincx).

by the attractive forces due to the faults and the dislocations tend to get wider apart, adopting an equilibrium configuration which is dependent on the distance between domain walls and the changes in lattice orientation. Refraction arises from a change in thickness of the crystal domain, from a difference in stacking fault energy or from a change in orientation of the structure at the domain wall. Pile-up against boundaries arises if dislocations have to generate stacking faults to pass from one domain to another. If the stacking fault energy required to do this is higher than the strain energy available, pile-up occurs at domain walls, with the dislocations tending to align parallel to the wall itself.

The scale $(0.01 \, \mu)$ upon which these dislocation phenomena are seen, although within the range of the electron microscope, lies outside the resolution range of X-ray topography. X-ray topographs of tin disulphide show two types of dislocation. These are the partials surrounding low energy stacking faults, which are evidently grown into the structure, and arrays of undissociated dislocations usually originating from a region of high strain within the crystal. Much of the tin disulphide crystals grown by iodine vapour transport prove to be perfect and completely free of dislocations. The average dislocation density falls as low as $100 \, \text{cm}^{-2}$. The undissociated dislocations include those having Burger's vectors parallel to $[1\bar{2}10]$. From their radius of curvature a lower limit of yield stress of $10^{-5} \, \text{dyn} \, \text{cm}^{-2}$ is estimated for tin disulphide. Other dislocations have Burger's vectors $\langle 11\bar{2}0 \rangle$ and vary from edge to screw along their length. Like the $[1\bar{2}10]$ dislocations they are thought to be generated by handling.

Tin diselenide grown by the iodine vapour transport technique is a further material which is sufficiently perfect for individual dislocations to be resolved by X-ray topography. In topographs of one sample of this material the dislocation arrays run between precipitate particles, probably of tin iodide, whose positions are associated with a prominent growth spiral. Crystals of TiS_2, ZrS_2, and HfS_2 grown by iodine vapour transport have a much higher dislocation density, and in the case of HfS_2 may be so curved by internal strain that divergent beam X-ray topography has to be used to obtain reasonable coverage of the superficial surface area. Experiments have indicated that dislocations in these dichalcogenide compounds are not moved by the application of an electric field.

Only one other compound of the layer type has received a detailed examination by X-ray topographic methods. This is indium sesquisulphide, In_2S_3 [133]. The material also crystallises in space group $P\bar{3}m1$ and may be grown by iodine vapour transport. The layer structure differs, however, from cadmium iodide. Indium lies in octahedral co-ordination between sulphur atoms in cubic close packing. The sulphur atoms lie in close-packed planes parallel to (0001), but the stacking sequence is triple, viz.

$$----- A\gamma BC\beta A\gamma BC\beta -----$$

where A, B, C are sulphur layers at $Z = 0$, $\frac{1}{3}$, and $\frac{2}{3}$, and γ, β are indium layers at $Z = 0.19$ and $Z = 0.81$. The crystals examined were sufficiently perfect to resolve individual dislocations by X-ray topography and the X-ray examination was supplemented by electron microscopy. As in dichalcogenide layer materials, most dislocations observed lie in the plane of easy glide (0001) and were curved. The most common Burger's vector was $b = [100]$ but dislocations having Burger's vectors parallel to $[010]$ and $[110]$ were also found. Growth layers, similar to those found in melt grown crystals were observed, but no stacking faults. If a stacking fault lies between sulphur layers B and C in the stacking sequences above, i.e. between anion-anion layers, the stacking sequence

$$----- A\gamma BC\beta A\gamma B \mid A\gamma B\alpha CA\gamma -----$$

results in the indium atom lying in the co-ordination of the cadmium iodide lattice.

The fault vector would be $R = \frac{1}{3}$ [210], but the high energy required makes this mechanism very unlikely. An anion-cation stacking fault would result in a stacking sequence

$$----- A\gamma BC\beta A \mid \beta CA\gamma B\alpha CA\gamma -----$$

and reduces the In−In distance to 3.02 Å, which is also improbable.

Dislocations observed in In_2S_3 do not split into partials. In the electron beam the compound decomposes with loss of sulphur, giving rise to stacking faults bounded by dislocations with Burger's vectors parallel to [100]. By electron microscopy, extensive six-sided nets of undissociated dislocations were observed, having Burger's vectors $b_1 = [100]$, $b_2 = [110]$, $b_3 = [010]$.

References

1. Landolt-Bornstein, vol. 1. part 4. 'Crystals'. p. 34. Springer-Verlag, Berlin. (1952).
2. R. W. G. Wyckoff, *Crystal Structures*, vol. 1. chap. 4. sect. IV.c.1. p. 266. Interscience Publishers, New York, 1960.
3. M. R. Tubbs, *Phys. Stat. Solidi* **49** (1972), 1.
4. J. A. Wilson and M. D. Yoffe, *Adv. Phys.* **18** (1969), 193.
5. D. L. Greenaway and R. Nitsche, *J. Phys. Chem. Solids* **26** (1965), 1445.
6. J. L. Verble and T. J. Wietling, *Solid State Commun.* **11** (1972), 941.
7. R. Huisman and F. Jellinek, *J. Less Common Metals* **17** (1969), 111.
8. R. C. Evans, *An Introduction to Crystal Chemistry*, 2nd. edn., Cambridge University Press, 1964.
9. F. Hulliger, *Structure and Bonding* (Springer-Verlag, Berlin) **4** (1968), 83.
10. F. Hulliger, *Helv. Phys. Acta*, **33** (1960), 959.
11. J. Guggenheim, F. Hulliger, and J. Mueller, *Helv. Phys. Acta* **34** (1961), 408.
12. R. B. Murray, R. A. Bromley, and A. D. Yoffe. *J. Phys. C* **5** (1972), 738, 746, 3038.
13. V. V. Sobolev, V. I. Donetskikh, A. A. Opalevskii, V. E. Fedorov, E. V. Labkov, and A. P. Mazhova, *Fiz. Tekh. Poluprov. (U.S.S.R.)* **5** (1971), 1025. (English translation in *Soviet Phys. Semicond. (U.S.A.)*).
14. R. A. Bromley, *Phys. Letters* A**37** (1970), 242.
15. P. G. Harper and D. R. Edmunson, *Phys. Stat. Solidi* A**4** (1971), 59.
16. J. V. Acrivos, W. Y. Liang, J. A. Wilson, and A. D. Yoffe, *J. Phys. C* **4** (1971), 118.
17. R. B. Somoano and A. Rembaum, *Phys. Rev. Letters* **27** (1971), 402.
18. B L. Evans and R. A. Hazelwood, *Phys. Stat. Solidi* A**4** (1971), 181.
19. L. H. Brixner, *J. Inorg. Nucl. Chem.* **24** (1962), 257.
20. E. Kadjik and F. Jellinek, *J. Less Common Metals* **19** (1969), 421.
21. R. Huisman, F. Kadjik, and F. Jellinek, *J. Less Common Metals* **23** (1971), 437.
22. V. I. Khalikman and A. G. Duksina, *Izv. Akad. Neorg. U.S.S.R.*, **7** (1971), 1127.
23. R. Huisman, F. Kadjik, and F. Jellinek, *J. Less Common Metals* **21** (1970), 187.
24. E. A. Antonova, K. V. Kiseleva, and S. A. Medvedev, *Fiz. Metallov. Metallovendenie (U.S.S.R.)* **27** (1969), 441.
25. G. E. Myers and G. L. Montet, *J. Appl. Phys.* **41** (1970), 4642.
26. J. Edwards and K. F. Frindt, *J. Phys. Chem. Solids* **32** (1971), 2217.
27. E. A. Antonova, S. A. Medvedev, I. Yu. Shebolin, *Zh. Eksper. Fiz. (U.S.S.R.)* **57** (1969), 329. (English translation in *Soviet Phys. J.E.T.P. (U.S.A.)*.)
28. F. Consadari, A. A. Fife, R. F. Frindt, and S. Gygax, *Appl. Phys. Letters* **18** (1971), 233.
29. G. Hahn and N. Schonberg, *Arkiv. für Kemi.* **7** (1954), 371.
30. A. H. Thompson, F. R. Gamble, and J. F. Revelli, *Solid State Commun.* **9** (1971), 981.
31. C. W. Chu and S. Huang, *Phys. Letters A* **36A** (1971), 93.
32. T. H. Geballe, A. Menth, F. J. Di Salvo, and F. R. Gamble, *Phys. Rev. Letters* **27** (1971), 314.
33. F. J. Di Salvo, R. Schwall, T. H. Geballe, F. R. Gamble, and J. H. Oseiki, *Phys. Rev. Letters* **27** (1971), 310.

34. B. Van Laar, H. M. Rietveld, and D. J. W. Ijido, *Acta Crystallogr.* **A25** (1969), 248. (Conference abstract: 8th. Intern. Congress, I.U.Cr., Buffalo, U.S.A. (1969).)
35. J. M. Voorhoeus-van den Berg and R. C. Sherwood, *J. Phys. Chem. Solids* **32** (1971), 167.
36. E. Ehrenfreund, A. C. Gossard, F. R. Gamble, and T. H. Geballe, *J. Appl. Phys.* **42** (1971), 1491.
37. H. J. Grimmeis, A. Rabenau, H. Hahn, and P. Ness, *Z. Elektrochem.* **65** (1961), 776.
38. R. Vilanova, *Compt. Rend. Hebd. Sean. Acad. Sci. (France), Series B* **271** (1970), 1101.
39. S. Ahmed and P. A. Lee, *J. Phys. D (Appl. Phys.)* **6** (1973), 593.
40. P. A. Lee, G. Said, R. Davis, and T. Lim, *J. Phys. Chem. Solids* **30** (1969), 2719.
41. P. A. Lee, G. Said, and R. Davis, *Solid State Commun.* **7** (1969), 1359.
42. S. G. Patil and R. H. Tredgold, *J. Phys. D.* **4** (1971), 718.
43. G. Said and P. A. Lee, *Phys. Stat. Solidi* **A15** (1973), 29.
44. S. G. Patil, *J. Phys. C* **5** (1972), 2881.
45. J. P. Gowers and P. A. Lee, *Solid State Commun.* **8**, (1970), 1447.
46. G. Mula and F. Aymerich, *Phys. Stat. Solidi* **51** (1972), K 35.
47. F. Aymerich, F. Melamin, and G. Mula, *Solid State Commun.* **12** (1973), 139.
48. L. A. Coward, *J. Non-Crystalline Solids* **6** (1971), 107.
49. T. Suntola, *Solid State Electronics* **14** (1971), 933.
50. M. Iida and A. Hameda, *Japan J. Appl. Phys.* **10** (1971), 224.
51. P. A. Lee and G. Said, *Brit. J. Appl. Phys. (J. Phys. D.) Series 2* **1**, (1968), 837.
52. B. L. Evans and R. A. Hazelwood, *Brit. J. Appl. Phys. (J. Phys. D.) Series 2,* **2** (1969), 1523.
53. V. V. Sobolev and V. I. Donetskikh, *Phys. Stat. Solidi* **42** (1970), K 53.
54. G. Busch, C. Frohlich, F. Hulliger, and C. Steigmeir, *Helv. Phys. Acta* **34** (1961), 359.
55. P. S. Popov and A. P. Zemlyanov, *Izv. Vuz. Fiz. (U.S.S.R.)* (1968), 121. (English translation in *Soviet Physics Journal (U.S.A.)*.)
56. G. Said, Ph.D. Thesis, Brighton Polytechnic, England 1971.
57. M. Y. A. Young and M. L. Cohen, *Phys. Rev.* **178** (1969), 1279.
58. C. Y. Fong, *Phys. Rev. B* **5** (1972), 3095.
59. W. Albers and J. Verbekt, *J. Materials Sci.* **5** (1970), 24.
60. W. Blitz, P. Ehrlich, and K. Meisel, *Z. Anorg. allg. Chem.* **234**, (1937), 92.
61. H. Hahn and B. Harder, *Z. Anorg. allg. Chem.* **288** (1956), 241.
62. H. Hahn and P. Ness, *Z. Anorg. allg. Chem.* **302** (1959), 17.
63. Y. Jeannin, *Ann. Chim.* **7** (1962), 1.
64. G. Hagg and N. Schonberg, *Arkiv für Khemi* **7** (1954), 371.
65. S. Bartram, *Diss. Abstr.* **19** (1958), 1216.
66. P. Ehrlich, *Z. Anorg. allg. Chem.* **260** (1949), 1.
67. P. Bernusset and Y. Jeannin, *Compt. Rend. Acad Sci. Paris* **255** (1962), 934.
68. M. Chevreton and F. Bertaut, *Compt. Rend. Acad. Sci. Paris* **255** (1962), 1275.
69. F. Raam, F. Gronvold, A. Kjekshus, and H. Haraldsen, *Z. Anorg. allg. Chem.* **317** (1962), 91·
70. F. Gronvold and F. J. Langmyhr, *Acta Chem. Scand.* **15** (1961), 1945.
71. E. Strozer, W. Biltz, and K. Meisel, *Z. Anorg. allg. Chem.* **242** (1939), 249.
72. P. Hahn, B. Harder, U. Mutschke, and P. Ness, *Z. Anorg. allg. Chem.* **292** (1957), 82.
73. H. Hahn and P. Ness, *Z. Anorg. allg. Chem.* **302** (1959), 37.
74. H. Hahn and P. Ness, *Z. Anorg. allg. Chem.* **302** (1959), 136.
75. F. K. McTaggart and A. B. Wadsley, *Austral. J. Chem.* **11** (1955), 445.
76. A. A. Al-Hilli and B. L. Evans, *J. Appl. Crystallogr.* **5** (1972), 221.
77. R. Nitsche, *J. Phys. Chem. Solids* **17** (1960), 163.
78. R. Nitsche, H. U. Bolsterli, and M. Lichtensteiger, *J. Phys. Chem. Solids* **21** (1961), 199.
79. B. E. Brown and D. J. Beernsten, *Acta Crystallogr.* **18** (1965), 31.
80. L. H. Brixner, *J. Inorg. Nucl. Chem.* **24**, (1962), 257. (28th. Intern. Congress Pure and Appl. Chem. Montreal, 1961. Conf. abstracts, p. 202.)
81. J. G. Smeggil and S. Bartram, *J. Solid State Chem.* **5** (1972), 391.
82. P. Bro, *J. Electrochem. Soc.* **109** (1962), 1110.
83. H. Schafer, *Chemical Transport Reactions*, Academic Press, New York, 1964 p. 55.
84. J. F. Revelli, W. R. Phillips, and R. E. Schwall, O.N.R. conference on the Physics and Chemistry of Layer Compounds. Abstracts. Monterey, California, 17th–18th. August, 1972.
85. H. P. B. Rimmington and A. A. Balchin, *Phys. Stat. Solidi* **A6** (1971), K 47.
86. C. R. Whitehouse, H. P. B. Rimmington, and A. A. Balchin, *Phys. Stat. Solidi* **A18** (1973), 623.

87. H. P. B. Rimmington and A. A. Balchin, *J. Crystal Growth* **21** (1974), 171.
88. H. P. B. Rimmington and A. A. Balchin, *J. Materials Sci.* **9** (1974), 343.
89. F. A. S. Alamy and A. A. Balchin, *Materials Res. Bull.* **8** (1973), 245.
90. H. P. B. Rimmington, A. A. Balchin, and B. K. Tanner, *J. Crystal Growth* **15** (1972), 51.
91. H. P. B. Rimmington, Ph.D. Thesis. Brighton Polytechnic, England 1973.
92. F. Hulliger and E. Mooser, *Progress Solid State Chem.* **2** (1965), 330.
93. F. Jellinek, International Conference on Electron Diffraction and the Nature of Defects in Crystals. Melbourne, 1965. Pergamon Press Inc., New York, (1966). Paper II.c.2.
94. S. Tengner, *Z. Anorg. allg. Chem.* **239** (1938), 127.
95. J. Benard and Y. Jeannin, *Adv. Chem.* **39** (1962), 191.
96. P. Ehrlich, *Z. Anorg. allg. Chem.* **234** (1937), 97.
97. L. J. Norrby and H. F. Franzen, *J. Solid State Chem.* **2** (1970), 36.
98. A. D. Wadsley, *Acta Crystallogr.* **10** (1957), 715.
99. I. Oftedal, *J. Physik. Chem.* **134** (1928), 301.
100. S. Brunix and M. Chevreton, *Materials Res. Bull.* **3** (1968), 309.
101. A. Gleizes and Y. Jeannin, *J. Solid State Chem.* **1** (1970), 180.
102. Y. Jeannin, *Bull. Soc. Fr. Miner. Crystallogr* **90** (1967), 528.
103. F. Jellinek, *Acta Crystallogr.* **10** (1957), 620.
104. W. Rüdorff, *Chimia* **19** (1965), 489.
105. E. Ehrenfreund and A. C. Gossard, *Phys. Rev. B.* **5** (1972), 1708.
106. A. H. Thompson, F. R. Gamble, and R. F. Koehler, *Phys. Rev. B.* **5** (1972), 2811.
107. M. Marezio, P. B. Dernier, A. Menth, and G. W. Hull, *J. Solid State Chem.* **4** (1972), 425.
108. C. B. Carter and P. M. Williams, *Phil. Mag.* **26** (1972), 393.
109. P. M. Williams, O.N.R. conference on the Physics and Chemistry of Layer Compounds, Monterey, California. August 17th.–18th, 1972. Conference abstracts.
110. J. M. Thomas, E. L. Evans, B. Barch, and J. L. L. Jenkins, *Nature (Phys. Sci.)* **235** (1972), 126.
111. H. Fernandez-Moran, M. Ohstuki, C. Hough, and H. Krebs, O.N.R. conference on the Physics and Chemistry of Layer Compounds, Monterey, California. August 17th.–18th., 1972. Conference abstracts.
112. F. R. Gamble, F. J. Di Salvo, R. A. Klemm, and T. H. Geballe, *Science* **168** (1970), 568.
113. H. Fernandez-Moran, M. Ohstuki, A. Hibino, and C. Hough, *Science* **174** (1971), 498.
114. R. Somoano, V. Hadek, and A. Rembaum, A.I.P. conference proceedings on Superconductivity in *d* and *f* band metals, Rochester, New York, U.S.A. 29th.–30th. October, 1971.
115. R. B. Somoano, V. Hadek, and A. Rembaum, *J. Chem. Phys.* **58** (1973), 697.
116. H. E. Barz, A. S. Cooper, E. Corenzwit, M. Marezio, B. T. Matthias, and P. H. Schmidt, *Science* **175** (1972), 884.
117. F. Jellinek, *Arkiv. für Kemi* **20** (1963), 447.
118. T. Tracey, P. S. Gentile, and J. T. Budnik, O.N.R. conference on the Physics and Chemistry of Layer Compounds, Monterey, California, U.S.A. August 17th.–18th., 1972. Conference abstracts.
119. A. R. Verma and P. Krishna, *Polymorphism and Polytypism in Crystals*, John Wiley, New York, 1966.
120. P. Krishna and A. R. Verma, *Phys. Stat. Solidi* **17** (1966), 437.
121. G. C. Trigunayat and G. K. Chadha, *Phys. Stat. Solidi* **A4** (1971), 9.
122. J. R. Gunter and H. R. Oswald, *Naturwissenschaften* **55** (1968), 177.
123. L. E. Conroy and K. R. Pisharody, *J. Solid State Chem.* **4**, (1972) 345.
124. A. J. Forty, *Phil. Mag.* **5** (1960), 787.
125. R. Prasad and O. N. Srivastava, *J. Phys. D. (Appl. Phys.)* **3** (1970), 91.
126. R. Prasad, G. Singh, and O. N. Srivastava, *Japan J. Appl. Phys.* **8** (1969), 810.
127. R. Prasad, *J. Crystal Growth* **15** (1972), 259.
128. B. K. Daniels, *Phil. Mag.* **14** (1966), 487.
129. J. I. Hanoka and V. Vand, *J. Appl. Phys.* **39** (1968), 5288.
130. S. Amelincx and P. Delavignette, in J. B. Newkirk and S. Wernick (eds.), *Direct Observation of Dislocations in Crystals*, Interscience Publishers, 1969.
131. S. Amelincx, *The Direct Observation of Dislocations*, Academic Press, New York, 1964, p. 271.
132. J. R. Gunter and H. R. Oswald, *J. Appl. Crystallogr.* **2** (1969), 196.
133. P. Buck, *J. Appl. Crystallogr.* **6** (1973), 1.

134. V. K. Agrawal and G. C. Trigunayat, *Acta Crystallogr.* **A25** (1969), 401, 407.
135. G. Lal and G. C. Trigunayat, *J. Crystal Growth* **11**, (1971), 177.
136. G. Lal and G. C. Trigunayat, *Acta Crystallogr.* **A26** (1970), 430.
137. V. K. Agrawal, *Acta Crystallogr.* **A28** (1972), 472.
138. G. C. Trigunayat, *Phys. Stat. Solidi* **4** (1971), 281.
139. N. Thompson, *Proc. Phys. Soc.* **66** (1953), 481.
140. P. Pratap and D. L. Bhattacharya, *Phys. Stat. Solidi* **A12**, (1972), 61.
141. P. Pratap and R. K. Gupta, *Phys. Stat. Solidi* **A9** (1972), 415.
142. R. E. Brown, *Acta Crystallogr.* **20** (1966), 264.
143. J. van Landuyt, G. Remaut, and S. Amelincx, *Materials Res. Bull.* **5** (1970), 731.
144. J. van Landuyt, G. Remaut, and S. Amelincx, *Phys. Stat. Solidi* **41** (1970), 271.

PHASE TRANSITIONS AND CHARGE DENSITY WAVES IN THE LAYERED TRANSITION METAL DICHALCOGENIDES

P. M. WILLIAMS

Dept. of Chemical Engineering and Chemical Technology, Imperial College, London*

Abstract. This present chapter deals with the occurrence within the d^1 metallic layered transition metal dichalcogenides of group Va, of Fermi surface driven structural instabilities. The primary aim here is to present the diffraction evidence from electron, X-ray, and neutron observations for the adoption of a charge density wave ground state coupled to a periodic lattice distortion and to show how this ground state may be perturbed by substitutional doping and intercalation. The electronic consequences of these effects are then briefly reviewed.

1. Introduction

The idealised structures of the transition metal dichalcogenides of groups IVa, Va and VIa have been fully discussed elsewhere in the present work. Likewise, the magnetic, optical and electrical properties of these materials have been described in detail; in this chapter, we will therefore reproduce only such data as is necessary. Briefly, the group IVa dichalcogenides, such as TiS_2, HfS_2 etc. adopt the CdI_2 structure, $p\bar{3}m$, (denoted hereafter as 1T) in which the metal is octahedrally coordinated by chalcogen atoms (Figure 1a) and on the basis of simple charge transfer arguments,

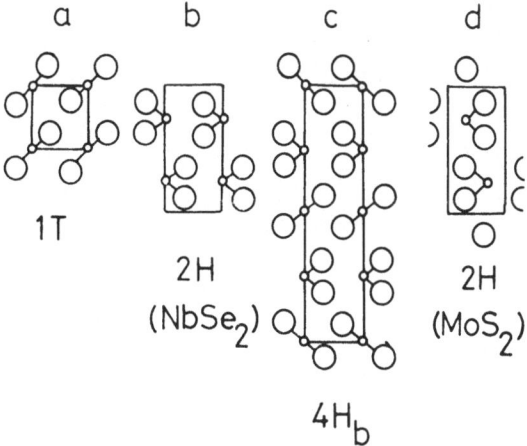

Fig. 1. (a) (11.0) projection of the unit cell of CdI_2, $P\bar{3}m$, showing $AAA...$ stacking sequence of layers. Large circles, chalcogen, small circles, metal. (b) (11.0) projection of unit cell of $NbSe_2$, Pb_3/mmc, showing $ABAB...$ stacking sequence and trigonal prism coordination. (c) (11.0) projection of unit cell of $4H_bTaS_2$ showing alternation of octahedral (cf. (a)) and trigonal prism (cf. (b)) layers. (d) (11.0) projection of unit cell of MoS_2, $P6_3/mmc$ (cf. – $NbSe_2$).

* Present address: V. G. Scientific, East Grinstead, Sussex, England.

F. Lévy (ed.), Crystallography and Crystal Chemistry of Materials with Layered Structures. 51–92. All Rights Reserved.
Copyright © 1976 by D. Reidel Publishing Company, Dordrecht-Holland.

are expected to be insulators with a filled 'p' like valence band, split by an energy gap from the empty metal 'd' states [1]. The sulphides and selenides of Hafnium and Zirconium conform broadly to this simple model, but the ditellurides of both metals, in common with the ditellurides throughout this group of materials are distorted [2]. Perhaps more surprisingly, the regular titanium chalcogenides are apparently metallic and exhibit tendencies towards structural distortions although there is some controversy, as we shall see later, over whether their metallic properties result from genuine band overlap or from impurity donation effects.

In group Va, however, genuine metallic properties result from the addition of the extra electron which enters the lowest energy band based on metal 'd' states. Again, the octahedrally coordinated CdI_2 structure is found (VSe_2, $1T-TaS_2$), but polytypes in which the metal atom is trigonal-prismatically coordinated by chalcogen are also common (e.g. $2H-NbSe_2$ $P6_3/mmc$, Figure 1b) as well as a range of mixed polytypes in which trigonal prismatic and octahedral coordination alternate (e.g. $4H_b$ Figure 1c). The ditellurides are distorted, and the structure of $NbTe_2$, for example, in which marked metal-metal pair bonding occurs, has been extensively investigated [2]. In group VI, the addition of a further electron results in semiconducting properties in MoS_2, which adopts a regular trigonal prismatic structure (Figure 1d) and again the ditellurides are exceptional.

Broadly speaking, the physical properties of the compounds (with the exception of the tellurides) through the series IVa → Va → VIa conform to this simple pattern, but within group Va where metallic properties are expected, the detailed behaviour of

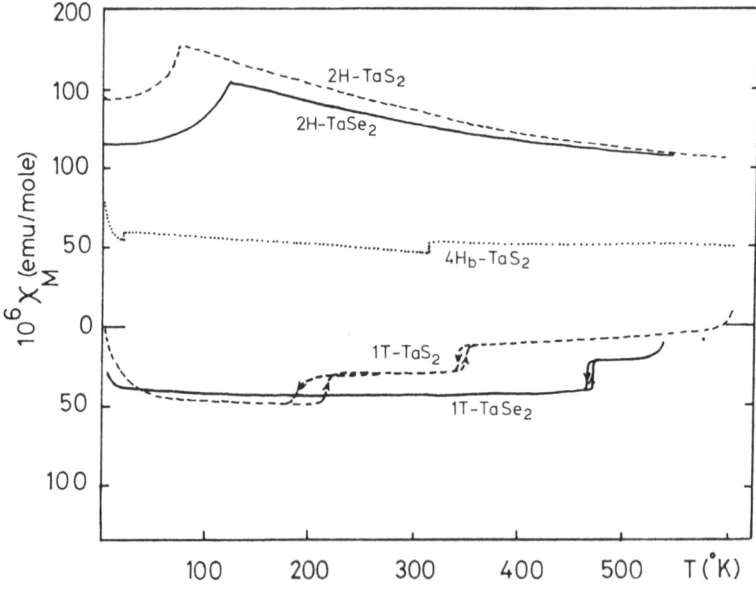

Fig. 2.   Molar susceptibilities for Ta chalcogenides (after Wilson *et al.* [3] – composite data [4, 5, 6]).

many of the materials has until recently been regarded as 'anomalous'. Thus in Figure 2, we reproduce magnetic susceptibility data for $2H-TaS_2$, $2H-TaSe_2$, $4H_b-TaS_2$, $1T-TaS_2$ and $1T-TaSe_2$, [3, 4, 5, 6]. From the simple band structure predictions of the W/Y model [1] and from calculated energy band schemes [7, 8, 9, 10] these Va materials should have a high density of 'd' states at the Fermi level, but only the susceptibilities of the 2H polytypes are consistent with this. Those for $4H_b$ and $1T-TaS_2$ and $1T-TaSe_2$ are anomalously low and furthermore, we encounter evidence in all three cases of phase transitions (at temperatures summarised in Table I) which

TABLE I

Temperatures at which the incommensurate – commensurate transition takes place for a range of transition metal dichalcogenides

		Td (K)
2H	TaSe₂	120
1T	TaSe₂	473
4H_b	TaSe₂	410
4H_b	TaS₂	316
1T	TaS₂	350 ($1T_1 \rightarrow 1T_2$)
		190 ($1T_2 \rightarrow 1T_3$)
1T	VSe₂	106
1T	TiSe₂	200

was not revealed in early X-ray studies of these materials [11]. Even more conclusive evidence for such transitions is revealed in resistivity measurements (Figure 3), where although the metallic behaviour of the 2H polytypes is only weakly perturbed near ~ 100 K, sharp first order discontinuities are observed for both the 1T materials. At low temperatures, $1T-TaSe_2$ apparently returns to a more normal metallic behaviour but the increase in resistivity with decreasing temperature in the sulphide is actually more reminiscent of a semiconductor, as pointed out originally by Thompson *et al.* [4] In fact the number of carriers in $1T-TaS_2$ measured by these authors is quite high and the resistivity behaviour is more properly described in terms of some enhanced carrier scattering, as pointed out by Wilson *et al.* [3]. Benda [12] in measurements of infra red reflectivity shows, furthermore, that the usual Drude type free carrier scattering is absent for the pure $1T-TaS_2$ phase and only appears for substantial Ti doping.

Such behaviour is clearly more complex than might be anticipated from the simple band models proposed [7, 8, 9, 10] for these group Va compounds and the strong evidence for first order transitions in the resistivity data for the 1T polytypes in particular has prompted, in recent years, a careful re-examination of the structures of such materials. Early X-ray studies [11] of $1T-TaS_2$ and $1T-TaSe_2$ suggested, in fact, that these compounds formed perfect CdI_2 structure crystals, but subsequent examination of $1T-TaS_2$ in transmission electron diffraction by Wilson and Yoffe [1] revealed complex patterns with extra periodicities not expected for the space group $P\bar{3}m$. At that time, these authors ascribed the diffraction effects to the possible formation of a shear structure (such as are observed in the titanium and vanadium oxides) or to the existence of an excitonic insulator phase (again following early investigations of the properties of $3d$ transition metal oxides [13]). Thompson and co-workers, in later diffraction studies associated [5] with measurements of the resistivity of $1T–TaS_2$ in fact observed first order transition behaviour in the c-axis dimensions (determined from X-ray powder diffractograms), an increase of 0.02 Å occurring at the lower temperature transition (in Figure 3). These latter observations have stimulated numerous more detailed investigations of the structures of single crystals of this and other

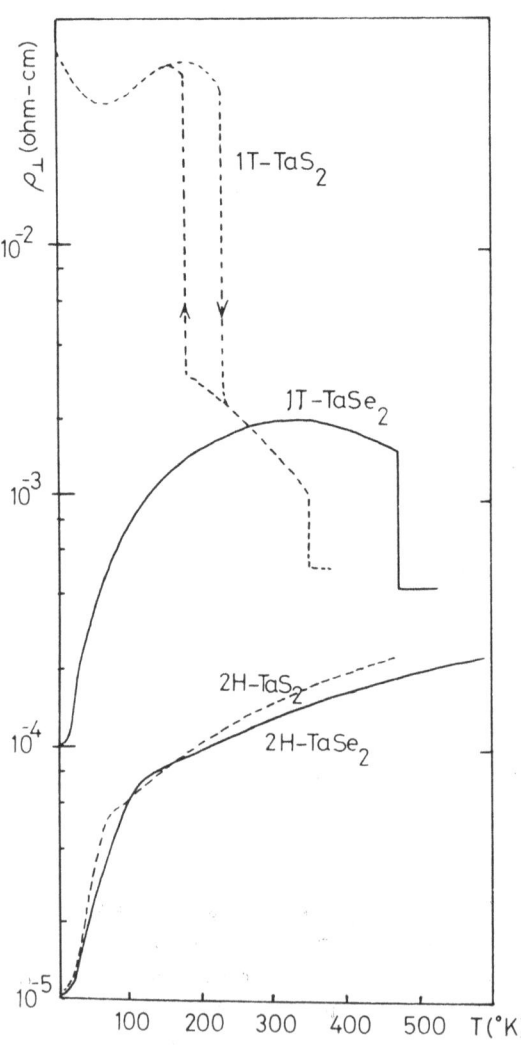

Fig. 3. Resistivity layer as a function of temperature for 1T and 2H tantalum dichalcogenides.

group Va materials using electron, neutron and X-ray diffraction techniques over temperature ranges from 5 to 500 K, and as we shall see below, complex diffraction effects are not limited to 1T–TaS$_2$ alone but are exhibited by nearly all of the group Va dichalcogenides. There is thus firm evidence linking the electronic anomalies observed for these materials with temperature dependent structural instabilities.

Again, in recent years, there has been considerable theoretical interest [7, 8, 9, 10] in the determination of the detailed features of the electronic energy bands in these materials. Numerous calculations have been carried out for the group Va materials, so that there is now a fair degree of consensus on the predicted form of the Fermi surface for such compounds; of particular interest within the present context is the highly anisotropic nature of these materials in which the relatively weak inter-layer interaction is expected to give rise to little dispersion $\parallel$ c*. The simple form of the energy bands for the 1T materials then gives rise to large flat areas of Fermi surface, a situation known to favour the formation of spin or charge density waves [14]. In either case the density wave in the conduction electron gas is then stabilised by coupling with a periodic lattice distortion [14], which thereby gives rise to the extra periodicities observed with diffraction techniques. We shall thus see that the structural instabilities displayed by these transition metal dichalcogenides are in fact related to the observations of spin density waves in chromium [14, 15] and to the Peierls-like distortions in the one dimensional metallic systems currently of interest [16]. We shall present here only a selection of the wealth of structural observations reported to date for the group Va materials and will attempt to show how these observations may be interpreted in terms of coupled CDW/PLD states and how these effects correlate with the 'anomalous' physical properties of the materials. For convenience, we will present the results for octahedral and trigonal prismatic polytypes in consecutive sections and will also deal separately with observations of both doped and intercalated materials.

2. The Direct Observation of Periodic Lattice Distortions in the Octahedrally Coordinated (1T) Transition Metal Dichalcogenides

The octahedral, 1T, polytypes related to the CdI$_2$ structure ($P\bar{3}m$) are most common amongst the $3d^1$ and $5d^1$ dichalcogenides of vanadium and tantalum, although DiSalvo and co-workers have recently reported the growth of 1T structures in crystals of NbSe$_2$ substitutionally doped with Ti [17]. We will consider first the disulphide and diselenide of tantalum and later the diselenide of vanadium. For convenience, we also present in this section the results for 4H$_b$–TaS$_2$ in which octahedrally and trigonally prismatically coordinated layers alternate (Figure 1); here, the effects most readily observed in diffraction may be directly attributed to distortions within the octahedral layers.

2.1. 1T–TaS$_2$

Since the earliest report of 'superlattice' effects in high energy electron diffraction

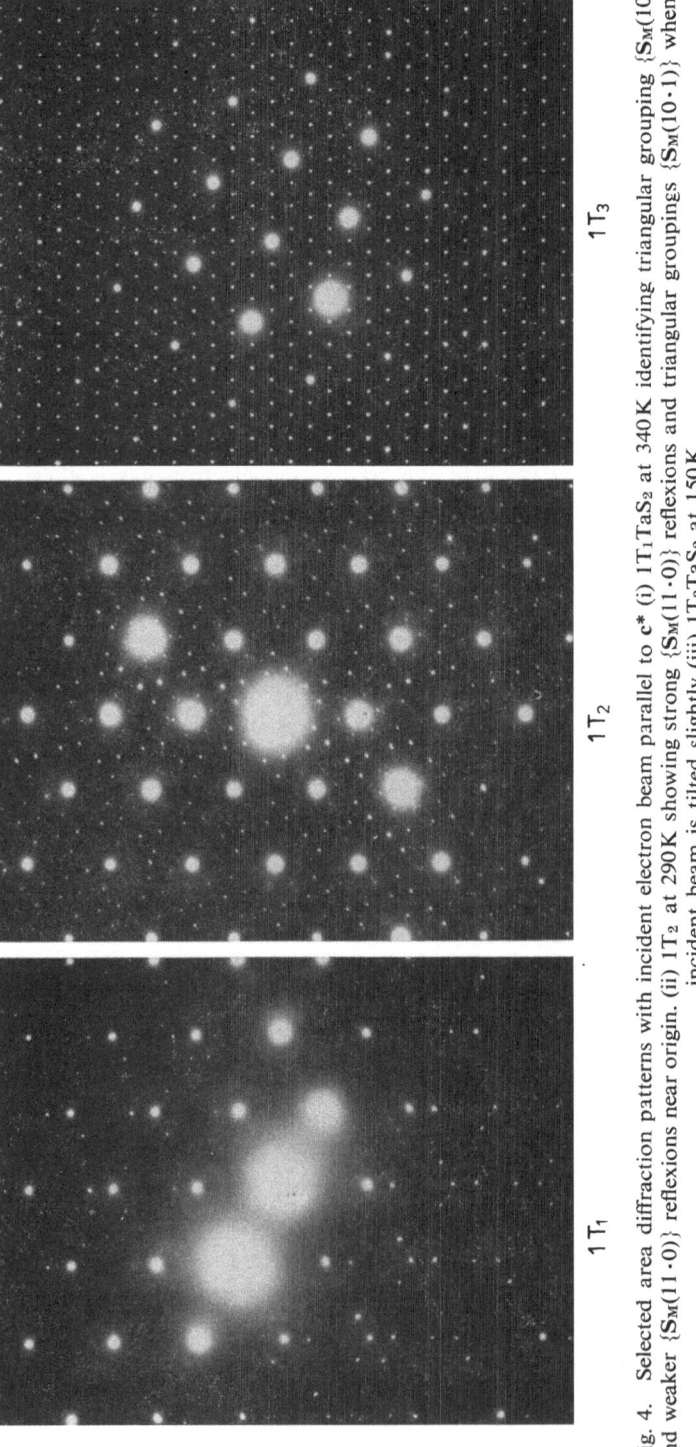

Fig. 4. Selected area diffraction patterns with incident electron beam parallel to **c*** (i) $1T_1TaS_2$ at 340 K identifying triangular grouping $\{S_M(10\cdot l)\}$ and weaker $\{S_M(11\cdot 0)\}$ reflexions near origin. (ii) $1T_2$ at 290 K showing strong $\{S_M(11\cdot 0)\}$ reflexions and triangular groupings $\{S_M(10\cdot 1)\}$ when the incident beam is tilted slightly (iii) $1T_3TaS_2$ at 150K.

$1T_1$ $1T_2$ $1T_3$

patterns of 1T–TaS₂ [1], extensive studies have been carried out by many groups [3, 18, 19] using electron and X-rays. Although there have been minor differences over interpretation of results, there is an almost precise reproducibility between the diffraction patterns recorded over a wide range of temperatures for meterials prepared in different laboratories. We reproduce in Figure 4 the results of Scruby *et al.* [18] showing ($hk \cdot 0$) transmission electron diffraction patterns for 1T–TaS₂ recorded at the temperatures indicated (using either heated or liquid nitrogen cooled stages in a JEM 7A electron microscope operating at 100 keV). The three phases are denoted $1T_1$, $1T_2$ and $1T_3$ (after Williams *et al.* [18]) to indicate regions in the $\varrho \to T$ relation-

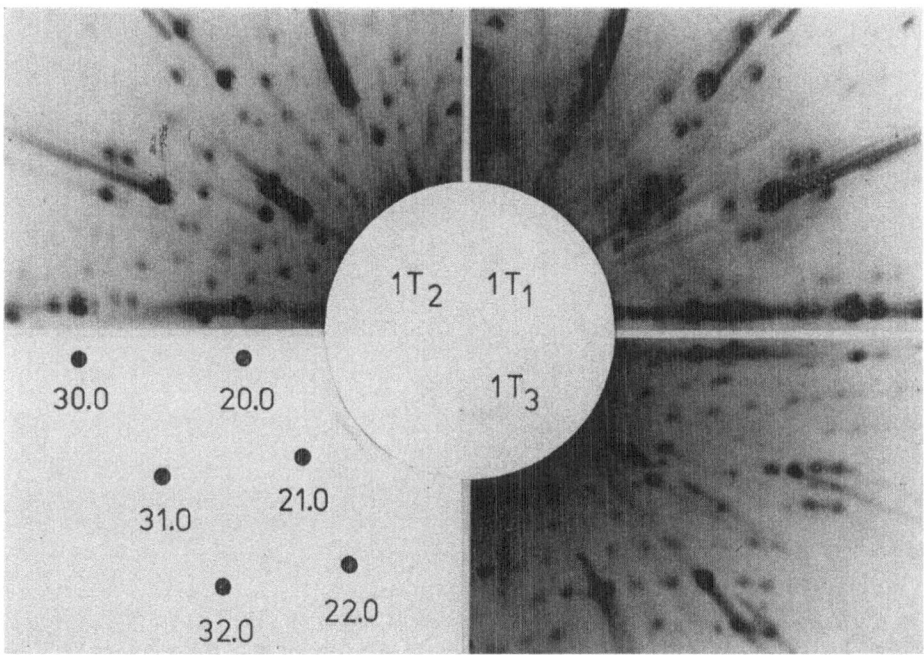

Fig. 5. Oscillation X-ray photographs with incident beam in similar orientation to Figure 4 for (i) $1T_1$ at 390 K, (ii) $1T_2$ at 290 K, (iii) $1T_3$ at 80 K, confirming the reciprocal space geometry given by electron diffraction. Distortion amplitude too small in $1T_1$ for $\{S_M(11 \cdot 0)\}$ to be observed.

ship of Figure 3 above the first transition, between first and second transitions and below the second transition in temperature respectively. The transformation temperature to the $1T_3$ phase we will denote T_d (after Wilson *et al.* [3]), that from $1T_1$–$1T_2$, T_d^1. Figure 5 reproduces these author's single crystal X-ray oscillation photographs, again for the three regimes $1T_1$, $1T_2$, $1T_3$. With both X-ray and electron diffraction the extra periodicities referred to above which are not consistent with the simple CdI₂ structure are clearly noted, the X-ray observations confirming that these periodicities seen perhaps most readily in selected area diffraction are characteristic of bulk single crystals and not of localised defects. We summarise this data schematically in Figure

6; thus the matrix CdI_2 structure reflexions, $\{M\}$, are surrounded by groups of extra reflexions, $\{S_M\}$, disposed with three or six fold symmetry about each 'parent' in $\{M\}$. If vectors Q_{in} define the disposition of the $\{S_M\}$ about each $\{M\}$ as shown then the following may be deduced for the three 'phases'.

(a) In both $1T_1$ and $1T_2$ the magnitude of vectors Q_{i1}, $|Q_{i1}|$, defining the first order reflexions in $\{S_M\}$, is an irrational fraction of the matrix structure. Only in $1T_3$ do Q_{i1} and a^* become *commensurate* (i.e. they are related on a rational coincidence mesh), the $\{S_M\}$ defining in the $(hk \cdot 0)$ plane a perfect $\sqrt{13} \times \sqrt{13}$ superlattice mesh.

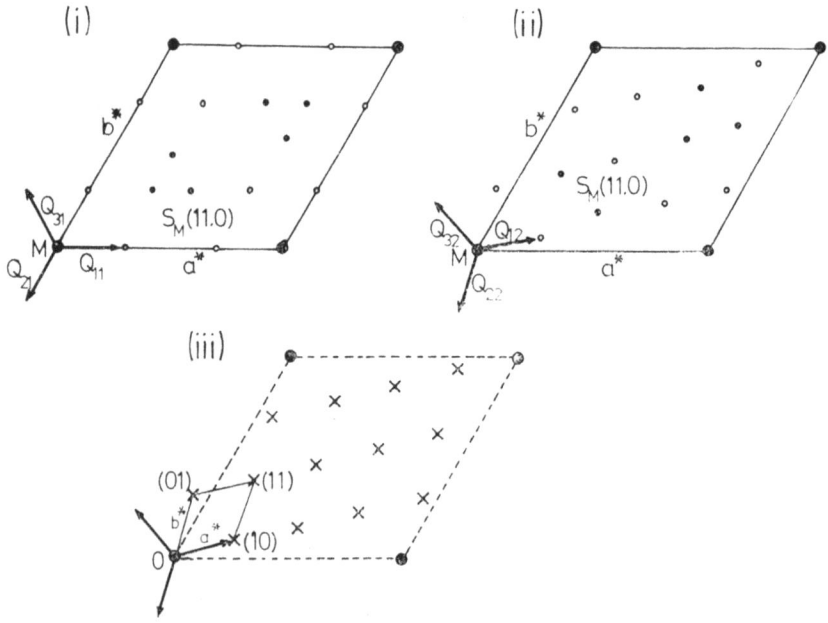

Fig. 6.   Reciprocal cells for each distorted 1T–TaS₂ phase, where Q_{i1}, Q_{i2}, Q_{i3} are distortion wave vectors. In (i) $1T_1$ and (ii) $1T_2$ the unit cell is that of the matrix with: ● – $\{S_M\}$ reflexions in $(hk \cdot 0)$ plane of $\{M\}$; ○ – $\{S_M\}$ reflexions at $\pm\frac{1}{3}c^*$ ($S_M(10 \cdot 1)$ and $S_M(01 \cdot \bar{1})$ lie at $+\frac{1}{3}c^*$ and $-\frac{1}{3}c^*$ respectively). In (iii) $1T_3$, the commensurate distortions define a reduced $\sqrt{13} \times \sqrt{13}$ reciprocal cell, shown in projection.

(b) The vector Q_{i1} in $1T_1$ is parallel to a^* but rotates by $11°6'$ away from a^* at the transition to $1T_2$. Q_{i2} then rotates gradually, as the temperature is lowered, to $\sim 14°$ at the transition to $1T_3$ to form the perfect $\sqrt{13} \times \sqrt{13}$ superlattice. At the same time $|Q_{i2}|$ varies with temperature as indicated in Table I.

(c) Second order reflexions of the type $Q_{12} \pm Q_{22}$ are observed in both $1T_1$ and $1T_2$ (and of course in the perfect superlattice mesh of $1T_3$); these are arrowed in Figure 4. Following the notation of Scruby *et al.*, we will adopt a system of coordinates local to each $\{M\}$ to describe the associated $\{S_M\}$; thus the first order reflexions we will denote $\{S_M(10 \cdot l)\}$, the second $\{S_M(11 \cdot l)\}$. This notation will be used throughout the text in the descriptions of all the materials.

(d) Within the $1T_1$ phase, in addition to the $\{S_M\}$, there is a diffuse background in the form of streaks passing through the $\{S_M(10 \cdot l)\}$ apparently in the form of 'circles' or 'bicycle chains' (Wilson). The intensity in these streaks increases with temperature until the irreversible 1T–2H inter-polytypic transition occurs near 450 K; Figure 7 shows a typical higher-temperature transmission electron diffraction pattern accentuating this streaking. The precise interpretation of this effect is the subject of some controversy, as we shall see later.

We have described these results for $1T$–TaS_2, in a degree of detail which cannot be matched for all the other materials without making the present text an unreadable catalogue, in order to establish the basic characteristics of the observations many of which are common to other layer materials within the group. Thus we shall see that in many respects, the incommensurate nature of the Q_{i1}, at high temperature is the

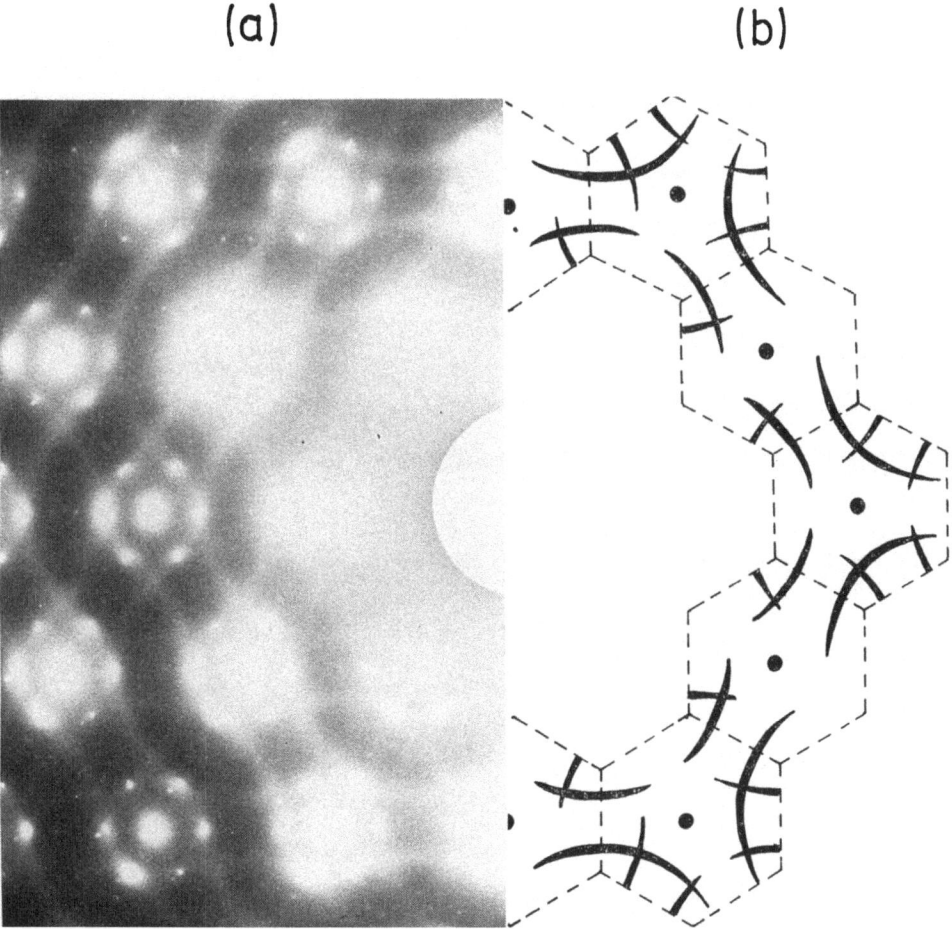

Fig. 7. Selected area diffraction pattern for $1T_1TaS_2$ at 420 K showing diffuse streaking which images the Fermi surface. All six $\{S_M(10 \cdot 1)\}$ reflexions are visible because of crystal buckling.

'signature' of the charge density wave. The diffracted intensities derive, however, not from scattering from the maxima in this charge density wave in the electron gas (this produces only a negligible contribution) but from the associated PLD. Here the effects in diffraction space are far less contentious in that the intensity variations in the $\{S_M\}$ for increasing h, k indices in M ($hk \cdot 0$) may be predicted exactly (Section 3 below). For the present, we will note that in Figure 8 which reproduces an Mo$K\alpha$ oscillation photograph with the beam incident perpendicular to $\mathbf{c}^*$, the intensities in the $\{S_M(10 \cdot l)\}$ (rows arrowed) increase dramatically as we pass from M (00.l), (01·l), through M (10·l), (11·l), to M (20·l), (21·l). This intensity variation, as has been pointed out by Comes and others [16] in the related case of the ID metals, is very specifically the 'signature' of a periodic modulation of the lattice and *cannot* be explained in terms of defect ordering or shear structures. This X-ray observation is thus, along with neutron data for the 2H polytypes, the single most conclusive piece of evidence for PLD/CDW behaviour in these materials. The intensities in the electron observations again show similar increases for $\{S_M(10 \cdot l)\}$ with increasing M ($hk \cdot 0$) but in this case, curvature of the Ewald sphere in reciprocal space precludes too literal an interpretation of the variations. Again with reference to Figure 8, careful study shows the reflexions $\{S_M(10.l)\}$ to be located out of the ($hk \cdot l$) planes at positions $l = \pm \frac{1}{3}c^*$, forming an octahedron of reflexions about each $\{M\}$ of local sym-

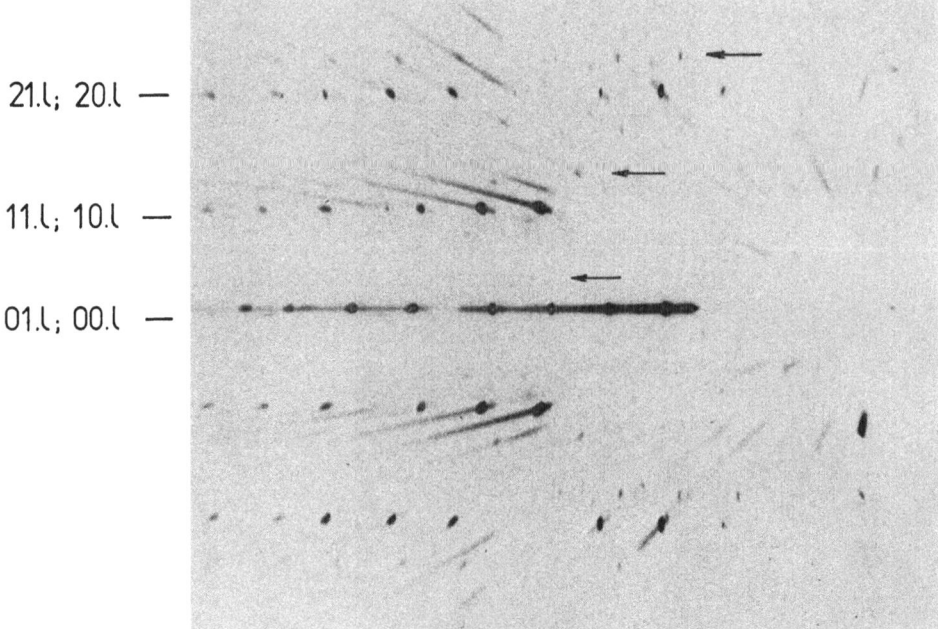

21.l; 20.l —

11.l; 10.l —

01.l; 00.l —

Fig. 8. Oscillation X ray photograph of 1T₂TaS₂ with incident beam perpendicular to $\mathbf{c}^*$ and recorded with a cylindrical camera. There is an increase in intensity of $\{S_M\}$ reflexions (indicated by arrows) relative to the adjacent matrix reflexions with increase in the component of scattering vector normal to $\mathbf{c}^*$.

metry $\bar{3}$ (a fact which may also be deduced from careful e.m. observations by tilting flat areas of sample). This observation is true for both the $1T_1$ and $1T_2$ phases and implies strong correlation between PLD's in neighbouring layers; the situation in $1T_3$ is more complex and is discussed fully below. Finally, with reference to TaS_2, we note that in both the $1T_1$ and $1T_2$ phases, the matrix appears to remain strictly trigonal (within the limits of error of these observations); in $1T_3$, there is some indication that this is no longer true [3, 18] the structure being triclinic with a streaking repeat of $13c_0$, as opposed to $3c_0$ in $1T_1$ and $1T_2$, for the CDW/PLD's.

2.2. $1T-TaSe_2$

The resistivity and susceptibility data for $1T-TaSe_2$ indicate only one first order transition. In diffraction, this is indeed shown to be the case, the results of Wilson *et al.* [3] revealing that a reversible transition between phases resembling the $1T_1$ and $1T_3$ regimes (using the notation for $1T-TaS_2$ – above) occurs near 465 K. The absence of the $1T_2$ phase in the selenide and the contrast in low temperature resistivity behaviour even when perfect $\sqrt{13} \times \sqrt{13}$ superlattices are observed for both the sulphide and the selenide (Figure 3) constitute major differences between what might otherwise have been expected to be very similar compounds. Thermodynamically, the enthalpy of transition of $1T_1-1T_2$ TaS_2 is comparable to that for $1T_1-1T_3$ $TaSe_2$, the $1T_2-1T_3$ transition enthalpy in the sulphide being nearly seven times smaller [5]. Within $1T_1$, Q_{in} is again parallel to $\mathbf{a}^*$ and strong diffuse scattering is observed as for the sulphide; the disposition of the $\{S_M (10.l)\}$ at $\pm\frac{1}{3}c^*$ again suggests a $3c_0$ stacking repeat for the PLD's in real space. $1T_3$ $TaSe_2$ exhibits the same perfect $\sqrt{13} \times \sqrt{13}$ superlattice below T_d with, once more, a $13c_0$ stacking repeat [20].

2.3. $4H_b-TaS_2$

This mixed polytype of TaS_2, in which trigonal and octahedral layers alternate, exhibits two first order discontinuities in resistivity, as shown in Figure 3. At the higher temperature, a transition between phases again related to $1T_1$ and $1T_3$ occurs at T_d, as is seen in Figure 9 which reproduces data of Williams *et al.* [21]. Above 415 K, (Figure 9a), extra reflexions $\{S_M(10 \cdot l)\}$ defined by incommensurate vectors Q_{in} are again accompanied by diffuse scattering in the form of streaks. The transition to the commensurate state (Figure 9c) results always in the formation of two superimposed $\sqrt{13} \times \sqrt{13}$ superlattices symmetry related by the mirror planes of the matrix structure normal to the $(hk \cdot 0)$ plane. Even for thin crystals the intensities in these two superlattices are equal leading to the suggestion [3] that the PLD's occur only in octahedral lamellae with alternating disposition of Q_{in} along c^* with respect to the matrix $\mathbf{a}^*$ axis. Note that the intensities of the $\{S_M(10 \cdot l)\}$ in these $(hk \cdot 0)$ projections both above and below the transition and their behaviour on tilting the samples (by up to 45°) suggest that they are located at $l=0$ [22], in contrast to the $l=\frac{1}{3}c^*$ position in $1T$ TaS_2 and $1T$ $TaSe_2$.

One further point of considerable theoretical interest emerges from careful studies of $4H_b$ TaS_2 in the vicinity of the 415 K transition. As is seen in Figure 9b, the transi-

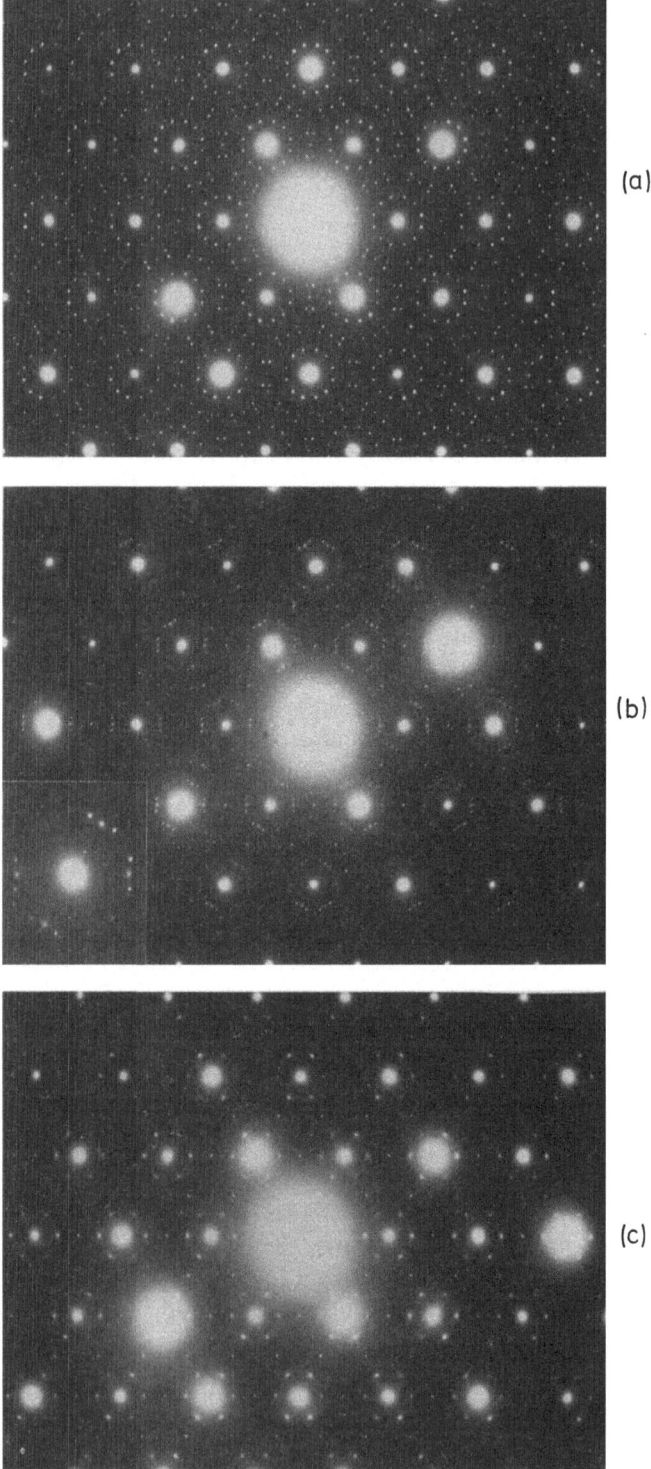

Fig. 9. (hk·0) selected area electron diffraction patterns for $4H_b$–TaS_2 (a) 290 K showing two orientations of a perfect $\sqrt{13} \times \sqrt{13}$ superlattice (b) At 416 K during sluggish transition to high temperature phase. Note increase in width of diffraction maxima for latter relative to low temperature phase by factor of approximately 5 (c) 450 K with reflexions $\{S_M\}$ now defined by vectors $\mathbf{Q}_{i1} \parallel \mathbf{a}^*$. A weak 2×2 superlattice overlays the patterns in all three cases.

tion is often sluggish and within the size of area defined by the selected area aperture of the electron microscope, both high and low temperature phases are found to co-exist [23]. However, as is readily observed in the inset in Figure 9b, the width of the diffraction maxima for the '$1T_1$' phase is approximately five times that for the '$1T_3$' phase at the same lattice temperature (as no significant temperature gradient can exist across such small areas under stable operating conditions). This observation suggests a dramatic increase in the coherence length of the CDW at the $1T_1 \rightarrow 1T_3$ transition, as is fully discussed below.

We note finally the appearance in all three plates of Figure 9 of a sharp 2×2 super-lattice which appears to be relatively temperature independent. The possible origins of this scattering have been discussed by Wilson *et al.* [3] and by Hughes and Liang [24] in terms of either chalcogen deficiency [3] or of a Fermi surface driven effect related to a different spanning vector from that producing the predominant $\sqrt{13}$ scattering [24]. From the sample dependence of this 2×2 superlattice, the author inclines towards the former interpretation, although the possible interplay between point defect ordering and the F.S. instability should undoubtedly be considered.

2.4. $1T$–VSe_2

The dichalcogenides of vanadium are again of octahedrally coordinated CdI_2 struc-ture; stoichiometry in the sulphide is however difficult to control, phases from VS_2 to VS being produced by incorporation of excess sulphur within the interlammellar region

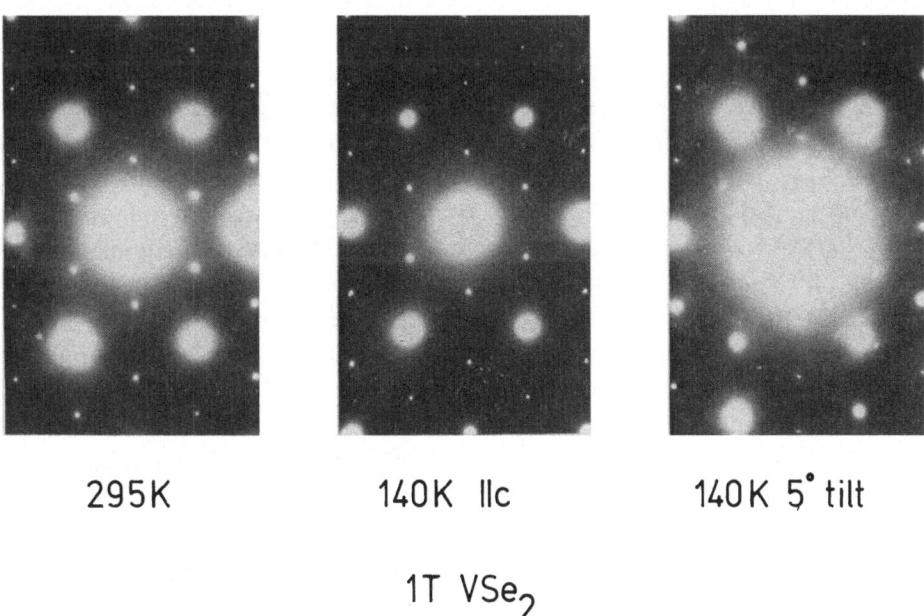

295K   140K ∥c   140K 5° tilt

1T VSe$_2$

Fig. 10.   ($hk \cdot 0$) selected area electron diffraction patterns for $1T$–VSe_2 at (i) 295 K (ii) 140 K with the beam incident exactly ∥ c^* (iii) 140 K with the crystal tilted by 5°, revealing strong $\{S_M(10 \cdot l)\}$ reflexions at $\pm c^*/3$ in addition to diffuse streaking.

(Figure 1). VSe$_2$ has been prepared and characterised recently by Thompson and Silbernagel [25] and, in common with the d^1 compounds, exhibits anomalies in resistivity and susceptibility. Figure 10 reproduces (hk.0) transmission electron diffraction patterns recorded at 300 K, 140 K and 140 K respectively and reveals once again at low temperatures the occurrence of sharp extra reflections $\{S_M(10 \cdot l)\}$, in addition to diffuse streaking which may be observed even at room temperature. At 140 K, Q_{in} is incommensurate with a^*, by about 2%, and strong diffuse scattering is present, but at 40 K, the projections of Q_{in} on the ($hk \cdot 0$) plane define a perfect 4×4 superlattice and the diffuse scattering is absent. We have not observed this incommensurate-commensurate transition directly (the effects in diffraction space being a relatively small change in Q_{in} and, we conjecture, the disappearance at the transition of the diffuse scattering as in 1T–TaS$_2$) but susceptibility data [25] suggests this occurs at 106 K. Tilting experiments in the electron microscope once more show the $\{S_M(10 \cdot l)\}$ to be located at $l = \pm \frac{1}{3}c^*$, as was the case with 1T–TaS$_2$ and 1T–TaSe$_2$.

2.5. The Titanium Dichalcogenides

TiS$_2$ and TiSe$_2$ share the same structure as the 1T polytypes of group Va, but the metal is now nominally in a di^0 configuration. Conductivity data, however, suggest weakly metallic properties [26] although band structure calculations [27] all indicate semi-conducting character, as is expected for the other d^0IVa compounds. In the latter case, donation of carriers from impurities or from (non-stoichiometric) excess of titanium is then thought to give rise to the metallic properties, except that we note that films of TiS$_2$ prepared by the reactive sputtering of Ti in H$_2$S show very low carrier density [28].

In general, we would therefore anticipate that TiSe$_2$ (and TiS$_2$) would have either no Fermi surface, or else one with very few carriers compared with 1T–TaS$_2$ for example. Under such circumstances, Fermi surface driven structural instabilities as described above for the d^1Va materials would not be expected to occur. Figure 11, however, shows the appearance of a well defined 2nd order structural transition near 250 K; thus at room temperature (11a), pronounced diffuse scattering in the form of streaking is observed along a^* directions, peaking in intensity around $\pm a^*/2$. By 120 K, this streaking has sharpened into well defined reflexions $\{S_M\}$ at $(\pm a^*/2, \pm c^*/n)$. From the variation in intensity of these $\{S_M\}$ across the plate and from the curvature of the crystal, we deduce that $n = 2$ and that the unit cell in fact doubles in dimensions in both the a and c axis directions following the transformation. Since the $\{S_M\}$ appear to be approximately equal in intensity to the $\{M\}$ in certain regions of reciprocal space, the magnitudes of the displacements 11c and 11a must be large. Nevertheless, resistivity data [25] indicate metallic behaviour right down to helium temperatures, so that this transformation is not simply driven by band filling and dia-magnetic pairing in a Peierl's-like sense [29]. Its possible relationship to the CDW/PLD anomalies in the d^1 compounds is discussed below.

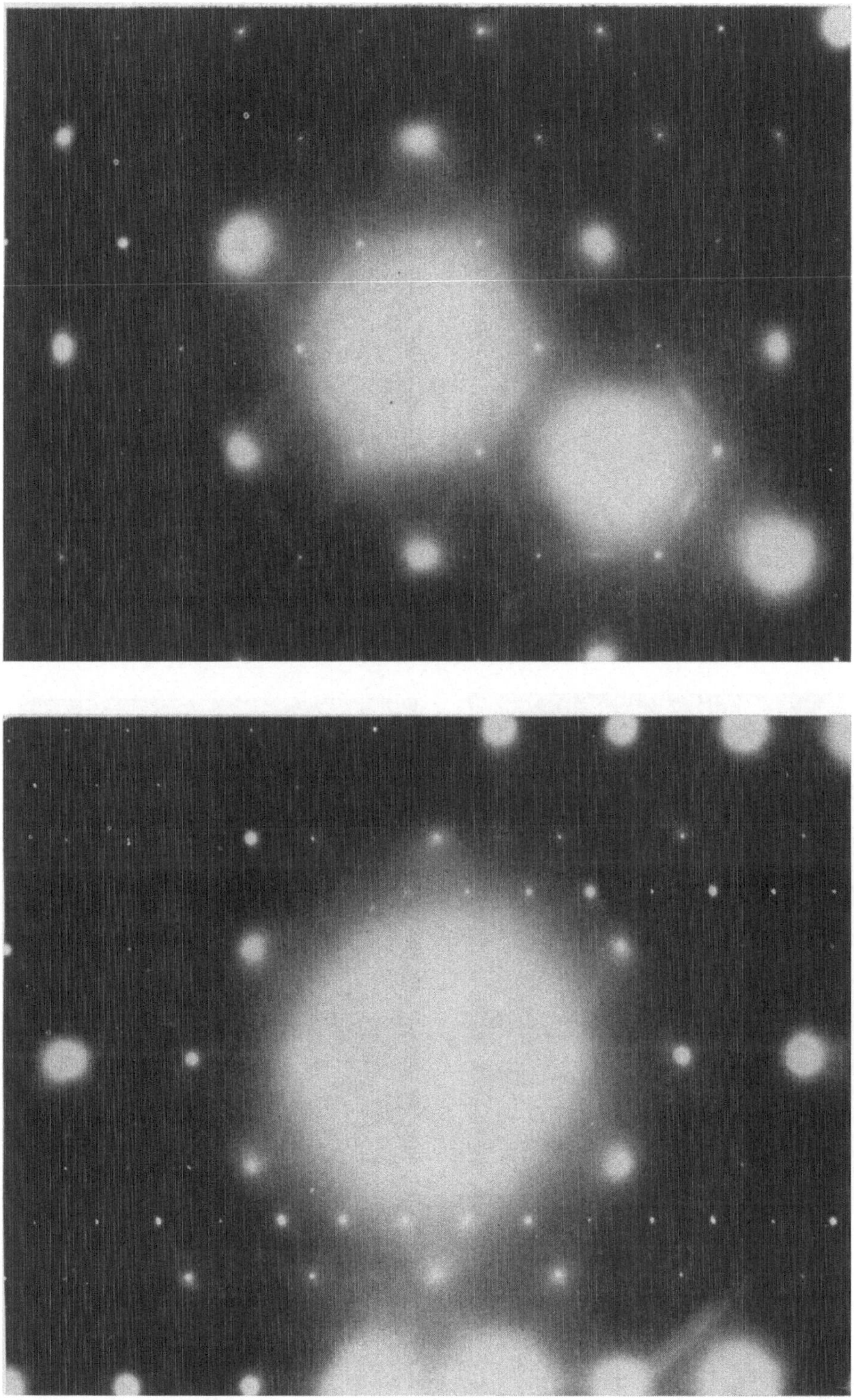

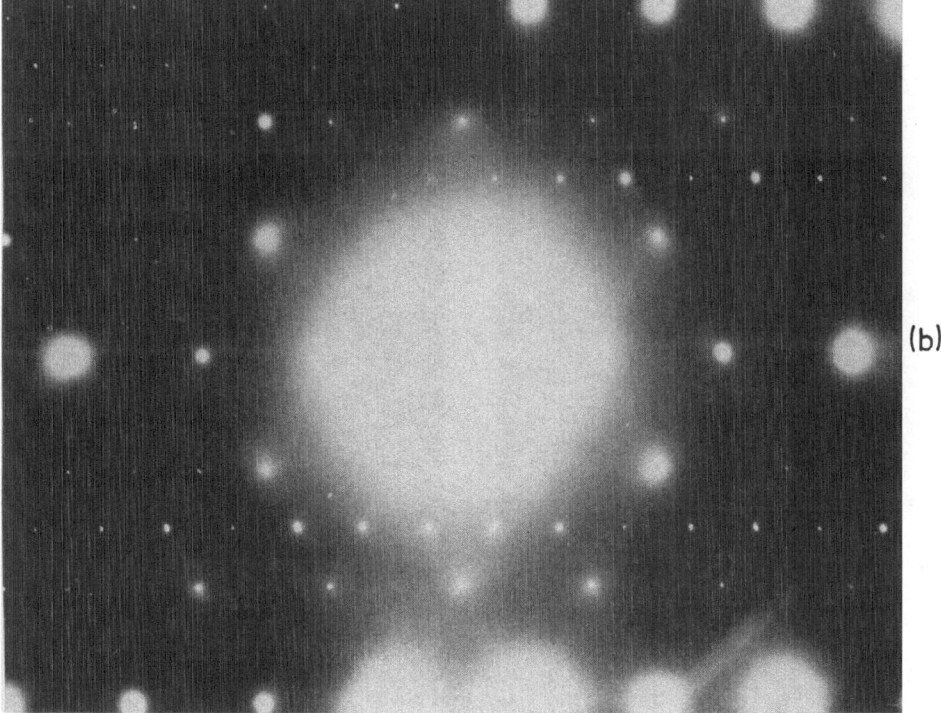

(a)

(b)

Fig. 11. ($hk \cdot 0$) selected area electron diffraction patterns for 1T–TiSe$_2$ at (a) 290 K showing streaks ∥ **a*** (b) 120 K showing sharp {$\mathbf{S}_M(10 \cdot l)$} reflexions at $\pm \mathbf{a}^*/2$, $\pm \mathbf{c}^*/2$.

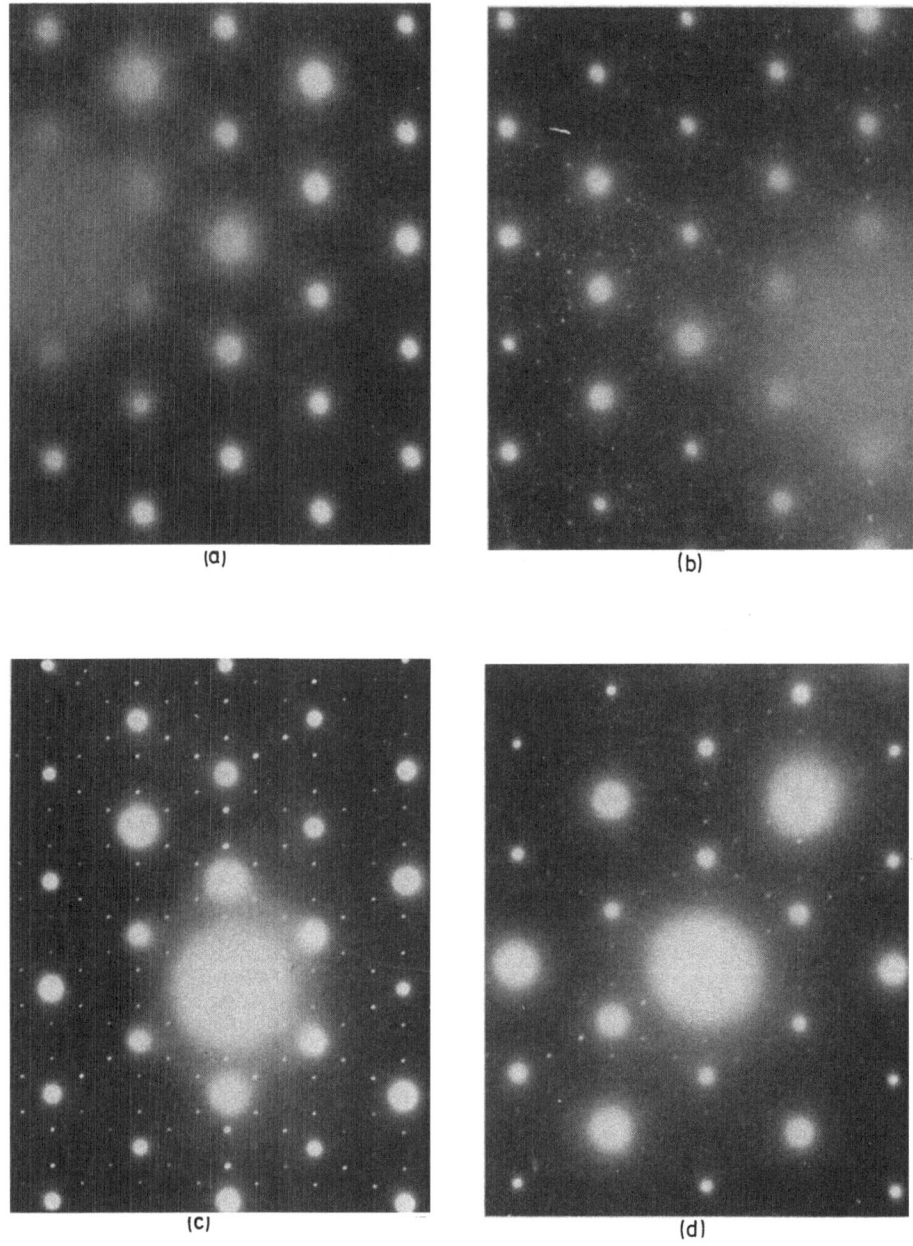

Fig. 12. $(hk \cdot 0)$ selected area electron diffraction patterns for (a) 2H TaSe$_2$ 300 K showing weak residual streaking $\parallel$ **a*** (which results from slight excess of Ta in crystals) (b) 2H TaSe$_2$, 140 K, immediately above T_d showing sharp $\{S_M(10 \cdot 0)\}$ reflexions with $Q_i < \frac{1}{3}$**c***, (c) 2H TaSe$_2$, 40 K, with Q_i now commensurate at $\frac{1}{3}$**c*** and showing second order $\{S_M(11 \cdot 0)\}$ reflexions. (d) 2H–NbSe$_2$, 17 K, showing sharp $\{S_M(10 \cdot 0)\}$ reflexions with Q_i incommensurate $(< \frac{1}{3}$**c***$)$; note the $\{S_M\}$ are weak when $\{M\}$ is strong for $h - k = 3n$, indicating antiphase PLD's in Nb and Se layers.

3. The Trigonal Prismatically Coordinated d^1 Transition Metal Dichalcogenides

3.1. 2H–TaSe$_2$ AND 2H–NbSe$_2$

We present results for both materials together because of their similarity in behaviour. Thus Figure 12 reproduces $(hk \cdot 0)$ electron diffraction patterns for 2H–TaSe$_2$ at 300, 110 and 40 K [30] and for 2H–NbSe$_2$ at 17 K [31]. In both materials, the electromagnetic anomalies described earlier are seen to be accompanied by the appearance at sufficiently low temperature of extra periodicities $\{S_M\}$ in diffraction, with vectors Q_{in} defining the $\{S_M\}$ close to but not coincident with $\pm a^*/_3$. The deviation, δ, from the commensurate 3×3 superlattice state is small at the onset of the appearance of the $\{S_M\}$ for both materials (1–2%) and this is amply demonstrated in recent neutron diffraction [32] observations reproduced in Figure 13. Thus, δ, is plotted as a function of temperature in Figure 14 from which it is seen that 2H–TaSe$_2$ transforms to the commensurate 3×3 superlattice state at 90 K. Below this temperature, Figure 12c shows that strong second order $\{S_M(11 \cdot l)\}$ reflexions, undetectable in the higher temperature patterns of Figure 12b, develop. Monckton et al. [32] have in fact measured the intensities in second order reflexions of the form $\{S_M(20 \cdot l)\}$ above T_d and deduce the presence of a periodic lattice distortion of wave vector $2Q_{in}$ from the anomalous magnitude of the $\{S_M(20 \cdot l)\}$ intensity. 2H–TaS$_2$ exhibits a similar incommensurate-commensurate transition [33] but 2H NbSe$_2$ remains incommensurate even down to 5 K (Figure 14) even though, as Monckton et al. point out [32], at T_0,

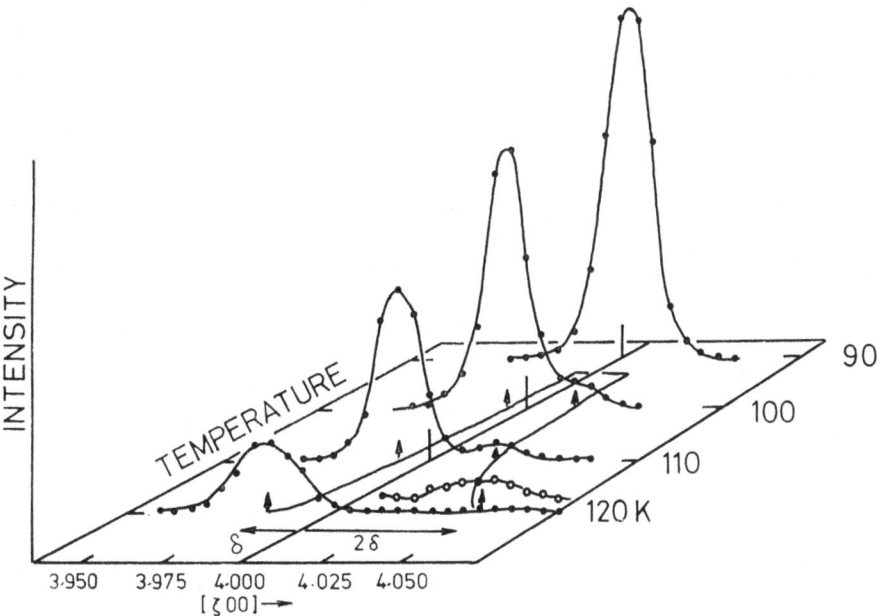

Fig. 13. Elastic scans along $\langle \zeta 00 \rangle$ showing the incommensurate primary peak in 2H TaSe₂ (cf. Figure 12) at $\zeta = (1-\delta) a^*/3$ and a secondary peak at $(1+2\delta) (a^*/3)$. Open circles indicate multiplication by ten.

the onset temperature of the incommensurate state (120 K for 2H–TaSe$_2$, 35 K for NbSe$_2$), δ is similar for both the selenides. Again, with reference to Figures 12 and 13, we note that the width of the diffraction maxima, and hence the coherence length of the CDW/PLD is very similar both above and below T_D in TaSe$_2$. Also with reference to these 2H polytypes, we note that the 2 × 2 superlattice observed in 4H$_b$ TaS$_2$ (Figure 9) is also commonly observed in crystals of 2H–TaS$_2$ particularly following transformation at high temperatures in vacuum from the 1T polytype. Once more, this effect may be interpreted in terms of chalcogen loss on heating, but the intensity in such 2 × 2 reflexions is larger than might be anticipated for such minimal losses. This point will be discussed further below.

Finally we note that observations on NbS$_2$ in the 2H-form [30] suggest that no structural anomalies occur in this material down to 17 K; this is in accordance with the absence down to He temperatures of such electronic perturbations as the sign reversal observed in 2H–NbSe$_2$ near 26 K.

4. Diffraction Studies of Intercalated and Doped Materials

The effects described above in the pure stoichiometric materials will be discussed in detail in Section 7 in terms of charge density wave instabilities. For the present, we

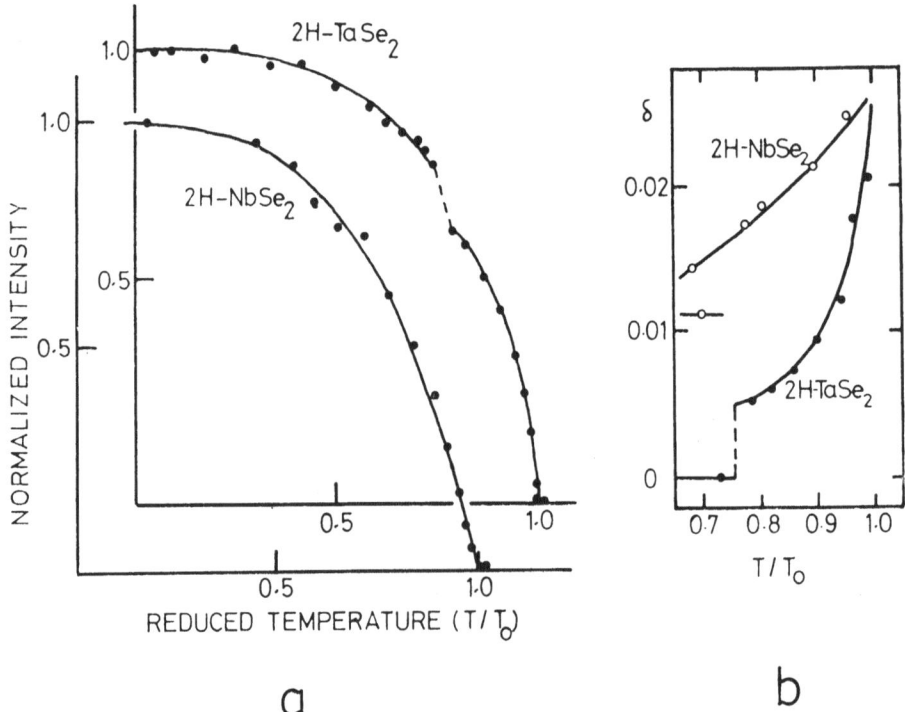

Fig. 14. (a) Normalised intensity of the primary superlattice Bragg peaks vs temperature. The break in the TaSe$_2$ curve marks the lock-in transition. $T_0 = 122.3$ K for 2H TaSe$_2$, 33.5 K for 2H NbSe$_2$. (b) Temperature dependence of δ (Figure 13) for 2H TaSe$_2$ and 2H NbSe$_2$. (After Monckton *et al.* [32].)

suggest only that these structural anomalies are related in some way to the Fermi surface in these materials; their behaviour following controlled changes in the size or shape of the Fermi surface should therefore be regarded as one of the most significant tests of this hypothesis. Within the layered compounds, such changes may readily be accomplished in one of two ways. Substitutional doping of the cation enables the electron density at and hence the size of the Fermi surface to be continuously varied, provided, of course, that it is possible to grow single-phase crystals of such mixed compounds. Alternatively, the layered compounds offer a unique means of controlling the intralamellar electron density via intercalation, that is the insertion between lamellae (Figure 1) of dopant species such as alkali metals and organic compounds, notably the amines. The third alternative of pressure induced Fermi surface changes is more complex and produces 'side effects' as we shall see in Section 7.

4.1. DOPED MATERIALS

Wilson, diSalvo and co-workers [3, 17] have extensively investigated the electronic properties and structures of doped group Va transition metal dichalcogenides, following the earlier work of Benda [12] and others on these systems. As we have seen above pure 1T–TaS$_2$ (or TaSe$_2$) exhibits perhaps the strongest PLD effects and is therefore a good candidate for such doping studies. Substitution of Ti for Ta effectively removes one electron per Ti atom from the Fermi surface whilst still producing single crystals of good quality and homogeneous 1T structure [3, 17]. Thus in Figure

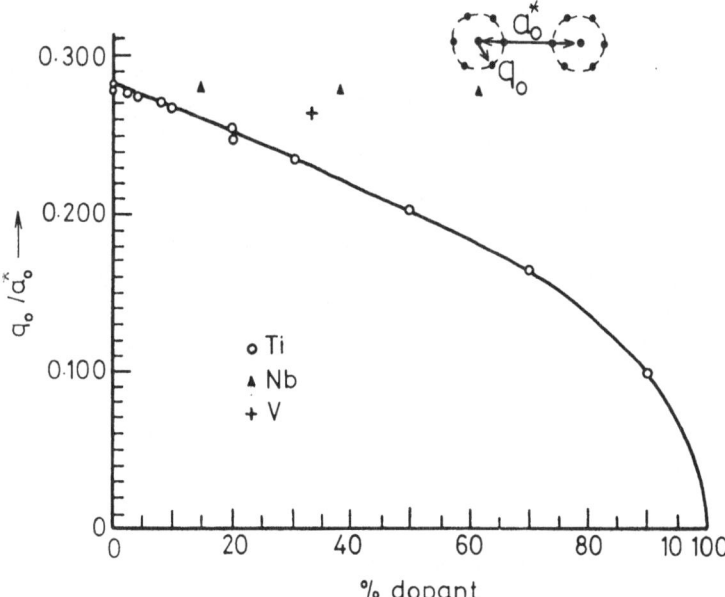

Fig. 15.   The ratio Q/a^* for the $\{S_M(11 \cdot l)\}$ reflexions in (i) Ti-doped 1T–Ta$_{1-x}$Ti$_x$S$_2$ (ii) 1T–Ta$_{1-x}$ $\times$ $\times$ Nb$_x$S$_2$ (iii) 1T–Ta$_{1-x}$V$_x$S$_2$, plotted against percentage dopant, x. For the Ti doping, Q/a^* falls approximately as $(1-x)$ (dashed line) indicating shrinking Fermi surface, whereas Nb(and V), $Q/a^* \simeq \sqrt{13}$ and does not alter up to 60% Nb doping. (After Wilson $et\ al.$ [3].)

15, we reproduce the data of Wilson *et al.* [3] for Ti doped $1T$–TaS_2 showing the variation of $\mathbf{Q}_{in}$ for the $\{S_M(11 \cdot l)\}$ reflexions plotted as the ratio of $|\mathbf{Q}_i|/|\mathbf{a}^*|$ in case of any slight changes in $\mathbf{a}^*$ on doping) with Ti concentration. The $\mathbf{Q}_{in}$ correspond essentially to the vector $\parallel \mathbf{a}^*$ in the $1T_1$ phase as Wilson *et al.* show that cation substitution acts to suppress the formation of the commensurate state in the selenide and both the $1T_2$ (which is 'near' commensurate) and $1T_3$ (commensurate) phases in the sulphide. Thus for $Ti_{0.06}Ta_{0.94}S_2$, T_d is reduced to room temperature (i.e. the high temperature $1T_1$ incommensurate state is stable over a wider range) whilst for $\gtrsim 15\%$ Ti doping, the commensurate state may not be attained at any temperature. Two discernible effects therefore manifest themselves in these observations. Firstly, the ratio $|\mathbf{Q}|/|\mathbf{a}^*|$ decreases for increasing Ti concentration, consistent, as we shall see below, with the removal of electrons from the Fermi surface. Secondly, the suppression of the commensurate $\sqrt{13} \times \sqrt{13}$ state and the lack of evidence for any adoption of other smaller wave vector superlattices suggests strong scattering of the CDW, as is considered further below. Confirmation of this latter conclusion follows from studies of TaS_2 doped with other group Va metals (Nb or V) where although charge transfer out of the Fermi surface does not occur, the commensurate state is still suppressed. Anion substitution, however, does not significantly affect either the commensurate or the incommensurate CDW state in the $1T$ Ta $(S/Se)_2$ system. DiSalvo *et al.* [17] have recently presented a comprehensive survey of results for such materials.

4.2. INTERCALATED MATERIALS

Phenomena associated with the intercalation of layered structures are diverse and have been considered in detail elsewhere in the present series, so that here we will discuss such effects only in so far as they interact with CDW formation. The insertion of atoms or molecules between layers may be accomplished by thermally activated diffusion from gas or liquid phase, or electrolytically, and a very wide range of intercalation complexes of the group Va transition metal dichalcogenides have been reported. Electrical or optical studies of these systems have often suggested [35] that the effects observed following intercalation may be interpreted in terms of charge transfer between the layers and the intercalate species, perhaps the classic example of this being the dramatic changes induced in the shape of the Fermi surface of graphite following the insertion of bromine to as dilute a stoichiometry as $C_{6000}Br$ [36]. Acceptors are not readily incorporated into layered TMD's but 'donor' intercalation complexes with organic amines [37] (and in general any organic compounds containing terminating $N:$, NH or NH_2 groups) metallic hydroxides, ammonia, alkali metals and earths, and rare earths have all been reported. Structurally, particularly with the organic intercalates of $2H$–TaS_2 [37], interest has often centred on the astounding anisotropy exhibited by such systems, powder diffractogram methods being commonly employed in the measurement of $\mathbf{c}$ and $\mathbf{a}$ axis changes following intercalation. More recently, single crystal diffraction studies have been reported, for example, of Na in $NbSe_2$, intercalated from ammonia solution [38], and pyridine in

$2H–TaS_2$ [39]. In both cases, interpretations of the diffraction effects have been given in terms of scattering from regular arrays of intercalate species between undistorted lamellae giving rise to extra periodicities not present in $\{M\}$. However, Wilson et al. have recently reported results for intercalation of ethylene-diamine into $4Hb–TaS_2$; in this case, the host lattice exhibits strong PLD/CDW effects (Figure 9), so that the new periodicities present in $(hk \cdot 0)$ electron diffraction patterns following intercalation may derive either from a modified PLD/CDW, or from an array of organic molecules,

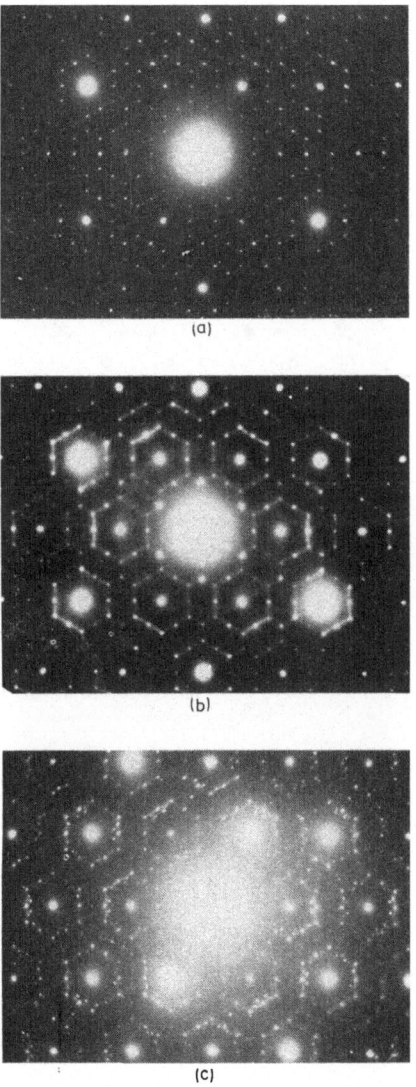

Fig. 16. $(hk \cdot 0)$ selected area electron diffraction patterns for $1T–TaS_2$ intercalated with (a) Na (b) K and (c) Eu from NH_3 solution, recorded at 290 K. Note geometric similarity in patterns (despite some disorder in Eu case).

or both. As these authors point out, however, simple charge transfer arguments leading to an expansion of the Fermi surface and hence of the $|Q_i|/|a^*|$ ratio, as is in fact observed, must be approached with caution in such a system where the gross structural changes following intercalation may in any case be expected to give rise to changes in the electronic energy bands and hence the form of the Fermi surface.

We consider here a slightly simpler case from the unpublished work of W. B. Clarke [40]. Figure 16, shows the $(hk \cdot 0)$ selected area electron diffraction patterns from $1T$–TaS_2 intercalated with sodium, potassium, and europium out of ammonia solution. Although complex, the patterns may be broken down into separate contributions as in Figure 17; from these two figures, the following simple observations may be made:

(a) Although the europium-intercalate is somewhat disordered (intercalation

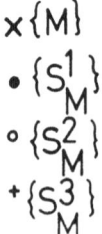

$$\times\{M\}$$
$$\bullet\,\{S_M^1\}$$
$$\circ\,\{S_M^2\}$$
$$+\,\{S_M^3\}$$

Fig. 17. Schematic breakdown of patterns in Figure 16 into three families of reflexions (a) $\{S_M^1\}$ which form a $\sqrt{3} \times \sqrt{3}$ superlattice and derive from scattering from ordered array of intercalated (b) $\{S_M^2\}$ at $Q = \pm\frac{1}{3}a^*$ from an increased spanning vector (c) $\{S_M^3\}$ at $Q = \pm 2a^*\sqrt{3}/9$.

times in solution were considerably longer for this metal compared with the other two with the consequent risk of contamination or partial decomposition), we see that the geometric form of the pattern is identical for each metal, suggesting the detailed effects are intrinsic to the TaS_2 lamellae, and not the intercalate.

(b) One family of reflexions, $\{S_M^1\}$, form an ordered array on a $\sqrt{3} \times \sqrt{3}$ superlattice, the intensities in $\{S_M^1\}$ falling off monotomically with increasing angles of scattering. Similar $\sqrt{3} \times \sqrt{3}$ superlattices are observed with Ni, Fe and Mn doped NbS_2 where the $(hk \cdot 0)$ diffraction pattern may be interpreted in terms of a regular array of inter-lamellar atoms. Here, for potassium, the $\{S_M^1\}$ are considerably more intense than for sodium.

(c) A second family of reflexions $\{S_M^2\}$ occur at $\pm \mathbf{a}^*/3$ and increase in intensity relative to $\{M\}$ with increasing scattering angle (cf. Plate 8).

(d) The third set, $\{S_M^3\}$, with $Q_{in} = 2\mathbf{a}^*\sqrt{3}/9$ again increase in intensity with scattering angle, and form a perfect $\sqrt{27} \times \sqrt{27}$ coincidence mesh with the $\{S_M^2\}$.

(e) The intensities in $\{M\}$, where reflexions are strong for $h - k = 3n$, are consistent with a change in the matrix stacking sequence from $AAA...(CdI_2)$ to $ABCABC...$ (rhombohedral).

(f) Tilting experiments at room temperature suggest that both the $\{S_M^2\}$ and the $\{S_M^3\}$ form rods in reciprocal space $\parallel \mathbf{C}^*$; this tentative conclusion remains to be confirmed by X-ray studies.

Cooling the sodium intercalated 1T–TaS_2 (Figure 16a) to liquid nitrogen temperatures produces an interesting transition as is seen in Figure 18a. The extra reflexions $\{S_M^2\}$ essentially 'merge' into the $\{S_M^3\}$, leaving only a faint residual streak, to produce a perfect 3×3 superlattice; once more, the intensity variation at high scattering angles in the $\{S_M\}$ relative to $\{M\}$ suggests PLD behaviour as opposed to a simple ordering of the intercalate, as was previously suggested for Na in $NbSe_2$.

Figure 18a was obtained by slow cooling of the intercalate in the electron microscope with the electron beam switched off; if, however, the sample is cooled with the electron beam on, the pattern in Figure 18b results. Here, although the $\{S_M^1\}$ on the $\sqrt{3} \times \sqrt{3}$ mesh are still sharp, the $\{S_M^2\}$ and $\{S_M^3\}$ form a disordered streak. This disordered pattern is identical to that reported by Wilson et al. for $4H_b TaS_2 \cdot EDA^1/_4$, again suggesting that the observed diffraction effects are intrinsic to the TaS_2 lamellae and do not merely result from a simple single centre scattering due to ordered intercalate arrays. Heating of the Na (or K and Eu) intercalated samples produces more complex effects than are observed in Figure 18b (W. B. Clarke, to be published) so that local electron beam heating cannot be the cause of the streak-like disorder. We believe therefore that the latter results from coulombic interactions between the energetic (100 KeV) electron beam and the potentially unstable layer of ionised intercalate.

Finally, the de-intercalation of the samples on controlled exposure to air produces an interesting result. The Na samples were stable for up to 24 hr in air, the potassium and europium intercalates for shorter times (20 min), but in each case the $(hk \cdot 0)$ diffraction pattern of Figure 18c results (that shown is for a de-intercalated K sample). This pattern results from the superposition of two commensurate $\sqrt{13} \times \sqrt{13}$ super-

(Fig. 18a)

Fig. 18.   (a) 1T–TaS₂–Na intercalated, cooled at 120 K, showing a perfect 3×3 superlattice. (b) 1T–TaS₂–Na intercalated cooled to 120 K with electron beam switched on showing sharp $\{S_M{}^1\}$ but disordered $\{S_M{}^2\}$ and $\{S_M{}^3\}$. This effect was reversible on warming to 290 K to give 16(a). (c) de-intercalated (K) 1T–TaS₂, showing 2 orientations of $\sqrt{13} \times \sqrt{13}$ superlattice and characteristic build up in intensity of $\{S_M\}$ for high $\{M\}$ of periodic distortion.

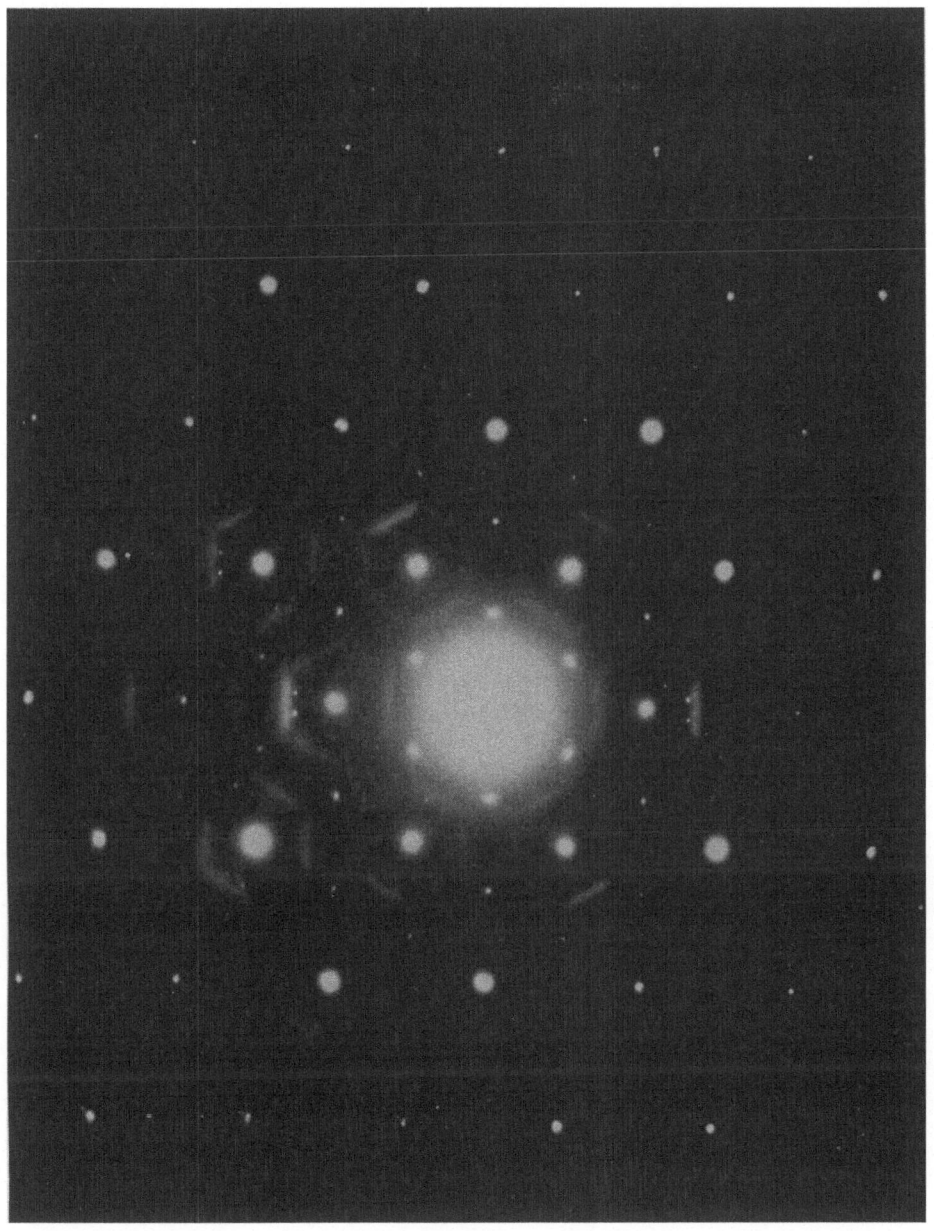

(Fig. 18b)

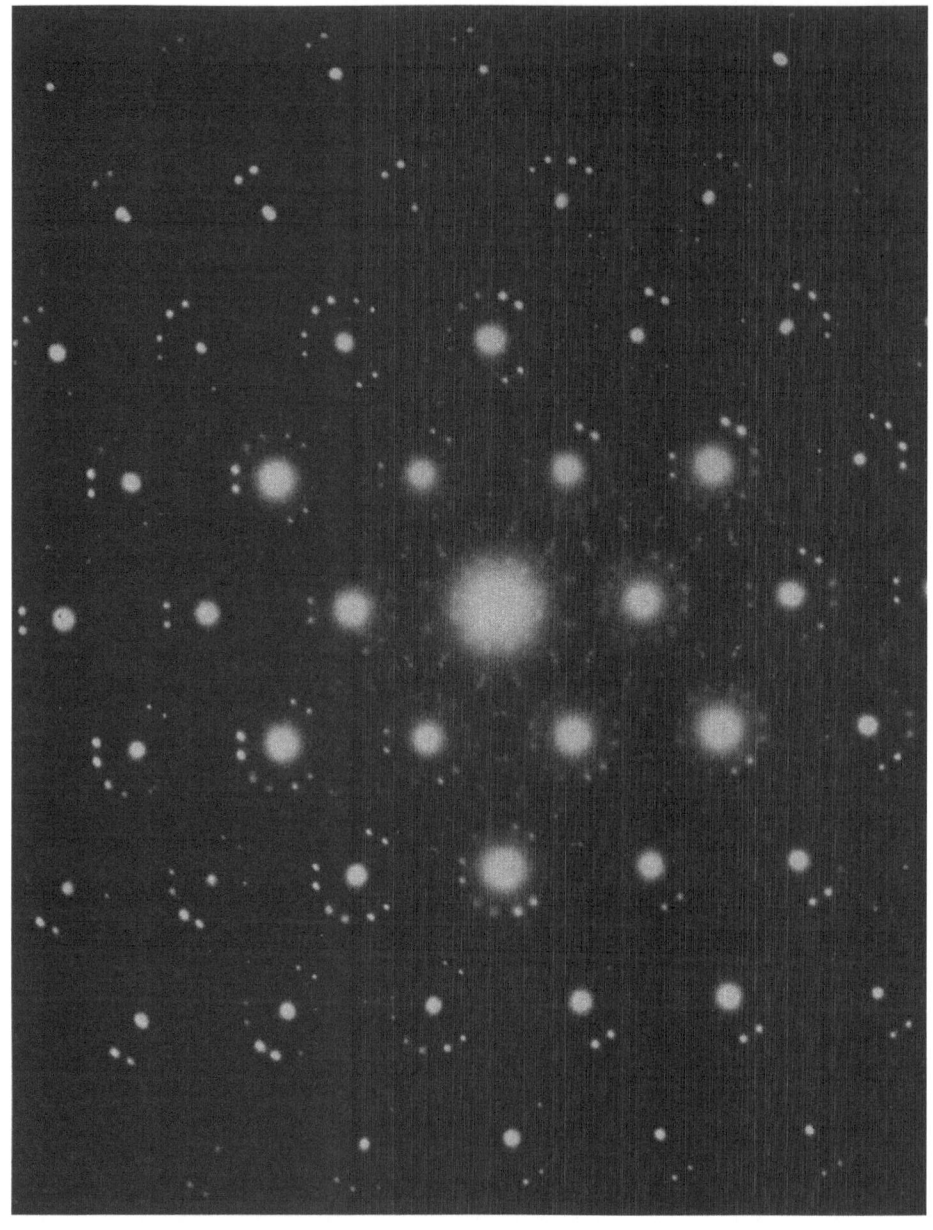

(Fig. 18c)

lattices and is identical to that observed for $4H_b$–TaS_2 (Figure 9). In fact, the deinter-calation leaves the crystals perfectly flat and strain free suggesting that such a drastic intra-layer coordination change may not have occurred ($4H_b$ samples are often dislocated heavily following the $1T$–$4H_b$ transformation). As a result of this flatness, the intensity behaviour in the $\{S_M\}$ relative to $\{M\}$ is dramatically highlighted in Figure 18c (see Section 4) and we conjecture that this pattern may derive from a PLD structure closely related to the low temperature $1T_3$ phase. Detailed discussion of all these intercalation observations is deferred until Section 7.

5. Diffraction by a Distorted Structure

The occurrence of extra periodicities $\{S_M\}$ in the diffraction observations in numerous laboratories over such a wide range of layered crystals, independent of their growth conditions, inevitably calls into question any interpretation of these phenomena in terms of impurity effects and suggests more fundamentally intrinsic origins. In view of the previous work [2] on the distorted tellurides, lattice distortions would seem to play a major role in the present observations and we examine this suggestion in some detail in this section.

Since for each of the hexagonal materials considered in the previous section, the extra reflexions $\{S_M\}$ are trigonally (or hexagonally) disposed about each matrix reflexion, three symmetry related distortion waves must be considered together. Each has a wave vector Q_i and amplitude U_Q, where $|Q_i| = Q_i = 1/\Lambda$, Λ being the periodicity of the distortion wave in direct space, and $|U_Q| = U_Q$ for $i = 1, 2, 3$. The layered structure of all the materials suggests that the distortion waves lie within each layer so that $Q_i \cdot c^* = 0$; experimental confirmation of this assumption for the case of $1T$–TaS_2 will be given later.

This restriction on the orientation of Q implies that the Fourier transform of the distortion wave system within a single layer is a family of rods in reciprocal space which are perpendicular to the layer and hence parallel to c^*. The relative phases of the distortion waves in adjacent layers modulate these rods so as to produce a set of extra reflexions $\{S_M\}$, where $M = ha^* + kb^* + lc^*$ is the scattering vector of a single matrix reflexion in the set $\{M\}$ and the scattering vector, relative to the origin of reciprocal space, of any extra reflexion, K, is given by $K = M + S_M$. The vector S_M has components parallel and perpendicular to the layer. Only the latter is always commensurate with the matrix i.e. $S_M \cdot c^* = 1/n$, n integral.

It is well known [41] that when a distortion is applied to a one dimensional lattice so that the displacement of each atom is a periodic function of its undisplaced distance from the origin, an infinite series of 'ghosts' or 'side bands' surrounds each diffraction maximum that arises from the undistorted structure. This theory is readily extended to three dimensions [14] and in the present case of three symmetry related distortions, the intensity of a matrix reflexion at M will be reduced by a factor

$$1 - \sum_i (\pi M \cdot U_Q)^2$$

and there will be six first order reflexions $S_M(10 \cdot l)$ with intensities proportional to:

$$\tfrac{1}{4}|2\pi(\mathbf{M}+\mathbf{Q}_i)\cdot\mathbf{U}_{Q_i}|^2 \quad \text{at} \quad \mathbf{M}\pm(\mathbf{Q}_i+nlc^*) \tag{1}$$

Strictly these expressions only hold when

$$(\mathbf{M}+\mathbf{Q}_i)\cdot\mathbf{U}_Q \ll \frac{1}{2\pi},$$

being approximations to the Bessel functions

$$J_0(\pi\mathbf{M}\cdot\mathbf{U}_Q)$$

and

$$J_1(2\pi(\mathbf{M}+\mathbf{Q}_i)\cdot\mathbf{U}_{Q_i})$$

which gives the true intensities for the $\{\mathbf{M}\}$ and $\{S_M(10 \cdot l)\}$ respectively. Nevertheless, since the approximation is valid for a distortion amplitude small compared with the lattice spacing and for the first few orders in $\{\mathbf{M}\}$, it is of use in considering the present results. To an equal level of approximation, there will be higher order reflexions $\{S_M(h0 \cdot l)\}$ with $-1 \leqslant l \leqslant +1$, with intensities proportional to

$$\left(\frac{1}{h!}\right)^2\left(\frac{1}{2}\right)^{2h}|2\pi(\mathbf{M}\pm h\mathbf{Q}_i)\cdot\mathbf{U}_Q|^{2h} \tag{2}$$

at $\mathbf{M}\pm h\mathbf{Q}_i+nlc^*$.

Furthermore, extra reflexions of the type $\{S_M(11 \cdot l)\}$ will have intensities proportional to

$$(\tfrac{1}{2})^4|2\pi(\mathbf{M}+\mathbf{Q}_1+\mathbf{Q}_2)\cdot\mathbf{U}_Q|^2|2\pi(\mathbf{M}+\mathbf{Q}_1+\mathbf{Q}_2)\cdot\mathbf{U}_{Q_2}|^2 \tag{3}$$

at $\mathbf{M}+\mathbf{Q}_1+\mathbf{Q}_2$.

This analysis was based on the assumption of only one atom per unit cell. Nevertheless, we may apply it to the interpretation of much of the data in Sections 2, 3, 4 with certain reservations. For TaS_2, the scattering amplitude of the Ta atom in the unit cell is considerably greater than that of the two sulphurs, so that we may, to a good approximation, apply the analysis directly. With the diselenides of Ta, Nb and V, the selenium now contributes significantly to the observed scattering intensity, particularly in VSe_2, so we must now take into account the distortion waves in the chalcogen layers as well as in the cation sheets. However, this leads to extra information concerning the relative phases of these two sets of distortion waves, as is seen for $NbSe_2$, for example.

5.1. THE 1T-POLYTYPES

The distortion waves in $1T–TaS_2$ have been described in detail by Scruby et al. [18]. For present purposes, we will merely summarise their conclusions, and note firstly that for a given wave vector $\mathbf{Q}$, there are three possible distortion modes within the TaS_2 layer

(i) Longitudinal, (LA); $\mathbf{U}_{Q_i} \parallel \mathbf{Q}_i$ and $\perp c^*$

(ii) Transverse, $(TA_{\parallel})$; $\mathbf{U}_{Q_i} \perp \mathbf{Q}_i$ and $\perp \mathbf{c}^*$

(iii) Transverse, $(TA_{\perp})$; $\mathbf{U}_{Q_i} \perp \mathbf{Q}_i$ and $\parallel \mathbf{c}^*$

By virtue of the dot vector product in Equation (1), we may readily deduce which of these possible modes contributes predominantly to the observations. The increase in intensity in Figure 8 of the $\{S_M\}$ surrounding $\{M(hk \cdot l)\}$ reciprocal lattice rows with increasing h is expected for the increase in $(\mathbf{M} + \mathbf{Q}_i)$ in Equation (1) and thus confirms the original supposition of a periodic distortion. Furthermore, since the $\{S_M\}$ increase in intensity with k but *not* significantly with l (for any given row with a fixed value of h) it must be concluded that $\mathbf{U}_{Q_i} \cdot \mathbf{c}^* \sim 0$, the distortions lying predominantly within the layers, as assumed above. Since the $\{S_M\}$ are also strongest for $\mathbf{Q}_i \parallel \mathbf{M}$ (see for example the X-ray data of Figure 5 and particularly the electron data of Figure 18c) and weakest for $\mathbf{Q}_i \perp \mathbf{M}$, $\mathbf{U}_Q$ must have a large component parallel to $\mathbf{Q}$ so that the distortions are predominantly LA. This conclusion is further reinforced in Figure 8 by the observations for $\mathbf{M}(00.l)$; since the $\{S_M\}$ for this row have finite intensity and $\mathbf{M} \cdot \mathbf{U}_Q = 0$, $\mathbf{Q}_i \cdot \mathbf{U}_Q \neq 0$, again implying strong LA contributions.

The modulation in intensity of the $\{S_M\}$ rods in reciprocal space to give maxima for $l = \pm \mathbf{c}^*/3$ in $1T_1$ and $1T_2$ implies strong phase correlations between PLD's in neighbouring layers, the PLD's stacking rhombohedrally. In $1T_3$, however, the diffraction evidence leads to the conclusion of a $13c_0$ repeat for the reduced zone 'distortion' cell, the stacking vector of the distortions being $\mathbf{c} + \mathbf{a}$. This interpretation is reinforced by recent neutron [20, 3] diffraction studies of $1T$–$TaSe_2$ below T_d which again favour a $13c_0$ repeat. In $1T_1$–$TaSe_2$ (i.e. above T_d) the $13c_0$ rhombohedral stacking sequence is once more adopted, as is also the case for $1T$–VSe_2. The significance in this $13c_0$ repeat will be discussed in Section 7. Only in $4H_b TaS_2$ of the 'octahedral' materials does the stacking vector change; here, however, the octahedrally coordinated layers are separated by trigonal prismatically coordinated layers, so that the constraints (see below) which favour rhombohedral stacking no longer apply.

The amplitudes, where measured, of the distortion waves in various materials are summarised in Table II. The coherence length of the distortions is more difficult to deduce. Comparison of the line widths of diffraction maxima in $\{M\}$ and $\{S_M\}$ for $1T$–TaS_2 using MoKα radiation [22], however, produce the somewhat surprising result that the coherence length of the PLD in this commensurate phase is comparable to the mosaic spread of the crystals used in the direction $\parallel \mathbf{a}$. In all materials above T_d, however, (T_d^1 in this case of $1T$–TaS_2) the coherence length is considerably reduced as a result of scattering; the observations of $4H_b TaS_2$

TABLE II

Amplitudes of distortion waves in the layered transition
metal dichalcogenides

2H	TaSe₂	0.09 Å	(5 K)
1T	TaS₂	0.03 Å	(400 K)
		0.1 Å	(295 K)

(Figure 8b inset) imply a reduction by a factor of five or so above T_d for this material. We note, however, that X-ray and electron diffraction are incapable of distinguishing between static and dynamic effects and that if the distortion modes become dynamic above T_d where the diffuse scattering also appears (but are static below) then we would anticipate some similar coherence length reduction. This point is discussed further below. The amplitude of the distortion wave in the $1T_2$ structure is sufficient (~ 0.1 Å) to produce considerable intensity in the $\{S_M(10 \cdot l)\}$ reflexions; for this case, it is then possible in the electron microscope to form a dark field lattice image of the distortion wave as is shown in Figure 19 (Scruby, to be published) where one of the three symmetry related wave vectors Q_i and a matrix $M(20 \cdot 0)$ reflexion have been used.

For the present, we will not comment further on the appearance of diffuse streaking above T_d except to identify (after Scruby et al). the scattering into the diffuse streaks as being predominately transverse in character, an observation self evident from the intensity distribution in Figure 7 for 1T–TaS$_2$ where radial streaks (i.e. $\parallel M$) are weak, peripheral streaks (i.e. $\perp M$) strong. Static X-ray photographs also reveal that there is no phase correlation in this diffuse background $\parallel c^*$ such as is observed for the $\{S_M\}$, a point favouring a dynamic interpretation of this effect.

5.2. The 2H polytypes

Qualitatively, the behaviour of the PLD/CDW effects in the 2H polytypes differs considerably from that in the 1T forms discussed above. The lower temperatures at which the effects manifest themselves in the 2H materials, coupled with the structural stability of the 2H trigonal prism coordination unit (below say 900 to 1000 K) facilitates observations of the 2nd order onset, at T_0, of the commensurate phase, in contrast to the 1T polytypes where irreversible transitions to the 2H phase usually set in below T_0. The distortion wave amplitudes are considerably less in the 2H forms (e.g. $\lesssim 0.09$ Å in 2H TaSe$_2$ at 5 K [32] compared with $\gtrsim 0.1$ Å in 1T–TaS$_2$ at 300 K [18]) and in all cases, observations indicate that the extra reflexions $\{S_M\}$ are located within the zero layer in reciprocal space, suggesting either rod-like diffraction maxima with no correlations between PLD's in neighbouring layers or alternatively a stacking repeat in the PLD's which is in phase with the matrix repeat.

Monckton et al. have shown in quantitative studies of TaSe$_2$ at 5 K that the static distortions in this material take the form of a predominant displacement of the Ta atoms $\parallel (\xi \cdot 00)$ directions, opposed by a smaller displacement of the Se atoms, also $\parallel (\xi \cdot 00)$ with a further 'TA$_\perp$'-type displacement $\parallel (00 \cdot \xi)$ of the latter. The electron observation of 2H NbSe$_2$ reinforce this argument (Figure 12d) since for reflexions where scattering from the Nb and Se atoms in the cell reinforce (for $h - k = 3n$) leading to strong $\{M\}$, the associated $\{S_M\}$ are weak, suggesting antiphase contributions from the PLD's in the Nb and Se layers. (For TaSe$_2$, $f(\text{Ta}) \gg f(\text{Se})$ so that this effect is not as readily observed). The distortion modes responsible for the $\{S_M\}$ are thus predominantly longitudinal in both the 1T and 2H forms, although by and large, this is the only point of similarity between the two classes of polytypes.

6. The Fermi Surfaces

Before discussing the origins of the PLD's in terms of coupled CDW effects, we will briefly review the expected form of the Fermi surface for both the 1T and 2H polytypes. Numerous calculations of the energy bands for these group Va materials have been reported, the overall features in the predicted Fermi surfaces being similar, with the possible exception of recent determinations of that for VSe_2 (Wexler, unpublished). We reproduce here in Figure 20a, b the energy bands within the 'd' manifold calculated by Mattheiss for 1T–TaS_2 (assuming an undistorted structure) and 2H–$NbSe_2$, with the resulting Fermi surfaces (sketched originally from Mattheiss's calculations by Wilson et al. [3]) reproduced in Figure 20c, d. That for 1T–TaS_2 is seen to be a simple single sheeted surface, with electrons located around M and L, and relatively little dispersion $\parallel$ c^*. Note that although the matrix structure is trigonal this surface is repeated with six-fold hexagonal symmetry here, although the angular photoemission studies of Smith and co-workers [42] reveal trigonal, not hexagonal symmetry in states close to E_f. For 2H $NbSe_2$, two surfaces result, degenerate in the ALH plane, since there are now two layers per unit cell. Here, we can imagine the electron pockets around LM in the 1T form to have touched, leaving closed hole-like surfaces around KH; we note, however, that this touching is determined critically by the energies of bands along ΓK and AH (Figure 19). Again, there is little band dispersion $\parallel$ c^*, with the exception of the inner hole surface of the upper band (Figure 20).

7. Discussion

The response of a system of conduction electrons in a metal to an applied spatial perturbation of wave vector $\mathbf{q}$ is given by the form of the generalised susceptibility function χ_q, where

$$\chi_q = \sum_k \frac{f_k(1 - f_{k+q})}{E_{k+q} - E_k}, \tag{4}$$

f_k, f_{k+q} being the Fermi distributions over states $|\mathbf{k}\rangle$, $|\mathbf{k}+\mathbf{q}\rangle$ of energies E_k, E_{k+q}. For a free electron gas Kohn [43] pointed out in 1959, that as a result of the form of the denominator in Equation (4), a logarithmic divergence in the static susceptibility χ^0_q may occur when $\mathbf{q}$ is close to spanning the Fermi surface i.e. $\mathbf{q} \sim 2\mathbf{k}_f$. This effect then gives rise to the familiar Kohn anomalies in the phonon spectra of metals [44]. Overhauser [14] and Lomer [15] later independently showed a similar instability in the spin susceptibility of the paramagnetic state of a free electron gas, leading to an interpretation of the phenomenon of spin waves in chromium (and to spin ordering in rare earths) [45]. The magnitudes of all these effects are determined by, amongst other things, the precise geometric form of the Fermi surface; in particular Afanaseev et al. [46] and later Fehlner and Loly [47] considered the increased contributions to the divergence in χ^0_q from cylindrical and nearly flat, parallel or 'nesting' pieces of Fermi surface compared with the spherical surface discussed by Kohn. The extreme

Fig. 19.   Dark field lattice image of PLD in $1T_2$–TaS_2 formed from one of the $\{S_M(10\cdot l)\}$ for
M(20.0). (After C. B. Scruby.)

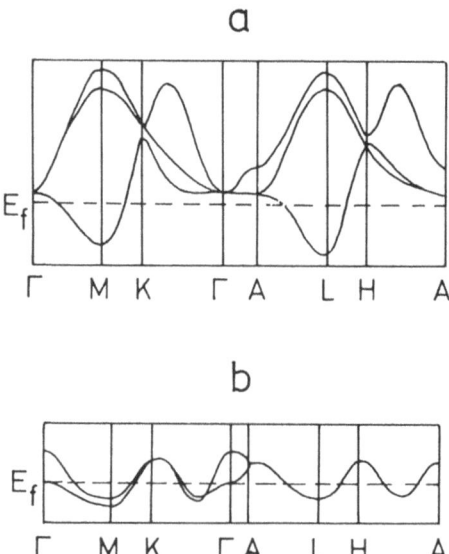

Fig. 20.   'd' manifold in APW calculations [8] for: (a) $1T$–TaS_2; (b) $2H$–$NbSe_2$; (c) Fermi surface
for $2H$–TaS_2 (i)l ower band (ii) upper band, deduced from average of APW and tight binding calcual-
tions; (d) Fermi surface of $1T$–TaS_2 showing simple predicted elliptical form of electron segments
cented on ML directions. (After Wilson et al. [3].)

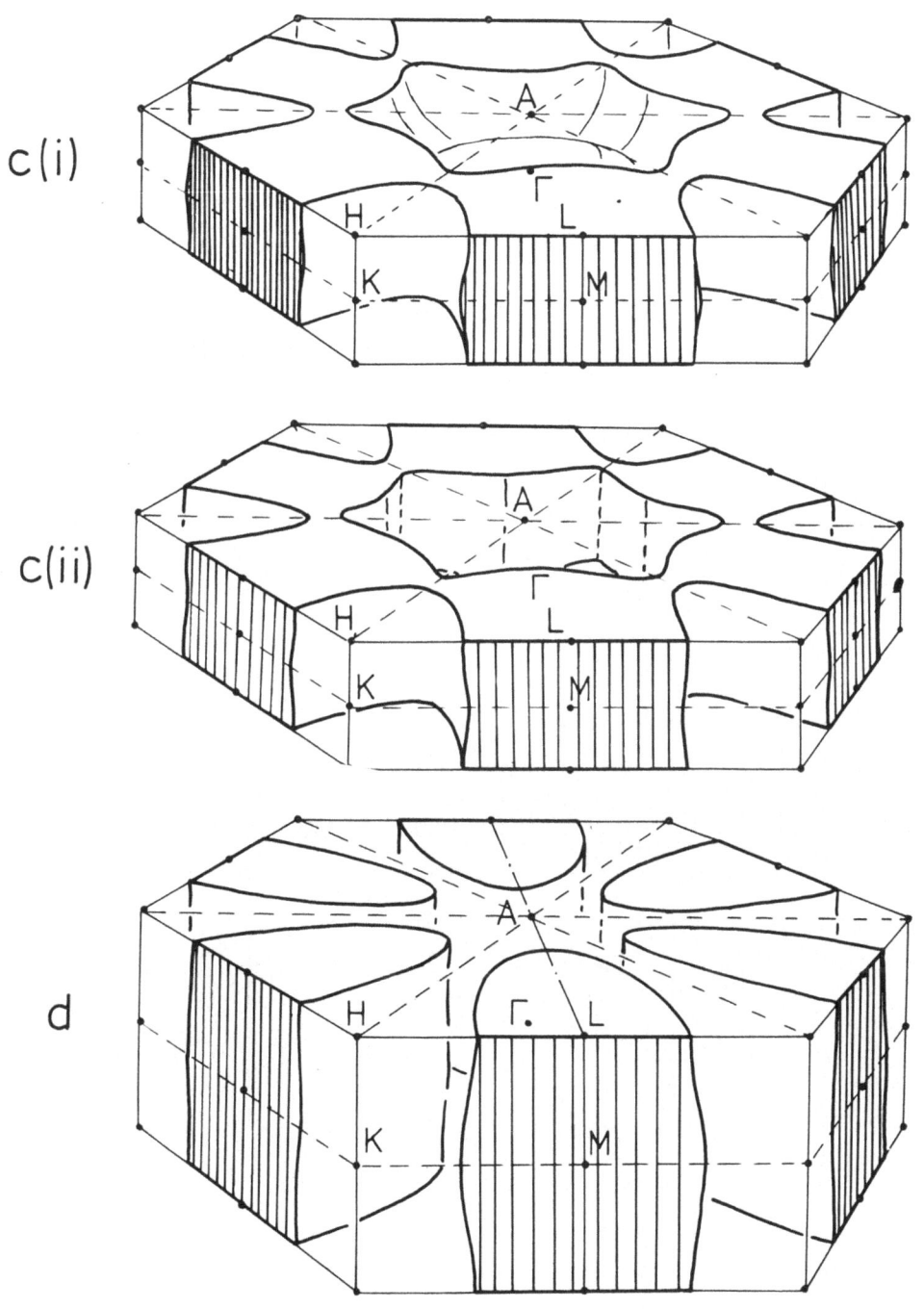

(Fig. 20)

case, of a flat surface is approximately realised in one dimensional metals such as KCP [16].

If electron-electron as well as electron-phonon interactions are then taken into account, then Overhauser [14] further postulated that below a suitable temperature, the divergence in χ°_q may lead to the adoption of a charge density wave ground state in which the spatial fluctuations in electron density are determined by the $2k_f$ wave vector. This CDW ground state is then stabilised by coupling to a periodic lattice distortion (PLD), the ion core displacements essentially screening out the inhomogeneities in electrictric field of the CDW, as considered by Overhauser [14] and Chan and Heine [48]. The lowering in electrostatic energy in the conduction electrons is thus balanced by the increased elastic strain energy of distortions in the lattice.

All of the diffraction observations of PLD's in the metallic layered transition metal dichalcogenides may be interpreted in terms of this coupled CDW/PLD model, and we again consider first the 1T polytypes. Clearly the near parallel walls of the Fermi surface in 1T–TaS$_2$ (Figure 20) satisfy the nesting criteria outlined above, although several perturbation wave vectors are geometrically possible. Empirically, Figure 21 compares the wave vectors Q_i measured for $1T_1$–TaS$_2$ near 350 K [18] with values of k_f inferred from the Mattheiss' APW scheme, projected onto the $(hk \cdot 0)$ (or $\Gamma MK\Gamma$) plane, from which it is seen that the measured Q_i just span segments of surface in

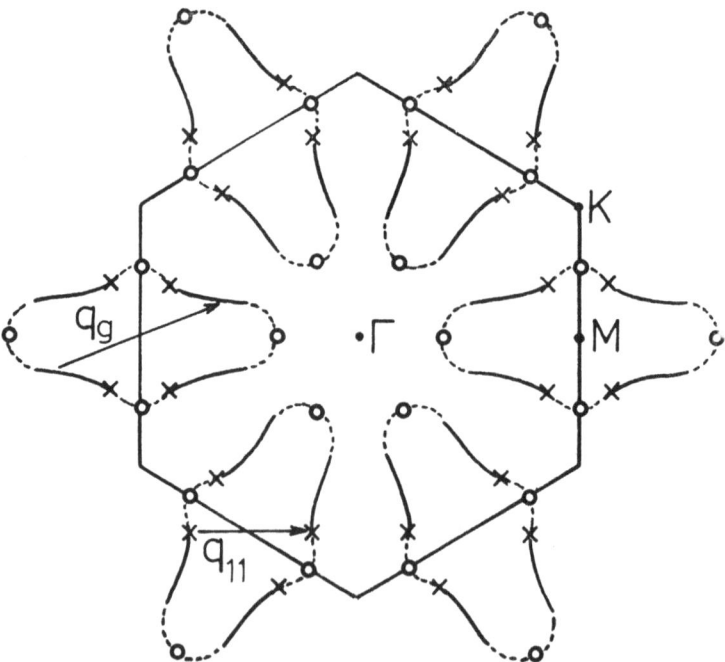

Fig. 21. Comparison of $\times$ measured vectors Q_i for 1T–TaS$_2$ with $\bigcirc$ k_f values inferred from APW calculations [8], projected onto ΓMK plane. Solid lines indicate parts of surface imaged in diffraction by TA$_\parallel$ phonons.

directions $\parallel \Gamma M$. A similar situation is expected to arise in 1T–TaSe$_2$ and 1T–VSe$_2$, although no detailed calculations of the form of the Fermi surface have been made for these materials; preliminary results for VSe$_2$ [49] indicate that a similar spanning vector to that determined in diffraction (Figure 10) is possible.

It seems reasonable to question, however, whether only the spanning vectors $\mathbf{Q}_i$, which give rise to the strong reflexions $\{\mathbf{S}_M\}$, may contribute to the divergence in χ°_q. Williams *et al.* [18] have in fact suggested that perturbations with general wave vectors $\mathbf{Q}_g$ (Figure 21) may lead to divergent contributions to χ^0_q, resulting in an enhanced Kohn-like anomaly in the phonon spectrum in these directions. Scattering of electrons from softened phonons in these modes then gives rise to the diffuse streaks

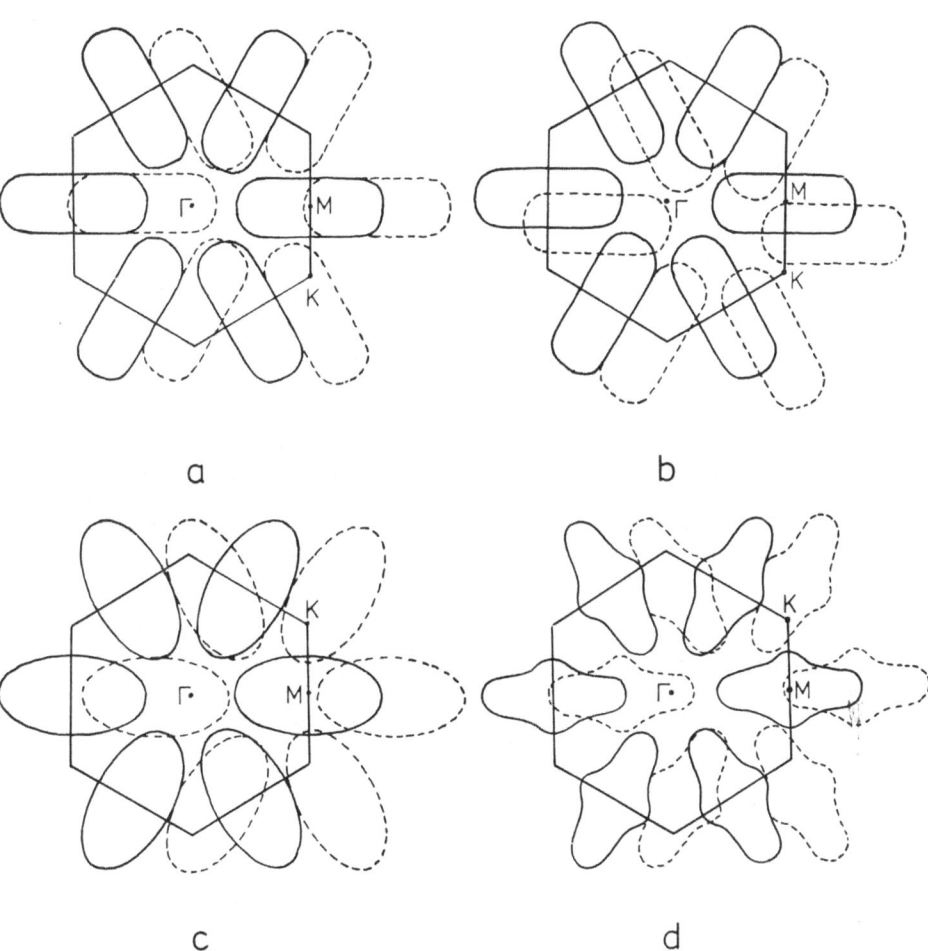

Fig. 22. (a) Demonstrating superior nesting of all six segments of Fermi surface for spanning vector $\mathbf{Q}_i \parallel \Gamma M$ for an ideal flat sided surface, compared with (b) nesting of only two segments of same surface for vector $\parallel \Gamma K$. (c) Elliptical Fermi surface showing how good nesting of (a) is destroyed by curvature of surface. (d) Shape of surface in Figure 21 proposed from image in phonon spectrum showing again how good nesting results for $\mathbf{Q}_i \parallel \Gamma M$.

prominent in diffraction patterns of $1T_1$–TaS_2, so that the form of the streak is actually an image of the Fermi surface. This conclusion is further supported by recent X-ray diffractometer investigations of the $1T$–$Ta_{0.9}Hf_{0.1}S_2$ [19], both studies leading to a form of Fermi surface with points of inflexion close to the Q_i spanning positions (Figure 21). This contrasts with the elliptical cross section proposed by Wilson *et al.*, but in fact is consistent with the deductions of these latter authors concerning the adoption of Q_i as the prominent wave vector. As they correctly point out for the unphysical case of perfectly flat parallel sided segments (Figure 22a) the Q_i adopted, parallel to ΓM in fact maximizes nesting of all six pieces of Fermi surface, whereas the other possibility parallel to ΓK results in nesting of only two out of the six segments (Figure 22b). However, as is readily seen in Figure 22c an ellipsoidal cross section destroys much of the nesting of Figure 22a and it is only for the proposed case of an inflected shape that optimum nesting occurs (Figure 22d). We note further that the disappearance of the diffuse streaks following the $1T_1$–$1T_2$ transition in which Q_{i1} rotates to Q_{i2} would be consistent with the destruction of much of the parent Fermi surface in the transition proposed again by Wilson *et al.* [3]. We therefore believe the form of Fermi surface of Figure 21 to be approximately correct for this material.

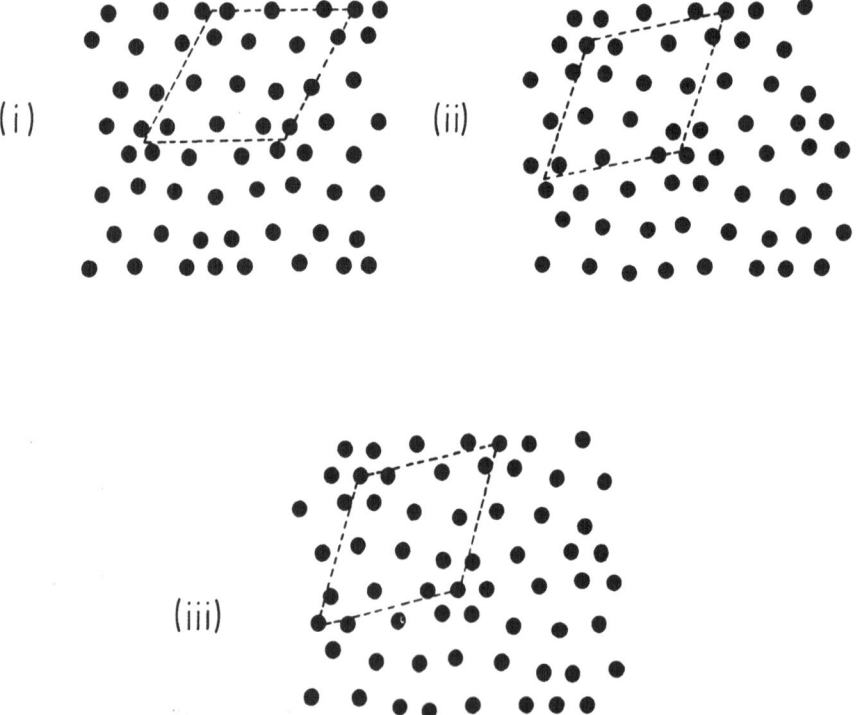

Fig. 23.   Model for single layer of Ta atoms in each distorted phase of $1T$–TaS_2 (for an arbitrarily large $U_Q \sim 0.1$ Å). Origin of distortion cell (broken lines) arbitrary in the incommensurate $1T_1$ and $1T_2$ phases, and at a metal site in the commensurate $1T_3$ phase.

The coupling of the CDW to the PLD raises inevitably the questions of phases and inter-layer interactions. Figure 23 reproduces for the case of $1T-TaS_2$ the form of modulated matrix structure for the incommensurate $1T_1$ and $1T_2$ phases, and the commensurate $1T_3$ phase (for an arbitrary LA modulation in amplitude). In the incommensurate phases, the origin of cell is clearly arbitrary, and as discussed by Williams et al. [18], within the reduced zone of the distortion cell in real space, no two atomic sites coincide. For the purposes of determining the repeat of the modulated structure along c^*, therefore, the situation is equivalent to the stacking of distortion waves in a continuum. Under these circumstances, electrostatic considerations suggest that rhombohedral stacking with a $3c_0$ repeat would minimise inter-layer interactions, as is observed in all the 1T polytypes. Only in the $4H_b$ case, where modulated 1T layers are separated by undistorted (at high temperatures that is) 2H layers, can such inter-layer interactions be neglected, the c^* repeat in this case being in phase with the matrix.

In all cases, screening considerations suggest [14, 48] that the CDW and the PLD will be exactly in phase, the maxima in electron density coinciding with maxima in the average cation density. The relative phases of PLD's within the anion layers may be less readily predicted, however, since the results [17] for anion substituted materials suggest relatively little interaction between the CDW and the anion PLD. The phase relationship of the latter with the cation PLD would seem to be determined purely by elastic strain energy considerations. For the cation sheets, it is further suggested [3, 18] that for the commensurate $1T_3$ states in TaS_2 and $TaSe_2$, the origins of the distortion cells within each layer should coincide with a cation site; this always ensures coincidence between the latter and the maxima in the CDW, thereby minimising electrostatic energy, and explains the drive towards the commensurate state observed in all materials. The transformation from the $3c_0$ incommensurate stacking repeat to the $13c_0$ repeat in $1T_3TaS_2$ [18] and $TaSe_2$ [20] is thus explained, since for a $\sqrt{13} \times \times \sqrt{13}$ cell origin on a cation site in one layer, the $3c_0$ rhombohedral repeat places the cell origins in the next two layers on intestitial sites adjacent to chalcogen atoms. Only for distortion wave stacking vectors $c+a$ or $c+2a$ can the cell origins coincide with cation sites in every layer, the stacking repeat being $13c_0$ in either case, as is observed. In $1T-VSe_2$ on the other hand, where the projection of the Q_i onto the $(hk \cdot 0)$ plane below Td defines a perfect 4×4 supercell no such change in stacking vector appears to take place following the incommensurate-commensurate transformation.

The above interpretation in terms of a Fermi surface driven instability may conveniently be tested by altering the electron density within the layer, as outlined above. The results of Wilson et al. [3] for Ti doped TaS_2 (Figure 15) confirm the Fermi surface origins of the diffraction effects since substitution of Ti with one fewer electron than Ta is expected to shrink the electron pockets back towards M and L, as seen in Figure 24, thereby reducing the magnitude of the spanning vector as is observed. This substitution, however, introduces random fluctuations into the cation site potentials, since it it not possible to anneal the Ti into a perfectly order substitutional sub-lattice. These potential fluctuations scatter the CDW, as discussed by Wilson et al. [3] and more recently McMillan [50] has shown how this scattering stabilises the incommensurate phase, suppressing lock-in to the commensurate superlattice state.

For the intercalated materials, however, the results reveal an expansion of the Fermi surface schematised in Figure 24, without scattering of the CDW. The comparison between intensities in the $\{S_M\}$ for the potassium and sodium intercalated materials in figures 16a, b, together with the persistence of sharp reflexions of the type $\{S_M^1\}$ following disorder of the $\{S_M^2\}$ and $\{S_M^3\}$ in Figure 18b strongly suggests an identification of this first family in terms of scattering from an ordered $\sqrt{3} \times \sqrt{3}$ array of intercalate ions, as is seen for Ni, Fe or Mn doped M NbS$_2$ (W. B. Clarke,

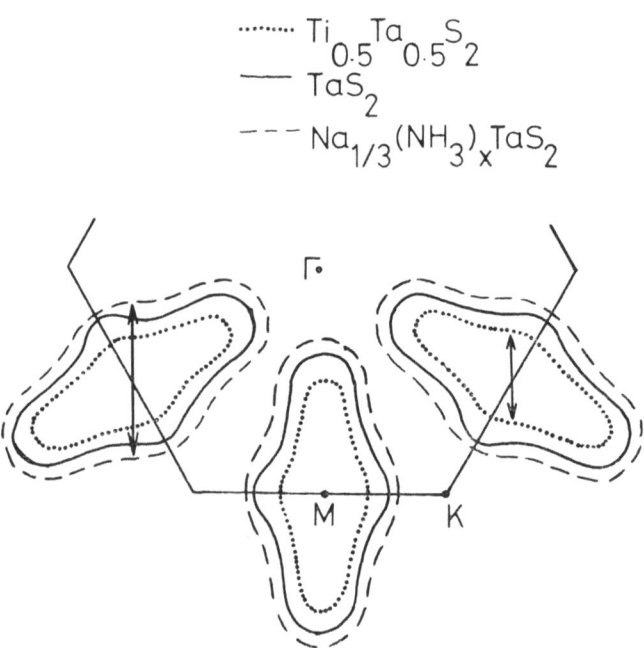

Fig. 24.   Schematic of one segment of Fermi surface for 1T–TaS₂ in ΓMK plane showing effect of (i) 50% Ti doping (ii) intercalation with Na in NH₃ solution to Na$_{1/3}$TaS₂. $\mathbf{Q}_i$ in both cases is still $\parallel \Gamma M$.

to be published). The $\{S_M^2\}$ situated at $\pm \mathbf{a}^*/3$ may then be identified with an increased $\mathbf{Q}_i^2$ spanning vector on the Fermi surface within the ΓMK plane and form the perfect 3×3 superlattice with the $\{S_M^1\}$ at low temperatures. Both the interlamellar site of the dopant cation and the coincident Fourier components for intercalated cation and CDW potentials favour an unattenuated CDW, as is observed from the width of the diffraction maxima for the $\{S_M^2\}$. The family $\{S_M^3\}$ behave in intensity in a similar manner to the $\{S_M^2\}$ and would appear to derive from a PLD coupled to an alternative $\mathbf{Q}_i^3$ CDW spanning vector. Since the insertion of charged cation sheets between layers of the host TaS$_2$ structure may be expected to give rise to considerably greater band dispersion along $\mathbf{c}^*$, it is possible that the $\mathbf{Q}_i^3$ vectors span the Fermi surface out of the ΓMK plane. An alternative which appears more likely is that the $\mathbf{Q}_i^3$ derive from a PLD within the intercalated cation layer; certainly following charge transfer to the TaS_2 layer from the alkali metal [35], even if neighbouring ions are partially screened by the incorporation of polar NH_3 groups during intercalation [51], the two dimensional ion sheets should be highly unstable with respect to a periodic distortion. Such effects have been detected in the related system of Cs-graphite intercalates. Under these circumstances, we would anticipate that the $\mathbf{Q}_i^3$ and $\mathbf{Q}_i^2$ vectors would in some way be related and form a coincidence mesh as is observed. The transformation to the perfect 3×3 structure at low temperatures then presumably reflects temperature

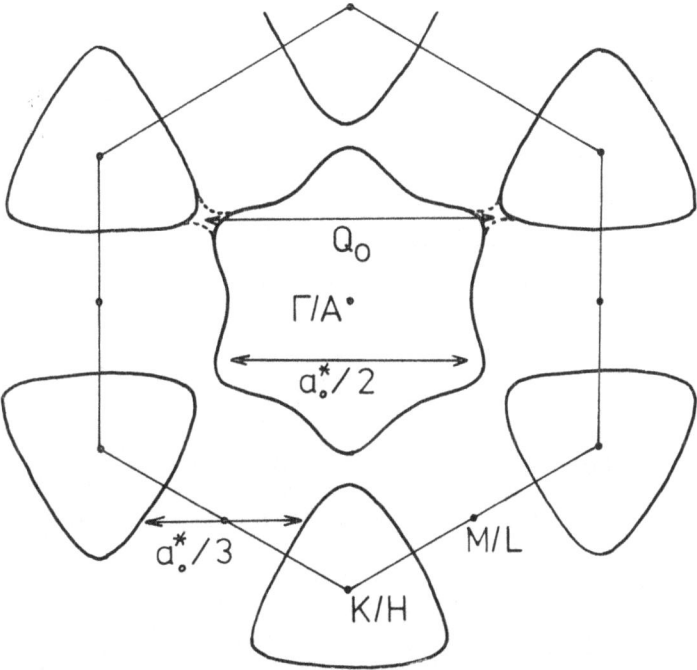

Fig. 25.   (a) ($hk \cdot 0$) projection of Fermi surface for 2H–NbSe₂ showing (i) $\mathbf{a}^*/3$ spanning vector suggested by Wilson *et al.* [3] (ii) $\mathbf{a}^*/2$ vector [3, 24] which may cause the $2a_0$ superlattice. (b) Alternative Fermi surface spanning vector proposed by Rice and Scott [53], joining saddle points which may give rise to the $3a_0$ superlattice in the 2H polytypes below T_0.

dependence in the degree of charge transfer. The reversible disordering of both the CDW's by coulombic scattering of the energetic electron beam is a phenomenon of interest both here and in the related case of $EDA_{1/4} \cdot 4H_b TaS_2$ [3] and further investigations of the interplay between CDW's and intercalation should prove fruitful. Acrivos [52] for example, has recently pointed out the coincidence between the reciprocal lattice vector of the $\{S_M(11 \cdot 0)\}$ superlattice reflexions in $Pyr_{1/2}2H \cdot TaS_2$ [39] and a possible spanning vector of the 2H Fermi surface, so that CDW effects may play a significant role in the determination of stoichiometry and superlattices in a wide range of intercalated systems.

If we return now to the related case of CDW/PLD effects in the 2H polytypes, then it is evident that much of the foregoing discussion can equally apply to these materials. Thus, Wilson *et al.* have proposed a Fermi surface spanning vector (Figure 25a) for the $3a_0$ CDW which is closely related to its counterpart in the 1T polytypes, although they point to the possible conflicts between the various sheets of Fermi surface as a likely cause for poorer nesting and hence the weaker CDW effects observed in the 2H case. More recently, Rice and Scott [53] have suggested an alternative mechanism for CDW formation in the 2H polytypes in which nesting in the usual sense does not apply. They propose (Figure 25b) a divergent contribution to χ_q^8 for $\mathbf{q} = \mathbf{Q}_0$ connecting

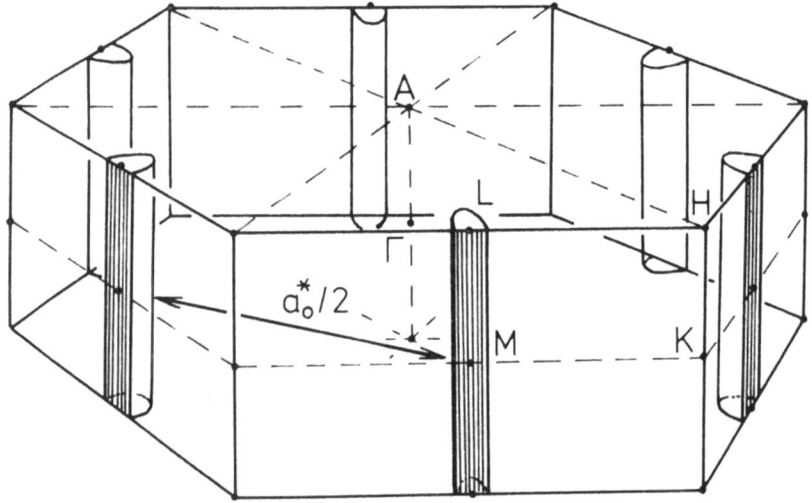

Fig. 26.   Schematic of 'Fermi surface' for 1T–TiSe₂ in form of narrow rods along *ML*, with possible spanning vector of $2a_0$ superlattice.

saddle points in the (2 dimensional) Fermi surface, a vector close to that which Wilson *et al.* [3] and Hughes and Liang [24] suggested may give rise to the $2a_0$ superlattice observed in some 2H material. At present, the precision in the determination of the geometry of the Fermi surface in the 2H polytypes does not permit a quantitative choice between either the **a***/2 or **a***/3 vectors. Nevertheless, it is interesting to note Rice and Scott's conclusions that their model does not destroy much Fermi surface following the $3a_0$ reconstruction, consistent with the improved metallic properties below T_0 in the 2H polytypes in contrast with the poorer conductivity following development of the incommensurate CDW state in the 1T materials where the nesting argument would seem more appropriate.

The occurrence of the $2a_0(\times 2c_0)$ superlattice in 1T–TiSe₂, nominally a d⁰ material, again calls into question the wide scale adoption of a nesting situation to explain the structural transformations in all the layer compounds. Here, the Fermi surface, whether produced by $d-p$ band overlap or impurity donation, would be expected by analogy with 1T–TaS$_2$ to comprise narrow diameter 'rods' about ML (Figure 26). Under such circumstances, large areas of Fermi surface are not available for nesting in the usual sense, although it is of interest to note that the vector spanning two neighbouring rods is very close to $\mathbf{a}^*/2$. Whether this transformation is closely related to those in the d^1 compounds, however, must remain the subject for future investigation.

8. Summary

The detailed discussion of the electronic consequences of the transitions outlined here is beyond the scope of the present chapter. We note briefly, however, that following the stabilisation of a CDW/PLD ground state, the destruction of large areas of nested Fermi surface in the 1T compounds below T_d is consistent with the first order discontinuities in resistivity observed for 1T–TaS$_2$ and 1T–TaSe$_2$, and their adoption of the 1T$_3$ $\sqrt{13} \times \sqrt{13}$ supercell reflects the observed drive towards diamagnetism (which may also explain the formation of the supercell in 1T–TiSe$_2$). The stacking of PLD's in both the incommensurate and commensurate states follows from a minimisation of electrostatic energies of interaction both between the CDW and PLD in the cation sheets of each individual layer and between CDW's in neighbouring layers, thereby explaining the adoption of the $3c_0$ and $13c_0$ stacking repeats observed crystallographically. Substitutional doping or intercalation, with associated charge transfer out of or into the d band confirms the Fermi surface origins of the instabilities and raises in the latter case the interesting question of the role of CDW effects in the formation of intercalate superstructures in general. For the 2H polytypes, the qualitative dissimilarities in the diffraction observations when compared with the 1T materials raises the possibility of a different Fermi surface mechanism supporting CDW/PLD formation. Electronically, the 2H polytypes, particularly NbSe$_2$, are fascinating in that they exhibit concurrently below T_c the apparently conflicting correlations responsible for superconductivity and for the observed Peierls-like distortions. In this context, it is of interest to note that the stiffening of the lattice of 2H–NbSe$_2$ under pressure [54] raises the critical temperature for superconductivity, possibly by reduction in amplitude of these distortions, whereas 2H–NbS$_2$ which does not exhibit CDW/PLD effects in diffraction shows no increase in T_c under pressure.

References

1. J. A. Wison and A. D. Yoffe, *Adv. Phys.* **18** (1969), 193.
2. B. E. Brown, *Acta. Crystallogr.* **20** (1966), 264.
3. This data is taken from J. A. Wilson, F. J. diSalvo, and S. Mahajan, *Adv. Phys.* **24** (1975), 117 and is in fact a composite of References 4, 5 and 6.
4. A. H. Thompson, F. R. Gamble, and J. F. Revelli, *Solid St. Commun.* **9** (1971), 981.
5. A. H. Thompson, K. R. Pisharody, and R. F. Koehler, *Phys. Rev. Letters* **29** (1972), 163.
6. F. J. diSalvo, B. J. Bagley, J. M. Voorhoeve, and J. V. Waszczak, *J. Phys. Chem. Solids* **34** (1973), 1357.
7. R. B. Murray, R. A. Bromley, and A. D. Yoffe, *J. Phys. C. Solid State* **5** (1972), 746.

8. L. F. Mattheiss, *Phys Rev. B.* **8** (1973), 3719.
9. C. Y. Fong and M. L. Cohen, *Phys. Rev. Letters* **32** (1974), 720.
10. R. V. Kasowski, *Phys. Rev. Letters* **30** (1973), 1175.
11. F. Jellinek, *J. Less Common Metals* **4** (1962), 9.
12. J. A. Benda, *Phys. Rev. B.* **10** (1974), 1409.
13. B. I. Halperin and T. M. Rice, in Seitz, Turnbull, and Ehrenreich (eds.), *Sol. St. Phys.*, Vol. 21, Academic Press, New York, 1968, p. 115.
14. A. W. Overhauser, *Phys. Rev.* **128** (1932), 1437; *ibid.* **167** (1968), 692; *ibid.* **133** (1971), 3173.
15. W. M. Lomer, *Proc. Phys. Soc.* **80** (1962), 489.
16. R. Comes, M. Lambert, H. Laundis, and H. R. Zeller, *Phys. Rev. B.* **8** (1973a), 571; *Phys. Stat. Sol.*, *B* **58** (1973b), 587.
17. F. J. diSalvo, J. A. Wilson, B. G. Bagley, and J. V. Waszczak, to be published (1975).
18. P. M. Williams, G. S. Parry, and C. B. Scruby, *Phil. Mag.* **29** (1974), 695.
    C. B. Scruby, P. M. Williams, and G. S. Parry, *Phil. Mag.* **31** (1975), 255.
19. Y. Yamada, J. C. Tsang, and G. V. Subba-Rao, *Phys. Rev. Letters* **34** (1975), 1389.
20. D. Monckton and J. D. Axe, unpublished work.
21. F. R. Shepherd, P. M. Williams, D. A. Young, and C. B. Scruby, in M. H. Pilkhum (ed.), *Proceedings XIIth International Conference on the Physics of Semiconductors*, Stuttgart, Teubner, Stuttgart, 1974.
22. C. B. Scruby, unpublished work.
23. J. van Landuyt, G. van Tendeloo, and S. Amelinckx, *Phys. Stat. Sol. A* **26** (1974), 359.
24. H. Hughes and W. Y. Liang, *J. Phys. C. Solid State* **7** (1974), L162.
25. A. H. Thompson and B. Silbernagel, private communication.
26. L. E. Conroy and K. C. Park, *Inorg. Chem.* **7** (1968) 459.
27. H. W. Myron and A. J. Freeman, *Phys. Rev. B* **9** (1974), 481.
28. R. S. Title, private communication.
29. R. E. Peierls, *Quantum Theory of Solids*, Clarendon Press, Oxford, 1955.
30. P. M. Williams and C. B. Scruby, unpublished work. See also Reference 3.
31. P. M. Williams, C. B. Scruby, and G. J. Tatlock, *Sol. St. Commun.* **17** (1975) 1197.
32. D. E. Monckton, J. D. Axe, and F. J. diSalvo, *Phys. Rev. Letters* **34** (1975), 734.
33. J. P. Tidman, O. Singh, A. E. Curzon, and R. F. Frindt, *Phil. Mag.* **30** (1974), 1191.
34. T. Forgan, Private communication.
35. J. V. Acrivos, W. Y. Liang, J. A. Wilson, and A. D. Yoffe, *J. Phys. C.* **4** (1971), L18.
36. D. A. Young and A. S. Bender, *J. Phys. C.* **5** (1972) 2163.
37. F. R. Gamble, F. J. diSalvo, R. A. Klemm, and T. H. Geballe, *Science N.Y.* **168** (1970), 568.
38. C. B. Carter and P. M. Williams, *Phil. Mag.* **26** (1972), 393.
39. G. S. Parry, C. B. Scruby, and P. M. Williams, *Phil. Mag.* **29** (1974), 601.
40. W. B. Clarke, unpublished work.
41. R. W. James, *The Optical Principles of the Diffraction of X-rays*, Bell, London, 1948.
42. N. V. Smith, M. M. Traum, and F. J. diSalvo, *Sol. St. Commun.* **15** (1974), 211; *Phys. Rev. Letters* **32**, (1974), 1241.
43. W. Kohn, *Phys. Rev. Letters* **2** (1959), 393.
44. First reported for lead: B. Brockhouse, *Phys. Rev. Letters* **7** (1961), 93.
45. S. C. Keeton and T. L. Loucks, *Phys. Rev.* **168** (1966), 672.
46. A. M. Afanaseev and Y. Kagan, *Sov. Phys. JETP* **16** (1963), 1030.
47. W. R. Fehlner and P. D. Loly, *Solid St. Commun.* **14** (1974), 653.
48. S. K. Chan and V. Heine, *J. Phys. F.* **3** (1973), 795.
49. G. Wexler, unpublished work.
50. W. L. McMillan, to be published.
51. F. R. Gamble, J. H. Osiecki, M. Cais, R. Pisharody, F. J. Di Salvo and T. H. Geballe, *Science* **174** (1971) 493.
52. J. V. Acrivos, to be published.
53. T. M. Rice and G. K. Scott, to be published.
54. P. Molinie, D. Jerome, and A. J. Grant, *Phil. Mag.* **30** (1974), 1091.

THE LAYERED STRUCTURES OF
TERNARY CHALCOGENIDES WITH ALKALI AND
TRANSITION METALS

W. BRONGER

Institut für Anorganische Chemie der RWTH Aachen, F.R.G.

1. Layered Structures of Binary Metal Chalcogenides as a Basis for the Construction of Ternary Compounds

The structure of binary fluorides and oxides with alkali and alkaline-earth metals can be largely understood by means of the supposition of electrostatic binding forces. Thus the sequence of structure types with changing co-ordination numbers for cations and anions depending on their radial quotients, can be nearly quantitatively predicted by a simple model using electrostatically attracting and repelling forces. Moreover, the lattice energies calculated by means of the electrostatic model agree satisfactorily with those which can be experimentally determined by means of the Born-Haber cycle. The dominant role of the electrostatic binding forces decreases considerably if one changes over to the fluorides and oxides of the remaining main group and transition metals, especially when the electropositive atoms are in higher oxidation states.

The binary chlorides and sulphides, bromides and selenides, as well as the iodides and tellurides of the alkali and alkaline-earth metals exhibit progressively increasing covalent binding character, especially as the atomic weight of the anionic partner increases. Analogous compounds with the other main group metals continue this tendency and the corresponding phases with the transition metals can already assume metallic properties. Above all, the systems with sulphur, selenium and tellurium, which will be discussed here in greater detail, can often be recognized as intermetallic systems. Here the existence of many phases, mostly having wide ranges of homogeneity, can be observed.

The transition from the more salt-like structured chalcogenides to those already having metallic properties can be seen clearly from the example of the largely stoichiometrically composed mono- and dichalcogenides – MX and MX_2 – of the metals of the 4th period (see Tables I and II). Included in the tables are the hitherto known high pressure and high temperature phases [1].

In the range of the MX compounds with the NaCl structure type, in which the cations occupy all the octahedral sites in a cubic close-packed structure of anions, mainly ionic bonds are found. Exceptions are the scandium chalcogenides, in which for scandium, the oxidation state $+3$ must be assumed, the electrons not required for the inert gas configuration of the X^{2-} giving rise to metallic properties. In the MX compounds with nickel arsenide structures, the metal atoms are situated in all octahedral sites of a hexagonal close-packed arrangement of the chalcogenide atoms. The relatively short distances between these sites along the hexagonal c-axis can lead to semiconductor or even metallic properties. The structural unity of the MX compounds with the nickel arsenide structure type is broken by the manganese compounds. Here, dependent on the half filled $3d$-shell of Mn^{2+}, the salt-like NaCl structure type is not only characteristic for the oxide, but also for the compounds with the heavier chalcogenides. The copper and zinc compounds limit the region of stability of the nickel

TABLE I

Crystal structures of the MX compounds

	Ca	Sc	Ti	V	Cr	Mn	Fe	Co	Ni	Cu	Zn
O	○	—	○	○	—	○	○	○	○	□	△
S	○	○	●	●	□	○ △	● ▲	●	● ■	▽	△
Se	○	○	●	●	●	○	● ▲	●	● ■	▽	△
Te	○	●	●	●	●	●	● ▲	●	●	▲	△

○ ≙ NaCl; △ ≙ zinc-blende or wurtzite; ● ≙ NiAs or variant; □ ≙ special types; ▲ ≙ CuTe type; ■ ≙ millerite; ▽ ≙ CuS type

TABLE II

Crystal structures of several MX_2 compounds

	Ti	V	Cr	Mn	Fe	Co	Ni	Cu
O	○	○	○	○	—	—	—	—
S	▲	—	—	●	● ■	●	●	●
Se	▲	▲	—	●	● ■	● ■	●	● ■
Te	▲	▲	▲	●	● ■	● ■	● ▲	● ▲

○ ≙ rutile or distorted rutile; ▲ ≙ CdI_2; ● ≙ pyrites; ■ ≙ marcasite structure.

arsenide structure type (see Table I) in the direction of the filled 3d-shell configuration. The compounds with the d^{10}-configuration on the extreme right of Table I possess largely ionic-covalent structures.

It is noteworthy that in the border regions of the structural fields, strong anisotropic atomic arrangements appear. Thus, the CuS type is a transition from a co-ordination structure to a layer structure. The CuTe type however is a pure layered structure, in which the iron chalcogenides also crystallize. It is a variant of the antitype of the PbO structure. Ideally, in this array, the anionic partners form a cubic close-packed arrangement in which the metal atoms occupy alternately every second layer of tetra-

hedral sites (see p. 109). The MX_2 structures listed in Table II lend themselves to
similar interpretation. Here the layer structure of the CdI_2 type forms the transition
from the more salt-like compounds to those with extensively metallic properties. In
the CdI_2 type, the anionic atoms form a hexagonal close-packed arrangement in which
the metal atoms occupy every second layer of octahedral sites (see p. 99).

The illustrated relationship in the structural sequence of the MX and MX_2 compounds of the 4th period metals can be extended firstly to the corresponding compounds of metals of the higher periods, and secondly, with respect to X, to the homologues of nitrogen. Layer structures always appear in the border areas between salt-
like and metallic structure types. It can be mentioned in passing that in the sequence
of element structures, the arrangements of the atoms in the layered structures of
graphite and black phosphorous indicate similar transitions.

When considering ternary chalcogenides with transition metals, the range of the
layered structures can wander and extend at random. In this article, only those chalcogenides which contain alkali metals in addition to transition metals will be dealt
with.

In these systems, new variations of layered atomic arrangements might be found,
firstly with host lattices of binary compounds – an example of this being the intercalation of alkali metal atoms into titanium dichalcogenides – and secondly, with
ternary layered structures, in which transition metal-chalcogenide arrangements appear, which are not known in the corresponding binary systems.

The sequence of the transition metals in the periodic table serves as a classification
for the ternary chalcogenides discussed in the following.

2. Layered Structures of $A_x M_y X_z$ Compounds with $A \cong$ Alkali Metal, $M \cong$ Transition Metal and $X \cong$ S or Se

2.1. THE SCANDIUM GROUP

Until now, no ternary chalcogenides with alkali metals and scandium were known.
However, with yttrium, lanthanum and the lanthanoids, a large number of ternary
sulphides with alkali metals has been synthesized and structurally investigated [2–5].
For the preparation, either mixtures consisting of the hydroxides of the transition
metals and the nitrates of the alkali metals, or mixtures consisting of oxides of the
former and carbonates of the latter, were converted in an H_2S flow. The reaction
temperatures were between 500° and 1200 °C. The alkali metal components were
generally added in large excess because of competitive side-reactions resulting in the
formation of volatile alkali-sulfur compounds. After reaction, the compounds obtained were purified in aqueous solutions. In several cases, single crystals could be
isolated. Analyses and structural investigations reveal that all the compounds have
the composition $ALnS_2$ ($A \cong$ alkali metal, $Ln \cong$ Y, La or one of the lanthanoids).
Phase widths were not observed. Because the radial quotients R_{Ln^+}/R_{A^+} are characteristic values for the chosen structure type of the $ALnS_2$ compounds, the quotients
and the ranges of the structure types observed are shown in relation to one another in

W.BRONGER

TABLE III

$R_{Ln^{3+}}/R_{A^+}$ values for the $ALnS_2$ phases obtained (the shaded area indicates the range of the NaCl structure)

	Ionic radii (Å)	Li^+ 0.68	Na^+ 0.97	K^+ 1.33	Rb^+ 1.47	Cs^+ 1.67
La^{3+}	1.06		1.09	0.80	0.72	0.63
Ce^{3+}	1.03		1.06	0.77	0.70	0.62
Pr^{3+}	1.01	1.49	1.04	0.78	0.69	
Nd^{3+}	1.00	1.47	1.03	0.75	0.68	
Sm^{3+}	0.96	1.41	0.98	0.72	0.65	
Eu^{3+}	0.95	1.39	0.97	0.71	0.65	
Gd^{3+}	0.94	1.38	0.97	0.71	0.64	
Tb^{3+}	0.92	1.35	0.95	0.69	0.62	
Dy^{3+}	0.91	1.34	0.94	0.68		
Y^{3+}	0.88	1.29	0.91	0.66		
Ho^{3+}	0.89	1.31	0.91	0.67		
Er^{3+}	0.88	1.29	0.91	0.66		
Tm^{3+}	0.86	1.26	0.89	0.65		
Yb^{3+}	0.85	1.25	0.88	0.64		
Lu^{3+}	0.84	1.24	0.87	0.63		

Table III. X-ray structure analyses indicated two structure types, depending on R_{Ln^+}/R_{A^+}. With larger values, the cations are statistically distributed in positions of the cations of the NaCl type. With smaller values, an ordered distribution of the cations according to the structure of the α-$NaFeO_2$ type is found. This layer structure may be described as a filled $CdCl_2$ type, in which the cations A and Ln alternately occupy every second layer of octahedral sites in a cubic close-packed arrangement of the anions, (see Figures 1 and 2). An equivalent description of the α-$NaFeO_2$ type layer structure can start with the NaCl type, in which the Ln and A atoms occupy every second cation layer $\perp$ [111]. The cell units and the symmetry of the ordered $ALnS_2$ compounds are hexagonal. In the space group $R\bar{3}m$ (No. 166) the following positions are occupied: A^+ in 3a, Ln^{3+} in 3b and S^{2-} in 6c. The z-parameters of the sulphur atoms were in some cases determined by least-squares refinements, in others, with the help of the ionic radii values. The lattice constants are listed in Table IV.

Up to the present time, no $ALnS_2$ phases with $R_{Ln^{3+}}/R_{A^+} < 0.62$ have been obtained.

The continuity of the observed lanthanoid contraction permits the assumption that all lanthanoids have the oxidation state $+3$. Hitherto existing measurements of the magnetic susceptibilities confirm this assumption [6]. To our knowledge, the electrical properties of the $ALnS_2$ phases have not, as yet, been published.

2.2. THE TITANIUM GROUP

The hitherto investigated ternary sulphides and selenides $A_xM_yX_z$ with $M \cong Ti$ or Zr, have distinctly different properties from those with $M \cong Y$, La or one of the lanthanoids. Here the so-called intercalation compounds were found.

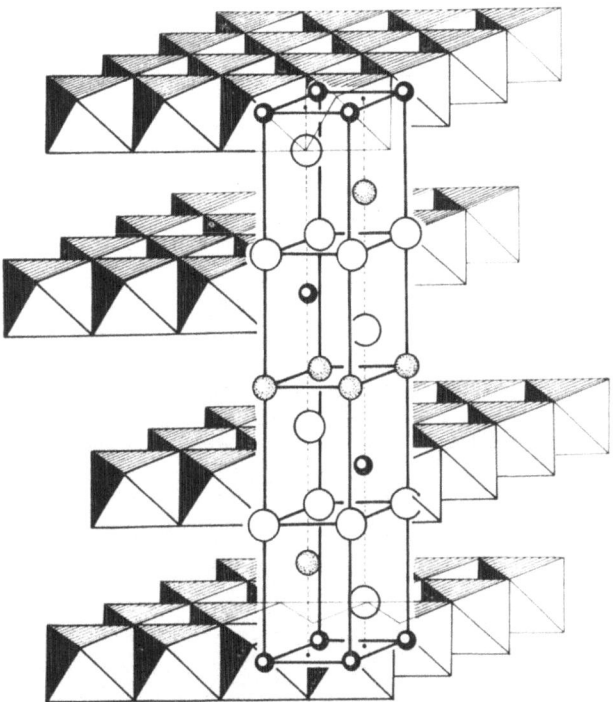

Fig. 1. The crystal structure of the α-NaFeO₂ type. The transition metal atoms are represented by black circles, the alkali metal atoms by dotted circles and the chalcogenide atoms by the white circles. This is valid for all the figures in this article.

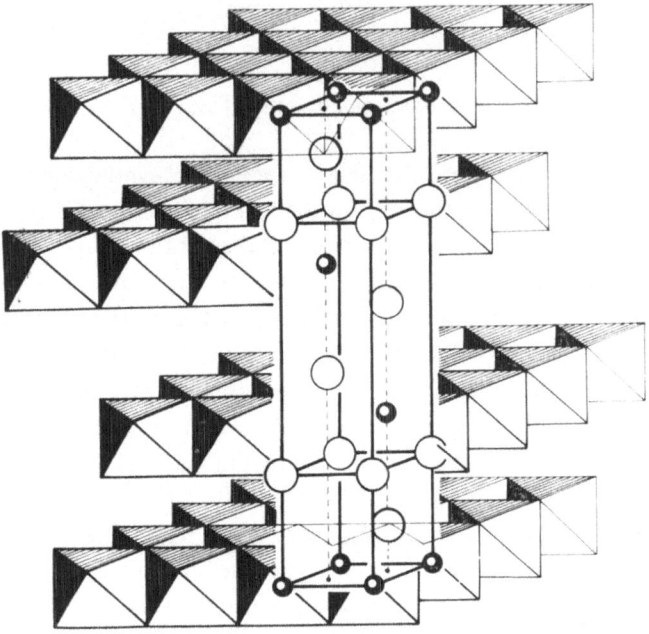

Fig. 2. The CdCl₂ structure.

TABLE IV

Lattice constants of the $ALnS_2$ compounds (Å) [2–5]

	a	c		a	c
$LiPrS_2$	5.686_5		$NaLaS_2$	5.881	
$LiNdS_2$	5.628_5		$NaCeS_2$	5.832	
$LiSmS_2$	5.588		$NaPrS_2$	5.777	
$LiEuS_2$	5.606		$NaNdS_2$	⎰5.803	
$LiGdS_2$	5.530			⎱4.100(4)	19.90(2)
$LiTbS_2$	5.505		$NaSmS_2$	4.056	19.87
$LiDyS_2$	5.474		$NaEuS_2$	4.042	19.92
$LiYS_2$	5.473		$NaGdS_2$	4.009	19.87
$LiHoS_2$	3.898	18.68	$NaTbS_2$	3.989	19.87
$LiErS_2$	3.875	18.63	$NaDyS_2$	3.978	19.92
$LiTmS_2$	3.829(9)	18.48(9)	$NaHoS_2$	3.949	19.86
$LiYbS_2$	3.842	18.54	$NaErS_2$	3.939	19.98
$LiLuS_2$	3.813(4)	18.41(2)	$NaYS_2$	3.968	19.89
			$NaTmS_2$	3.915(3)	19.88(2)
$KLaS_2$	4.264	21.89_5	$NaYbS_2$	3.902(2)	19.91(1)
$KCeS_2$	4.223	21.80	$NaLuS_2$	3.885(3)	19.87(2)
$KPrS_2$	4.185	21.75			
$KNdS_2$	4.160_5	21.83	$RbLaS_2$	4.292	22.89
$KSmS_2$	4.107	21.76	$RbCeS_2$	4.246	22.80
$KEuS_2$	4.093	21.85_5	$RbPrS_2$	4.222	22.87
$KGdS_2$	4.075	21.89	$RbNdS_2$	4.189	22.89
$KTbS_2$	4.051	21.87	$RbSmS_2$	4.141	22.86
$KDyS_2$	4.030	21.83	$RbEuS_2$	4.119	22.84
KYS_2	4.022	21.85_5	$RbGdS_2$	4.098	22.88
$KHoS_2$	4.009_5	21.80	$RbTbS_2$	4.070	22.80
$KErS_2$	3.993	21.77			
$KTmS_2$	3.997(4)	21.84(2)	$CsLaS_2$	4.306	24.08
$KYbS_2$	3.964	21.82	$CsCeS_2$	4.262	23.99
$KLuS_2$	3.947(2)	21.79(2)			

They could be synthesized by reactions of TiS_2, $TiSe_2$ or ZrS_2 with alkali metals in liquid ammonia.

Rüdorff and co-workers were the first to succeed in synthesizing titanium compounds [7]. The following formulae show the maximum content of alkali metals obtained for each compound under his experimental conditions: $Li_{0.6}TiS_2$, $Na_{0.8}TiS_2$, $K_{0.8}TiS_2$, $Cs_{0.6}TiS_2$ and $Na_{0.9}TiSe_2$. Preliminary structure investigations on the sodium compounds show a cation distribution which corresponds to that in the α-$NaFeO_2$ type. This means, that in the $CdCl_2$ type the unoccupied octahedral sites in the anion array were partly occupied by sodium atoms. Thus, the $[X-Ti-X]$ slabs in the TiS_2 or $TiSe_2$ host structures are shifted, so that the hexagonal c-axis now includes three $[X-Ti-X]$ slabs instead of one and three planes of sodium atoms, (see Figure 1).

Recently reports of further investigations of the A_xTiS_2 intercalation compounds have been published [8–11]. Under varied experimental conditions it was possible to obtain phases extending to the limiting compositions $ATiS_2$. In detail, the following phase sequences were published:

In the system with lithium, only one phase exists. The lithium atoms are intercalated into the structure of TiS_2 with values of $0<x<1$. The lattice constants of the hexagonal unit cell increase from $a=3.412$ Å and $c=5.695$ Å for TiS_2, to $a=3.435$ ±0.005 Å and $c=6.18\pm0.02$ Å for $LiTiS_2$. Assuming the same space group $P\bar{3}m1$ (No. 164) as for the obviously isotypic compounds $LiVS_2$ [12] and $LiCrS_2$ [12, 13], the following positions are occupied: Ti in (a) (000); Li in (b) $(00\frac{1}{2})$; 2S in $(d)\pm(\frac{1}{3}\frac{2}{3}z)$. When $x=1$, the variable z takes the value 0.228. The layer structure of $LiTiS_2$ described here can thus be considered as a filled CdI_2 type (see Figures 3 and 4). An equivalent description may start with the NiAs type, in which the cation layers $\perp$ [001] are alternatingly occupied by only lithium or only titanium atoms.

For the systems with heavier alkali metals, the phase sequences were published and are listed in Table V.

Only those phases indicated by Ia and Ib were characterized by X-ray investigations, and then only by powder patterns according to the Debye-Scherrer method. According to this, the Ia structure found in the Na_xTiS_2 system corresponds to the atomic arrangement of the α-$NaFeO_2$ type with partially filled alkali metal layers; it thus corresponds to the structure found by Rüdorff for the composition $Na_{0.8}TiS_2$. For the Ib

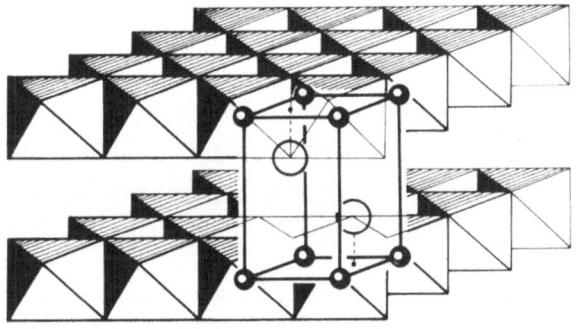

Fig. 3.   The CdI₂ structure.

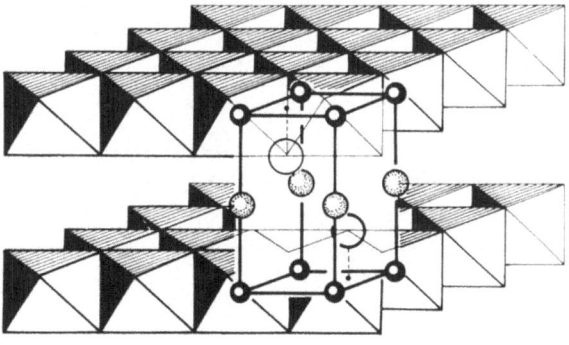

Fig. 4.   The crystal structure of LiTiS₂.

TABLE V

Phase sequences in $A_x TiS_2$ systems. The lattice constants are given in Ångström units. The standard deviations are given in brackets

		Ib		Ia
$Na_x TiS_2$		$0.17 < x < 0.33$	$0.38 < x < 0.72$	$0.79 < x \leqslant 1$
		$Na_{0.33}TiS_2$:	$Na_{0.55}TiS_2$:	$Na_1 TiS_2$:
		$a = 3.406\ (7)$	$a = 3.433\ (5)$	$a = 3.469\ (5)$
		$c = 38.2\ (1)$	$c = 20.94\ (5)$	$c = 20.58\ (5)$
$K_x TiS_2$	$0.06 < x < 0.08$	$0.14 < x < 0.16$	$0.28 < x \leqslant 1$	
	$K_{0.08}TiS_2$:	$K_{0.15}TiS_2$:	$K_1 TiS_2$:	
	$a = 3.405\ (10)$	$a = 3.410\ (7)$	$a = 3.488\ (5)$	
	$c = 75.5\ (3)$	$c = 41.1\ (1)$	$c = 22.83\ (5)$	
$Rb_x TiS_2$	$0.04 < x < 0.06$	$0.12 < x < 0.32$	$0.42 < x \leqslant 1$	
	$Rb_{0.05}TiS_2$:	$Rb_{0.26}TiS_2$:	$Rb_1 TiS_2$:	
	$a = 3.400\ (10)$	$a = 3.404\ (7)$	$a = 3.427\ (5)$	
	$c = 75.4\ (3)$	$c = 41.3\ (1)$	$c = 24.30\ (5)$	
$Cs_x TiS_2$	$0.03 < x < 0.04$	$0.08 < x < 0.10$	$0.56 < x \leqslant 1$	
	$Cs_{0.04}TiS_2$:	$Cs_{0.09}TiS_2$:	$Cs_1 TiS_2$:	
	$a = 3.404\ (10)$	$a = 3.410\ (7)$	$a = 3.500\ (5)$	
	$c = 76.4\ (3)$	$c = 43.3\ (1)$	$c = 24.51\ (5)$	

type, a stacking variant was found in which the alkali metal atoms have a trigonal-prismatic sulphur co-ordination. The interrelationships between the different atomic arrangements may be seen in the projections $\perp$ [100] (see Figure 5).

In the systems $A_x ZrS_2$, largely analogous relationships were found (see Table VI) [11, 14].

The particular reactivity of the intercalation compounds of types $A_x TiS_2$ and $A_x ZrS_2$ is particularly noticeable in reactions with acids. Even with water as a weak acid, reactions occur in which H_2 is formed, accompanied by a reformation of the disulphides which may still contain small amounts of alkali metals.

In some cases, the magnetic moments of the intercalation compounds mentioned above have been determined. Thus, for example, for $Na_{0.8}TiS_2$, a μ_{eff} value of 0.61 μ_B at $T = 293$ K is given. For $NaTiS_2$, the published susceptibility value $(T = 297$ K$)$ corresponds to a $\mu_{eff} = 0.80\ \mu_B$. The potassium compounds have the following values: $K_{0.44}TiS_2$: 0.46 μ_B; $K_{0.56}TiS_2$: 0.60 μ_B; $K_{0.74}TiS_2$: 0.80 μ_B; $KTiS_2$: 0.93 μ_B. In some cases, susceptibility values which were almost temperature-independent were found. This fact probably indicates Pauli-paramagnetism, and consequently a metallic behavior. This character seems to diminish with increasing alkali metal content. According to preliminary conductivity measurements, $Na_{0.8}TiS_2$ was found to be a conductor [7].

For the $A_x ZrS_2$ phases, the magnetic susceptibility values can also be interpreted ui terms of Pauli-paramagnetism [11, 14].

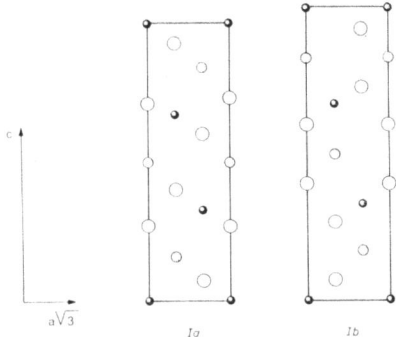

Fig. 5. Projections of the atomic arrangements of the Ia and Ib types ⊥ [100] (see text).

TABLE VI

Phase sequence in A_xZrS_2 systems. The lattice constants are given in Ångström units. The standard deviations are given in brackets

	Ib			Ia
Li_xZrS_2				$0.25 < x \leqslant 1$ Li_1ZrS_2: $a = 3.603$ (5) $c = 18.75$ (4)
Na_xZrS_2	$x = 0.32$ $Na_{0.32}ZrS_2$: $a = 3.640$ (7) $c = 39.84$ (10)			$0.64 < x \leqslant 1$ Na_1ZrS_2: $a = 3.666$ (5) $c = 20.35$ (5)
K_xZrS_2	$0.18 < x < 0.30$ $K_{0.196}ZrS_2$: $a = 3.644$ (7) $c = 39.83$ (10)	$0.42 < x < 0.71$ $K_{0.58}ZrS_2$: $a = 3.637$ (5) $c = 23.22$ (5)	Iab $x = 0.86$ $K_{0.86}ZrS_2$: $a = 3.666$ (5) $c = 67.9$ (1)	$x = 1$ K_1ZrS_2: $a = 3.718$ (5) $c = 22.20$ (5)
Rb_xZrS_2	$x = 0.33$ $Rb_{0.33}ZrS_2$: $a = 3.643$ (7) $c = 48.85$ (10)	$0.54 < x < 0.75$ $Rb_{0.75}ZrS_2$: $a = 3.657$ (5) $c = 24.05$ (5)		
Cs_xZrS_2	$x = 0.37$ $Cs_{0.37}ZrS_2$: $a = 3.636$ (7) $c = 50.9$ (1)	$0.56 < x < 0.66$ $Cs_{0.64}ZrS_2$ $a = 3.644$ (7) $c = 25.23$ (5)		

2.3. THE VANADIUM GROUP

$LiVS_2$ [12] was obtained by reaction of mixtures consisting of Li_2CO_3 and V_2O_3 in an H_2S atmosphere at temperatures of 500 °C and, subsequently, 700 °C. Investigations of the crystal structure by means of X-ray and neutron diffraction experiments indicate a filled CdI_2 type. In the space group $P\bar{3}m1$ (No. 164), the following positions are occupied: V in (a) (000); Li in (b) $(00\frac{1}{2})$ and S in (d) $\pm(\frac{1}{3}\frac{2}{3}z)$ with $z = 0.235$.

The lattice constants of the hexagonal unit cell at 293 K are: $a = 3.3803$ (2) and $c = 6.1381$ (5) Å. A neutron diffraction experiment at 4.2 K shows no magnetic spin structure. The results show that $LiVS_2$ is isotypic with $LiTiS_2$ [10] and $LiCrS_2$ [12, 13] (see Figure 4).

Similar structural relationships can be found for the sodium compounds: $NaVS_2$ and $NaVSe_2$ crystallize in the α-$NaFeO_2$ structure type (see Figure 2) and are isotypic with $NaTiS_2$ and $NaCrS_2$ (see also p. 106).

A large number of intercalation compounds of the dichalcogenides of the elements niobium and tantalum have already been investigated. Besides phases with the general formulae T_xNbS_2 and T_xTaSe_2 with $T \cong Ti$, V, Cr, Mn, Fe, Co, Ni, Rh [15–18], there are also phases with the compositions A_xMS_2 and A_xMSe_2 with $A \cong Li$, Na or K and $M \cong Nb$ or Ta [19].

It was possible to synthesize the last mentioned compounds by reactions between the elements in question, or between mixtures of the alkali metals and the transition metal dichalcogenides in evacuated quartz tubes at 800 °C.

The disulphides of niobium and tantalum – each of which exist in several forms and in which the Nb or Ta atoms may have trigonal-prismatic or octahedral co-ordination [20, 21] – were used here in the so-called 2s-modification (see Figure 6). In this structure, the layer sequence can be described by the scheme $AcA\,BcB$, with the capital letters denoting the chalcogenide layers and the small letters the metal atom layers. The atomic arrangement mentioned here is closely related to a second basic type of dichalcogenide, the 2s-MoS_2 form which displays the layer sequence $AbA\,BaB$ (see p. 108, Figure 9).

The A_xMX_2 phases with $M \cong Nb$ or Ta which are thermally stable over a wide temperature range, but which are not stable in air, generally show homogeneous phases only for $x \approx \frac{2}{3}$; for some sulphides with lithium and sodium a range of homogeneity exists for $0.4 < x < 0.7$. According to X-ray investigations on powder samples, the compounds obtained crystallize in hexagonal layered structures (see Table VII) in which the alkali metal atoms are intercalated between $[X-M-X]$ slabs. The coordination of the Nb and Ta atoms by sulphur or selenium atoms is, in all cases, trigonal

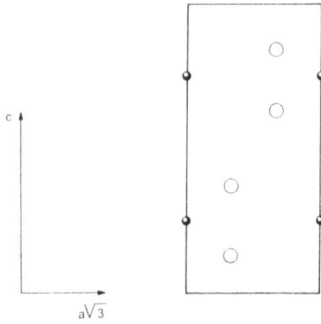

Fig. 6.   Projection of the atomic arrangement of the 2s-NbS_2 structure $\perp$ [100].

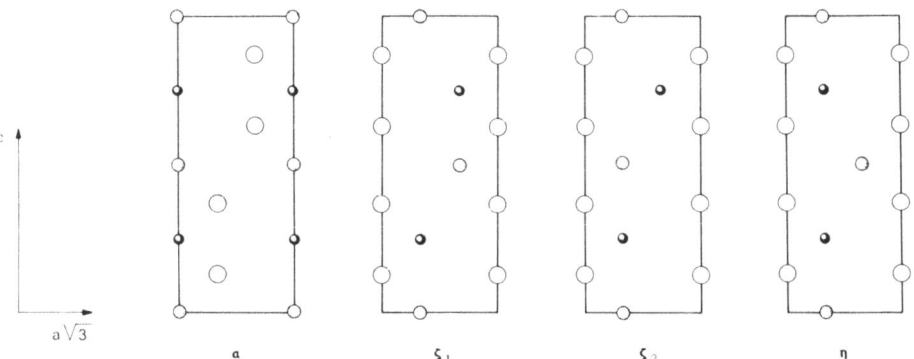

Fig. 7. Projections $\perp$ [100] of the hexagonal atomic arrangements of A_xMX_2 compounds ($A \triangleq Na$, K; $M \triangleq Nb$, Ta; $X \triangleq S$, Se) (see text).

prismatic. The co-ordination of the Na and K atoms is also trigonal-prismatic in the structures of the selenides and most of the sulphides. In some modifications of the sulphides, however, an octahedral environment is possible. The structure types are illustrated in Figure 7. From these, only the model which is derived from the 2s-form of the dichalcogenides (α) allows octahedral co-ordination of the alkali metal atoms. The models ζ_1 and ζ_2 could not be distinguished by means of an X-ray analysis. According to MO-calculations, the trigonal-prismatic co-ordinations for the d^1–d^2 configurations assumed for the niobium and tantalum atoms appear to be particularly stable [22].

Closely related to the alkali metal ions A^+ is the Tl^+ ion. For this reason, a number of results concerning the synthesis and structure of thallium intercalation compounds are mentioned at this point [23]. The synthesis of compounds having the composition $Tl_{0.5}MX_2$ with $M \triangleq Nb$ or Ta and $X \triangleq S$ or Se, was achieved by reactions between either MX_2 and Tl, or between the elements in question placed in corundum crucibles and sealed in quartz ampules at temperatures between 380° and 600 °C. Potential phase widths of the ternary compounds obtained were not investigated. The lattice constants of possible hexagonal unit cells derived from the X-ray powder patterns are listed in Table VIII.

Assuming that the thallium compounds have crystal structures which correspond to those of the alkali metal compounds, and taking the thickness of the $[X-M-X]$ slabs from the binary compounds, the displacements per intercalated metal layer $\perp$ [001] are shown in Table IX.

It is surprising that the displacements for the thallium intercalation compounds are consistently somewhat smaller than those for the corresponding potassium compounds. A substantially larger displacement – of the order of that for say, rubidium compounds – is expected here, because the ionic radii of Tl^+ and Rb^+ have the same value of $r = 1.47$ Å, whereas the radius of K^+ is only 1.33 Å.

TABLE VII

Structures and lattice constants (Å) of the dichalcogenides MX_2 of niobium and tantalum ($2s$-forms) as well as of the alkali metal intercalation compounds (the lattice constants of the A_xMX_2 phases refer to the maximum alkali metal content with $x \approx \frac{2}{3}$; standard deviations are given in brackets)

Compound	Structure	a	c
$2s$-NbS$_2$	α	3.31	11.89
$2s$-TaS$_2$	α	3.315	12.10
Li$_x$NbS$_2$)^a	α	3.331 (1)	12.90 (1)
Li$_x$TaS$_2$)	α	3.333 (1)	12.89 (1)
		3.35	12.88
Na$_x$NbS$_2$	α (or η)	3.366 (1)	14.52 (1)
	ζ	3.363 (1)	14.525 (5)
Na$_x$TaS$_2$	α (or η)	3.345 (1)	14.54 (1)
	ζ	3.337 (1)	14.59 (1)
K$_x$NbS$_2$	α (or η)	3.345 (5)	16.22 (2)
K$_x$TaS$_2$	ζ	3.332 (1)	16.20 (1)
$2s$-NbSe$_2$	α	3.442	12.54
$2s$-TaSe$_2$	α	3.437	12.72
Na$_x$NbSe$_2$	ζ	3.476 (1)	15.366 (5)
Na$_x$TaSe$_2$	ζ	3.458 (1)	15.403 (5)
K$_x$NbSe$_2$	η	3.480 (2)	17.04 (1)
K$_x$TaSe$_2$	η	3.463 (1)	17.05 (1)

[a] Here the lithium positions are unknown.

TABLE VIII

Lattice constants of hexagonal unit cells of $Tl_{0.5}MX_2$ compounds

Compound	a (Å)	c (Å)
Tl$_{0.5}$NbS$_2$	3.35	16.08
Tl$_{0.5}$TaS$_2$	3.33	8.08
Tl$_{0.5}$NbSe$_2$	3.47	8.34
Tl$_{0.5}$TaSe$_2$	3.46	8.36

TABLE IX

Thickness of the Tl-layers (Å) in the dichalcogenide intercalation compounds

	NbS$_2$	TaS$_2$	NbSe$_2$	TaSe$_2$
Li	0.5	0.4		
Na	1.3	1.2	1.4$_1$	1.3$_4$
K	2.1$_6$	2.0$_5$	2.2$_5$	2.1$_7$
Tl	2.0$_9$	2.0$_3$	2.0$_7$	2.0$_0$

2.4. THE CHROMIUM GROUP

The synthesis of the phase $LiCrS_2$ [12, 13] was carried out by reactions of Li_2CO_3 with Cr_2O_3 in an H_2S atmosphere at 500 °C, or in a CS_2 saturated argon stream for a short period at 800 °C, or 1100 °C. $LiCO_3$ was usually added in excess and graphite crucibles were used.

The crystal structure of $LiCrS_2$ was determined from X-ray and neutron diffraction experiments on powder samples, as well as from X-ray structure investigations on single crystals. Earlier investigations have been shown to be incorrect [24]. As was the case for the isotypic compounds $LiTiS_2$ and $LiVS_2$, a layered structure was found for $LiCrS_2$, which may be described as a filled CdI_2 or ordered NiAs type (see Figure 4). The published lattice constants and position parameters are as follows: a(Å): 3.456(1) [13]; 3.4637(3) [12] for 293 K or 3.4515(3) [12] for 4.2 K. c(Å): 6.020(2) [13]; 6.0369(11) [12] for 293 K or 6.0212(10) [12] for 4.2 K.

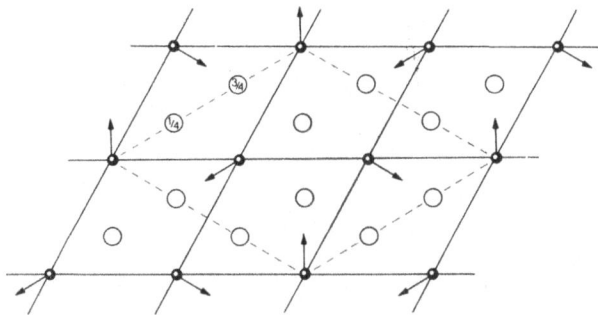

Fig. 8. Spin arrangement in $LiCrS_2$ within one hexagonal chromium layer. In adjacent layers the moments are reversed. The chemical cell is shown in dark outline, the magnetic cell is indicated by the broken lines.

The space group is $P\bar{3}m1$ with Cr in (a), Li in (b) and S in (d) with $z=0.2247(7)$ [13] or 0.220(2) [12] for 293 K and 0.224(1) [12] for 4.2 K. The neutron diffraction experiments reveal an ordered spin structure at 4.2 K in the slabs $\perp$ [001] (see Figure 8) and an antiparallel coupling of neighbouring slabs. The Néel point is ≈ 55 K.

The compounds $NaCrS_2$ and $KCrS_2$ have been known for almost 100 years [25÷28]. The synthesis of these compounds can be achieved by means of melting mixtures of alkali salts like carbonates or rhodanides, and chromium compounds like chromium oxide or potassium chromate in an oxygen free atmosphere at temperatures of approximately 800°–900 °C. The crystalline compounds can be obtained from the melts after extraction with pure water. Investigations of the crystal structure revealed a layered structure of the α-$NaFeO_2$ type for both compounds [28, 29] (see Figure 2). The lattice constants of the hexagonal unit cells are: $a=3.53$ Å and $c=19.49$ Å for $NaCrS_2$ and $a=3.618$ Å and $c=21.16$ Å for $KCrS_2$ (for $NaCrS_2$ the values $a=3.549$ and $c=19.44$ Å have been also reported [30]).

As in the case of the lithium compounds, isotypic structures exist for the $NaMS_2$ phases with $M \cong Ti$, V or Cr. In Table X some properties of the $NaMS_2$ compounds are listed.

TABLE X

Properties of $NaMS_2$ compounds [7]

	$NaTiS_2$	$NaVS_2$	$NaCrS_2$
Colour	black	black	orange
Lattice constants (Å)	$a=3.47$	3.57	3.53
	$c=20.6$	19.65	19.49
Magnetic moment μ_{eff} at $T=293$ K (μ_B)	0.80	2.1	3.7
$\mu_{calculated}$ (μ_B)	1.8	2.8	3.8
Specific resistance $[\Omega \cdot cm]$	low	2.2	$>10^8$
Reaction with water	H_2-development	slow hydrolysis	stable

A comparison between the three isostructural compounds whose lattice constants differ only slightly from one another, clearly shows that with an increasing number of electrons in the $3d$ level, the tendency towards delocalization of the electrons decreases. In the chromium compound, practically no delocalization is to be found; in the titanium compound, however, metallic properties can already be found, although the metal-metal distances are only slightly shorter. For $NaCrS_2$, no phase width has been observed; Na_xTiS_2, on the other hand, exists for $0.79 < x < 1$.

Measurements of the magnetic properties of $NaCrS_2$ and $KCrS_2$ show paramagnetic behaviour at higher temperatures; the moments calculated from the Curie-Weiß law are $\mu = 3.7$ μ_B for $NaCrS_2$ and $\mu = 4.0$ μ_B for $KCrS_2$, which agree satisfactorily with the value expected for Cr^{3+} ions [29]. Moreover, the magnetic properties of $NaCrS_2$ were measured on single crystals at temperatures down to 4.2 K [30]. The data obtained agree with an antiferromagnetic helix model, in which the spin moments lie in a plane $\perp [001]$. Further neutron diffraction experiments were carried out [31].

The compound $NaCrSe_2$ is isotypic with $NaCrS_2$. The following values are given for the lattice constants of the hexagonal unit cell: $a=3.708$ Å and $c=20.29$ Å [32] or $a=3.729$ Å and $c=20.48$ Å [30]. The magnetic properties are similar to those of $NaCrS_2$ [30, 31].

$ACrS_2$ compounds with $A \cong Rb$ or Cs have already been mentioned in several publications. Up to now, however, exact structure determinations have not been carried out. It is probable that no pure compounds were obtained [24].

Ternary sulphides, with alkali metals and molybdenum or tungsten, were obtained as intercalation compounds in a way similar to that used for the neighbouring elements niobium and tantalum [7, 23]. However, the synthesis could not be achieved by reactions of MoS_2 or WS_2 with molten potassium or sodium, because at higher temperatures the formation of Mo or W and alkali metal sulphides were observed. However, ternary sulphides and selenides can be obtained if the metal sulphides react with alkali metals dissolved in liquid ammonia. Black to blue-black compounds are formed, which are pyrophoric and which react violently with water, forming H_2 and alkali hydroxides. The results of the investigations on the compounds obtained are listed in Table XI below.

It is worthy of note that ammonia is incorporated in the lithium compounds, in the ratio of 1 NH_3 to 1 Li.

The structure of the initial metal chalcogenides is evidently only slightly changed by the intercalation. It can be assumed that the slabs of the $2s$-MoS_2 structure (see Figure 9) are preserved as a whole, with the hexagonal a-axis – corresponding to the distances $M-M$ or $S-S$ within the layers – increasing by only a few hundredth's of an Ångström. The displacements in the direction of the c-axis, i.e. the increased spacings between neighbouring $[X-M-X]$ slabs produced by the intercalated alkali metal layers, correspond to those obtained for the intercalation compounds of the niobium and tantalum dichalcogenides as shown in Table IX.

Preliminary measurements of the magnetic behavior of the A_xMoX_2 and A_xWX_2 intercalation compounds indicate metallic behavior [7].

It is possible that 1:1 compounds of the general formula AMS_2 exist here too. It must be mentioned, however, that the phases obtained were not investigated thoroughly enough. Structural investigations on single crystals were not made [24].

TABLE XI

Molybdenum and tungsten chalcogenide
intercalation compounds

Compounds	c-axes of the hexagonal unit cells (Å)	Displacement per alkali metal layer (Å)
MoS_2	12.3_0	–
$Li_{0.8}(NH_3)_{0.8}MoS_2$	19.0_0	3.35
$Na_{0.6}MoS_2$	15.0_0	1.35
$K_{0.6}MoS_2$	16.2_0	1.95
$Cs_{0.5}MoS_2$	17.7_8	2.74
$MoSe_2$	12.9_3	–
$K_{0.5}MoSe_2$	17.1_4	2.22
WS_2	12.3_6	–
$Li_{0.5}(NH_3)_{0.6}WS_2$	19.0_8	3.36
$Na_{0.5}WS_2$	15.0_8	1.36
$K_{0.5}WS_2$	16.3_8	2.01
$Rb_{0.5}WS_2$	17.1_8	2.41
$Cs_{0.5}WS_2$	17.8_8	2.76
WSe_2	12.9_7	–
$K_{0.4}WSe_2$	–	–

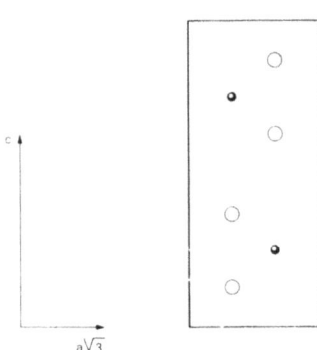

Fig. 9.   Projection of the atomic arrangement of the 2s-MoS₂ structure ⊥ [100].

2.5. THE MANGANESE GROUP

A compound having the composition $K_2S \cdot 3MnS$ was first reported in 1846 [34]. Recent investigations have confirmed this stoichiometry. Moreover, apart from $K_2Mn_3S_4$, the compounds $Rb_2Mn_3S_4$ and $Cs_2Mn_3S_4$ have been synthesized and their crystal structures have thus been determined [35]. The synthesis was carried out by reactions between alkali carbonates, pure manganese and sulphur in a molten state at temperatures of 700° to 900 °C. Pure, red plate-shaped crystals were isolated from

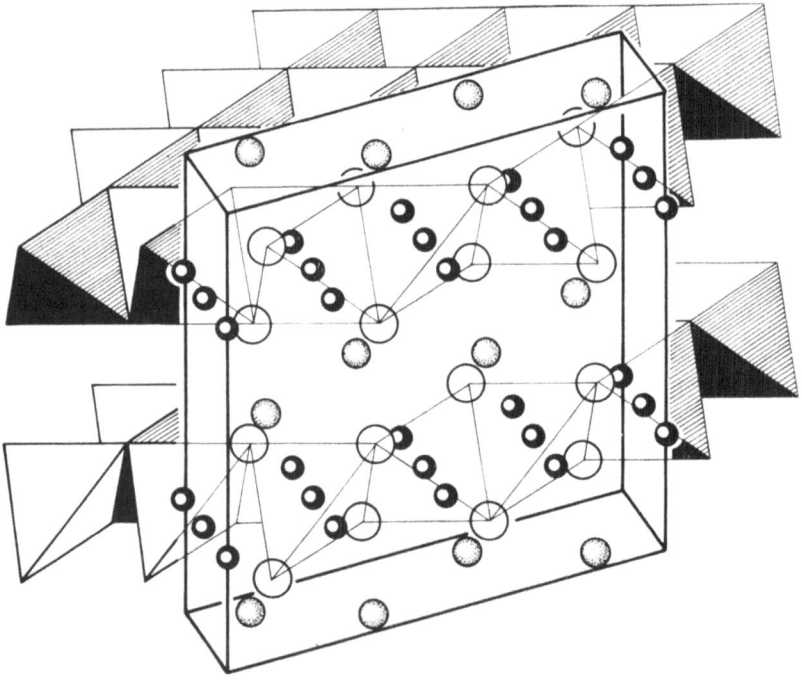

Fig. 10.   The $Cs_2Mn_3S_4$ structure.

the melts by purification in an aqueous solution. The crystals are only stable over long periods in an inert gas atmosphere. A structural investigation on single crystals revealed a layered structure type for $Cs_2Mn_3S_4$, in which alternating manganese and alkali metal layers are separated by sulphur layers (Figure 10). Here, the structure can again be derived from a binary layered structure type, which, by intercalating with alkali metal atoms, leads to the atom arrangement observed in the ternary phase: the order of the manganese and the sulphur atoms corresponds to the CuTe type (see Figure 11). The chalcogenide atoms form layers of tetrahedra connected by their edges in two dimensions. The metal atoms occupy the centres of the tetrahedra. The varying stoichiometry of the ternary compounds is explained by the fact that only three quarters of the tetrahedra centres are occupied ($Cs_2Mn_3\square S_4$). X-ray analyses show an occupation differing from the statistical distribution (see Figure 12). This has not as yet been fully explained. In detail, the atoms occupy the following positions in the space group *Ibam* (No. 72): 1 Mn in 4a, 4 Mn in 4b, 7 Mn in 8g with $y=0.234$, 8 Cs in 8j with $x=0.243$ and $y=0.123$, 16 S in 16k with $x=0.236$, $y=0.972$ and $z=0.153$. The orthorhombic unit cell has the following dimensions: $a=5.920$, $b=11.47$ and $c=14.16$ Å. For the caesium atoms, the co-ordination number 8 results, where the 8 sulphur atoms occupy the corners of a distorted cube. Structure investigations on X-ray diagrams of powder samples reveal a structure for $Rb_2Mn_3S_4$ similar to that of $Cs_2Mn_3S_4$. This is also valid for the non-statistical distribution of the manganese atoms. The unit cell has the dimensions: $a=5.845$, $b=11.21$ and $c=13.66$ Å. For $K_2Mn_3S_4$ the atomic positions are not yet known, however the lattice constants suggest considerable similarity with the layered structure of the $Cs_2Mn_3S_4$ type.

Preliminary investigations of the magnetic behavior show analogies with the behavior of the MnS forms. Possible Néel-points lie within the same temperature range, and the paramagnetic moments correspond to those calculated for the d^5 state of an Mn^{2+}.

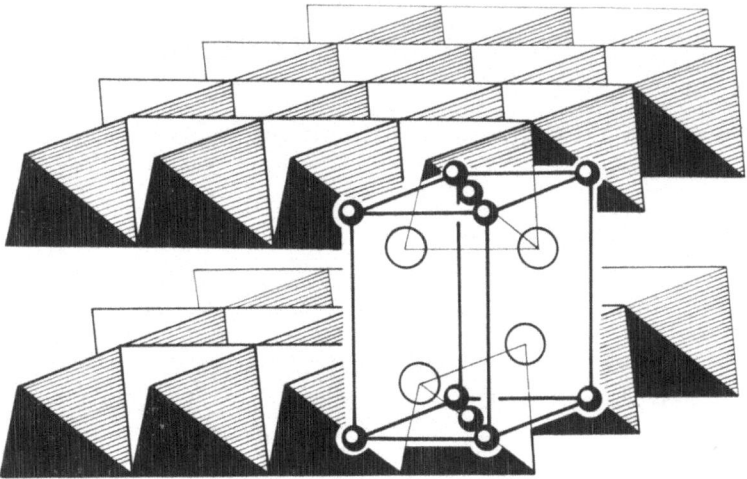

Fig. 11. The CuTe structure.

Up to now, ternary chalcogenides with technetium have not been reported, whereas intercalation compounds of rhenium do exist, and closely correspond to those of molybdenum and tungsten [7]. This is confirmed by their reactivity and by the results of structure investigations. The results of X-ray investigations known up to now are listed in Table XII, and are arranged in the same way as for the molybdenum and tungsten compounds in Table XI.

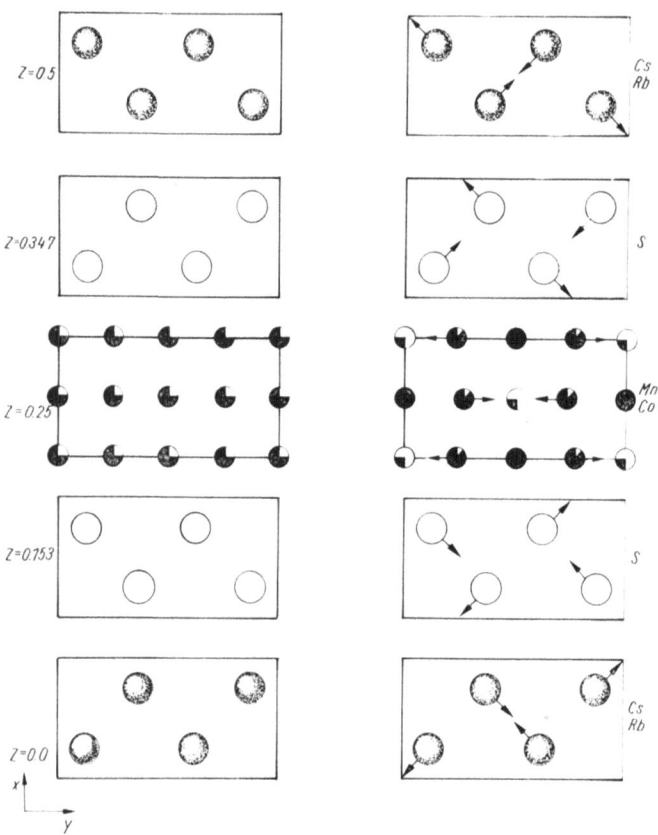

Fig. 12.   The sequence of the layers in the $Cs_2Mn_3S_4$ structure.

TABLE XII

Rhenium chalcogenide intercalation compounds

Compound	c-axes of the hexa-gonal unit cells (Å)	Displacement per alkali metal layer (Å)
ReS_2	12.1_9	–
$K_{0.8}ReS_2$	15.5_0	1.65
$Rb_{0.6}ReS_2$	16.3_6	2.05

A comparison between the results in Tables XI and XII shows that the displacements between neighbouring [X–M–X] slabs decrease with increasing alkali metal content. Compare for example: $K_{0.5}WS_2 \cong 2.00$ Å, $K_{0.6}MoS_2 \cong 1.95$ Å and $K_{0.8}ReS_2 \cong 1.65$ Å.

2.6. THE IRON GROUP

Of the systems A/Fe/S or A/Fe/Se, only the compounds $KFeS_2$, $RbFeS_2$, $CsFeS_2$, $KFeSe_2$, $RbFeSe_2$ and $CsFeSe_2$ are definitely known at present [29, 36–40]. These compounds crystallize with chain structures instead of layered structures. In every case, tetrahedra of chalcogen atoms are connected by their edges in one dimension. The iron atoms occupy the centres of the tetrahedra and the alkali metal atoms separate these arrangements.

The thallium compounds, however, have layered structures. They have the same stoichiometry as the alkali metal compounds, namely $TlFeS_2$ and $TlFeSe_2$ [41, 42]. The synthesis of these compounds was achieved by a reaction between the elements in evacuated quartz ampules at 300° to 500 °C. The structure is characterized by the fact that the chalcogen tetrahedra surrounding the iron atoms are connected by their corners in two dimensions. The thallium atoms are intercalated in layers between the [X–Fe–X] slabs (see Figure 13). Here also, the structure of a binary compound – namely the layered structure of the red HgI_2 – may serve as a basis for the atomic arrangement of these ternary compounds. In the HgI_2 structure, the metal atoms are tetrahedrally co-ordinated by non-metal atoms in a way similar to the iron atoms in the sulphides mentioned above. The tetrahedra are connected by their corners in two dimensions (see Figure 14). The following values have been published: $TlFeS_2$: $a = 3.753$ Å, $c = 13.342$ Å; $TLFeSe_2$: $a = 3.881$ Å, $c = 13.965$ Å. In the tetragonal space

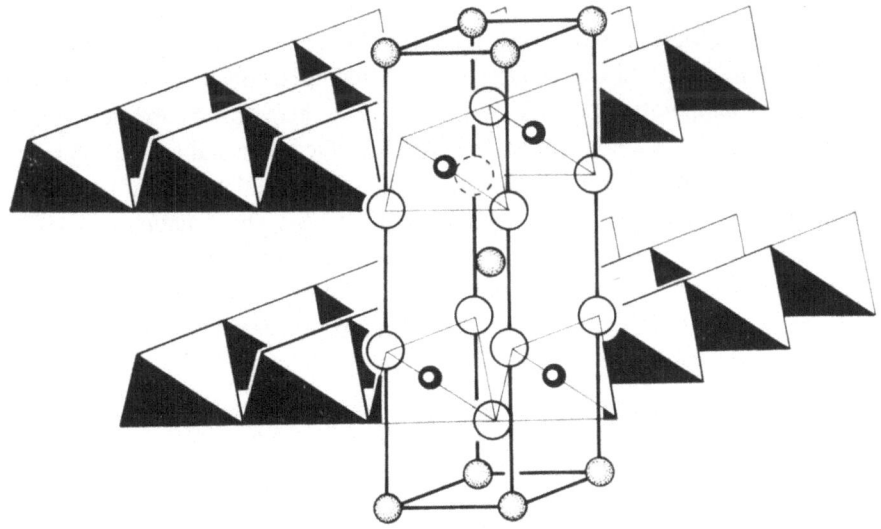

Fig. 13. The $TlFeS_2$ structure.

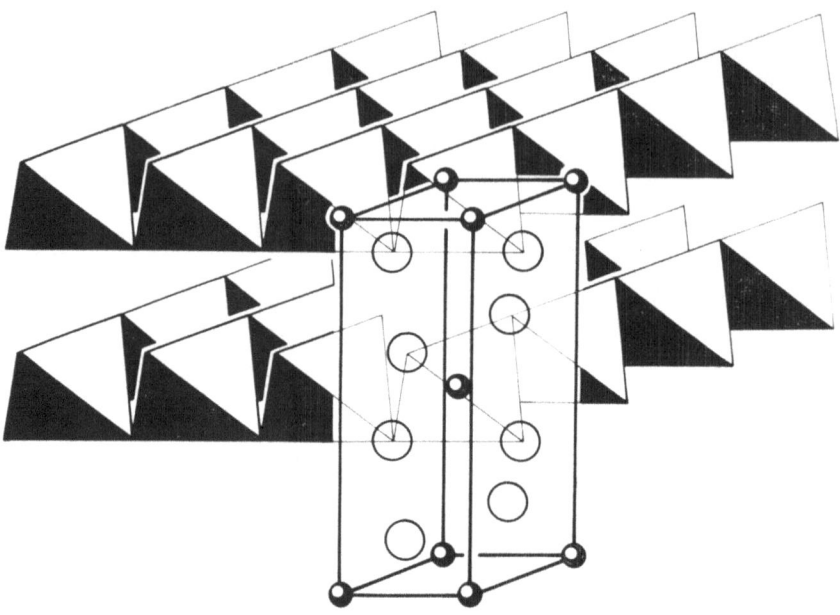

Fig. 14. The structure of HgI₂ (red form).

group $I\bar{4}m2$ (No. 119), the position (2a) (000) is occupied by Tl, (2c) $(0\frac{1}{2}\frac{1}{4})$ by Fe and (4e) $\pm$ (00z) by S with $z = 0.3513$. Each thallium atom occupies the centre of a slightly distorted cube, the corners of which are occupied by sulphur atoms. Ternary chalcogenides with alkali metals and ruthenium or osmium are not known at present.

2.7. THE COBALT GROUP

For cobalt, the compounds $Rb_2Co_3S_4$ and $Cs_2Co_3S_4$, which are analogous to the phases with manganese, could be synthesized [35]. The synthesis was again carried out by reactions of alkali carbonates with sulphur and cobalt in a pure nitrogen atmosphere at temperatures of 700° to 900 °C, and thereafter by extraction of the melts with water. The plate-like crystals have a metallic lustre and are unstable in air. The corresponding potassium compound could not be synthesized in a pure form, since the products obtained were contaminated by CoS_2. The structure of $Cs_2Co_3S_4$ and $Rb_2Co_3S_4$ corresponds to that of the manganese compounds mentioned above (see Figure 10, p. 108). Here too, the sulphur tetrahedra surrounding the cobalt atoms form layers $\perp$ [001]. The lattice constants of the orthorhombic unit cells (formula units: $z = 4$; space group: $Ibam$) are:

$$Rb_2Co_3S_4: \ a = 5.64_4, \quad b = 11.04, \quad c = 13.37 \ \text{Å}$$
$$Cs_2Co_3S_4: \ a = 5.73_0, \quad b = 11.23, \quad c = 13.88 \ \text{Å}.$$

Ternary chalcogenides with alkali metals and rhodium or iridium are not known at present.

2.8. THE NICKEL GROUP

Ternary chalcogenides could be prepared with nickel [43–45]. Analyses show that they have the same stoichiometry as the manganese and cobalt compounds. The compounds $K_2Ni_3S_4$, $Rb_2Ni_3S_4$ and $Cs_2Ni_3S_4$ could be isolated as plate-like crystals having a golden lustre, whilst $K_2Ni_3Se_4$ has a metallic green lustre [45]. In a humid atmosphere, the crystals are unstable and lose their lustre. Because of their plate-like habit these compounds may be assumed to crystallize in a larger structure. Structure investigations have not been successful up to now due to characteristic stacking faults of the crystals. It is certain that the nickel phases are not isotypic with the manganese and cobalt compounds. Moreover, because of the weak temperature-independent paramagnetic behavior, we may assume that each nickel atom has a co-planar co-ordination of four sulphur or selenium atoms [45].

The palladium compounds could be synthesized by similar fusion reactions [45, 46]. The compounds $K_2Pd_3S_4$, $Rb_2Pd_3S_4$ and $Cs_2Pd_3S_4$ form plate-like crystals having a violet metallic lustre; in transmitted light, however, they appear red. Like the nickel compounds, they exhibit a non-magnetic behavior. Structure investigations on single crystals of $Cs_2Pd_3S_4$ [47] reveal a monoclinic unit cell with $a = 6.275$ Å, $b = 14.02$ Å, $c = 6.166$ Å and $\beta = 120.6°$. In the space group with the highest possible symmetry

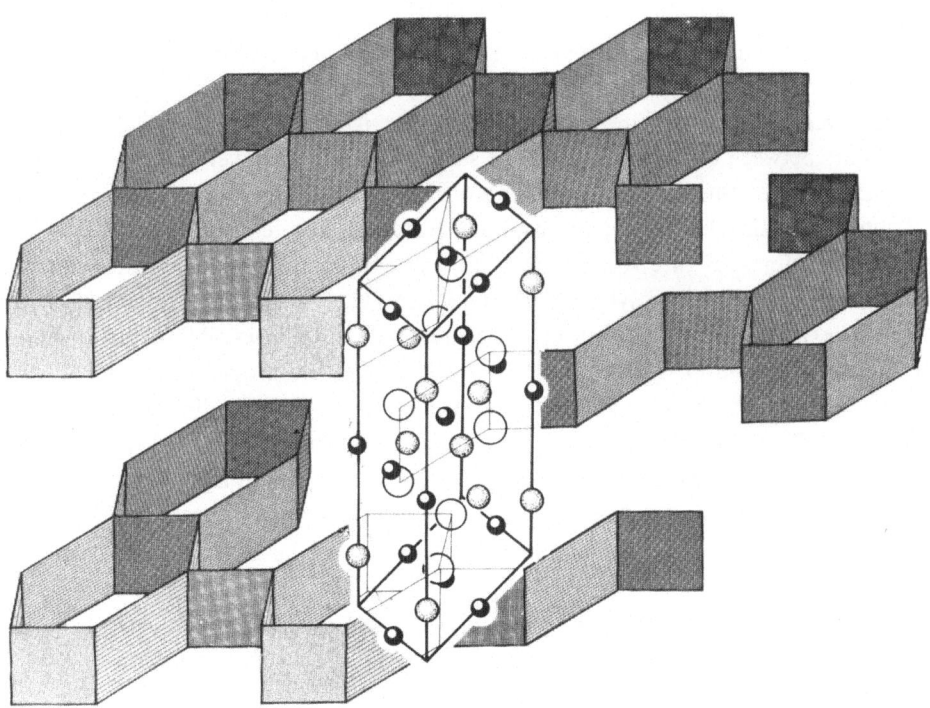

Fig. 15. The $Cs_2Pd_3S_4$ structure.

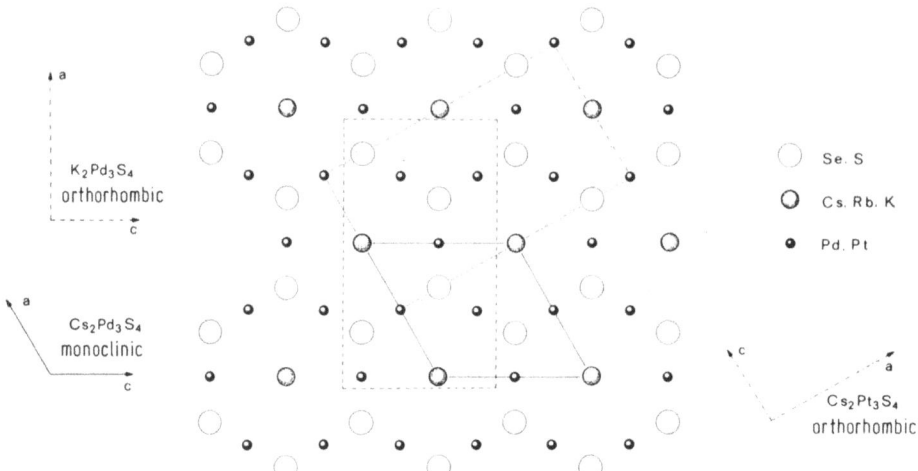

Fig. 16.   Layer of palladium atoms. Where alkali metal and chalcogenide atoms are shown, one
atom always lies above and one below the projection plane (see text).

$C2/m$ (No. 12), the following positions are occupied: 2 Pd_I in 2b, 2 Pd_{II} in 2c, 2 Pd_{III} in 2d; 4 Cs in 4g with $y=0.163$ and 8S in 8j with $x=0.164$, $y=0.393$ and $z=0.337$. The atomic arrangement of the $Cs_2Pd_3S_4$ type can be described as a layer structure (see Figure 15). The sequence of the layers $\perp$ [010] is as follows: a palladium layer, a sulphur layer, a double caesium layer, a sulphur layer and lastly a palladium layer. If neighbouring caesium and sulphur layers are considered as a single layer, a sequence of close-packed planes of the type ...$ABBAAB$... is obtained. The palladium atoms which are between two layers having the same symbol in the layer sequence, obtain a co-planar co-ordination of four sulphur atoms. Thus, the palladium atoms form layers of interconnecting hexagons; the caesium atoms are located above and below the center of each (see Figure 16). Caesium has the co-ordination number 8 with reference to sulphur. The sulphur polyhedron does not completely screen the caesium atom, an opening existing in the direction of the nearest palladium layer. Below this opening, another caesium atom is situated. The most characteristic property of this structure type is certainly the co-planar sulphur co-ordination of the palladium atoms, which, as in the PdS structure, forms a special atomic arrangement. However, whilst in the PdS and PtS structures, the connection of the co-planar sulphur arrangements is three dimensional, a previously unknown two dimensional connection along the sides of the rectangles formed by the sulphur atoms is found in the $Cs_2Pd_3S_4$ structure type (see Figure 15).

$K_2Pd_3S_4$ apparently crystallizes in two forms [48]. The one which can be characterized by a hexagonal unit cell ($a=6.11_7$ Å, $c=25.57_5$ Å) has an unknown atomic arrangement, as synthesis of single crystals has not yet been achieved. For the other form, structure investigations on single crystals reveal an orthorhombic unit cell with $a=10.65_2$ Å, $b=25.61_4$ Å, $c=6.09_5$ Å.

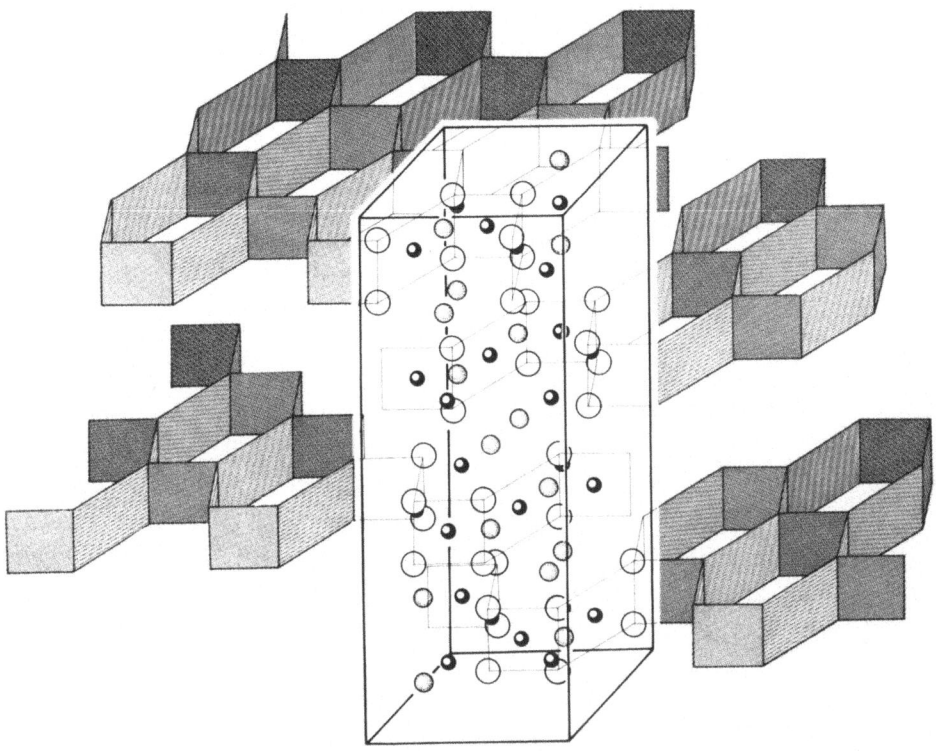

Fig. 17. The $K_2Pd_3S_4$ structure.

The space group with the highest possible symmetry is *Fddd* (No. 70). The unit cell contains eight formula units, and in it the following positions are occupied: Pd_I in 8a, Pd_{II} in 16f with $y=0.375$; K in 16f with $y=0.702$; S in 32h with $x=0.291$, $y=0.065$ and $z=0.125$. The $K_2Pd_3S_4$ structure type is a stacking variant of the above mentioned $Cs_2Pd_3S_4$ layer structure (see Figure 17). Whilst the latter has a pattern of two [S − Pd − S] slabs, the former has a pattern of four. If neighbouring caesium and sulphur layers are again considered as one layer, a sequence of close-packed planes of the type ...$DAABBCCDDA$... is obtained.

$Rb_2Pd_3S_4$ crystallizes in two forms; one has the $Cs_2Pd_3S_4$ structure, the other, the $K_2Pd_3S_4$ structure [48]. The lattice constants of the two modifications are: $Rb_2Pd_3S_4$ (monoclinic) $a=6.22_4$ Å, $b=13.32_0$ Å, $c=6.11_7$ Å and $\beta=120.58°$; $Rb_2Pd_3S_4$ (orthorhombic) $a=10.71_5$ Å, $b=26.62_3$ Å and $c=62.22_4$ Å. Structure investigations on single crystals reveal the following positions of the atoms in the monoclinic form: Pd_I in 2b, Pd_{II} in 2c, Pd_{III} in 2d; Rb in 4g with $y=0.159$ and S in 8j with $x=0.166$, $y=0.384$ and $z=0.336$.

The selenides $Cs_2Pd_3Se_4$, $Rb_2Pd_3Se_4$ and $K_2Pd_3Se_4$ could be obtained by the above mentioned fusion reactions at approximately 850°C [45]. The compounds could be isolated as plate-like crystals having a metallic green lustre. Structure investigations again reveal layered structure types: $Cs_2Pd_3Se_4$ is isotypic with $Cs_2Pd_3S_4$ ($a = 6.49_3$ Å, $b = 14.20_4$ Å, $c = 6.41_4$ Å and $\beta = 120.4_1°$). $Rb_2Pd_3Se_4$ and $K_2Pd_3Se_4$ crystallize in the orthorhombic $K_2Pd_3S_4$ structure (see Figure 17).

The lattice constants are:

	a (Å)	b (Å)	c (Å)
$Rb_2Pd_3Se_4$	11.07_6	27.09_2	6.35_1
$K_2Pd_3Se_4$	10.95_4	26.19_6	6.31_7

The compounds K_2PtS_2 and Rb_2PtS_2 have a chain structure type in which the co-planar sulphur co-ordinations of the platinum atoms are connected by their sides in one dimension [50]. Moreover, the compounds $Cs_2Pt_3S_4$ and $Rb_2Pt_3S_4$ which have the same stoichiometry as the palladium compounds mentioned above, are known to exist [51]. In this case X-ray structure investigations on single crystals reveal an orthorhombic layer structure which is very similar to that of $Cs_2Pd_3S_4$ (see Figure 18). The unit cells contain four formula units and have the following lattice constants: $a = 10.86$ Å, $b = 13.60$ Å, $c = 6.39$ Å for $Cs_2Pt_3S_4$ and $a = 10.84$ Å, $b = 13.30$ Å

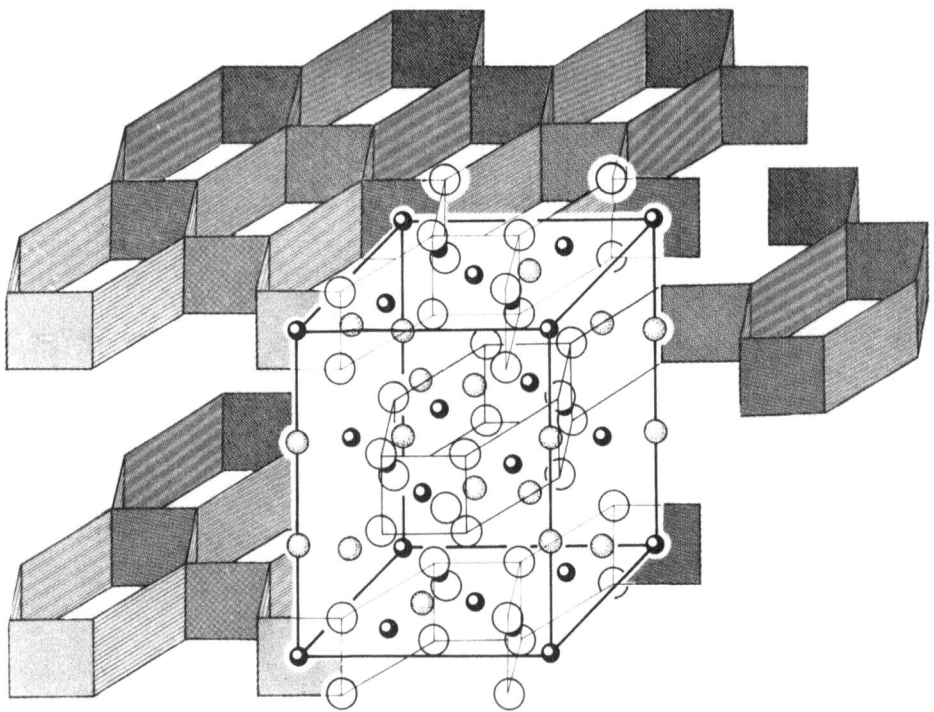

Fig. 18.   The $Cs_2Pt_3S_4$ structure.

TABLE XIII

Atom positions for $Cs_2Pt_3S_4$ and $Rb_2Pt_3S_4$

| | $Cs_2Pt_3S_4$ | | | $Rb_2Pt_3S_4$ | | |
	x	y	z	x	y	z
Cs, Rb in 8h	0	0.342	0	0	0.343	0
Pt in 4b	0	0	0	0	0	0
8d	0.25	0	0.25	0.25	0	0.25
S in 160[a]	0.165	0.116	0	0.17	0.12	0

[a] An accurate determination of the sulphur positions in the rubidium compound was not possible because of the poor quality of the crystals obtained.

and $c = 6.39$ Å for $Rb_2Pt_3S_4$. In the space group $Fmmm$ (No. 69), the positions are occupied as shown in Table XIII.

Besides the ternary platinum chalcogenides already mentioned, which contain platinum in the oxidation state $+2$, phases in which the platinum atoms have the oxidation state $+4$ are also known. The most accurately investigated compound up to now is $K_2Pt_4S_6$ [52]. Again a layered structure was found. Investigations on single crystals indicate that the space group is $R\bar{3}m$ (No. 166). The hexagonal unit cell has the constants $a = 7.01$ Å and $c = 19.14$ Å. The following positions are occupied 3 Pt(IV) in 3b, 9 Pt(II) in 9d, 18 S in 18h with $x = 0.167$ and $z = 0.104$ and 6 K in 6c with $z = 0.300$. In the structure found, $[Pt_4S_6]$ slabs are arranged in such a manner that in each slab the platinum atoms form a layer of interconnecting triangles (see Figure 19). The sulphur atoms, too, form layers of interconnecting triangles, but in these, one quarter of the positions are unoccupied. Two differently coordinated platinum atoms can be recognized. One quarter of the atoms (Pt(IV)) have an octahedral sulphur co-ordination, the rest (Pt(II)) have a rectangular one. The potassium atoms form double layers between the $[Pt_4S_6]$ slabs. $K_2Pt_4S_6$ probably crystallize in a second modification which also has hexagonal symmetry and the same a-axis, but a c-value which is one third smaller ($c = 12.77$ Å).

According to X-ray investigations on powder samples, $A_2Pt_4S_6$ phases with $A \cong Rb$

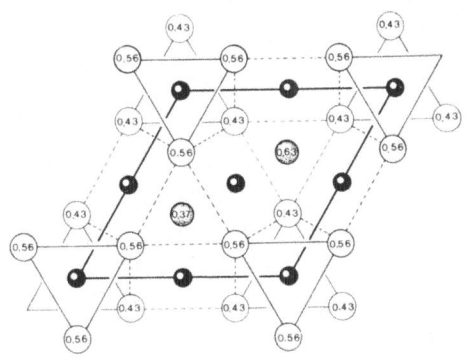

Fig. 19. $[Pt_4S_6]$-slab of the $K_2Pt_4S_6$ structure (All Pt-atoms have the z-Parameter 0.5).

TABLE XIV

Lattice constants of
$A_2Pt_4X_6$ compounds

Compound	Lattice constants	
	a (Å)	c (Å)
$K_2Pt_4S_6$	7.01	19.14
$Rb_2Pt_4S_6$	7.02	20.08
$Cs_2Pt_4S_6$	7.03	21.13
$K_2Pt_4Se_6$	7.36	19.35
$Rb_2Pt_4Se_6$	7.35	20.20

TABLE XV

Lattice constants of
$A_2M_3^{II}M^{IV}X_6$ compounds

Compound	Lattice constants	
	a (Å)	c (Å)
$K_2Pb_3IrS_6$	7.02	19.19
$K_2Pd_3PtS_6$	6.97	20.17
$K_2Pt_3SnS_6$	7.08	12.83
$Rb_2Pt_3SnS_6$	7.14	13.59
$K_2Pd_3SnS_6$	7.09	13.02

TABLE XVI

Lattice constants of the thallium
compounds $Tl_2M_3^{II}M^{IV}S_6$

Compound	Lattice constants	
	a (Å)	c (Å)
$Tl_2Pt_4S_6$	12.26	17.58
$Tl_2Pt_3SnS_6$	12.35	17.96
$Tl_2Pt_3TaS_6$	11.84	18.91
$Tl_2Pt_3ZrS_6$	12.24	17.78
$Tl_2Pd_3PtS_6$	12.07	18.26
$Tl_2Pd_3SnS_6$	12.33	18.18
$Tl_2Ni_3PtS_6$	12.36	17.84

and Cs and some selenides, seem to have the same structure as $K_2Pt_4S_6$ [52]. The lattice constants of the hexagonal unit cells are listed in Table XIV.

Moreover, the platinum atoms which have the oxidation state $+2$, as well as those having the oxidation state $+4$, could be replaced by other elements. The quaternary compounds known up to now are listed in Table XV [52].

Some compounds with thallium have recently been synthesized. Their lattice constants are shown in Table XVI above [53].

It was surprising to find that the hexagonal a-axes of the thallium compounds had larger values than those of the corresponding alkali metal compounds. The following approximate relationship was found: $\sqrt{3} \cdot a_A \approx a_{Tl}$, where $A \triangleq$ alkali metal compound and $Tl \triangleq$ thallium compound.

This result might possibly be explained by a lower symmetry of the atomic arrangement in the thallium layers which perhaps corresponds to that in the Tl_2S structure. For the latter, a hexagonal layer structure type was found with $a = 12.20$ Å. If the atomic arrangement of the thallium atoms had had a higher symmetry (anti CdI_2-type), the a-value would have diminished to the value $a/\sqrt{3}$.

2.9. THE COPPER GROUP

Of the ternary sulphides with alkali metals and copper – most of which have been known for a long time [54, 55] – only the structures of the compounds KCu_4S_3 and $RbCu_4S_3$ have been solved [56].

Shiny, dark blue crystals of KCu_4S_3 and $RbCu_4S_3$ crystallize as square platelets after melting a mixture of copper, alkali carbonates and sulphur. X-ray investigations on single crystals reveal tetragonal unit cells with the following lattice constants: KCu_4S_3: $a = 3.908(8)$ Å, $c = 9.28(1)$ Å; $RbCu_4S_3$: $a = 3.928(8)$ Å, $c = 9.43(1)$ Å. In the space group with the highest possible symmetry $P4/mmm$ (No. 123), the following positions are occupied: 1 K or 1 Rb in 1b; 4 Cu in 4i with $z = 0.159$; 1S in 1a and 2S in 2b with $z = 0.298$ for KCu_4S_3 and $z = 0.292$ for $RbCu_4S_3$. The structure is illustrated in Figure 20. It is characterized by $[S-Cu-S-Cu-S]$ slabs $\perp [001]$. These slabs are separated by alkali metal layers. The sulphur atoms form almost regular tetrahedra. Each sulphur atom in the outer layers belongs to four tetrahedra, whereas the sulphur atoms in the inner layers belong to eight. The copper atoms are displaced a small amount from the centres of the tetrahedra in the direction of the alkali metal layers; this displacement is a little more for the rubidium compound than for the potassium one. The alkali metal atoms are surrounded by eight sulphur atoms located at the corners of a cube. As all the copper atoms occupy the same crystallographic position, it is impossible to distinguish between copper atoms in the oxidation states $+1$ and $+2$ suggested by the formula A $Cu_3^{+1}Cu^{+2}S_3$. Specific conductivity measurements (40 $\Omega^{-1} \cdot cm^{-1}$ for KCu_4S_3 and 60 $\Omega^{-1} \cdot cm^{-1}$ for $RbCu_4S_3$) show a metallic behaviour which is contrary to the formulation of different oxidation states of the copper atoms [56].

The synthesis of a hitherto unknown copper compound of composition KCu_3S_2 has recently been successfully carried out by means of a fusion reaction [57]. Structure investigations on single crystals reveal a monoclinic symmetry for KCu_3S_2. The space group with the highest possible symmetry is $C2/m$ (No. 12). In the unit cell ($z = 4$), all the atoms occupy the position 4i, the corresponding position parameter values being listed in Table XVII and the lattice constants in Table XVIII. An illustration of this complicated structure type is given in Figure 21. The atomic arrangement can be described as a layer structure. The layers are made up of copper-sulphur arrangements which are $\parallel (001)$, and which are separated by layers of potassium atoms.

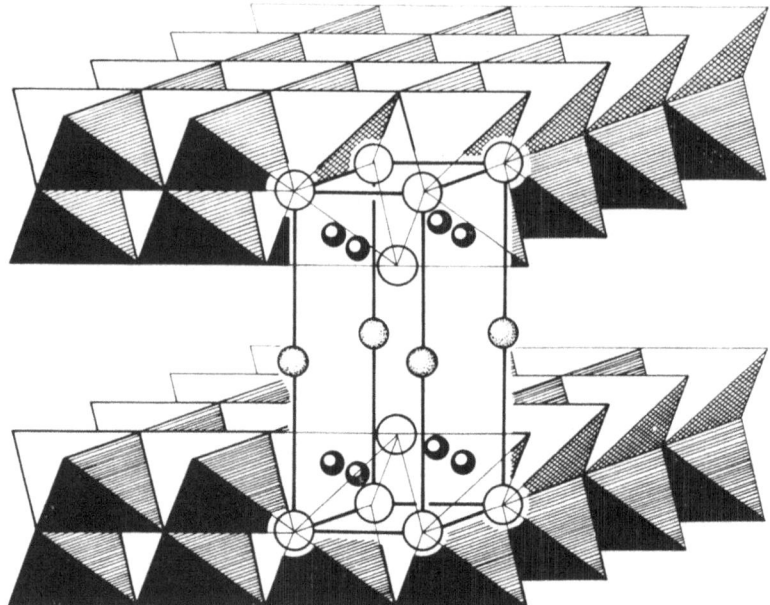

Fig. 20.   The KCu_4S_3 structure.

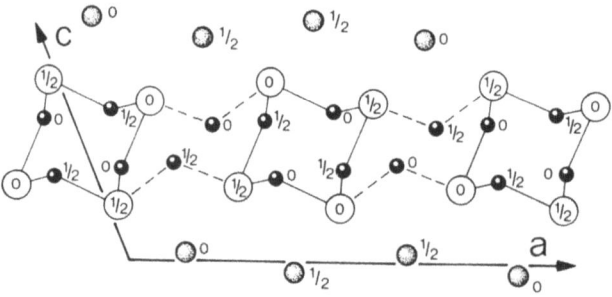

Fig. 21.   Projection $\perp$ [010] of the monoclinic atomic arrangement of the KCu_3S_2 structure.

Of the three crystallographically different copper atoms, two (Cu_I and Cu_{III}) possess a similar sulphur co-ordination: Three nearest sulphur atoms form an almost equilateral triangle; the copper atom lies only slightly above the plane of this triangle, thus forming a very flat, only slightly distorted trigonal pyramid, with the copper atom at the apex. The sulphur triangles are connected together to form channels or tubes in the direction of the b-axis. The Cu_{II} atoms connect these tubes thus forming layers (see Figure 21). Each Cu_{II} atom has two closely situated neighbouring sulphur atoms, and two further away. These four sulphur atoms form a slightly distorted tetrahedron.

$RbAg_3S_2$ and $CsAg_3S_2$ are isotypic with KCu_3S_2 (see Table XVII and XVIII) [57]. In the system K/Ag/S, the compound $K_2Ag_4S_3$ was found [57]. The same composition exists with rubidium. Structure investigations show the same layer structure for both compounds. The space group is $C2/m$ (No. 12). In the unit cell ($z=4$), all the atoms occupy the position 4i. The position parameter values for $K_2Ag_4S_3$ are listed in Table XIX. The isotypical layer structure of $Rb_2Ag_4S_3$ was only determined from X-ray diffraction experiments on powdered samples. The lattice constants are listed in Table XVIII.

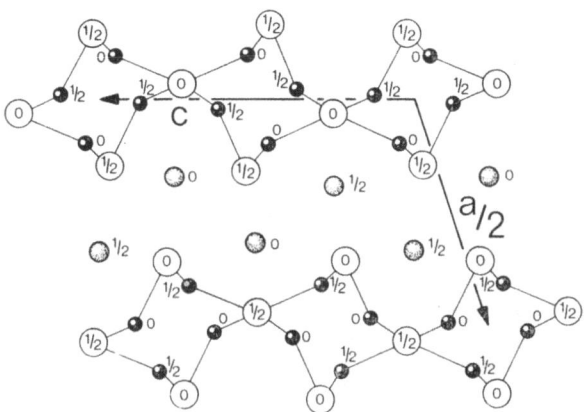

Fig. 22.   Projection $\perp$ [010] of the monoclinic atomic arrangement of the $K_2Ag_4S_3$ structure.

The structure of the $A_2Ag_4S_3$ compounds is similar to that of the AAg_3S_2 phases. The silver atoms also have three nearest sulphur neighbours, the sulphur triangles again being connected together to form tubes. In this case, however, they are connected by common sulphur atoms. The alkali metal atoms separate these layers, and have six nearest sulphur neighbours which form a distorted octahedron (see Figure 22). Similar silver sulphur arrangements to those in the above mentioned compounds were found in Ag_2S, while very close relationships were found with $BaCu_4S_3$ [58].

Concluding Remarks

Those ternary chalcogenides with alkali and transition metals which are now known reveal characteristic anisotropic arrangements. Apart from a few chain structure types, the structures observed are mainly layered. In the latter, $[X-M-X]$ slabs, or in some copper compounds $[X-M-X-M-X]$ slabs, are separated by alkali metal layers.

Two different methods of synthesizing these compounds are mainly used at present: Firstly, the intercalation of alkali metal atoms into the dichalcogenides of the transition metals, mostly by reaction in liquid ammonia; secondly, the reactions of alkali metal compounds – mostly carbonates – with sulphur and transition metals by fusion reactions. A systematic examination of the ternary chalcogenides obtained by these methods reveals that for certain transition metals, for example, Ru, Rh, Os and Ir, no ternary sulphides with alkali metals could be obtained. It is thus apparent that other synthesizing methods need to be developed.

TABLE XVII

Position parameters for atoms in the AM_3S_2
phases. All the atoms occupy the position 4i in
the space group $C2/m$. For each atom, both
the x and the z values are given.

	KCu_3S_2	$RbAg_3S_2$	$CsAg_3S_2$
A	0.1338	0.1337	0.1356
	0.0462	0.0437	0.0428
M_I	0.5957	0.5984	0.5954
	0.6426	0.6545	0.6473
M_{II}	0.6886	0.6880	0.6886
	0.4275	0.4215	0.4248
M_{III}	0.0610	0.0565	0.0564
	0.4034	0.3967	0.3972
S_I	0.1950	0.1977	0.1924
	0.6882	0.6976	0.6833
S_{II}	0.5204	0.5174	0.5215
	0.2398	0.2240	0.2342

TABLE XVIII

Lattice constants for ternary sulphides with copper and
silver. All the compounds crystallize in the
space group $C2/m$.

Compounds	a (Å)	b (Å)	c (Å)	β (Å)
KCu_3S_2	14.77_3	3.92_5	8.18_2	113.5
$RbAg_3S_2$	16.12_8	4.30_6	8.77_3	114.6
$CsAg_3S_2$	16.17	4.32_7	8.93	113.0
$K_2Ag_4S_3$	17.36_4	4.29_6	11.60_3	108.3
$Rb_2Ag_4S_3$	17.88_4	4.33_1	11.84_9	108.6

It is interesting to note that although the structure of these compounds have been
well investigated, the physical properties are not well known. A deeper understanding
of the bond structure of these compounds can only be obtained when more accurate
measurements of the magnetic, electric and optical properties are available.

Finally, the question may be asked whether the layer structures discussed here can
be obtained with other metals and/or nonmetals. By replacing the alkali metals by the
alkaline-earth metals, only compounds with barium could be synthesized [59, 60].
As published results have already shown, the tendency here is towards three dimen-
sionally connected atomic arrangements, perhaps initiated by the higher charge of
the ionic partner. Thus, in the $BaMnS_2$ structure [61] for example, the $[S-Mn-S]$
slabs are no longer in one plane, but form wavy slabs which create an atomic arrangement similar to that of the crystobalit type. Initial results on the replacement of sulphur
or selenium by phosphoruos or arsenic, show that again layer structures exist. By syn-

TABLE XIX

Position parameters for atoms in the compound $K_2Ag_4S_3$. All the atoms occupy the position 4i in the space group $C2/m$. For each atom, both the x and the z values are given.

Atom	Position parameter	Atom	Position parameter
K_I	0.3318 0.1597	Ag_{IV}	0.5947 0.5130
K_{II}	0.1869 0.3423	S_I	0.1450 0.0419
Ag_I	0.0105 0.8781	S_{II}	0.1599 0.6125
Ag_{II}	0.0203 0.6333	S_{III}	0.4656 0.7203
Ag_{III}	0.5944 0.0915		

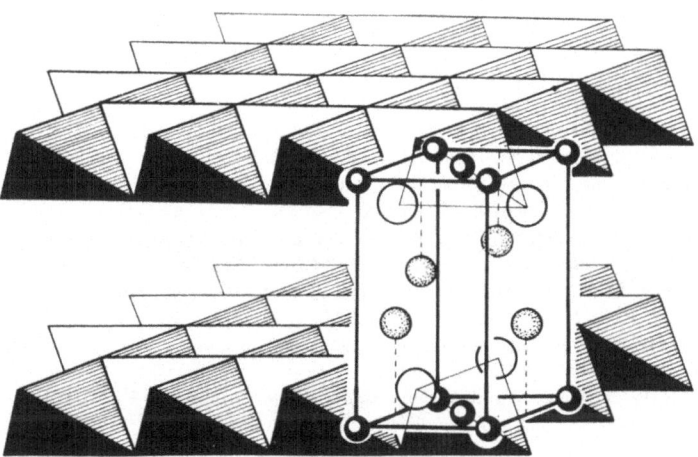

Fig. 23.   The KMnP structure.

thesis of the elements, it was possible to obtain the compounds KMnP and KMnAs [62]. Both crystallize in a layer structure similar to $Cs_2Mn_3S_4$ (see Figures 23 and 10). The phosphorous or arsenic atoms form layers of tetrahedra connected by their edges in two dimensions. The manganese atoms occupy the centers of the tetrahedra. The [P−Mn−P] or [As−Mn−As] slabs were separated by the alkali metal atoms. Thus one characteristic feature of the ternary sulphide structure types is imitated in a ternary phosphide and a ternary arsenide. Future experiments will show if it is possible to synthesize ternary phosphides or arsenides with the other layered structures known from the chalcogenides.

References

1. F. Hulliger, *Struct. Bonding* **4** (1968), 83.
2. R. Ballestracci and E. F. Bertaut, *Bull. Soc. Fr. Mineral.-Cristallogr.* **87** (1964), 512.
3. R. Ballestracci, *Bull. Soc. Fr. Mineral.-Cristallogr.* **88** (1965), 207.
4. M. Tromme, *Compt. Rend. Acad. Sci.* **273** (1971), 849.
5. W. Bronger, R. Elter, E. Maus, and T. Schmitt, *Rev. Chim. Miner.* **10** (1973), 147.
6. W. Bronger and T. Schmitt: unpublished.
7. W. Rüdorff, *Chimia* **19** (1965), 489.
8. M. Danot, A. Le Blanc, and J. Rouxel, *Bull. Soc. Chim. Fr.* **8** (1969), p. 2670.
9. J. Rouxel, M. Danot, and J. Bichon, *Bull. Soc. Chim. Fr.* **11** (1971), p. 3930.
10. J. Bichon, M. Danot, and J. Rouxel, *Compt. Rend. Acad. Sci.* **276** (1973), 1283.
11. A. Leblanc-Soreau, M. Danot, L. Trichet, and J. Rouxel, *Mat. Res. Bull.* **9** (1974), 191.
12. B. van Laar and D. J. W. Ijdo, *J. Solid State Chem.* **3** (1971), 590.
13. J. G. White and H. L. Pinch, *Inorg. Chem.* **9** (1970), 2581.
14. J. Rouxel, J. Cousseau, and L. Trichet, *Compt. Rend. Acad. Sci.* **273** (1971), 243.
15. J. M. van den Berg and P. Cossee, *Inorg. Chim. Acta* **2** (1968), 143.
16. B. van Laar, H. M. Rietveld, and D. J. W. Ijdo, *3rd Intern. Conf. Solid Comp. Trans. Elem.*, Oslo, 1969, p. 79.
17. K. Ansenhofer, J. M. van den Berg, P. Cossee, and J. N. Helle, *J. Phys. Chem. Solids* **31** (1970), 1057.
18. B. van Laar and H. M. Rietveld, *J. Solid State Chem.* **3** (1971), 154.
19. W. P. F. A. M. Omloo and F. Jellinek, *J. Less Common Metals* **20** (1970), 121.
20. F. Jellinek, *Arkiv Kemi* **20** (1963), 447.
21. F. Kadijk, R. Huisman, and F. Jellinek, *Rec. Trav. Chim.* **83** (1964), 768.
22. C. Haas, Paper presented at Conf. on Crystal Structure and Chemical Bonding in Inorg. Chem., Wageningen, The Netherlands, 1974.
23. V. Schmidt, Dissertation Tübingen, 1971.
24. M. Sergent and J. Prigent, *Compt. Rend. Acad. Sci.* **261** (1965), 5135.
25. M. Gröger, S.-B. *Akad. Wiss. Wien, Abt. II* **83** (1880), 749.
26. R. Schneider, *J. Prakt. Chem.* (2) **56** (1897), 401.
27. J. Milbauer, *Z. Anorg. Chem.* **42** (1904) 438,
28. J. W. Boon and C. H. MacGillavry, *Rec. Trav. Chim. Pays-Bas* **61** (1942), 910.
29. W. Rüdorff and K. Stegemann, *Z. Anorg. Allg. Chem.* **251** (1943), 376.
30. P. F. Bongers, C. F. van Bruggen, J. Koopstra, W. P. F. A. M. Omloo, G. A. Wiegers, and F. Jellinek, *J. Phys. Chem. Solids* **29** (1968), 977.
31. F. M. R. Engelsman, B. van Laar, G. A. Wiegers, and F. Jellinek, *3rd Intern. Conf. Solid Comp. Trans. Elem.*, Oslo, 1969, p. 72.
32. W. Rüdorff, W. R. Ruston, and A. Scherhaufer, *Acta. Crystallogr.* **1** (1948), 196.
33. W. Rüdorff and H. H. Sick, *Angew. Chem.* **71** (1959), 127.
34. A. Völker, *Liebigs Ann. Chem.* **59** (1846), 35.
35. W. Bronger and P. Böttcher, *Z. Anorg. Allg. Chem.* **390** (1972), 1.
36. K. Preis, *J. Prakt. Chem.* **107** (1869), 12.
37. R. Schneider, *J. Prakt. Chem.* **108** (1869), 16.
38. W. Bronger, *Z. Anorg. Allg. Chem.* **359** (1968), 225.
39. J. Meyer and H. Bratke, *Z. Anorg. Allg. Chem.* **135** (1924), 297.
40. W. Bronger, *Naturwissenschaften* **53** (1966), 525.
41. R. Wandji and J. Kamsu Kom, *Compt. Rend. Acad. Sci.* **275** (1972), 813.
42. A. Kutojlu, *Naturwissenschaften* **61** (1974), 125.
43. R. Schneider, *Poggendorffs Ann.* **151** (1874), 437.
44. J. Bellucci and L. Bellucci, *Gazz. Chim. Ital.* **38** (1908), 635.
45. W. Bronger, J. Eyck, W. Rüdorff, and A. Stössel, *Z. Anorg. Allg. Chem.* **375** (1970), 1.
46. R. Schneider, *Poggendorffs Ann.* **141** (1870), 532.
47. W. Bronger and J. Huster, *J. Less Common Metals* **23** (1971), 67.
48. J. Huster and W. Bronger, *J. Solid State Chem.* **11** (1974), 254.
49. J. Huster and W. Bronger, *Z. Naturforsch.* **29b** (1974), 594.
50. W. Bronger and O. Günther, *J. Less Common Metals* **27** (1972), 73.

51. O. Günther and W. Bronger, *J. Less Common Metals* **31** (1973), 255.
52. W. Rüdorff, A. Stössel, and V. Schmidt, *Z. Anorg. Allg. Chem.* **357** (1968), 264.
53. V. Schmidt and W. Rüdorff, *Z. Anorg. Allg. Chem.* **397** (1973), 51.
54. R. Schneider, *J. Prakt. Chem.* **108** (1865), 16.
55. J. Milbauer, *Z. Anorg. Allg. Chem.* **42** (1904), 440.
56. W. Rüdorff, H. G. Schwarz, and M. Walter, *Z. Anorg. Allg. Chem.* **269** (1952), 141.
57. C. Burschka, Dissertation Aachen, 1975.
58. J. E. Iglesias, K. E. Pachali, and H. Steinfink, *Mat. Res. Bull.* **7** (1972), 1247.
59. I. E. Grey and H. Steinfink, *J. Amer. Chem. Soc.* **92** (1970), 5093.
60. R. A. Gardener, M. Vlasse, and A. Wold, *Acta Crystallogr.* **B25** (1969), 781.
61. D. Schmitz and W. Bronger, *Z. Anorg. Allg. Chem.* **402** (1973), 225.
62. L. Linowsky and W. Bronger, *Z. Anorg. Allg. Chem.* **409** (1974), 221.

STRUCTURAL ASPECTS OF NON-STOICHIOMETRY
IN MATERIALS WITH LAYERED STRUCTURES

R. J. D. TILLEY

School of Materials Science, University of Bradford, Bradford BD7 1DP, Yorkshire, England

1. Introduction

Materials with layered structures constitute a large class of solids which possess interesting physical and chemical properties and which have been intensively investigated in recent years. The materials span a range from pure elements, through relatively simple binary and ternary inorganic compounds to complex layer silicates, clay minerals and organic crystals. This vast number of compounds presents a practical difficulty in writing an account of non-stoichiometry in layer structured materials. If all compounds are included, the account can degenerate into a catalogue of literature references and the more important relationships which may exist between structure and non-stoichiometry may well be obscured or lost altogether, while a consideration of only a few materials will be insufficient to account for the observed phenomena in all the variety of chemical and structural types encountered.

In this chapter, the latter approach has been chosen as the lesser of two evils. Hence, in the following pages, an attempt has been made to classify and describe a fairly small number of materials, which, although all are layer structures, differ appreciably from each other in the way in which they accommodate stoichiometric variability. The materials chosen have been included for a variety of reasons. Some were chosen principally because they were and are of some considerable interest in applied research. This has meant that there is a fair amount of active research on them at present, and that the literature is fairly extensive and the phases relatively well characterised. Others have been included because the way in which they accommodate non-stoichiometry is well understood. Indeed, it is remarkable how many materials are supposed to have composition ranges, but how few have had this variation characterised precisely in terms of crystal chemistry. In general, therefore, the materials considered are those in which a large body of structural information is available, and this leads, in the main, to the oxides and chalcogenides of the transition metals.

Within this general framework, three groups of materials have been considered, viz. (i) those in which the layer structure is quite closely associated with a stoichiometric formula, and in which the layer structure is effectively destroyed by changes in stoichiometry, (ii) layer structures which are tolerant to stoichiometric variability, and (iii) those layer structures which can transform more or less continuously into a non-layer structure as the composition varies. There are, of course, no hard and fast divisions between these groups, and it may be that the grouping chosen is not particularly well suited for all purposes, or that some compounds discussed could well be considered in either of two groups if they are somewhat borderline in character. This is

F. Lévy (ed.), Crystallography and Crystal Chemistry of Materials with Layered Structures. 127–184. *All Rights Reserved.*
Copyright © 1976 by D. Reidel Publishing Company, Dordrecht-Holland.

particularly so of the first two groups, as one has to set some sort of arbitrary limit upon the degree of composition variation allowed for a material before it becomes tolerant or intolerant to variation.

The arrangement has an advantage that rather unrelated chemical species can be placed in juxtaposition, to reveal their similarities while the fact that they have different chemical formulae or constituents becomes irrelevant. It is also hoped that this presentation will avoid merely duplicating the information contained in other chapters which closely parallel this one, and with which a certain amount of overlap must necessarily occur. Finally, such a grouping has fitted in well with three chemically distinguishable types of materials which allows each of the following sections to preserve some measure of coherence, and which has simplified the task of reviewing the existing literature. The examples chosen for each group can then be listed as (i) layer structures which contain readily distinguishable fragments of the ReO_3 structure, (ii) the β-alumina family of phases, which are of some interest for technological reasons, and (iii) some of the transition metal sulphides. Within each group, there are a large number of omissions, and particularly in group (iii) only a small number of the many non-stoichiometric layer chalcogenides have been selected as illustrative.

This account of the non-stoichiometry of layer structures is limited to describing the structural aspects of the topic and as such is certainly a restricted and unbalanced viewpoint. The nature of the chemical bonding in these phases is of interest, and to many research workers, the associated physical properties, such as magnetic and electrical behaviour are of paramount importance. Similarly, the thermodynamics of non-stoichiometric compounds is an aspect of the subject which merits attention. The chapter is, therefore, to be considered as only one facet of the topic of non-stoichiometry as a whole, although other chapters in this and the other volumes partly redress this balance. It is, however, true to say that from a crystal chemical viewpoint, neither thermodynamic measurements or measurements of physical properties have, to date, revealed the complexity of the structural chemistry of non-stoichiometric materials, and in some cases have obscured it completely. Some of the reasons for this are summarised in the following section, which consists of a brief review of the more important aspects of non-stoichiometry in general, both for materials with and without a layered structure. The remaining sections consider the structural aspects of the groups of materials mentioned above and the final section summarises the more important conclusions to be drawn from these considerations.

2. Non-Stoichiometry

2.1. NON-STOICHIOMETRY AND STRUCTURE

The stoichiometry of a compound is the ratio of the number of anions to cations, and following traditional chemical thought, this ratio should be a fairly small integer. The concept of non-stoichiometry arose when it was appreciated that many compounds, particularly the solid compounds of the transition metals, did not altogether obey this principle. Indeed, many of them appeared to tolerate quite wide departures of composition on either side of the integral ratio while others never achieved any composition which was integral. Thus, for example, titanium monoxide spans a composition range from approximately $TiO_{0.73}$ to $TiO_{1.23}$ and titanium disulphide, TiS_2, appears to have a real composition of between $TiS_{1.81}$ and $TiS_{1.92}$, never achieving the ideal composition of TiS_2. It is these compounds, and others with a similar behaviour, that have been termed non-stoichiometric compounds, while those in which the anion to cation ratio is a small integer, sodium chloride, for example, are referred to as stoichiometric compounds.

Clearly, the basis upon which a compound is declared to be non-stoichiometric or not is an operational one and rests upon an ability to declare that only one compound is present during the observed composition variation. Thus, characterisation of non-stoichiometry has always been closely linked with experimental procedures which allow the number of phases present in a particular sample to be determined. The most commonly used experimental technique for this is undoubtedly X-ray diffraction, particularly from powder samples. Hence, if we consider a binary compound between a metal A and a non-metal Z of formula A_aZ_z, it is regarded as non-stoichiometric in nature if the X-ray diffraction pattern remains unchanged, apart from small lattice parameter variations, while the anion to cation ratio, z/a, changes measurably. In the case of ternary and more complex phases, the same operational definition applies, so that a compound $A_aB_bZ_z$ is considered to be a non-stoichiometric material if the anion to cation ratio, $z/(a+b)$ varies while the structure, from the point of view of X-rays, remains unchanged. The compounds formed by isomorphous replacement of one cation by another in solid solution, $Cr_xAl_{2-x}O_3$ for example, are not considered to be non-stoichiometric as the anion to cation ratio remains constant.

This definition must be modified slightly to encompass a large number of phases which are often called non-stoichiometric phases in the literature, but which have no experimentally discernable composition range. These are the members of the growing number of homologous series of oxides which have rather complex formulae rather than simple ones; $W_{20}O_{58}$ is an example. In the main, many of these phases appear to be strictly stoichiometric, although the ratio of anions to cations is not a small integer. Such materials will be termed line phases throughout this article, implying that, on a phase diagram they show no extended composition range. This definition of non-stoichiometry is entirely structural and has been emphasised as it accords well with the theme of the present article. However for some purposes, particularly theoretical considerations of non-stoichiometric solids from the standpoint of thermodynamics, a definition in terms of the phase rule is to be preferred. Detailed considerations of such definitions and their implications will not be included here, but for a full account reference can be made to articles by Anderson [1, 2] and Fender [3].

Besides X-ray diffraction, a variety of other chemical and physical properties are also used to define the composition range over which a non-stoichiometric compound is stable, particularly if thermodynamic quantities are of importance in the study. Thermogravimetric techniques coupled with the simultaneous observation of the partial vapour pressure of one of the components of the compound are frequently employed, as free energy measurements can be derived from the data. Electrical properties such as conductivity and Seebeck coefficient are also often chosen and solid state galvanic cells are becoming increasingly used for free energy measurements.

It is therefore clear that the range of composition over which a non-stoichiometric phase exists, its homogeneity range, may well appear to differ according to the experimental technique employed. Moreover, not only the type of technique employed, but also its sensitivity must be carefully evaluated, and the number of data points recorded must be sufficient to eliminate the possibility of mistakenly confusing a multiphase region with a continuum. Careful studies by Hyde and coworkers on the titanium-oxygen system reveal this necessity clearly [4, 5].

Once such a non-stoichiometric phase has been found, it is of some interest to ascertain its structure and attempt to determine how the homogeneity range is accounted for. In the past such structural considerations have been dominated by the concept of the point defect in its various aspects. In the main this is due to the elegance and approachability of the theoretical basis of point defect theory, originally formulated by the application of statistical thermodynamics to crystalline solids by Schottky and Wagner in the 1930s [6] and to the ease with which point defects in a crystal structure can be visualised. The original ideas put forward by Schottky and Wagner were extended to other non-stoichiometeric phases by Anderson [7] and more recently have been applied extensively by many authors. Much of this literature has been correlated and expanded by Kröger [8].

A further reason for the almost universal application of point defect arguments was to be found in the rather low resolution or discrimination of the experimental techniques employed. Diffraction techniques and the other experimental methods commonly employed to study non-stoichiometry in crystals are more or less constrained by the fact that they present data which is an average result from a large number of unit cells of the material under examination. This has often obscured the true complexity of the defect structures present, so that until recently more elaborate theories of non-stoichiometry have seemed unnecessary.

Certainly the introduction of high resolution focussing powder cameras has added a great deal to the knowledge of the real extent of homogeneity ranges of non-stoichiometric compounds, and with these cameras a large number of compounds once thought to have broad composition ranges have been shown to consist of numbers of very closely related line phases, each with a precise composition. Despite this success, the shortcomings of the method, and more particularly of the low-resolution Debye-Scherrer technique have frequently been ignored. They were stated explicitly by Wadsley in 1963 [9] and more recently have been reiterated by Andersson and Wadsley [10]. They will not be included here except to emphasise one aspect, the fact that X-rays in particular do not see all atoms equally. This becomes important when compounds of heavy cations are being studied, for in this case the X-ray diagram mainly represents the spatial arrangement of the metal ions. Non-stoichiometry which principally affects the anion sublattice may go completely undetected provided that the cation sublattice retains its original integrity, even if a considerable degree of ordering exists among the defects present in the anion array. Neutron diffraction to some extent corrects this bias. However, as fairly substantial volumes of crystal are required to obtain diffractograms, these still present average data from a large number of unit cells.

Two diffraction techniques have emerged in the last few years which are able to give information about local atomic disorder. The first of these is an extension of neutron diffraction, although in principle the technique is also applicable to X-ray diffraction, and relates to the diffuse scattering present when the crystal structure is disordered. The theory of the method has been given by Willis [11, 12] and a review of some recent results by Fender [3] and will not be reported here. The results of the theoretical analysis may be summarised thus; the scattering from an imperfect crystal consists of two components, elastic Bragg scattering, and diffuse scattering. The Bragg scattering is that most commonly measured, and the scattering referred to by implication above. This yields data about the contents of the average unit cell of the crystal structure. The diffuse scattering, on the other hand, gives information about the local defect structure of the material. Hence, when such defects are responsible for stoichiometric variability, it gives direct information about the structural arrangements responsible. Analysis of diffuse scattering of neutron beams has shown that point defects are not responsible for non-stoichiometry in a number of phases with anion deficient and anion excess fluorite structures and in wüstite. Instead, volumes of the structure, usually 2 or 3 unit cells in extent contain a structural rearrangement of atomic species different from the matrix. These regions are frequently referred to as defect clusters. In the compounds studied they are arranged in a random fashion throughout the crystal. These materials do not have a layered structure and will not be described further. At the time of writing, no layered structure materials appear to have been studied by this technique, although it would appear an extremely promising one to use.

The second technique to yield information about local order and disorder in crystals is that of high resolution electron microscopy, in which the crystal structure is, within the capabilities of the instrument, directly resolved. The theory and practice of this technique have been discussed in detail by Allpress and Sanders [13]. A summary of the theory, and a discussion of recent results and applications of the technique are given by Anderson and Tilley [14]. The central feature of this technique is that a large number of diffracted beams as well as the transmitted beam are allowed to contribute to the formation of the image. Under these circumstances, provided that the crystal examined is very thin, of the order of 10 nm, the contrast of the micrograph is a direct representation of the charge density of the structure in the particular orientation examined. There are a number of limitations to the technique, and some care must be taken in ensuring that a number of variable factors are correct before this interpretation of the micrograph is permissible. However, in cases where such a simple interpretation is valid, the results are very impressive, as not only is detail of 2–3×10^{-10} m resolved, but also non-periodic disorders of the crystal structure are recorded. This technique is most successful in yielding information on crystals containing coherent intergrowths of several structural types at a level unresolved by conventional X-ray crystallography. In this way, a number of supposed non-stoichiometric oxides have been found to be comprised of intergrowths of stoichiometric line phases of similar structure but in varying proportions. The technique is also sensitive enough to reveal the presence of filled tunnels or interstices in a number of rather open framework or cage-like structures. Details of these results are given in [14]. This technique will undoubtedly be improved and lead to more structural information at an atomic level in compounds of variable composition, and thus directly clarify the nature of the mode by which non-stoichiometry is structurally accommodated in these phases.

2.2. CHEMICAL DEFECTS; POINT DEFECTS AND EXTENDED DEFECTS

From an accumulation of experimental evidence of the type detailed above it is becoming clear that the simple point defect model applies best to the most ionic solids, halides and oxides, which have very small homogeneity ranges. These solids are, in fact, virtually stoichiometric and the introduction of stoichiometric variability either by chemical methods such as doping with altervalent ions, or by physical means such as irradiation with X-rays or electrons, is fairly difficult to accomplish.

Once the degree of stoichiometric variation becomes appreciable, which in practical terms means that one X atom in one or two thousand has been removed in the case of reduction, defect interaction becomes important and the point defect model breaks down. Structurally two possibilities arise. As defect populations are imagined to increase, they can either be considered to interact to form clusters in which the original point defect structure is to some extent distinguishable from the regular crystal structure, or else the structure of the crystal can rearrange in a sufficiently significant way to eliminate the element of point defect character. Further increases in composition change will result in an increase in the concentration of these larger structural elements which can be thought of as extended defects. Eventually, further interactions result in ordering, or structural rearrangement which is physically detected as a 'new phase'. This new structure may be totally unrelated to the parent or may contain the extended defects in an ordered array to produce a superstructure of the parent.

This scheme of changes has been discussed in the past by Anderson [2]. Table I summarises this evolution from a point defect model, via differing degrees and types of interaction, to the formation of a second phase. This table, which summarises the progression from a solid containing point defects to one in which the structure is changed substantially into a different phase, is not intended to suggest that point defects are a precursor in any such transformation, or that structurally, point defects are the sole means of accommodating stoichiometric changes in some ranges of composition. Instead, it is rather better regarded as indicating how simpler theories must be successively refined to apply realistically to the more typical non-stoichiometric solid which has an appreciable stoichiometry range. It also indicates the various structural possibilities open to a non-stoichiometric phase to allow it to retain a neutral structure. Thus, in crystals which, for example, accommodate their non-stoichiometry by way of crystallographic shear, any simple point defect model must be considered largely redundant, as interactions within the solid make planar faults the preferred structural defect type.

TABLE I

Schematic relationships between non-stoichiometry and the defect structure of crystals

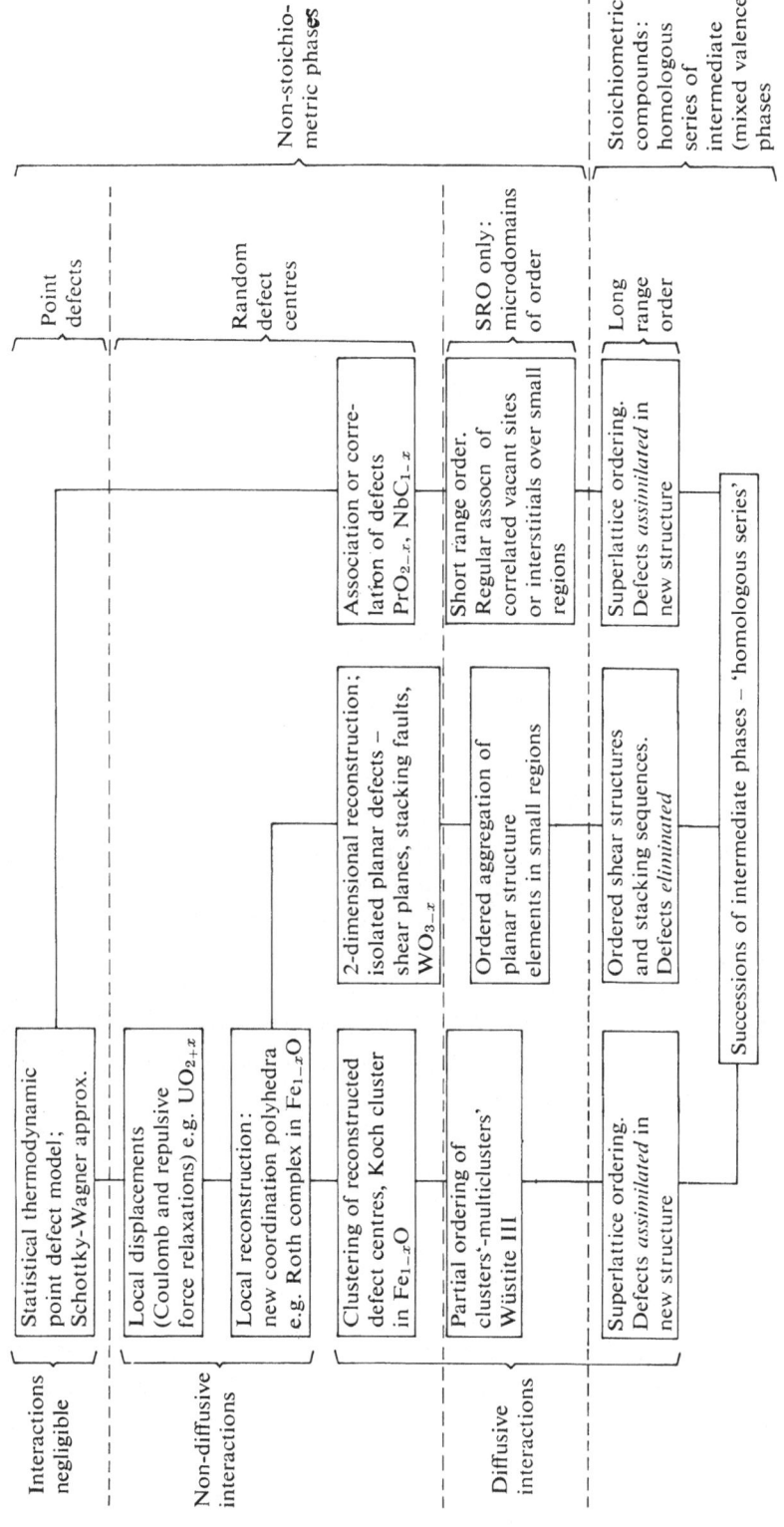

Taken from J. S. Anderson [2]

The existence of extended defects which are structurally more complex than point defects, and which are responsible for stoichiometric variability in crystals, does not mean to say that the point defect model is wrong or too simple to apply to real solids. Neither does the question as to the importance or existence of point defects in such solids loose its meaning. It suggests, instead, that experimental data must be treated with a great deal of caution. The assumption that nonstoichiometry is structurally accommodated by point defects should not be made automatically and the emphasis of empirical study must be to prove their existence instead. If such evidence is not found, other structural models should also be considered.

One advatage of point defect concepts is that they are sufficiently simple to be used as the physical basis for statistical thermodynamic calculations. If ones aim is to correlate non-stoichiometry and structure rather than to calculate thermodynamic properties of crystals, point defect concepts become far too restricting. Wadsley [9] described a number of structural ways in which composition of a crystal could become variable without recourse to point defects or clusters of point defects. Of these, the two to have received most attention recently have been crystallographic shear and intergrowth between similar structures. However, the other ways in which composition can be altered which were tabulated by Wadsley, viz. substitution of one atom type by another, interpolation of additional atoms into a structure, and subtraction of atoms from a structure, are also independent of any restriction to isolated sites within the crystal.

This structural viewpoint suggests that for some purposes crystals which are able to tolerate substantial composition ranges can best be thought of as host structures containing chemical defects. The size of these chemical defects will no longer be restricted to single atom sites within the crystal, but may well be extensive in volume. Such chemical defects cease to be defects and are assimilated into the crystal as necessary structure elements as soon as they become ordered. Chemical defects as a group can then be envisaged as a parallel to those faults which principally influence physical and mechanical properties, notably dislocations and stacking faults. Both of these groups may influence the ease of chemical reaction, indeed, enhanced chemical reactivity at dislocations is well known, but chemical defects are distinguished by the fact that their presence in a crystal alters the composition of that material.

The structural changes in the neighbourhood of such chemical defects may merely represent relaxations of the parent lattice. In other cases, however, their structure can take on a discernable similarity to a known structure type with a differing stoichiometry. This has an importance for theoretical treatments of non-stoichiometry. If the structural fragments are small, and if the positions of at least one atomic type in them are by and large, continuous, and not severely disrupted on passing from the matrix to the defect, the defect units can be termed microdomains of structure *A* in structure *B*. While this idea of explaining stoichiometric variation by the structural incorporation of microdomains of one stoichiometric phase in another is initially appealing, it is difficult to express in either structural or statistical mechanical formulations. Anderson has attempted such an analysis [15] and his conclusions are that microdomains, if they exist, would only be several unit cells in volume. This is the size of the defects which are found in some oxides, notably $Fe_{1-x}O$ and UO_{2+x} [16, 17] and at this extreme microdomains may be indistinguishable from what are often referred to as clusters of point defects.

When the structural units become larger, they must take the form of coherent intergrowths at a unit cell level or they would be revealed as separate phases by diffraction techniques. For such coherent formations to occur, the two structural types must be similar at least along the plane which unites them. This restraint is not a serious one. Many compounds can be considered as a close packing of larger atoms with smaller atoms or ions in interstices. Others are composed of sheets, which can be interleaved one with another to provide variable stoichiometry reflecting the numbers of each type of component sheet present. Intergrowths between materials whose structure consists of a packing of columns of a simpler structure is geometrically more constrained, but even here, coherent intergrowths of two or more line phases can yield apparent stoichiometry ranges, as the remarkable structures found in the higher niobium oxides reveal [10, 14].

So far this summary of the structural aspects of non-stoichiometry has been concerned with structures encountered in crystals examined at room temperature. The interaction between chemical defects within crystals will, of course, be temperature dependent. One can envisage that moderate temperatures will have only a small effect which in some cases may cause a certain amount of disordering, but in others may allow sufficient atomic diffusion for some degree of ordering to take place. Higher temperatures will lead to disorder among the defects responsible for the changes in stoichiometry. The extent to which this happens will clearly vary considerably from one phase to another. Operationally, therefore, line phases may take on the appearance of broad non-stoichiometric compounds at high temperatures when disorder of the chemical defects present is not readily detected. Very little evidence concerning these high temperature states is available, due to the difficulty of obtaining experimental data.

2.3. NON-STOICHIOMETRY IN THE LAYERED STRUCTURES

The general principles briefly described above apply equally well to any structural type. However, the particular geometrical and chemical nature of the layer structures will impose some constraints and tend to favour particular structural modes of accommodating composition variation over others. From a crystal chemical point of view, these structures have fairly densely packed layers, separated by rather open interlayer regions. The bonding in these two regions is rather different in character. Bonding within layers is quite strong, while the layers themselves are held together by weaker long-range forces which are often considered to be of Van der Waals type. These are broad based generalisations, and as pointed out earlier, the wide range of chemical types encountered as layer structures means that they can only be used as a guide. Nevertheless, one would expect that a consideration of non-stoichiometry in these systems would be concerned with two radically different crystallographic regions, these being the interlayer and intra-layer parts of the compound. Thus composition variation can be concerned with the alteration of the chemical constitution of either or both of these regions of the compound.

The most easily visualised mode of stoichiometric variation in the layered structures is interpolation of additional material between the layers. Provided that geometrical considerations allow it, one would expect neutral molecules or atoms to be accommodated easily between layers and in very variable proportions. Such is the case for many hydrates and also for the recently prepared organic intercalation compounds of the layered transition metal sulphides. These molecules can be ordered or disordered, and usually cause the layered structure to expand considerably. In the case of the introduction of charged ions between the layers, some degree of charge compensation must also occur for the crystal to remain neutral overall. The simplest way that this can happen is by the incorporation of a suitable number of oppositely charged species into the interlayer region, and such a situation seems to prevail in silver β-alumina (cf. Section 4). If this is not possible, then charge compensation must occur in the intra-layer structure. This can be accomplished if the layer contains cations which can adopt several valence states, but it may also result in structural rearrangements which destroy the layer structure. Examples of both these consequences are found in Sections 3 and 5.

There are two other ways in which the layer structure can readily support variation of composition. The first of these is by collapse of the layers onto one another so as to eliminate the interlayer region; the process of crystallographic shear. It does not appear to occur commonly in layer structures although in Section 3 a few examples are given in which the collapse takes place at an angle to the layers and results in an eventual destruction of the layered nature of the crystal.

The other likely way in which layer structures will modify composition is by coherent intergrowth. Many layered structures are composed of close packed sheets of anions stacked in cubic or hexagonal arrays. Geometrically different structure types can readily intergrow into such an arrangement which would allow very subtle changes in stoichiometry to take place. This takes place in a number of minerals with layered structures and in some inorganic oxides which do not posses a layer structure, notably the niobium oxides related to Nb_2O_5 but few examples are to be found in the materials chosen for this chapter, although coherent intergrowths between the β- and β''-alumina structures is illustrated in Section 4.

To obtain composition changes in this way the component elements of the structures which intergrow coherently must have differing chemical compositions, although their structures must be similar along the planes common to each unit. If these units of structure have identical compositions, the materials so formed are called polytypes. These interesting compounds lie outside the scope of this article, but are discussed at length in another chapter of this volume. However, there is clearly a close relationship between non-stoichiometry, intergrowth and polytypism in layered materials. Some aspects of this have recently been discussed by Caro [18] and the problem is also considered further in Section 5 of this chapter.

3. Layer Structures Related to the ReO₃ Structure Type

The ReO_3 structure is possessed by only a few compounds, which are generally regarded as being fairly ionic. The structure is not a layer structure, but a 3 dimensional array of MX_6 octahedra joined by corner-sharing. It is, however, a useful starting point from which to discuss a number of layer structures which may be considered to be made up of readily distinguishable fragments of an ReO_3 type parent. Not all compounds which can be described in this way will be considered, but only a selection, to illustrate the various ways in which structure and stoichiometry interact in this class of

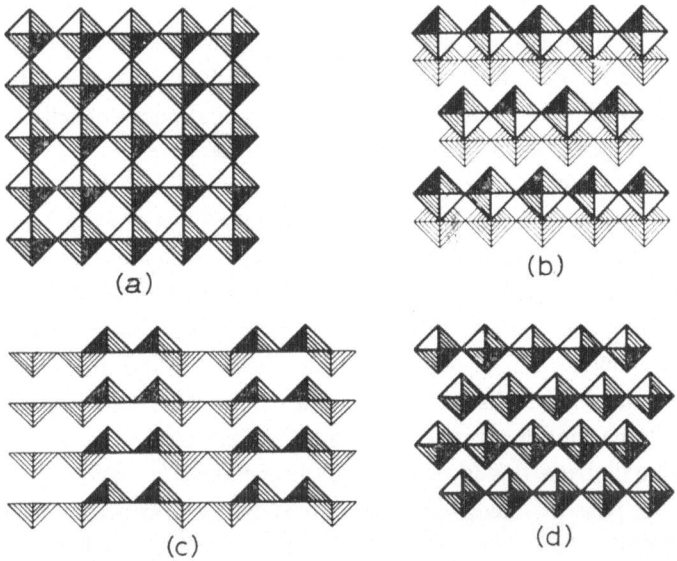

Fig. 1. Projections of ReO₃ related structures. (a) ReO₃; (b) MoO₃; (c) V₂O₅; (d) SnF₄. (a) (b) and (d) are comprised of metal-oxygen octahedra, shown as shaded squares in projection while (c) is composed of square pyramids, shown as half squares.

materials. Those chosen are related to the binary oxides MoO_3 and V_2O_5, the fluoride SnF_4 and the ternary oxides of the perovskite type. Figure 1 shows the structures of the binary compounds. Perovskite is derived from the ReO_3 structure, simply by interpolating large A cations between the BO_6 octahedra to yield the ABO_3 formula associated with these oxides.

These layer compounds have been grouped together not only because of their structural relation to the ReO_3 structure type but because, in most cases, stoichiometric changes result in a destruction of the layer structure, although new and different layer structures may be the result of these changes. These transformations are often accompanied by a drastic change in electrical properties. In the so called 'bronzes' of molybdenum and vanadium, high conductivities are often found, while in the perovskite related phases, dielectric behaviour and ferroelectric properties are sensitive to stoichiometry. These are, of course, a consequence of the changes in bonding which are also the cause of the destruction of the layer structure itself. Although of considerable interst, both from the point of view of academic research and technology, these changes in bonding lie rather outside the scope of this chapter, and will only be touched upon when of direct relevance to the problems of structure and stoichiometry.

3.1. DERIVATIVES OF THE MoO_3 STRUCTURE

The structure of MoO_3 is shown in Figure 1. It consists of layers of corner and edged shared MoO_6 octahedra which lie parallel to the (010) plane [19]. Each octahedron has one free apex which gives the oxygen atom at this point a somewhat different chemical reactivity to the others, as it is bound to only one metal atom. These MoO_6 octahedra are, however, far from regular and the structure can also be described as built from chains of corner shared MoO_4 tetrahedra running parallel to the c direction. This description is rather less suitable for our purposes, and does not reflect the cleavage properties of the oxide, so will not be considered further here.

The component octahedra in the structure become less distorted if some of the oxygen atoms are replaced by other anions, notably hydroxyl or halogen groups, or by water. Two phases have been characterised in the system when oxygen is partly replaced by fluorine [20]. These can be written as $MoO_{3-x}F_x$ with $x = 0.2$ and 0.6. In the $MoO_{2.8}F_{0.2}$ phase the layer structure is preserved and the structure is essentially the same as that of MoO_3. It differs from it in that the metal-anion octahedra are considerably less distorted than in the parent oxide, and this no longer allows the structure to be considered in terms of metal-anion tetrahedra. A further substitution of fluorine destroys this layer structure altogether. The octahedra become more regular, and the phase formed, $MoO_{2.4}F_{0.6}$, has the ReO_3 structure. The structure of MoF_3 is also of the ReO_3 type. Clearly a very delicate balance of forces decides which of these similar structure types forms.

A substitution of oxygen by hydroxyl ions leads to a series of molybdenum hydroxides $MoO_{3-x}(OH)_x$ with the degree of substitution, x, able to take values of 0.5, 1.0, 1.6 and 2.0 [21]. The compound $MoO_{2.5}(OH)_{0.5}$ has been studied in detail, and a structure determination [22] has shown that the material is also similar to MoO_3, but with more regular metal oxygen octahedra. $MoO_{2.8}F_{0.2}$ is, in fact, isostructural with $MoO_{2.5}(OH)_{0.5}$ if the hydrogen atoms in the latter compound are disregarded. These hydrogen atoms have not been located with certainty, as the structure determination was by X-ray diffraction, but were placed in the structure in positions compatible with possible hydrogen bonding between the layers of MoO_6 octahedra. These hydrogen atoms may well be ordered in view of the discrete series of phases formed, although any such ordering would be sensitive to temperature and long range order would probably give way to short range order at higher temperatures. If such ordering does take place, it will be of interest to determine whether the ordering takes place over the whole structure or whether it is confined to the hydrogen atoms within any one sheet.

Crystal structures of the other members of the $MoO_{3-x}(OH)_x$ series have not yet been determined. It is of some interest to determine if the structure changes to the

ReO$_3$ type at a certain degree of substitution of hydroxyl for oxygen. If this were so, a comparison with the results of fluorine substitution might allow some conclusions to be drawn about the relationship between bonding and the relative stabilities of the MoO$_3$ and ReO$_3$ structure types.

The structures of the molybdic acids are also of interest from this point of view, but in these materials there has been an overall oxidation of the structure causing it to open into layers of the SnF$_4$ type. These phases are therefore considered in Section 3.3.

The examples quoted above can be considered to be analogous to oxidation, and because of this a layer structure tends to be preserved. Reduction, on the other hand, can be accommodated by lattice collapse and it is this which is more likely to lead to a destruction of the layered structure. In the binary system, very small degrees of oxygen loss from MoO$_3$ seem to be structurally accommodated by the formation of domains of composition MoO$_{2.9975}$ in the crystals of the parent oxide [23]. The way in which these domains coexist with the MoO$_3$ matrix has been established by electron microscopy, and a possible structure for them has been proposed on the basis of electron diffraction patterns. This structure consists of an ordered array of vacant oxygen positions in the MoO$_3$ layers. Although this structure must be regarded as tentative, because three-dimensional data were not available for a complete structure determination, it does reveal that at these small degrees of reduction the layer structure of MoO$_3$ is preserved.

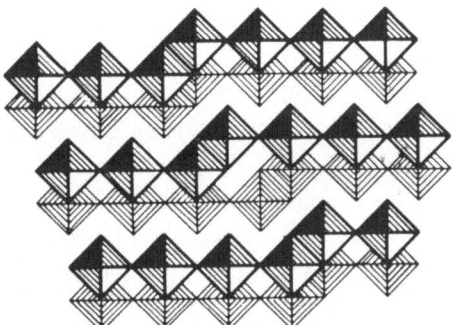

Fig. 2.   Diagrammatic representation of the formation of crystallographic shear planes on (120) in MoO$_3$ due to small degrees of oxygen loss.

In general, this domain structure of ordered vacancies is a low temperature phase. At higher temperatures these empty positions appear to aggregate on (120) planes in the MoO$_3$ structure, which collapse to form crystallographic shear (CS) planes and so eliminate vacancies. Figure 2 shows this collapse diagrammatically. Hence the course of reduction at these very small changes in stoichiometry results in collapse of the MoO$_3$ lattice on random (120) planes. Further reduction results in a repeat of this cycle and an increase in the density of the CS planes. The composition range over which this behaviour takes place is from MoO$_3$ to MoO$_{2.95}$ and the extent to which these planar defects destroy the layer structure depends upon their density. The crystals become blue and other physical properties are also likely to be different than in the parent unreduced phase.

Further reduction, to compositions down to $MoO_{2.88}$ lead to the formation of a number of CS phases, but based upon collapse along $(35\bar{1})$ planes. When these CS planes are ordered, an oxide is formed with formula Mo_nO_{3n-2}. Oxides with n equal to 18, 19, 20, 21 and 22 have been observed by electron microscopy [24]. In these the layer structure is considerably interrupted as the CS planes are not parallel or normal to the layers of the MoO_3 structure. Such oxides can thus be regarded as having lost the original layer structure which has been destroyed by reduction to approximately $MoO_{2.90}$. Any further reduction leads to a total reorganisation of the MoO_6 octahedra, and at low temperatures, tunnel compounds form in which all semblance of the layer structure is lost [19]. At somewhat higher temperatures the structure changes to the ReO_3 type with the degree of reduction again taken up by the insertion of planar faults which consist either of CS planes, in which the octahedral co-ordination of oxygen around the Mo atoms is preserved, or planes of MoO_4 tetrahedra [19]. It can be seen, therefore, that the tolerance of the MoO_3 layer lattice to reduction is small, and is greatest at lower temperatures.

The MoO_3 structure can also be reduced by reaction with other metals, notably the alkali metals, to form 'bronzes' of formula M_xMoO_3. These can be prepared by a variety of techniques, but electrolysis of fused mixtures of MoO_3 and alkali metal compounds is, at present, one of the most successful methods to use. This indicates that the reaction is not a simple one in which the alkali metal atoms enter the MoO_3 layer structure, but instead, that complex structural rearrangements are necessary.

The structures of two of these bronzes are shown in Figures 3 and 4. $Cs_{0.25}MoO_3$, with a metal to oxygen ratio of $1:2.4$, has alkali metal atoms lying between layers made up of groups of edge sharing MoO_6 octahedra linked by corners [25]. Clearly all structural resemblance to the original MoO_3 lattice has been lost. The same can be said of the silver bronze, $Ag_6Mo_{10}O_{33}$, which has a metal to oxygen ratio of $1:2.063$. Once again the molybdenum retains its octahedral co-ordination, and the silver atoms

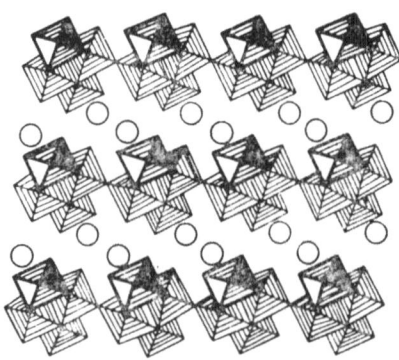

Fig. 3.   The structure of $Cs_{0.25}MoO_3$ projected onto (010). The circles represent positions partially occupied by caesium ions and the squares MoO_6 octahedra.

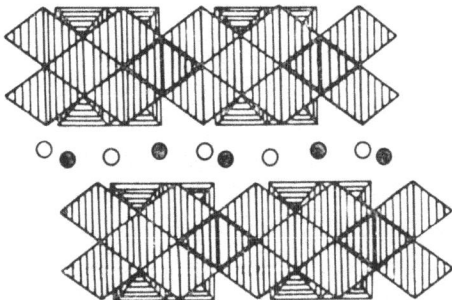

Fig. 4. The structure of $Ag_6Mo_{10}O_{33}$ projected onto (001). The positions partially occupied by silver ions are represented by filled or open circles and the MoO_6 octahedra by hatched diamonds.

lie in sheets between layers of fairly close packed MoO_6 octahedra linked by edges and corners [26]. It would be expected that both of these compounds would be able to tolerate some degree of stoichiometric variability. This is likely to take the structural form of a variation in the stoichiometry of the metal atoms between the layers with charge compensation coming from the variable valence of the molybdenum in the MoO_6 octahedra. At present, the range of composition over which these compounds can exist has not been determined.

3.2. DERIVATIVES OF THE V_2O_5 STRUCTURE

The reduction of V_2O_5 is similar to MoO_3 in that the layer structure is soon destroyed. Reduction at temperatures of 500–600 K results in the formation of a phase similar to that found in MoO_3 in which ordered arrays of oxygen vacancies are presumed to exist [27]. Higher temperatures, or more substantial degrees of reduction, below approximately $VO_{2.49}$ yield a two phase material. The crystal structure of this second phase is so far unknown, but is probably related to the other vanadium oxides found in the phase range between VO_2 and V_2O_5, namely V_3O_7 [28], V_4O_9 [29] or V_6O_{13} [30]. None of these are layered structures, and, as in the case of MoO_3, the layer structure is soon destroyed by loss of oxygen.

Replacement of the oxygen in V_2O_5 by fluorine is possible to a limited extent and the compounds $V_2O_{4.987}F_{0.013}$, and $V_2O_{4.975}F_{0.025}$ have been characterised [31]. These compounds have the same structure as V_2O_5 and behave in a very similar fashion. The total extent to which fluorine substitution can take place, and the resultant effects upon the layer structure are unknown, but may well depend upon the tolerance of the V_2O_5 structure to the presence of V^{+4} ions, which are necessary in order to preserve charge neutrality.

As in the case of MoO_3, a large number of metal 'bronzes', $M_xV_2O_5$, and related compounds have been prepared in recent years [32], and have been studied largely because of their interesting electrical properties. Not all of them are layered structures, and not all can be prepared by the direct reaction of metal and V_2O_5. The so-called α phases, however, are described as possessing a structure only slightly different from

V_2O_5. The metal atoms M are inserted between V_2O_5 sheets in an apparently disordered fashion, between the available positions indicated in Figure 5. The formula of a bronze with complete occupation of these sites would be MV_2O_5, but in reality only small degrees of occupation are found, up to about $M_{0.1}V_2O_5$. This appears to be closely related to size effects, as it is the larger metal ions which have smaller total degrees of insertion. Many metal atoms have been inserted into V_2O_5 to form these α phases, among which one can list the alkali metals, Ag and Cu, and a number of the 3d transition metals, notably Cr, Fe, Co. No doubt many other metals will enter the V_2O_5 lattice in small amounts to form similar phases.

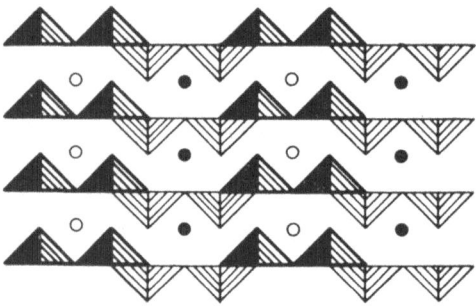

Fig. 5.   The structure of the α-bronzes, $M_xV_2O_5$. The metal ions M occupy some of the positions
indicated by filled and open circles, while the V_2O_5 matrix remains intact.

An increase in metal to V_2O_5 ratio results in the formation of a number of other bronze structures, not, in general, related directly to the structure of V_2O_5, but containing fragments of the V_2O_5 sheets linked to form a 3-dimensional structure. Of these phases richer in inserted metal, we will describe three which are of interest in this present context.

The phase $Na_xW_2O_5$ ($0.7 < x < 1.0$), denoted as α', appears to be topologically identical to the α phases [33] and differs from them only in slight distortions of the VO_5 pentagonal pyramids which make up the structure. However, this phase is not formed by a continuous insertion of sodium into the α phase, and between the limits of this latter material, $Na_{0.02}V_2O_5$, and the high sodium compound $Na_{0.7}V_2O_5$, a totally different (β) structure is found. Why this should be so is not clear. The α' phase seems able to tolerate considerable stoichiometric variability with little structural consequences. The range of allowed sodium insertion is large, from $x = 0.7$ to 1.0, and the structure appears to be a fairly stable one. Why the other vanadium bronzes do not adopt this structure is uncertain, but ionic size may have an important role to play here, and possibly temperature of preparation will also influence the structure type found.

The α' structure has a substantial tolerance, not only for variation of sodium concentration, but also for substitution of oxygen by fluorine [34]. In this case, the structure is stable over the composition range $0.95 < x < 1$ for the series $Na_xV_2O_{5-x}F_x$, with a limiting composition of NaV_2O_4F. NaV_2O_5 and NaV_2O_4F also form a complete solid solution [35], $NaV_2O_{5-x}F_x$ with $0 < x < 1$, in which only very small structural changes seem to accompany the insertion of the fluorine atoms at the expense of the oxygen atoms. The major change seems to be in the distribution of the two types of vanadium atom which the altervalent cations Na^+ and F^- require to maintain charge neutrality. In this case, the ions involved appear to be V^{+5} and V^{+4}. In NaV_2O_5, the V^{+5} and V^{+4} atoms are completely ordered, and in the compound NaV_2O_4F, only the V^{+4} species is present. Considerations of both the structures and electrical and magnetic properties of the intermediate $Na_xV_2O_{5-x}F_x$ phases has shown that in these the V^{+4} and V^{+5} ions are disordered over the available sites [35].

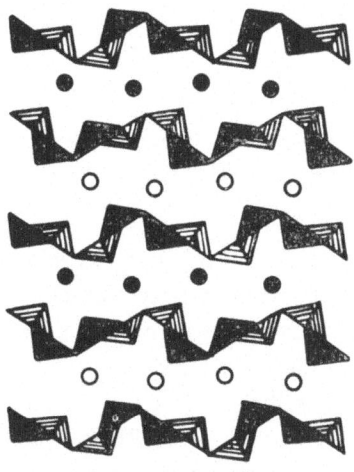

Fig. 6. The structure of γ-$Li_xV_2O_5$ projected onto (010). The positions partially occupied by the lithium atoms are represented by filled or open circles.

The effect of ionic size is shown when one compares the behaviour of lithium with sodium. When bronzes of composition close to LiV_2O_5 are prepared, a layer structure results which is somewhat different than the α' phase of NaV_2O_5 [32]. This structure is shown in Figure 6. It has sheets of V_2O_5 pyramids which are rather like folded fragments of the V_2O_5 sheets between which the lithium atoms are inserted. The composition range of this material, from $Li_{0.88}V_2O_5$ to LiV_2O_5, is quite considerable. The structure appears to tolerate this because of the ability to compensate for the Li^{+1} charge by V^{+4} ions. Whether these two different species of vanadium atoms are ordered is not certain.

The final example chosen further reveals the complexity of these bronze systems and indicates that simple insertion of metal atoms into the layer structure of V_2O_5 is not a true reflection of the reaction mechanisms involved. This example is chosen from the $Ag_xV_2O_5$ system. At relatively high silver concentrations, between $Ag_{0.67}V_2O_5$ and $Ag_{0.86}V_2O_5$, the structure shown in Figure 7 is formed [36]. The layers are composed of groups of four distorted VO_6 octahedra linked by edge sharing. The silver atoms lie between these sheets and this geometry allows for the latitude of the silver

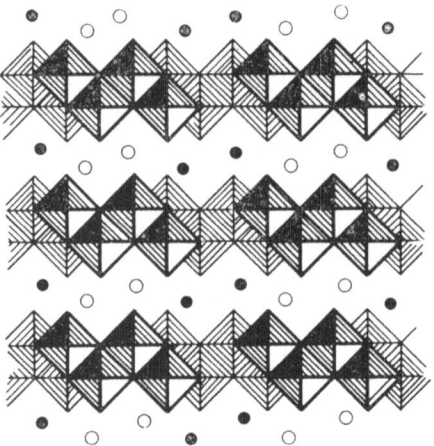

Fig. 7.   The structure of $Ag_xV_2O_5$ $(0.67 \leqslant x \leqslant 0.86)$. The silver atoms occupy some of the positions
indicated by the filled and open circles.

concentrations that the structure will tolerate. Outside these limits, new structural
types are formed which are unrelated to the layer structures.

A very closely related structure is formed in the copper $- V_2O_5$ system between the
limits $Cu_{0.85}V_2O_5$ and CuV_2O_5. The structure of this compound differs from the
structure of $Ag_{0.86}V_2O_5$ only in that the sheets of VO_6 octahedra are displaced somewhat on passing from one layer to the next [37]. So in many ways the two structures
are topologically identical. The valence state of the copper ions in these phases has
been shown to be $+1$ from a consideration of their magnetic and electrical properties,
and the variable amount of copper incorporated into the materials, again reflects the
stability of the compensating V^{+4} ions in these oxides.

Other examples of compounds related to these higher vanadium bronzes will be
found in the literature, but all tend to display the same characteristics. Layers of
linked vanadium-oxygen polyhedra, usually distorted octahedra, but sometimes tetrahedra or even trigonal prisms are separated by sheets of additional metal atoms. The
ability of the structure to tolerate such variations in composition is then due to the
ability of the cations in the $V-O$ layers to change their valence without significant
changes in their oxygen coordination polyhedra.

3.3. Derivatives of the SnF_4 structure type

As described in the introduction to this section, the ReO_3 structure type can be 'oxi-
dised' by opening the sheets of corner sharing MX_6 octahedra to produce a structure
of the SnF_4 type, composed of a stack of individual sheets of MX_6 octahedra linked
by four of the available six vertices (Figure 1). The stoichiometry of each layer is MX_4.
If the same formal procedure is applied to the perovskite family, layers of stoichiometry A_2MX_4 are formed, where A represents the large cations. In each case these
layers have a fixed anion to cation ratio, but stoichiometric flexibility can be achieved

by uniting several layers to form slabs of ReO_3 type or perovskite type to give compounds with a variety of chemical formulae. In this section, some of these will be described.

The fluorides, SnF_4, PbF_4 and NbF_4 are isostructural. The structure of these compounds is shown in Figure 1, the sheets of MF_6 octahedra being held together by electrostatic forces. One can imagine that reduction of these phases by removal of fluorine will result in an unstable anion co-ordination around the cations unless some lattice reconstruction is able to take place. By analogy with WO_3 [14, 38] one could suspect that a lattice collapse is likely, with the formation of CS planes parallel to the MX_4 sheets. This would result in a homologous series of compounds M_nX_{4n-1} should the collapse be regular. Such a collapse would preserve the layer structure of the parent until total collapse lead to the MX_3 structure. However, collapse at an angle to these layers would appear to be equally likely, and this would interrupt the layer sequence as has been illustrated in the case of MoO_{3-x} CS phases.

Fig. 8. The idealised structure of $SmZrF_7$, emphasising its relationship to the SnF_4 and ReO_3 structure types.

These possibilities have not yet been fully explored experimentally. A study of the ZrF_4-SmF_3 system has, however, shown that the compound $SmZrF_7$ (Figure 8) has rather a similar idealised structure in which $n=2$ in the formula above [39]. The co-ordination of the samarium atoms by fluorine is not octahedral and the CS plane is not produced by a normal collapse of the SnF_4 type of parent lattice or expansion of the ReO_3 type structure. Nevertheless, similar principles apply, and replacement of Sm in this phase by cations which are smaller and do prefer octahedral co-ordination may produce a number of related CS phases. In particular a study of the structures to be found in the systems formed by MF_3 fluorides of the ReO_3 type (i.e. MoF_3, TaF_3, FeF_3, AlF_3) and SnF_4 or NbF_4 would be of interest.

Stoichiometric variation can be obtained in a different fashion by interpolating cations between the MF_4 layers. A number of such phases AMF_4 are known, where A is usually a large alkali metal cation and M is a smaller cation, typically Al, Ga or

Fe [40, 41, 42]. The compounds with this structure which are already known do not seem to possess any degree of non-stoichiometry. In the $MF - FeF_3$ system, which has been studied structurally in some detail, a change in the ratio of MF to FeF_3 results in the formation of totally different structures. The AMF_4 compounds with this structure do not appear to tolerate any change in anion to cation ratio, although formally a large number of structures could be built up merely by uniting some MF_4 slabs to others by corner sharing, as indicated above.

The lack of these structures may well be associated with the fact that the fluorides are generally unstable at high, or even moderate, temperatures. There are a number of closely related oxides in which such series do form, although these can also be considered to be derivatives of the perovskite structure. If the alkali metal cations in the AMF_4 structure described above are placed between the MF_4 octahedra, that is, within the layer, and additional ions added to make up the formula A_2MF_4, a sheet of the perovskite structure is obtained which is one unit cell thick. This structure is

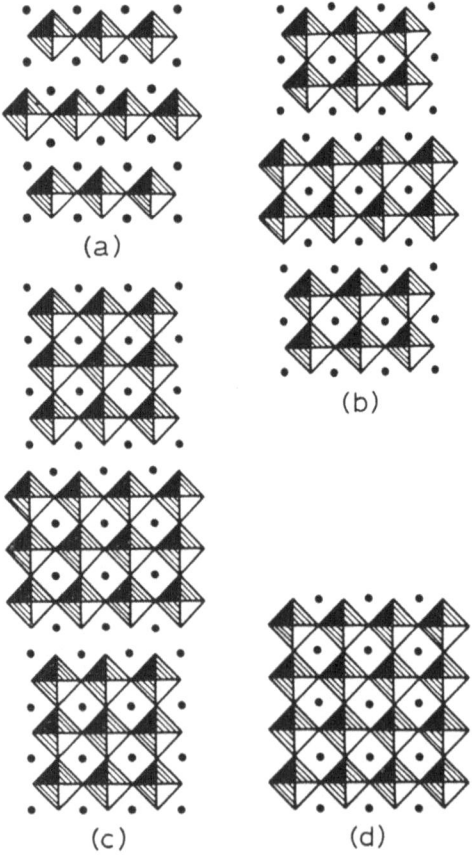

(a)

(b)

(c)

(d)

Fig. 9.   The structures of the perovskite related phases (a) Sr_2TiO_4; (b) $Sr_3Ti_2O_7$; (c) $Sr_4Ti_3O_{10}$ and (d) $SrTiO_3$ (perovskite structure). (b) and (c) are intergrowths between units of (a) and (d). The shaded squares represent TiO_6 octahedra and the filled circles strontium atoms.

typified by the compound K_2NiF_4 which has the MX_4 layers of octahedra displaced one with another, as in the SnF_4 compound compared to the ReO_3 structure, Figure 1. A number of oxides possess this structure [40] and we can mention here the material Sr_2TiO_4 in which the TiO_6 octahedra form layers and the Sr atoms are in the open positions between these octahedra [43]. A change in the Sr to Ti ratio in the direction of the parent $SrTiO_3$ perovskite produces $Sr_3Ti_2O_7$ with double sheets of TiO_6 octahedra linked by corner sharing, and $Sr_4Ti_3O_{10}$ with treble sheets of TiO_6 octahedra linked by corner sharing [43]. These structures, shown in Figure 9, form part of a homologous series of oxides $Sr_{n+1}Ti_nO_{3n-1}$, which, of course, yields the perovskite $SrTiO_3$ when $n=\infty$. Whether such compounds are formed in all related oxide systems is not known, or the extent to which the $Sr_{n+1}Ti_nO_{3n+1}$ homologous series can be extended beyond the $n=4$ member. It would seem likely, however, that many of these hypothetical structures could form if the preparation conditions are suitable.

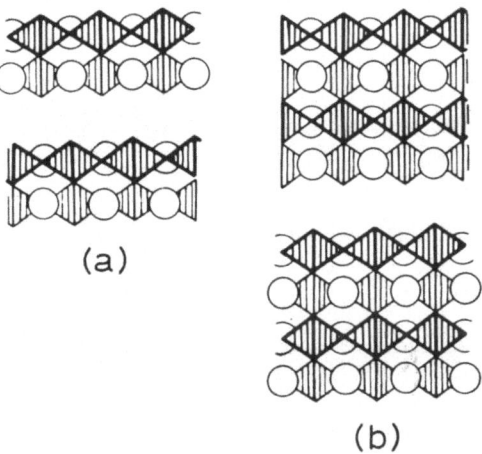

(a)

(b)

Fig. 10. The structures of the perovskite related phases (a) $NaNbO_2F_2$, projected onto (010) and (b) $Ca_2Nb_2O_7$, projected onto (010). The sheets are parallel to (100) in both structures.

An alternative way for the ReO_3 structure to break up into layers is not by separating the MX_6 octahedra along (100) planes, but along (110) planes. A number of fluorides possess such a structure, as do the oxides $\beta - BiNbO_4$ [44] and $NaNbO_2F_2$ [45]. As the anions in these layers are in a distorted hexagonal array these materials can be related to the PdF_3 structure type, but for the present purposes, they are represented as slabs of a perovskite such as is shown in Figure 10. As before, a potential homologous series of oxides is possible if one imagines the thickness of the ReO_3 sheets to increase in a regular fashion. So far, only one member of this series is known, $Ca_2Nb_2O_7$ also shown in Figure 10 [46, 47]. Potentially a large number of these oxides may exist.

The final structures to be considered in this group are the molybdic acids. These materials, with structures closely related to the SnF_4 structure type, are typical of a large number of hydrated layered minerals. The interpolation of variable amounts of water in between the layers is relatively simple, and complete dehydration leads to a collapse of the structure to another type. This latter reaction is often a topotactic one.

There is some confusion in the literature about the structural chemistry of the molybdic acids, but the determination of the crystal structure of H_4MoO_6 ($MoO_3 \cdot 2H_2O$) by Krebs [48] and the crystal-chemical study of its dehydration through H_2MoO_4 ($MoO_3 \cdot H_2O$) to anhydrous MoO_3 by Günter [49] has allowed some order to be introduced into this complexity. The crystal structure of H_4MoO_6 is rather similar to that of SnF_4 (see Figure 1) and isolated sheets of MoO_6 octahedra form the dominant structural motif. In reality, these layers are of formula H_2MoO_4, as one oxygen in each MoO_6 octahedron has been replaced by the oxygen of a water molecule, in an ordered fashion. Moreover, the layers are much further apart than in SnF_4, and in this space the second set of water molecules are positioned, as shown in Figure 11. The structure is thus more accurately written as $H_2MoO_4 \cdot H_2O$. The tungstic acid $H_2WO_4 \cdot H_2O$ is apparently isostructural [50]. In common with a large number of layer structure hydrates, there is a certain flexibility of the structure to the amount of interlayer water

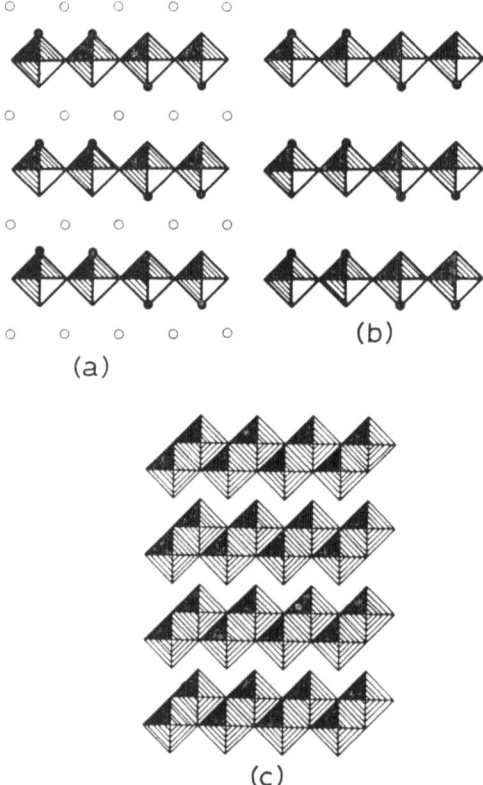

Fig. 11. The structures of the molybdic acids. (a) $H_2MoO_4 \cdot H_2O$; (b) H_2MoO_4. The non-bonded water molecules are representeed by open circles and the water molecules bonded to Mo atoms by filled circles. Dehydration of (a) leads to (b) and then to (c), the MoO_3 structure projected onto (100).

accommodated, and this may be the cause of the varying compositions reported for the molybdic acids. In the present case, water molecules can be removed from this interlayer region completely, leaving the skeleton of H_2MoO_4 layers intact (Figure 11). This type of behaviour is common among the clay minerals and reflects the fact that these latter water molecules are not an essential part of the crystal structure. The water molecules which are bound to the molybdenum atoms are, however, structurally important. Further dehydration removes these, and must result in some lattice collapse in order to preserve stoichiometry. This appears to happen in a topotactic fashion, and the structure collapses to form MoO_3 in the manner shown in Figure 11. This process is very closely related to the process of crystallographic shear found in some of the tungsten and molybdenum oxides [38]. It can formally be described as crystallographic shear in an MO_4 parent to produce an MO_3 structure type. From a geometrical standpoint, therefore, if the dehydration and subsequent lattice collapse took place on ordered planes, rather than at random, a homologous series of phases would be formed between H_2MoO_4 and MoO_3. Collapse on 1 plane in every n would lead to a phase $H_{2n-2}Mo_nO_{4n-2}$ and ultimately collapse on every second plane to a hypothetical material $H_2Mo_2O_6$ with the MoO_3 structure. Crystallographic shear phases of this type have only been found in a small number of oxides, and the reason for the preference of such ordered structures over other possibilities is still obscure. Therefore, it is not possible to predict if such phases would form if experimental conditions were favourable.

4. Layer Structures Related to β-Alumina

The name β-alumina is frequently applied indiscriminately to a group of non-stoichiometric layered materials formed in the Na—Al—O system. Originally, as the name implies, the compound β-alumina was believed to be a binary aluminium oxide, due to the fact that the presence of sodium in the material was undetected. However, at an early stage, X-ray structure determinations and associated chemical analysis showed that the formula of β-alumina was approximately $NaAl_{11}O_{17}$. Since that time a number of isostructural compounds have been characterised in which sodium has been replaced by other monovalent ions or aluminium by other trivalent ions. In addition, a number of other phases have been characterised with structures closely related to the β-alumina structure. In the systems based upon alumina four structures are known, which have been labelled β, β″, β‴ and β⁗. These too can also be prepared from alkali metal cations other than sodium, and some appear to be stabililised by additions of a few percent of divalent cations, particularly magnesium.

Interest in these phases has been great in recent years, as β-alumina has been successfully employed as a solid electrolyte for the passage of sodium ions in the sodium-sulphur battery. This device, first announced by the Ford Motor Co. in 1966, makes use of the extremely high mobility of sodium ions in β-alumina, a mobility equalling that of many ionic species in liquid solutions. The descriptive term 'super-ionic conductors' is often applied to these and similar materials. Investigations by many research groups have led to the conclusion that, at present, the sodium β-alumina sulphur combination provides one of the most viable types of storage battery and it is this which has stimulated a good deal of research into the properties and structure of the various β-aluminas. Kummer has recently reviewed this research and has described the properties and uses of β-alumina in some detail [51], while the earlier literature on these materials has been critically evaluated by De Vries and Roth [52].

Before describing the non-stoichiometry in these phases, the idealised structures will be described. The general outline of the structure was determined as long ago as 1931 by Bragg et al. [53] while Beevers and coworkers improved upon this and proposed the formula $NaAl_{11}O_{17}$ for the compound [54]. More recent structural studies on β-alumina [55], β''-alumina [56] and β'''-alumina [57] have confirmed the general validity of the earlier studies. In all of them, the dominant structural motif is that of sheets of cubic close packed oxygen layers. In the β and β'' forms, these sheets are four oxygen atom layers thick, while in the β''' and β'''' forms they are six oxygen atom layers thick. The non-alkali metal ions are predominantly located in some of the octahedral and tetrahedral interstices within these layers, giving them a structure similar to that of spinel, $MgAl_2O_4$. They are often referred to as spinel 'blocks' although they are, in reality, two dimensional sheets. The structures of these sheets are shown in Figure 12.

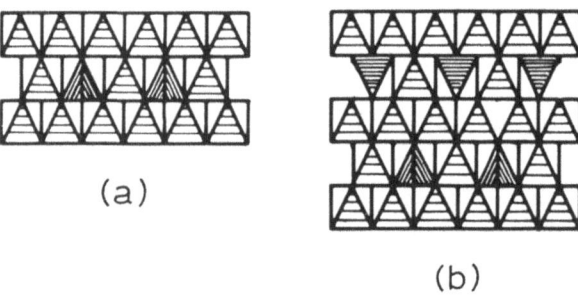

(a)

(b)

Fig. 12.   The infinite slabs of spinel structure type found in the β-alumina phases. (a) containing four oxygen layers, (b) containing six oxygen layers. The structures are represented by packing of MO_6 octahedra (lightly shaded) and MO_4 tetrahedra (heavily shaded).

Between these layers, the alkali metal atoms, and a few oxygen atoms and trivalent cations are located. These latter serve to provide weak links between the slabs of spinel but as the population of atoms in such layers is fairly low, this binding is weak. The β-aluminas therefore have an easy cleavage parallel to the spinel layers. There is also a lack of strong bonding between the spinel layers and the alkali metal ions which is revealed by the ease with which these latter ions may be substituted when crystals are immersed in molten salts of other cations, and, of course, in their very high mobility. Both of these features are discussed below.

The spinel layers can be stacked in one of two ways, as a consequence of their symmetry. Two adjacent layers can be related by a two-fold axis or a three-fold axis to give a hexagonal unit cell of either two layers or three layers in the repeat distance normal to the layers themselves. This direction is taken as the c direction. The material β-alumina is usually considered to be the phase with a unit cell consisting of two of the thinner 4 oxygen layer spinel sheets, and a c dimension of about 2.27 nm. Figure 13 shows this crystal structure. The idealised formula for β-alumina is $NaAl_{11}O_{17}$. The analogous phase containing three spinel layers per unit cell is the β'' phase. This has a c dimension of about 3.38 nm. The β''' and β'''' forms are the analogues of β and β'' alumina but based upon the thicker spinel slabs which are made up of six close packed oxygen layers. Their c dimensions are 3.18 nm and 4.77 nm respectively. They have never been found in the ternary Na-Al-O system, but they are formed in the presence of small amounts of divalent oxides, notably MgO, CoO and ZnO [51, 58].

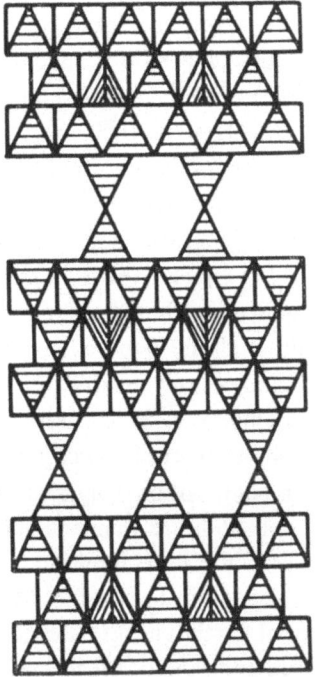

Fig. 13.   The structure of β-alumina shown as a packing of spinel layers. The alkali metal ions, which are not shown, are to be found in the empty spaces between the spinel layers.

Isomorphous structures have been formed by ions which have similar size and charge to Na^+ and Al^{+3}. Of the alkali metals, K^+ and Rb^+ have been successfully used in place of Na^+, while Ga^{+3} and Fe^{+3} are able to react in place of aluminium. Not all of these systems are able to form both β and β'' structures. For example Ga^{+3} forms β and β'' phases with Na^+ but only a β phase with K^+. In the case of Fe^{+3}, β phases are formed with Na^+, K^+ and Rb^+ while the existence of a β'' phase in the potassium system is not yet conclusively proved. The reaction of oxides of smaller divalent ions such as Mg^{+2} or Ni^{+2} yields the well known spinel structure in which the spinel layers are now infinite in thickness. Reaction with the larger divalent ions, such as Ba^{+2} or Sr^{+2} results in the magnetoplumbite structure $PbFe_{12}O_{19}$. This structure is very similar to the β-alumina structure, the only difference being that the open Na containing layers of the β-aluminas are now stoichiometrically filled with a packing of oxygen atoms and Pb^{+2} ions.

While these broad structural details are undoubtedly correct, there is much confusion in the literature concerning the true formulae of these phases and their relative stabilities. Most work has concentrated on the ternary Na—Al—O system, by virtue of its commercial importance, and the other systems described above will not be considered further. Even in the case of the sodium compounds, there is a great deal of uncertainty surrounding the β-alumina region of the phase diagram and a number of somewhat different versions can be found in the literature [51, 52, 58, 59]. In general it seems that the β phase has a composition range of between $Na_2O \cdot 5.33\ Al_2O_3$ and $Na_2O \cdot 8.5\ Al_2O_3$ at 1400°C. This composition range changes with temperature but remains considerable. The β'' phase appears to have a much smaller composition range of from approximately $Na_2O \cdot 7Al_2O_3$ to $Na_2O \cdot 8.5\ Al_2O_3$. It should be noted that there is certainly a measure of disagreement in the literature about these composition ranges and both should be accepted with caution. One of the difficulties of completely characterising these phases is that the β'' form is not formed alone, but only intergrown with the β phase. It converts to the β phase above 1500°C, but whether the reverse is true below that temperature, indicating that β'' alumina is the stable phase at lower temperatures is not yet clear, although the favoured view at the moment is that β is the stable phase at all temperatures and β'' is metastable.

The way in which the stoichiometric variations are accommodated structurally is open to some doubt. For the β phase, charge balance is achieved by having a composition of $NaAl_{11}O_{17}$. The known composition ranges of neither the β nor the β'' phase encompass this composition, and both are richer in sodium. For the β phase, this excess is of the order of 29% over the stoichiometric formula. The many structural studies leave no doubt that this sodium lies in the open layers between the spinel blocks. The problem is to account for the charge balance. One possibility, structurally attractive, is merely to introduce additional oxygen ions into this open region of the structure, to compensate for the charge imbalance. The other possibility is to introduce some sort of disorder, Al^{+3} vacancies for example, into the spinel layers.

Experimentally, the problem centres upon the difficulty of interpretation of the X-ray results. The most careful study in the literature to date is that of Peters, Bettman, Moore and Glick [55]. These authors showed that while the atoms in the spinel blocks could be assigned positions with confidence, the same could not be said for the atoms in the layer between the spinel blocks and containing the sodium atoms. It was concluded that the sodium atoms were 'smeared out in a complex pattern' and only their average position is centred at the expected sites, shown in Figure 14. The reason for this displacement was considered to be due to site distortion caused by electric fields due to Al^{+3} vacancies. These were assumed to be present in Al^{+3} sites adjacent to the sodium planes and were introduced to account for charge balance due to the excess sodium in the lattice over the stoichiometric formula of $NaAl_{11}O_{17}$. They suggest that other alternatives, particularly the introduction of extra oxygen into the interspinel layer region is less likely. Thus, Peters *et al.* [55] came down in favour of a structural model in which the extra sodium is distributed in a disordered fashion over some of the available sites in the interspinel regions and aluminium vacancies are distributed at random over one set of the octahedrally co-ordinated Al^{+3} sites which should be fully occupied. Because of the postulated interaction between these aluminium vacancies and the sodium ions, neither can be regarded as point defects, but

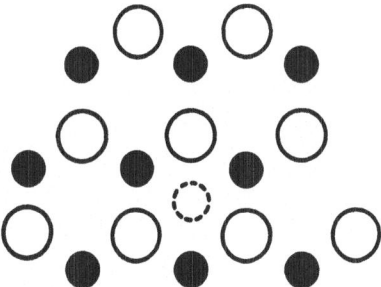

Fig. 14. The idealised structure of the alkali metal containing plane in β-alumina. The oxygen atoms are represented by large open circles. The filled circles represent the idealised positions occupied by alkali metal atoms. In the real structure, this site is only partially occupied, and excess metal is distributed between oxygen atoms in the case of sodium β-alumina or near to the sites shown by a dotted circle in the case of silver β-alumina.

rather as defect clusters. The simplest of such clusters will be formed from one vacant Al^{+3} site combined with three sodium ions, and such an entity fits well with the X-ray data available. The clusters may, of course, be more complex, but at present no data is available which is able to clarify this point. The stoichiometric variation in these materials will therefore be associated with the number of these complexes present, and the composition range will reflect the maximum and minimum numbers of complexes which the crystal is able to tolerate. The model implies that the clusters will not be excessively mobile, as the Al^{+3} vacancies will be restricted in movement to the regions close to the spinel layer upper and lower surfaces. Moreover Al^{+3} diffusion at room temperature in spinel crystals is extremely slow.

An alternative structural model, which is rejected by Peters *et al.* [55] on the basis of their results is that the sodium ions are thermally disordered. Line phases frequently become non-stoichiometric at high temperatures, when increasing mobility of atoms in the solid state becomes important (c.f. Section 2). Room temperature is, of course, merely an arbitrary temperature in this respect, and the order-disorder transition for the sodium ions may well be below this temperature. Many non-stoichiometric solids are known where disorder takes place at only a few hundreds of degrees centigrade; the FeS_x system discussed in Section 5 being one of these. Even if the mobility of sodium were not excessive at room temperature, the long exposure times necessary to register diffracted intensities would obscure the sodium positions. However, electron microscopy of β-alumina shows no trace of ordering in specimens under normal operating conditions, in which photographic exposure times of only one or two seconds prevail. Low temperature X-ray diffraction or electron miscroscopy should allow this hypothesis to be checked.

If the sodium positions are constantly changing at room temperature it is conceivable that an extended defect cluster involving Al^{+3} vacancies may not be the best model possible, and recourse may well have to be made to one in which extra oxygen is inserted into these alkali metal planes. This oxygen could be associated with the sodium to form extended two dimensional clusters, or could be totally independent of the sodium positions. The observation that the c parameter of β-alumina does not vary in a linear fashion across the whole of the stoichiometry range [60] suggests that one defect structure may give way to another as the sodium content varies, or that defect interactions may change as the stoichiometry alters. Diffuse neutron scattering might be of use in this respect as it might well reveal the presence or absence of defect clusters in these crystals and their structures. Finally, the problem of the number of oxygen atoms in the alkali metal planes could also be attacked by using neutron diffraction, as this is another piece of information which is badly needed in solving the structure of these phases.

These models do not indicate why the β and β'' modifications of the sodium compound form or whether the growth of one or other of these structures depends upon physical parameters such as temperature alone, or whether chemical composition is also important. The β''-alumina phase does appear to be somewhat richer in sodium than the β-phase. The two structures are able to integrow coherently and Figure 15 shows such a disordered crystal. This is one alternative method of accommodating some degree of non-stoichiometry, into crystals grown at lower temperatures where

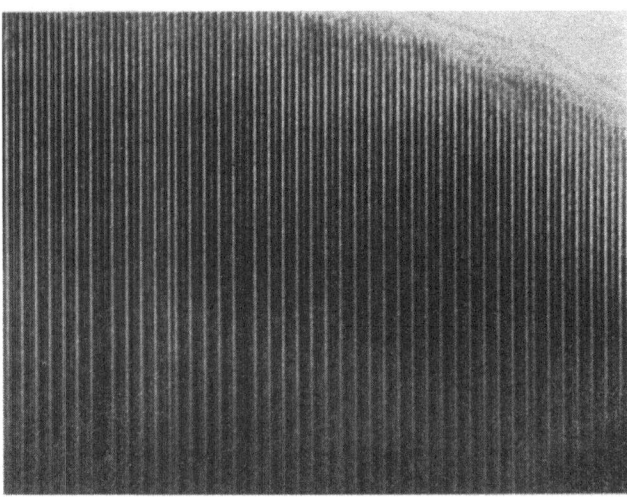

Fig. 15. Electron microcraph showing a disordered fragment of a β-alumina crystal comprised of a coherent intergrowth of the β and β'' forms. The intergrowth is best seen away from the edge of the crystal. The fringes at the crystal edge are approximately 1.1 nm apart and correspond to the spinel layer spacing.

the β'' form is stable but such a mechanism does not explain the sodium excess over the stoichiometric $NaAl_{11}O_{17}$ formula in crystals of monophasic β-alumina, and will not, therefore, negate the questions posed above.

Stoichiometric variability of the other phases mentioned earlier, the β''' and β'''' structures, and the analogous phases which can be formed from alternative monovalent and trivalent cations, have not been examined in detail, except for the sodium gallium oxides, which appear to have comparable phase ranges to the aluminium isomorphs [61]. One cannot, therefore, turn to these for further clarification of these problems. The many isomorphs which can be prepared from sodium β-alumina by partial or complete ion exchange of the sodium in molten salt baths [51] have also not been studied in sufficient detail structurally to help in this respect, with the exception of the silver compound.

Silver β-alumina, prepared from single crystals of the sodium compound by immersion in molten silver nitrate for several days, is more attractive than the sodium compound for structural studies because of the much higher relative scattering factor of the silver atoms. This should allow the heavier silver cations to be located with more precision than the lighter sodium. Roth [62] has studied the structure of single crystals of this silver substituted β-alumina, and has come to some very interesting conclusions. He found, as expected, that all the silver atoms lie between the spinel blocks, which themselves are somewhat distorted and very similar to the spinel blocks in sodium β-alumina. Secondly he reported that the compound contained about 25% more silver than the stoichiometric formula of $AgAl_{11}O_{17}$, again similar to the case of the sodium compound. However, the experimental results suggested that the excess charge was compensated for by extra oxygen atoms in the interspinel layer region, and not by Al^{+3} vacancies. Finally, it was concluded that, although the silver ions were disordered over the sites shown in Figure 14, the statistical distribution was quite different than that for sodium. These last two features suggest that from a structural viewpoint, non-stoichiometry in the silver compound is quite different from that in the sodium phase. Suggestions as to this difference have not been put forward although the larger size and more easy polarisability of the silver atoms may be important here. It would be of interest to study the continuous series of phases (Na, Ag) β-alumina which can be prepared by partial replacement of sodium to attempt to obtain some correlation between silver content and the preference of the structure for Al^{+3} vacancy clusters versus interlayer oxygen clusters.

To account for the stoichiometry, Roth has put forward a very elegant structural model, the essence of which is shown in Figure 16. His suggestion is that the β-alumina materials are stoichiometrically perfect $MAl_{11}O_{17}$ phases of the layer structure type already described within small enough volumes of the crystal. The M^{+1} ions are able to occupy either of the available sites a or b in Figure 16 but not both, either fully or statistically. Because of the very likely small energetic differences between these two structures, either is capable of forming. Hence one can imagine the silver containing layers being composed of domains of one sort or the other, a or b. Where such domains meet, in a domain wall, the concentration of M atoms goes up. Two sorts of domain wall are shown in Figure 16. In one, the concentration of M is 1.5 times that in the domains, and in the other it is twice that in either domain. A comparison of the observed composition with this model suggests that domains of between 5 and 10×10^{-10} m wide are required. These, of course, are too small to be resolved by X-ray techniques and even electron microscopy would be unlikely to be able to reveal them unless the domains were well ordered from one layer to another. As the spinel layers are compact and fairly thick, this would seem to have only a low probability of

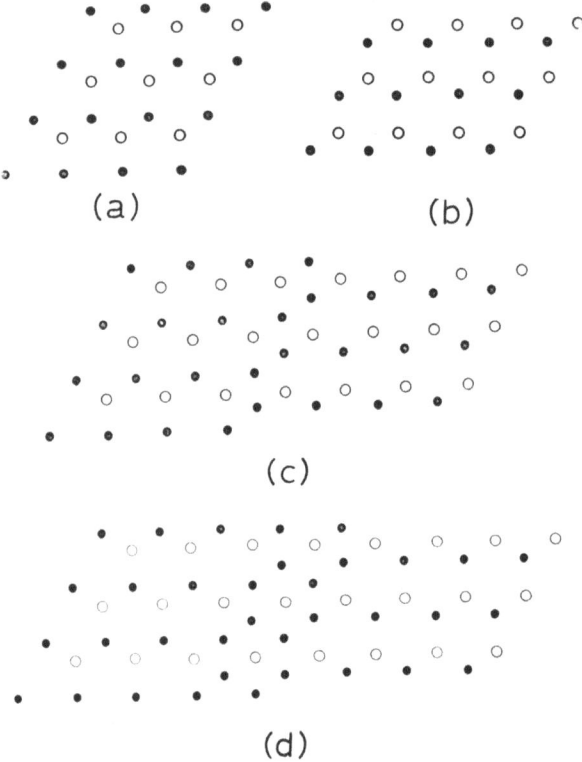

Fig. 16. Possible arrangements of alkali metal ions in the β-alumina structure. (a) and (b) show two alternative cation distributions. (c) and (d) show possible domain walls which could form between these alternatives. Each domain wall is richer in alkali metal ions than the domains themselves, and leads to an enhancement of the alkali metal composition. Alkali metals are represented by filled circles and oxygen atoms by open circles.

holding, and it is rather more likely that the domain structure of any one layer of M atoms is unrelated to the domain structure of any other layer of M atoms.

The domain structures within each layer are likely to be influenced by the thermal history of the specimen and hence the rather uncertain stoichiometries of these phases can be at least partially accounted for. The high diffusion rate of the monovalent ions in these structures would also be related to domain wall migration. Electrostatic calculations by Van Gool and Bottelberghs [63] indicate that this suggestion is not un-reasonable. However, if such domains are viable, careful annealing should eventually lead to the formation of crystals with at least some large domains, even if single domain $MAl_{11}O_{17}$ crystals are not formed. The model also implies that the domain walls are non-equilibrium structures, and again, therefore, that the stoichiometric $MAl_{11}O_{17}$ structure should be obtained, even if only rarely. So far there is no evidence for this, and further experiments would be needed to throw more light on this hypothesis. In particular, as the boundaries are closely linked with the degree of non-stoichiometry in the crystal, the total length of boundary should remain sensibly constant as the temperature changes, unless of course, a parallel change in alkali metal content is also recorded.

In this respect a paper by Le Cars, Comes, Deschamp and Théry [64] is of interest. They have studied the nature of diffuse scattering from crystals of silver β-alumina. At temperatures of 750 K this scattering is extremely diffuse, indicating that the silver ions are disordered, but as the temperature falls to 77 K the diffuse scattering condenses into more discrete spots. These are still rather broad, and do not have the same quality of sharpness as the 'normal' reflections, but nevertheless, they indicate an ordering of the silver atoms in the structure into microdomains. The results have not yielded a definitive model for the ordering within the microdomains, but they do suggest that at 77 K their size is of the order of 5 nm. This decreases to a half this value at room temperature. A few parallel results on sodium β-alumina also mentioned in this latter paper serve to indicate that this case seems more complex than that of silver, and no structural models were derived.

These results, which indicate the formation of ordered domains of silver ions at low temperatures lends support to Roth's model for the accommodation of non-stoichiometry in these materials. However, in terms of this model, any change in total domain wall length would lead to a change in stoichiometry which, experimentally, has not been observed. This difficulty can be overcome by assuming that the domain walls are rather mobile at higher temperatures, so that cooling freezes in a certain domain wall configuration rather than alters domain sizes. Other explanations may well be possible, and structurally very small microdomains may be equivalent to small 'defect clusters' of silver and oxygen atoms or fragments of domain wall. Further experimental results will be necessary to answer these questions satisfactorily.

Very little is known about the structural accommodation of non-stoichiometry in the quaternary β-aluminas. The β''-alumina structure can be prepared free of β-alumina at high temperatures, and is able to retain this structure on cooling if the reactants contain a few percent of the ions, Mg^{+2}, Co^{+2}. Ni^{+2}, Zn^{+2} or Li^{+1}. The crystal structure of a β''-alumina of idealised formula $Na_2MgAl_{10}O_{17}$ has been determined by Bettman and Peters [56]. The true formula of the material studied was found to be $Na_{1.67}Mg_{0.67}Al_{10.33}O_{17}$. However, no information concerning the structural nature of the non-stoichiometry or the phase-range, if any, of this material was included although there is some evidence to suggest that the material does not have a very appreciable composition range [58]. The curious feature of why the addition of such small amounts of Mg^{+2} favours the rhombohedral stacking of the spinel sheets is also unexplained as yet.

It is understandable that an addition of a spinel forming divalent ion such as Mg^{+2} to the β-aluminas would encourage thicker layers of spinel to form. Materials with spinel layers six oxygen layers in thickness rather than four, the β''' and β'''' modifications, have been found in the quaternary $Mg-Na-Al-O$ and related zinc and cobalt systems as mentioned previously. These phases also appear to have compositions of $Na_{1.36}Mg_{1.88}Al_{14.96}O_{25}$ (β''') and $Na_{1.67}Mg_{2.67}Al_{14.33}O_{25}$ (β''') and very little stoichiometry range [58]. The structure of the β''' phase has been determined by Bettman and Terner [57] but, as in the case of the other studies, the structure of the alkali metal containing plane is uncertain and the structural mode of accommodating the non-stoichiometry is not completely resolved.

Structurally, the β-aluminas are very closely related to the hexagonal ferrites. Although these are not layered structures, and thus a consideration of these materials falls outside the scope of this chapter, it is perhaps worth making a small digression to consider their structures in order to attempt to understand the β-alumina family in more depth.

The parent structure of the hexagonal ferrites, magnetoplumbite $PbFe_{12}O_{19}$, is very similar to that of β-alumina. The difference between the two structure types is that the open alkali metal containing layer of the β-alumina phases becomes a close packed PbO_3 layer in magnetoplumbite. It is such a mixed anion-cation sheet as is found in the perovskites, and it destroys the layer structure of the β-aluminas by extending the close-packed oxygen array of the spinel blocks throughout the whole of the crystal.

When one considers the structurally related phases produced by replacing Pb by Ba, a remarkably complex group of materials are encountered (see, e.g. [65, 66, 67, 68, 69] and references therein). A large number of different stacking sequences of the same type of spinel-like and perovskite-like layers appear to form. The unit cells of these materials, in a direction perpendicular to the spinel layers, varies from about 2.3 nm in magnetoplumbite and its isomorphs, to values of the order of 160 nm. One of the most prolific family of phases is formed by stacking units of magnetoplumbite type (designated M) and a phase identical to magnetoplumbite in all respects except that the BaO_3 perovskite sheet is now twice as thick and consists of two layers. This latter structure, of composition $Ba_2Me^{II}Fe_{12}O_{22}$, is designated as Y. Series of oxides M_pY_n exist, which fall into subdivision M_2Y_n, M_4Y_n, M_6Y_n and so on. Two typical long unit cell examples are the phases represented by the stacking $MY_6MY_{10}MY_7MY_{10}$, with $c=157.7$ nm and $M_3Y_4MY_7MYMYMY_6MY_8$, with $c=145.5$ nm. Besides this series in which the perovskite layers are doubled, another group of phases form in which the M type intergrows with a structure analogous to that in β'''-alumina, in which the spinel layer is 6 oxygen layers thick and the alkali metal sheet is replaced by a BaO_3 sheet. This structure is referred to as the W type. Only a relatively few M_nW phases are known, the largest being M_5W with a unit cell length of 22.3 nm. Many more of these phases are given in the references listed above.

There are a number of questions which arise from a consideration of these hexagonal ferrites in parallel with the aluminates, particularly with respect to the possible proliferation of phases in the $Ba-Al-O$ system or in mixed $Ba-Al-Fe-O$ oxides. These, however, fall outside the scope of this article, as they are not layered structures. However, we can consider the likelihood of such proliferation in the Fe analogues of the β-aluminas, and the β-aluminas themselves.

Considering firstly the M_pY_n series; in the β-aluminas, this would involve a juxtaposition of two Na containing sheets at intervals throughout the structure. Such an arrangement seems inherently unattractive. The Na^+ ions would be brought into close proximity and would not have the countercharge compensation of an almost close-packed layer of oxygen atoms to confer stability. Should such a juxtaposition of two alkali metal containing sheets be achieved, the material would be expected to have Na^+ ion conductivities even higher than in the known β-aluminas.

Analogues of the M_nW series should be more likely to form, as the β'''-alumina phase is the analogue of the W compound just as magnetoplumbite is the analogue of β-alumina. Therefore, there appears to be a possibility of obtaining $\beta_n\beta'''$ intergrowths structurally analogous to the M_nW series of ferrites. Whether these would be found in both the $Na-Al-O$ systems and $Na-Fe-O$ systems with equal probability would be uncertain, but experiments to check this idea would be of some interest.

It is interesting to note that both in the hexagonal ferrites and in the β-aluminas, the greatest thickness of spinel layer encountered is equal to 6 layers of close-packed oxygen. By drawing analogies with the layered perovskite derivatives discussed elsewhere in this chapter, one is able to speculate on the possibilities of forming series of compounds in which the spinel layers increase in thickness on passing from one member to another. Thus the formation of phases with 8, 10 and so on oxygen layers in the spinel slabs can be envisaged. In the systems studied to date, such structures appear to be rejected in favour of two-phase regions where spinel coexists with β-alumina phases of the type described earlier, but should they form such intergrowth structures would provide very subtle ways of altering the stoichiometries of crystals.

5. Layer Structures of Some Transition Metal Sulphides

Many of the transition metal sulphides have layer lattices, and indeed, this class of compounds frequently springs to mind when layer structures are discussed, rather than the clay minerals or some of the compounds mentioned elsewhere in this chapter. Nevertheless, the layer-like characteristics of a number of these materials, TiS_2 for example, are not so prominent as those of, say, β-alumina. This reflects the changed bonding characteristics of the sulphides, which are frequently of a metallic nature rather than ionic and results in electron delocalisation across layers of structure as well as within layers.

Structures of many of these phases will be found elsewhere in this volume, and as indicated earlier, not all of these structural types will be dealt with in this section. Instead, a group of sulphides has been chosen to illustrate the most common way in which these compounds tend to accommodate stoichiometric variation. In general, these materials, with compositions between MS_2 and MS, take up additional metal atoms or ions into the MS_2 phase.

There are two principle layer structure types which can be considered as examples, those in which the cations in the MX_2 parent lie in octahedral sites, and those in which they lie in trigonal prismatic sites. In both of these types, the structures are made of stackings of MX_2 layers in a sequence... $XMX\ XMX\ XMX$.... The anion layers are in hexagonal, cubic or mixed close packing. Interpolation of extra cations into the structure between the layers eventually leads to the sequence ... $XMXMXMX$.... As the density of these additional atoms increases, so the regions between the layers are filled up. The layer nature of the lattice is then lost, but in a gradual way, so that it is rather difficult to be precise about which compositions are best described as layered structures and which are not.

MX_2 materials in which the cations are octahedrally coordinated are often of the C6 (CdI_2) structure type or slightly distorted variants of it if the cations are relatively small compared to anions [9, 40, 70]. As the cation size increases, the anion sheets above and below them are pushed further apart and allow the cation coordination to become trigonal prismatic. The sulphides of the 4d and 5d transition metals are often of this type. The relationship between cation size, bonding and these two structure types has been discussed by Huisman et al. [71] and by Hulliger [70].

The structural changes taking place in both of these groups of compounds are very similar, and either one can be chosen as illustrative. Despite the fact that there has recently been a great deal of interest in the trigonal prismatic niobium and tantalum sulphides because of their super-conducting properties, the C6 structure type has been investigated in most detail and only this will be discussed here. Further information concerning the non-stoichiometry of materials with trigonal prismatic coordinated cations will be found in references [72, 73, 74, 75].

The C6 structure type can be described as being made up of hexagonal layers of sulphur atoms in an infinite array. The metal atoms are situated in octahedral sites and are localised to lie on alternate planes of available positions. If the unoccupied sites are also filled, the layered character is destroyed and the resulting structure is of the B8 (NiAs) type. These are illustrated in Figure 17. The transition between these two structure types has been well documented [9, 40, 76] and one would expect that a continuous transition between one extreme and the other would be possible in principle. At compositions slightly more metal rich than MX_2, the interpolated cations can be termed interstitials, while at compositions approaching MX it is more usual to regard the stoichiometric variation to be due to subtraction of atoms from the parent B8 phase, in which case they are usually termed vacancies. These expressions, however, are poor ones to apply over much of the phase range, and to phases prepared at low temperatures, where ordered structures prevail and where the occupied and empty sites are part of the structure of the phase as a whole. 'Vacancies' and 'interstitials' are then totally assimilated into the structure.

When the interpolated cations are limited to every other empty layer of the C6 parent, only a limited number of ordered permutations are possible. Some of these arrangements, applicable to the B8 end of the composition range, have been tabulated by Chevreton, [76] and are illustrated later in this section. This behaviour is found in the chromium and iron sulphides, described below, although the complete C6-B8 transformation is not found in either case. If the interpolated atoms situated on alternate planes are not completely ordered in all phases, the situation existing in the titanium sulphides is found. In this case a further complexity is introduced, because the stacking of the sulphur layers also changes from hexagonal to a complex sequence

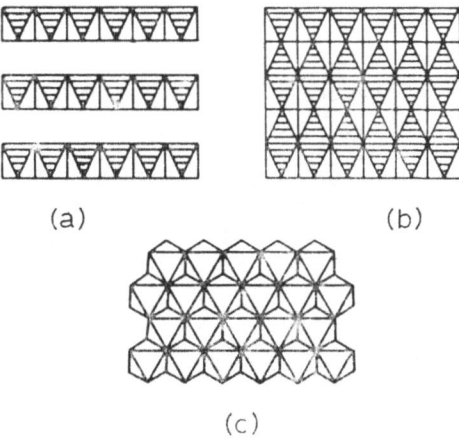

(a)   (b)

(c)

Fig. 17.   The structures of (a) TiS_2 (C6, CdI_2 type) and (b) TiS (B8, NiAs type) shown as packings of MS_6 octahedra. The hexagonal c axis is vertical. (c) The same structures viewed along [001] to show the packing of MS_6 octahedra within each sheet.

of hexagonal and cubic layers. Hence, once again the idealised C6-B8 transition is not found, although it does take place in some ternary metal titanium sulphides described below. At higher temperatures, two alternate structural possibilities occur, either the interpolated atoms can be disordered within each alternate layer, or if temperatures are high enough, the whole cation sublattice can become disordered. Such behaviour is found in the chromium and iron sulphides.

While a discussion of the titanium, chromium and iron sulphides is able to include all these major structural alternatives, and were chosen for this purpose, one further group of non-stoichiometric sulphides needs to be mentioned. These are the recently discovered organic intercalation compounds. In these materials, organic molecules, often with a basic character such as amines, are interpolated between the MX_2 layers. The main interest in these phases is in their electrical properties, as a number of these intercalates are superconducting. Very little detailed work has yet been carried out on their structures and the major structural modification caused by the change in composition is an expansion normal to the layers. In view of this paucity of structural detail, these interesting materials are also omitted from the following discussion, but reference can be made to Hooley's Chapter 1 in volume 1 for preparative details and some structural information.

TABLE II

The structure of titanium sulphides in the composition range TiS to TiS$_2$

Nominal composition	Phase	Symmetry	Unit cell dimensions (nm)		Number of sulphur layers per unit cell	Occupancy of metal sites	Homogeneity range at 1000 °C
			(a)	(c)			
TiS_2	TiS_2	trigonal, $P\bar{3}m1$	0.3408	0.56912 [1]	2	h 1.0 h 0.05 [1]	$TiO_{1.04}S_2$ — $Ti_{1.105}S_2$ [1]
$Ti_{1.156}S_2$	Ti_2S_3	rhombohedral, $R\bar{3}m$	0.342	13.77 [2]	48	(c 0.15 c 1.0 c 0.15 h 1.0 h 0.15 c 1.0 c 0.15 h 1.0 h 0.15 c 1.0 c 0.15 c 1.0 c 0.15 h 1.0 h 0.15 c 1.0)$_3$ [2]	
$Ti_{1.170}S_2$	Ti_2S_3	hexagonal, $P\bar{3}m1$	0.342	2.288 [2]	8	c 0.17 h 1.0 c 0.17 h 1.0 h 0.17 c 1.0 h 0.17 c 1.0 [2]	
$Ti_{1.198}S_2$	Ti_2S_3	hexagonal, $P6_3mc$	0.343	3.438 [2]	12	(c 1.0 c 0.20 h 1.0 h 0.20 c 1.0 h 0.17)$_2$ [2]	
$Ti_{1.6}S_2$	Ti_2S_3	hexagonal, $P\bar{3}m1$	0.342	2.87 [2]	10	c*h 1.0 c*h 1.0 h*h 1.0 h*c 1.0 h*c 1.0 (a) [2]	
$Ti_{1.25}S_2$	Ti_2S_3	hexagonal, —	0.343	3.43 [1]	12	—	$Ti_{1.325}S_2$ — $Ti_{1.34}S_2$ [1]
	Ti_5S_8	rhombohedral, $R\bar{3}m$	0.3418	3.436 [3]	12	c 1.0 c 0.2 h 1.0 h 0.2 [9]	—
$Ti_{1.33}S_2$	Ti_2S_3	hexagonal, $P6_3mc$	0.3426	1.1433 [1]	4	c 1.0 h 0.33 [1]	$Ti_{1.26}S_2$ — $Ti_{1.45}S_2$ [1]
	Ti_2S_3	hexagonal, $P6_3/mmc$ (b)	0.5936	2.2866 [4]	8	c 1.0 h 0.33 [4]	—
$Ti_{2.45}S_4$		hexagonal, $P6_3mc$	0.34198	1.1444 [5]	4	c 1.0 h 0.227 [5]	—
$Ti_{1.43}S_2$	Ti_2S_3 (c)	hexagonal, —	0.342	1.144 [6]	4	—	—
			0.342	6.88 [6]	24	—	—
			0.342	11.46 [4]	40	—	—

Table II (Continued)

Nominal composition	Phase	Symmetry	Unit cell dimensions (nm) (a)	(c)	Number of sulphur layers per unit cell	Occupancy of metal sites	Homogeneity range at 1000 °C
$Ti_{1.50}S_2$	Ti_3S_4	rhombohedral, **R3̄m**	0.3441	6.048 [7, 8]	21	c 0.95 h 0.7 c 0.7 h 0.9 c 0.5 h 1.0 h 0.5 [7]	$Ti_{1.53}S_2$–$Ti_{1.56}S_2$[9]
$Ti_{1.60}S_2$	$6Ti_4S_5$	hexagonal, P6₃/mmc	0.3439	2.893 [7, 10]	10	c 0.9 h 0.9 c 0.6 h 1.0 h 0.6 [7] c 0.875 h 0.875 c 0.65 h 1.0 h 0.65 [10]	–
$Ti_{1.71}S_2$	Ti_8S_9	rhombohedral, **R3̄m**	0.3425	2.6493 [4]	9	c 0.83 h 1.0 h 0.83 [4]	$Ti_{1.67}S_2$–$Ti_{1.78}S_2$[9]
Ti_2S_2	TiS	hexagonal, P6₃/mmc	0.3299	0.6380 [4]	2	h 1.0 h 1.0 [4]	$Ti_{1.885}S_2$–$Ti_{2.03}S_2$[9]

(a) composition of partly filled layers (*) uncertain
(b) Pseudohexagonal. The true cell is monoclinic
(c) There are a number of other phases in this range with undetermined structures [2]

References

1. Y. Jeannin, *Ann. Chim.* 7 (1962), 57.
2. E. Tronc and M. Huber, *J. Phys. Chem. Solids* 34 (1973), 2045
3. E. Flick, G. A. Wiegers, and F. Jellinek, *Rec. Trav. Chim.* 85 (1966), 869.
4. S. F. Bartram, Dissertation Abstracts 19 (1958), 1216.
5. L. J. Norby and H. F. Franzen, *J. Solid State Chem.* 2 (1970), 36.
6. E. Tronc and M. Huber, *Compt. Rend. Acad. Sci. Paris* 272C (1971), 1018.
7. G. A. Wiegers and F. Jellinek, *J. Solid State Chem.* 1 (1970), 519.
8. Y. Jeannin, *Compt. Rend. Acad. Sci. Paris* 256 (1963), 3111.
9. Y. Jacquin and Y. Jeannin, *Compt. Rend. Acad. Sci. Paris* 256 (1963), 5362.
10. O. Beckmann, H. Boller, and H. Novotny, *Monatsh. Chem.* 101 (1970), 945.

5.1. The binary titanium sulphides

TiS_2 possesses the C6 ($Cd(OH)_2$) structure and TiS the B8(NiAs) structure, and it would seem likely that in this system a complete transition from one phase to the other would be possible by the interpolation or subtraction of titanium atoms, as the case may be. However, this is not found, and the system exhibits a considerable degree of complexity. The phases found up to the present in this system are listed in Table II. The rather curious feature of these materials is that the packing of the sulphur atoms does not remain hexagonal through the whole of the phase range, but complex sequences of hexagonal and cubic close packed layers are formed. Clearly the relationship between structure and non-stoichiometry in these sulphides is not straightforward, and the extent to which the phases between TiS_2 and TiS possess truly layered structures is a moot point.

The TiS_2 phase, does, however, possess the physical properties associated with layered materials. It appears to be always a little rich in titanium compared to the ideal formula, and can be written as $Ti_{1+x}S_2$. The value of x is quite large, varying from 0.04 to 0.105. This composition range, and the conclusion that structurally it is due to interstital titanium atoms located between the TiS_2 layers, was derived from a consideration of density and lattice parameter measurements [77]. Thus, at the titanium rich end of this phase, about one in every ten possible octahedral sites in the empty sheets between the TiS_2 layers is occupied. This is quite a high proportion of filled sites and the proximity of some titanium atoms in this layer must be appreciable. Because of this one would expect that ordering may take place at some temperatures. Electron microscope observations of typical vapour grown TiS_2 crystals prepared below 1100 K occasionally showed diffuse scattering on the basal plane of the reciprocal lattice, in addition to the usual sharp spots expected [78]. These indicate that short range order is present, and this is likely to be between these extra titanium atoms. No attempt has been made to correlate the extent of the diffuse scattering with expected domain size of such ordered regions. It is, however, strongly indicative of the fact that ordering may well take place, probably at lower temperatures in this system.

According to the earlier literature, the next phase below TiS_2 is of formula close to Ti_2S_3, although Jellinek, in 1962, already noted a compound Ti_5S_8 and also lists another phase which he tentatively labelled as Ti_3S_5 [72]. In more recent reviews, the Ti_5S_8 phase is considered to be the one nearest in composition to TiS_2 [73, 79] following the structure determination by Flick, Wiegers and Jellinek [80]. This material does not have a reported composition range, and its structure is made up of a regular sequence of cubic and hexagonal sulphur layers (see Table II).

Rather more recent investigations [81, 82, 83] have indicated that many more phases exist in the composition range between $Ti_{1.105}S_2$, the lower end of the TiS_2 phase, and the Ti_2S_3 composition. A study of the literature pertaining to the Ti_2S_3 phase shows considerable variation for the composition of this material and its supposed stoichiometry range [9, 72, 73, 77, 82]. It is now apparent that these reports merely reflect the true complexity of this part of the phase diagram. Within the homogeneity range originally reported by Jeannin [77] at least 16 different structures form and maybe many more. Moreover, electron microscopy [82] has indicated that crystals prepared in this composition range are frequently severely disordered and consist of random intergrowths of various structures, many of which have not been prepared in the form of well ordered single crystals. Figure 18 shows such a crystal fragment.

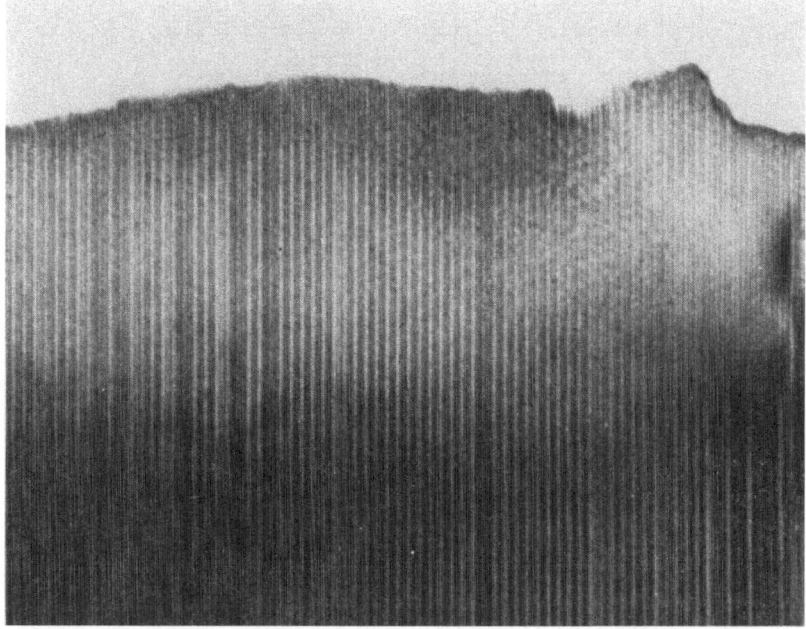

Fig. 18. Electron micrograph of a disordered titanium sulphide crystal obtained from a sample of overall composition $Ti_{1.25}S_2$. The narrowest fringes present the elementary sulphur layer spacing, 0.28 nm, while the intergrowths are more clearly revealed in thicker parts of the flake.

Figure 19 shows the structures of these phases. They are all composed of packings of sulphur layers in complex cubic or hexagonal sequences. The titanium atom sequence consists of alternately filled and partially filled sheets of octahedral sites, as in the TiS_2 structure. In all the structures determined to the present time, the titanium atoms in the partially filled layers are distributed at random over the available sites, and the degree of occupation of these sites is near to 0.25. How close this fraction can approach to 0.5 before the structure type radically alters is, as yet, uncertain.

All the structures so far determined have unique compositions which are recorded in Table 2. Tronc and Huber [83] refer to these materials as polytypes. This must be considered to be not yet proven, as polytypism implies that composition is invariant from one structural modification to another (see Chapter VI). Many other systems with non-layered structures exhibit a succession of complex phases within a very small composition range, and one can refer to the niobium oxides [14, 38, 84] the tungsten tantalum oxides [84, 85, 86] the barium ferrites, (already referred to in Section 4) and

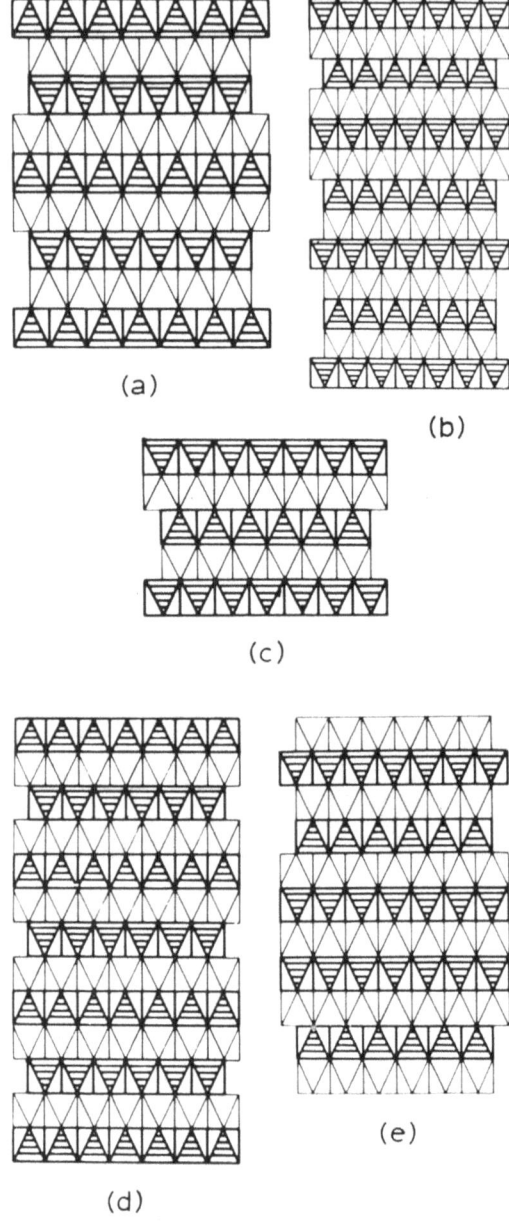

Fig. 19.   Sections through the structures of some titanium sulphides falling within the Ti_2S_3 composition range. (a) 8H $Ti_{1.17}S_2$, (b) 12H $Ti_{1.20}S_2$, (c) 4H $Ti_{1.225}S_2$, (d) 12R $Ti_{1.25}S_2$, (e) 10H $Ti_{1.25}S_2$.
The hexagonal axis is vertical in each drawing, and each structure is shown as a packing of MS_6
octahedra. The alternate layers of unfilled MS_6 octahedra are drawn unshaded.

the bismuth tellurides occurring in the composition range BiTe to $BiTe_{1.325}$ [87] for examples. Several other materials which show a similar complexity of structural types over small composition ranges are discussed by Anderson [84]. The point to be emphasized is that a complexity of structural types occurring within very narrow compositional limits, such as has been found in the Ti—S system between TiS_2 and Ti_2S_3 does not, of itself, indicate that polytypism is occurring or is the only way of labelling the complexity. Indeed, the fact that the phases so far characterised by X-rays have differing compositions supports the view that they are not polytypes in the true meaning of the term. Many polytypic forms of substances with no stoichiometric variation are known, as reference to Chapter VI and the literature cited therein will make apparent.

It is clear, therefore, that while some major features of the titanium sulphides with compositions close to $Ti_{1.25}S_2$ have been clarified, a great deal of careful investigation is now required to elucidate the finer details. One question already alluded to is whether polytypism is prevalent in these compounds, or whether a discrete set of structural types with very close compositions is found. Another question of some importance is to firmly establish the importance of temperature in these investigations. The literature cited shows clear evidence that in the titanium sulphides, temperature has an appreciable effect upon both the apparent existence ranges of the non-stoichiometric phases such as TiS_2 and the structures of the complex phases formed. This is not at all a new finding. It was stated explicitly by Wadsley in 1964 [9] who said then that a careful study of one such C6-B8 system in depth, taking composition and temperature as variables, was necessary. The Ti-S system, which has received a great deal of attention since that time, would still benefit from such a careful study.

From this point of view, two possibilities seem to present themselves. At high temperatures the homogeneity range of the TiS_2 phase might be very extensive, and the hypothetical C6-B8 transformation might exist over an appreciable part of the range between TiS_2 and Ti_2S_3. At the lowest temperatures, a sequence of ordered phases might be present, with a complex arrangement of close packed sulphur layers reflecting the composition of each structure. The phases so far reported could lie between these two extremes. Such a scheme of things naturally leaves one to consider whether the disordered partially occupied layers of titanium atoms in the phases with compositions close to $Ti_{1.25}S_2$ are totally disordered, or whether each sheet is partially ordered into domains of structure, with the partial ordering within each sheet being totally unrelated to ordering within other such fractionally occupied sheets. These questions are, of course, entirely analogous to those posed when considering the β-alumina phases. The second possibility is that the phases found so far are stable over the majority of the temperature range under which the solid exists and that the disordered nature of every alternate sheet of titanium atoms is also the typical state of these structures. Further experimental evidence is needed to clarify these issues.

When the phases with a composition between TiS and Ti_2S_3 are considered, the layer description breaks down and the phases resemble metallic alloys rather than micaceous compounds. For completeness they will be considered here, although in a strict sense they are best regarded as deriving from the non-layered TiS (NiAs type) parent rather from TiS_2.

The main difference between these phases and those described above is that the succession of layers of titanium atoms no longer falls into the repeating pattern of sheets of alternately full and fractionally filled octahedral sites. The structures of these materials are summarized in Table 2, where it can be seen that the succession of sulphur layers is again a complex sequence of hexagonal and cubic packing. The degree of occupancy of the octahedral sites between each pair of sulphur layers does not accord, at first sight, with any apparently logical sequence. This is not strictly so, as Wiegers and Jellinek [88] have shown that the occupation of the titanium layers can be regarded as possessing a wave-like periodicity with a repeat distance of an integral number of wavelengths in the repeat unit of the crystal. In the phases listed in Table II, there are 0, 1, 2 and 3 wavelengths in the compounds TiS, Ti_8S_9, Ti_4S_5 and Ti_3S_4 respectively. An exact correlation of the type of sulpur packing with the fraction of the titanium sites occupied seems not to be possible, although a fully occupied layer of titanium sites is always found between two hexagonal (h) type sulphur layers. This is not so above $Ti_{1.50}S_2$.

The discussion of the phases so far has been largely based on diffraction evidence and, despite the complexity reported in the literature, it is certain that a large number of structures exist within the $TiS-TiS_2$ region of the Ti–S phase diagram. It is, therefore, of interest to consider, in this light, the results obtained by Delmaire and Le Brusq [89] from a study of the electrical properties of these compounds. In their experiments the electrical resistance and the Seebeck coefficient of $Ti_{1+x}S_2$ samples were measured as a function of the changing composition of the sulphide phase under known vapour pressures of sulphur. The temperatures of the experiments fell in the region 700–1000°C. All the isotherms were smooth curves, and did not show sharp breaks at the crystallographically known structures. The authors conclude that the structure of the region $TiS_2 - Ti_2S_3$ is that of a solid solution with a formula $Ti_{1+x}S_2$. The diffraction results show that this is not the case. One must question, therefore, the degree to which the variation of electrical properties of these materials are structure sensitive. If they are insensitive to the stacking of the sulphur layers, then the phase range $TiS_2 - Ti_2S_3$ does indeed behave as a single phase $Ti_{1+x}S_2$ region. Indeed, if diffraction method were insensitive to such variations in sulphur packing, they would arrive at precisely this conclusion, and earlier studies using techniques with a somewhat lower resolution than modern ones, did, in fact, arrive at these same inferences.

A consideration of the processes taking place at an atomic level would suggest that electrical properties should be fairly insensitive to the changes from hexagonal to cubic packing. The use of polycrystalline samples also obscures the layer nature of these compounds. Hence, one is lead to conclude that the two sets of data are in no way contradictory, merely that they record different aspects of the materials. Similarly, it is difficult to be certain that the variation of sulphur partial pressure over these phases will reflect the sulphur layer packing. The results of electron microscopy [82] and X-ray studies [83] suggest that considerable disorder in the sulphur layer packing is a more usual state of affairs even after crystals have been annealed for considerable periods of time. In experiments in which the sulphur partial pressure is varied over the surface of crystals, and equilibrium judged by weight changes is achieved fairly rapidly, such disorder is likely to be persistent. Once again, the results support this suggestion, and once again the interpretation in terms of a solid solution is accurate, but only in the sense of it being a 'low-resolution' result.

From a structural viewpoint, the complex packing sequences ellucidated by X-ray diffraction can be derived from a hexagonal packing of anions by a process of slip which involves the introduction of stacking faults into the matrix. Mechanisms for these transformations have been described for metals because of the practical importance of the phenomena and are contained in many textbooks of metallurgy. Alternatively, the hexagonal sequence of suphur layers in TiS_2 can be altered by a collapse of the lattice to produce a crystallographic shear plane. The stacking sequence of the metal layers in the TiS_2-Ti_2S_3 phase region cannot readily be generated by crystallographic shear, but the introduction of stacking faults preserves the alternating metal layer sequence of full and partially full sites. However, whether any of the intermediate titanium sulphides can be directly produced from TiS_2 by reduction is unknown, and such a reaction, especially at low temperatures, may well produce a different sequence of structures. This consideration must also be remembered when considering the results of experiments such as those of Delmaire and Le Brusq [89] described above.

An alternative formal description of the titanium sulphide structures becomes apparent if they are compared with the zirconium sulphides. The ZrS_2 phase is isostructural with TiS_2, but the ZrS phase differs considerably and is related to the cubic rocksalt structure type. It apparently has empty sites on both the anion and cation sublattices and there is evidence that these are sometimes ordered [90]. The composition range of this material is broad enough to encompass the formal composition Zr_2S_3. It would seem likely that low temperature annealing of ZrS_x preparations within this composition range might well yield a large number of ordered phases. Thus the zirconium sulphides extend the trend seen in the titanium sulphides towards a more cubic packing of the anion layers. In some respects, therefore, the complex $Ti_{1+x}S_2$ packings can be considered to be intergrowths between TiS_2 of the C6 type and a hypothetical 'Ti_2S_3' or 'TiS' phase with cubic packing which could be considered to be analogous to the $Zr_{1+x}S_2$ counterpart.

5.2. TERNARY TITANIUM SULPHIDES

The packing sequences found in the titanium sulphides are rather surprising at first sight and cannot easily be explained. From an experimental point of view, reduction with other metal cations, to form M_xTiS_2 compositions rather than $Ti_{1+x}S_2$, might well be chosen to throw some light on the possible reasons underlying the structural complexity found. Quite an extensive literature exists on these ternary phases [70, 73, 74, 75, 79, 91] and here we will only very briefly consider two small groups of results, those in which the third element is a 3d transition metal and those in which it is an alkali metal.

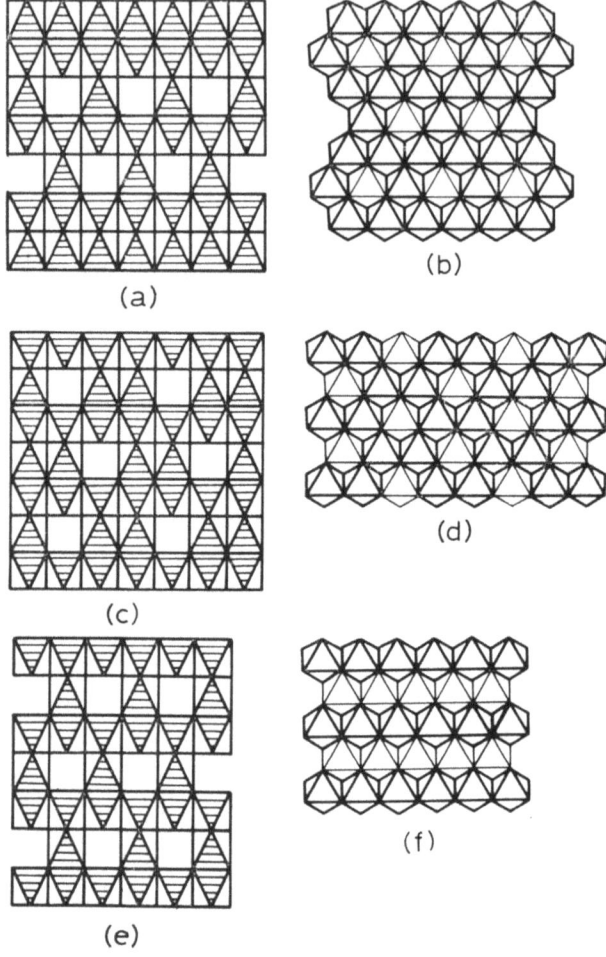

Fig. 20. Sections through the structures of the ordered chromium sulphides, shown as packings of MS_6 octahedra. (a), (b), Cr_7S_8; (c), (d), Cr_5S_6; (e), (f), Cr_3S_4; (g), (h), Cr_2S_3 (trigonal); (i) Cr_2S_3 (rhombohedral); (j) Cr_5S_8. In (a), (c), (e), (g) and (i) the hexagonal c acis is vertical and octahedra not occupied by cations are omitted. In (b), (d), (f), (h), (j), the unoccupied octahedra are drawn with light lines. The arrangement of unoccupied sites in (b) is the same as in 3C and 4C pyrrhotite, Fe_7S_8, and the arrangement in (h) applies to both forms of Cr_2S_3. The structure of Cr_5S_8 normal to the layer shown in (j) is the same as in Cr_3S_4 (e), and the structure of 3C pyrrhotite the same as in Cr_7S_8 (a).

Among the ternary compounds which contain small transition metal cations, the MTi_2S_4 phases are prominent, because of their interesting and potentially useful physical properties. Some of this information, as well as the structures of these materials has been given by Jellinek [73] and Iglesias and Steinfink [92]. In the analogous transition metal oxides AB_2O_4 the almost universally adopted structure is the spinel type. This structure is built up of a cubic close packed array of oxygen ions. The cations are ordered over a fraction of the available octahedral and tetrahedral positions in this array (cf. Section 4). The spinels themselves tend to be stoichiometric compounds

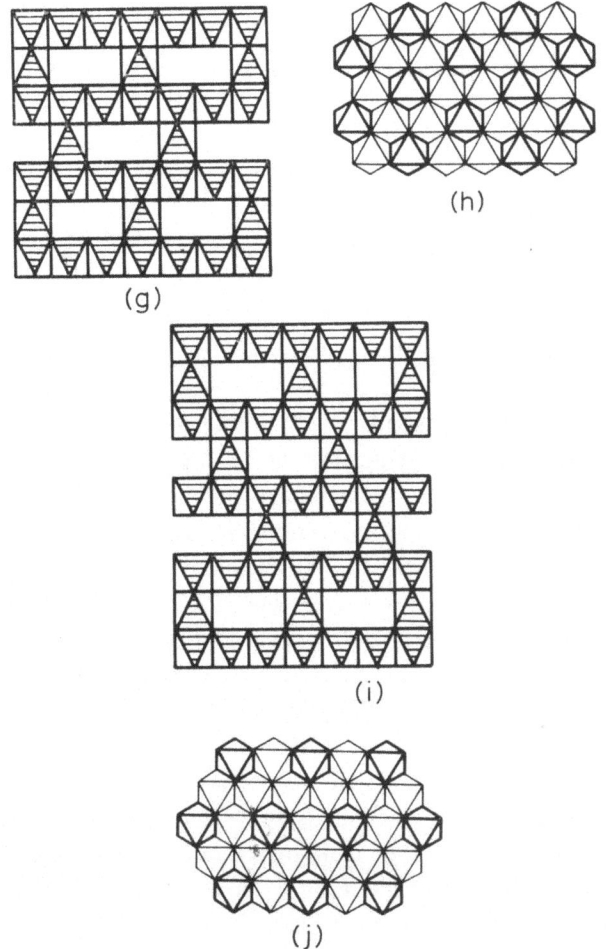

Fig. 20g–j.

with little tendency to variable composition, and the structure is adopted by binary oxides, M_3O_4 as well as ternary systems.

There seems to be no theoretical reason why this structure will not be favoured by the MTi_2S_4 compounds [92] but in point of fact the Cr_3S_4 structure type is the one preferred. This structure, shown in Figure 20, is considered in more detail in Section 5.4. At this point, we only need to note that it is composed of hexagonally close packed layers of anions, and the cations are to be found in some of the available octahedral positions and are, as in spinel, distributed in an ordered fashion. Why this structure is not found for compounds of composition close to Ti_3S_4 in the binary system is still unexplained.

From among the other ternary systems M_xTiS_2 containing a transition metal cation M, only two will be considered. Of these the Fe_xTiS_2 system has been studied in some detail because these materials are potentially useful in applications calling for ferro-magnets. The $FeTi_2S_4$ phase has the Cr_3S_4 structure, as mentioned above [93, 94, 95] and it seems that the ordered iron atoms in this structure can disorder with facility at temperatures of the order of $450\,°C$ [95]. In the rest of the composition range, x values of $\frac{1}{4}$, $\frac{1}{3}$ and $\frac{2}{3}$ have been investigated, as well as the $x = \frac{1}{2}$ sample, which corresponds to $FeTi_2S_4$. All of these materials have structures consisting of hexagonally packed sulphur layers with the iron and titanium atoms in octahedral positions [94]. The iron and titanium layers are distinct and alternate with each other. The titanium layers are complete and the iron atoms are ordered over the available octahedral sites, as required by the stoichiometry. These materials, therefore, behave in the manner expected for the C6-B8 transition, and complex alternations of cubic and hexagonal anion layers are not found.

The Ni_xTiS_2 system is very similar [96]. Within the approximate limits TiS_2 to $Ni_{0.2}TiS_2$ a phase structurally identical to the $Ti_{1+x}S_2$ compound forms. Thereafter, a series of ordered structures results, with close similarities to the iron and chromium sulphides described in the following two sections. The compositions of these phases are $Ni_{0.25}TiS_2$, $Ni_{0.33}TiS_2$, $Ni_{0.40}TiS_2$, $Ni_{0.50}TiS_2$ and $Ni_{0.75}TiS_2$. All have the hexago-nal packing of the sulphur layers preserved. The nickel atoms are to be found only in the empty sites in the TiS_2 parent, and they are generally ordered. The last four of these compounds are isomorphic with the chromium phases Cr_2S_3 (rhombohedral), Cr_2S_3 (trigonal), Cr_3S_4 and Cr_7S_8 (ordered), which are described in the following section. Once again, the complexity of the binary system has been extinguished and the ternary system behaves in a rather similar way to that expected for the C6-B8 transition.

Besides these transition metal compounds, some studies are also found in the litera-ture which deal with the insertion of alkali metals into the TiS_2 structure. They differ from the previous studies in that the reaction temperatures are low, as reactions are frequently carried out in liquid ammonia. The alkali metals appear to be interpolated between the TiS_2 layers of the parent compound. These compounds fall into two groups, depending upon ionic size [97, 98]. For the small Li^+ ion, a continuous range of compositions from TiS_2 to $LiTiS_2$ is found for the phase Li_xTiS_2. The lithium ions occupy octahedral sites in the empty sheets between the TiS_2 layers to present an example of a perfect C6-B8 transformation in which ordered intermediate phases are not found.

The larger alkali metal ions, potassium, rubidium and caesium are topologically similar and structures are found with wide ranges of stoichiometric variability; from $K_{0.28}TiS_2$ to $KTiS_2$, $Rb_{0.42}TiS_2$ to $RbTiS_2$ and $Cs_{0.56}TiS_2$ to $CsTiS_2$. They differ from the case of lithium only in that the alkali metal ions are now in trigonal prismatic coordination. As all of these compounds can be prepared from single crystals of TiS_2, to provide trigonal prismatic sites the individual TiS_2 layers must slip one anion position as they are forced apart. This results in a certain degree of lattice disorder which is revealed in the X-ray photographs. In the case of Na_xTiS_2, the situation lies between that of Li_xTiS_2 and K_xTiS_2. The trigonal prismatic coordination of the $Na_{0.38}TiS_2$ to $Na_{0.72}TiS_2$ compositions gives way to octahedral coordination for the highest sodium content phases, $Na_{0.79}TiS_2 - NaTiS_2$. Thus, $NaTiS_2$ has the same B8 type of structure as $LiTiS_2$ with an ordered arrangement where Na and Ti layers alternate.

In these alkali metal rich phases, there does not appear to be an ordering of alkali metal ions and a series of ordered phases does not seem to form. At lower compositions, two other structure types are able to form. In these, the layers of alkali metal, although remaining full, are separated by slabs of unchanged TiS_2 structure. This suggests that once one $TiS_2 - TiS_2$ pair has opened sufficiently to admit some alkali metal atoms, the layer is filled in preference to a partial occupancy of all such layers. However, it does not explain how these layers are able to order, as they apparently do.

Only the briefest details have been given of these systems, and there are other features of interest which can be found in the original literature. Nevertheless, they serve to illustrate the fact that the C6-B8 transition is observed in principle for these two widely differing groups of interpolated cations. The structural influence of the interpolated cations is small, and size appears to dominate any changes in the hexagonal layer sequence of the sulphur atoms. The ordering found in the iron and nickel ternary systems, and also found in the binary chromium and iron systems discussed below indicate that, as expected, co-operative d electron interactions have a role to play in this phenomenon, and it is perhaps not so surprising that the titanium ions, with fewest d electrons, are disordered. However, the complex way in which the sulphur packing in the binary titanium-sulphur system varies with stoichiometry is still unexplained.

5.3. THE BINARY CHROMIUM SULPHIDES

The various phases formed in the chromium-sulphur system can be regarded as almost filled $M_{1+x}X_2$ structures of the C6-B8 transition, and consist of hexagonally close packed anion layers with alternate layers of completely and partly filled octahedral metal atom sites. The chromium atoms in these sites are ordered, at least at some temperatures. In contrast to the Ti-S system, the highest sulphide formed is the Cr_2S_3 phase. Between this and the CrS composition four structures are found. The classic structural study on these compounds was that of Jellinek [99] and we will discuss these results here. Full details of references to other studies of the chromium sulphides prior to 1972 will be found in the reviews by Jellinek [72, 73, 75], Flahaut [79] and Hulliger [70]. Many of these earlier reports are concerned with the electrical and magnetic properties of these materials which are not only of academic interest but also of great industrial potential. Studies since 1972 have also mainly been concerned with these physical properties and the fewer structural investigations have only served to confirm Jellinek's original conclusions. Two of these papers [100, 101] are of note, however, and the results contained therein will also be discussed.

Jellinek's samples [99] were cooled slowly from a preparation temperature of 1000°C with the exception of the CrS phase, which was also annealed at 300°C in order to improve the reproducability of the results. X-ray diagrams showed the phases CrS, Cr_7S_8, Cr_5S_6, Cr_3S_4, (trigonal) and Cr_2S_3 (rhombohedral) to have formed. At compositions falling between those of these compounds, two phases were noted. All structures have narrow stoichiometry ranges, with both the Cr_2S_3 modifications and Cr_5S_6 being effectively line phases. Table III includes structural data and the composition ranges of these materials. In all phases except Cr_7S_8 the distribution of the chromium atoms in the partly filled layers was found to be ordered. In the Cr_7S_8 material the chromium atoms in these layers were found to be distributed in a random fashion over the available positions. However, Jellinek [99] pointed out that ordering was slow in those phases which are rather more rich in chromium, and longer annealing times or alternative preparative conditions might lead to an ordered structure. This has been confirmed by Popma and van Bruggen [102]. They have reported in detail upon the CrS and Cr_7S_8 phases, and also include results on the Cr_5S_6 phase. The Cr_7S_8 structure is typical of their findings. Samples annealed at low temperatures, below about 500 K, show an ordered structure not reported by Jellinek, with a complete ordering of the chromium atoms in every alternate layer, similar to the other Cr-S phases. The structure is of the Fe_7Se_8 type. Above 590 K the ordering of the chromium atoms in the partially filled layers is destroyed, but the sequence of full and partly full layers of the C6-B8 type is maintained and this is the structure originally found by Jellinek. These ordered structures are shown in Figure 20, as a packing of CrS_6 octahedra. Recently the phase Cr_5S_8 has also been prepared under high sulphur vapour pressures [103]. This material differs from the Ti_5S_8 phase in that the anion layers remain hexagonal. The cations are arranged in an ordered fashion such that every alternate layer of cation sites is completely occupied, as shown in Figure 20. Thus it is structurally analogous to the lower chromium sulphides.

To a first approximation Jellinek's results indicate that all phases are line phases. In contrast to this, two papers [100, 101] which report the results of equilibrium sulphur partial pressure measurements over these sulphides suggest that the phases have rather broader stoichiometry ranges. The results of Young, Smeltzer and Kirkaldy [101] were obtained at 700°C. They find all the phases that Jellinek reports, but find rather large phase ranges for Cr_7S_8 and Cr_3S_4. In a similar study Igaki, Ohashi and Mikami [100, 104] have suggested that the phases described by Jellinek [99] do not occur, and that instead a broad composition range exists for CrS_x. They do indicate that there are two phases with a boundary at $CrS_{1.463}$, that is, the phase boundary of the trigonal and rhombohedral 'Cr_2S_3' phases of Jellinek, and also that there are two separate phase regions below this composition with a phase gap between $CrS_{1.390}$ and $CrS_{1.420}$. These experiments were all carried out by studying the sulphur vapour pressure equilibrium over the suphide samples at temperatures in excess of 1270 K.

Here we see a similar pattern to that in the titanium sulphides, where diffraction evidence points to a succession of ordered phases, while studies of partial pressure and physical properties present a less complex picture. As in the previous discussion of this problem the apparent inconsistency of these results can be reconciled if one recalls that temperature is as important an experimental parameter as composition [9]. The Fe-S system (vide infra) emphasises this in a particularly striking way. Thus one would expect a transition from ordered structures at low temperatures to disordered B8

TABLE III

The structures of chromium sulphides in the composition range above CrS

Phase	Symmetry	Unit cell dimensions (nm)			β	Ref.	Homogeneity range		
		a	b	c			Jellinek [1]	Igaki [3]	Young [4]
CrS	monoclinic, C2/c	0.3826	0.5913	0.6089	101.6°	[1]	narrow	–	–
Cr_7S_8	trigonal, P3̄m1	0.3460		0.5762		[1]	1.136–1.149	1.200–1.390	1.130–1.142
Cr_5S_6	trigonal, P31c	0.5982		1.1509		[1]	narrow		1.190–1.203
Cr_3S_4	monoclinic	0.5964	0.3428	1.1272	91.5°	[1]	1.265–1.316		1.286–1.377
Cr_2S_3	trigonal, P31c	0.5941		1.180		[1]	narrow	1.420–1.463	1.423–1.463
Cr_2S_3	rhombohedral, R3̄	0.5938		1.6675		[1]	narrow	1.463–1.480	1.463–1.500
Cr_5S_8	monoclinic, P2/m	1.1783	0.6789	1.1063	90.82°	[2]	–	–	–

References

1. F. Jellinek, *Acta Crystallogr.* **10** (1957), 620.
2. A. W. Sleight and T. A. Bither, *Inorg. Chem.* **8** (1969), 566.
3. K. Igaki, N. Ohashi, and M. Mikami, *J. Phys. Soc. Japan* **31** (1971), 1424.
4. D. J. Young, W. W. Smeltzer, and J. S. Kirkaldy, *J. Electrochem. Soc.* **120** (1973), 1221.

structures at high temperatures. In the titanium sulphur system the completely disordered B8 structures have not been observed and must form only at very high temperatures, perhaps above the decomposition temperatures of the phases themselves. In the iron sulphides, the disordering temperature is relatively low, between 400 and 600 K, depending upon composition. The chromium sulphides would then be anticipated to fall between these two extremes. The partial study of Popma and van Bruggen [102] published some time before the studies of Igaki *et al.* [100, 104] and Young *et al.* [101] is in complete agreement with this suggestion. As mentioned above, the Cr_7S_8 structure partially disorders at 590 K. Above 800 K even this vestigial order is lost, and the structure becomes the B8 type with a random distribution of chromium atoms over all the available octahedral sites, some of which are necessarily not occupied. The B8 structure could be retained to room temperature by quenching. The CrS and Cr_5S_6 phases behave similarly, with a B8 type of structure existing above about 800 K, and for Cr_5S_6, an intermediate disordered C6-B8 type of phase occurring between approximately 600 and 800 K.

The results described above which credit wide stoichiometry ranges to the chromium sulphides at high temperatures are therefore concerned with these disordered structures and their conclusions do not apply to the low temperature ordered phases.

5.4. THE BINARY IRON SULPHIDES

In moving two places to the right in the 3d transition metal series, from titanium to chromium, the most notable changes found were the restricted composition range of the chromium sulphides of the B8-C6 type, and the relatively simpler ordering patterns found in the structures. A move a further two places to the right to iron apparently continues this trend. If experiments are carried out to synthesise iron sulphides from the component elements, using the usual sealed ampoule techniques, B8-C6 type phases only exist between the approximate compositions of Fe_2S_2 and $Fe_{1.75}S_2$. The B8-C6 phase range is thus becoming narrower as the number of d electrons on the metal atoms increases. Above $Fe_{1.75}S_2$ an FeS_2 phase of the pyrite structure is in equilibrium with lower sulphides. This material does not have a layer structure, and will not be considered further here. The other modification of FeS_2 which possesses the marcasite structure is also not a layered structure. Hence the crystal chemistry of the iron-sulphur system has departed considerably from the starting point of the titanium sulphides.

The lower end of the $FeS-FeS_2$ region is of the greatest interest from a crystal-chemical viewpoint within the context of this chapter. Although the layer structure is almost completely lost, the formal analogy with the B8-C6 transformation still exists, and for that reason this region will be described. A brief survey of the literature reveals that the $Fe_2S_2-Fe_{1.75}S_2$ region of the phase diagram is extremely complex, with a large number of structures occurring within a small composition range [70, 73, 79, 105]. Stoichiometric FeS itself has the B8 NiAs structure above about 140°C and in general this structure is retained over a composition range from $Fe_{1.8}S_2$ to Fe_2S_2 at temperatures of the order of 300°C. At higher temperatures this composition range would appear to be wider still [105]. In it, the non-stoichiometry seems to be accommodated structurally by total disorder among the octahedral positions occupied by the Fe atoms. This would agree with the other systems of this type we have discussed. However, whether these filled sites are distributed completely at random or not is open to some speculation. Certainly microdomains of ordered material which themselves are disordered may exist at some temperatures.

At temperatures below about 300°C, a large number of structures are found which are derived from the FeS (B8) parent and are generally regarded as structures containing Fe atoms in ordered arrays. The crystal structures of only two of these have been refined carefully FeS, (troilite) [106] and Fe_7S_8 (4C pyrrhotite) [107] and these recent determinations confirm the orginal structures described by Bertaut in the early 1950's [108, 109]. The B8 FeS structure becomes rather distorted at temperatures below 140°C and the c axis is doubled to form troilite. This is due to the iron atoms moving together in groups of three in the (001) plane and is accompanied by small displacements of the sulphur atoms. The transformation is martensitic in type and diffusion of atoms is not required as the transition temperature is passed. Structurally these changes are very small, but the transformation is accompanied by changes in the electrical and magnetic properties of the FeS phase, and it has been studied considerably from that aspect [73, 105]. The troilite phase is stoichiometric or very close to it.

The other structure to be carefully refined is that of pyrrhotite, Fe_7S_8. This is also derived from the B8 structure, and is closely related to the chromium sulphides described earlier. It consists of hexagonally close-packed sulphur layers, with iron atoms in octahedral sites (Figure 22). Every alternate layer of Fe atoms is completely full, the remaining set of layers is not full, and the occupied sites are ordered in the way shown in Figure 21. The true cell is monoclinic, and the cell dimensions are multiples of the NiAs lattice parameters A and C given by $a = 2\sqrt{3}A$, $b = 2A$, $c = 4C$ and $\beta = 90.3°$ This is often referred to as the 4C phase in the literature of the iron sulphides, the troilite then being called the 2C phase ($a = \sqrt{3}A$, $c = 2C$) while the NiAs high temperature form is often known as the 1C modification. At temperatures of above 350°C the 4C pyrrhotite structure also reverts to the NiAs type of disordered structure [110] and then closely resembles the analogous (disordered) Cr_7S_8 compound.

Thus, to recapitulate, at the two extremes of composition FeS and Fe_7S_8 we have at temperatures below 140°C (FeS) to 310°C (Fe_7S_8) ordered structures related to the NiAs (B8) parent by ordering of the iron atoms, while above these temperatures the ordering is lost and the B8 structure alone persists. Between these limits of temperature and composition, a very large number of other structures exist. They are all derived from the B8 parent by ordering of the iron atoms and the structures formed appear to be extremely sensitive both to temperature of formation and to composition. To make matters more complex, the literature also indicates that the superstructures of the B8 type found may also critically depend upon small amounts of impurities and upon the thermal history of the samples (see, for instance, the references cited in [105]). It is hardly surprising, therefore, that the recent literature of the Fe-S system in the pyrrhotite region abounds with reports of new superstructures.

The study of Chevreton *et al.* [111] on the phase range $FeS-Fe_{0.85}S$ ($Fe_{1.70}S_2$) is of interest, because each different composition they reported for their preparations yielded different hexagonal superstructures derived from the B8 type. Some of these axes are very long, particularly in the *c* direction, where cell lengths of up to 27 times the B8 values were recorded. They note that the different structures formed depended upon the rate of sample cooling and the time of reaction as well as the method of sample preparation.

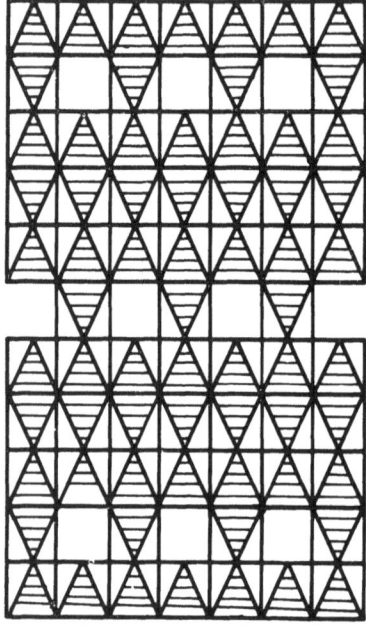

Fig. 21.   The structure of 4C pyrrhotite shown as a packing of FeS_6 octahedra. The octahedra unoccupied by cations are omitted. The arrangement of filled octahedra within the partially full layers is shown in Figure 20b.

Morimoto and co-workers have, for a number of years, made careful studies on the $FeS-Fe_7S_8$ phase range, and have described in most detail the phases characterised to date [112, 113, 114]. Their studies have been mainly single crystal studies of both synthetic and natural pyrrhotites. All phases discovered can be regarded as superstructures of the B8 type. A series formula has been allocated to a number of these structures, designated by $Fe_{n-1}S_n$ with *n* taking integral values of up to 8. These phases appear to be stoichiometric, and to have no large composition ranges.

Besides these structures, there is also found a series of superlattice phases which have a continuous variation of the *c* parameter with composition. The reflections on the X-ray photographs defining the *c* repeat are found to be diffuse, but if the maxima of these diffuse spots are determined, a smooth variation with changing composition is found. The reciprocal lattice parameters in the *c* direction are given by $1 \pm \delta l$, where δl is positive and non integral, and varies continuously from $\frac{1}{3}$ to $\frac{1}{2}$ as the composition changes from approximately $Fe_{0.95}S$ ($Fe_{1.90}S_2$) to $Fe_{0.90}S$ ($Fe_{1.8}S_2$). The *c* parameter therefore varies continuously between the limits 3C to 6C. These continuously varying structures can be quenched to room temperature. It would appear reasonable to assume that given sufficient annealing time, a series of ordered and integral c values would result, but with very long c repeat units. One study of Morimoto et al. [114] has indeed used long annealing times. After annealing for two months at 600°C the samples were cooled over a further two months to room temperature, and then stored for a year before examination. In these samples non integral c repeat units were still found. However, such annealing times are still extremely short when one bears in mind geological annealing times. It is therefore not surprising that geological samples of pyrrhotites, which frequently contain a variety of impurities, do show a confused and complex pattern of behaviour.

The variable non-integral c parameter is not the sole feature of this type in these phases. For some time now a variety of pyrrhotite with a hexagonal cell and a repeat in the c direction of 3C (where C is equal to the B8 type of repeat distance) has been known. Its structure has been refined by Fleet [107] and is shown in Figure 20 (a), (b). It has been found by Morimoto [112, 114] that a continuous variation in the a parameter of this phase can occur so that structures with cell dimensions $a = NA$, $c = 3C$ can be observed. In these materials N can take continuous values from 40 to 90.

Finally, another long and apparently continuously variable super-structure has been found for materials with compositions close to $Fe_{1.76}S_2$ at temperatures close to 300°C. In these, the cell dimensions are expressed as $a = 2A$ and $c = MC$. M can take all values between 3.0 and 4.0 continuously, but the patterns of the subsidiary maxima are somewhat different from those of the NC type described earlier and therefore imply that significantly different structural types are present. A phase diagram showing the occurrence of these modifications is shown in Figure 22.

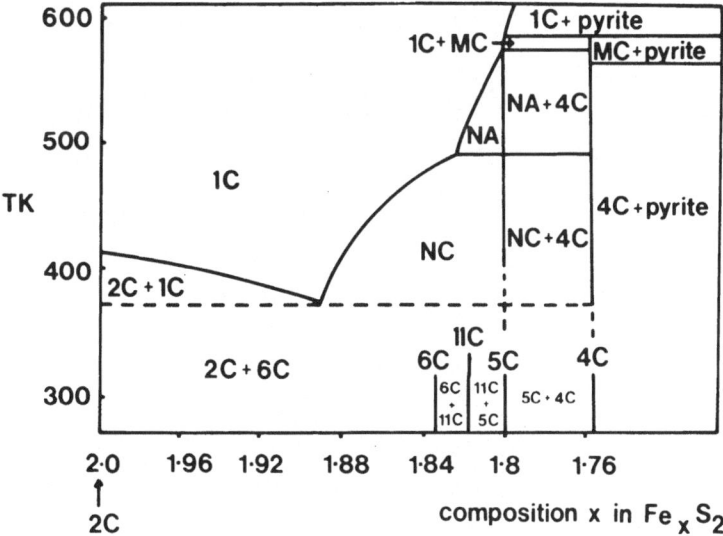

Fig. 22. Phase diagram of the iron-sulphur system in the region of the pyrrhotite phases (after Nakazawa and Morimoto [110]).

The comments made above concerning the NC type of phases are also germain here. However, until more structural studies are available it will only be possible to speculate on the structural types found. One feature of recent studies which seems to emerge from the literature is that the ordered phases appear to be stoichiometric, and that each prepared composition is either monophasic or else disordered, implying that the stable structural modifications were unable to form under the experimental conditions employed. The structures of all these phases are unknown although they are probably based upon a B8 type of subcell. Nevertheless, Chevreton *et al.* [111] have stated that it is uncertain if the sulphur stacking is always hexagonal, and it is possible that in some of these structures, an alternation of cubic and hexagonal stacking of sulphur layers may be found as in the Ti-S system. Similarly, the structural sequence of the iron atoms in which full and partly full layers of octahedral sites alternate may break down in some regions of the phase diagram. The existence of the mineral smythite is an indication of this. This material has the composition Fe_3S_4, and its layered structure is revealed by its basal cleavage, which is perfect. The structure of this phase differs from that expected after consideration of the chromium and titanium sulphides in that the succession of iron layers is not an alternation of full and partly full layers. Instead three full layers of iron sites alternate with one completely empty layer [115]. Thus the

Fig. 23.   The structure of smythite, Fe_3S_4, shown as a packing of FeS_6 octahedra. The empty sheets of octahedra are shown unshaded.

structure can be regarded of slabs of B8 structure, four sulphur layers thick, separated by an empty layer of octahedral metal sites of the sort found in the C6 structure type. The phase cannot be considered as an ordered intergrowth of C6 and B8 structures because each B8 unit is displaced with respect to the ones on either side of it to form a cubic sequence, as shown in Figure 23. It is possible, therefore, that some of the structural complexity of the pyrrhotite phases may reflect the occurrence of structural units other than the B8 type, although the existence of smythite itself is unlikely, as it has only been synthesised from solution in the laboratory, [105].

This brief survey of the iron sulphur system close to FeS concludes this section as further progression across the periodic table results in the formation of non-layered structure derivatives. It is necessary only to emphasise once again that the chromium and iron systems have been treated with extreme brevity, and many of the questions posed in the discussion of the titanium sulphides apply to these other phases as well. They should be borne in mind in these latter sections although they have usually not been explicitly restated.

6. Conclusions

Despite the fact that this chapter must be regarded as incomplete in its coverage it is hoped that sufficient information has been presented to allow a general picture of the inter-relationship of non-stoichiometry and structure in these compounds to be drawn. One conclusion which can be reached from the systems presented here is that an un-critical interpretation of non-stoichiometry in terms of point defect populations is likely to be far from the truth. The structures obtained as the stoichiometry varies are invariably far more complex than that suggested by simply introducing hypothetical vacancies or interstitial atoms. Stoichiometric change is accommodated by much larger volume elements than the point defect concept implies. These spatially larger chemical defects can be planar faults or small volumes of distorted crystal matrix, 'defect clusters'. In some systems, even these defects are unnecessary, and a change of ordering of one or other of the atomic species present takes place when the composition changes slightly, to yield a set of structurally related but ordered compounds.

The second conclusion to be drawn from the evidence presented is that the observable evidence for non-stoichiometric behaviour must be treated with caution. The resolution limits of each technique used should be carefully considered in interpretation of data. Evidence from two different techniques, or taken under widely differing conditions should only be equated if there is sufficient reason for doing so. For example data collected at high temperatures should not be regarded as reflecting structures found at room temperature unless evidence suggests that this extrapolation is justified.

Finally, it can be stated that far more experimental work is necessary on these layered structure materials before the structural aspects of non-stoichiometric behaviour can be understood, or incorporated into a complete theory of non-stoichiometry. It is hoped that this account has indicated some ways in which this knowledge is still incomplete.

References

1. J. S. Anderson, in *The Chemistry of Extended Defects in Non-Stoichiometric Solids*, (ed. by L. Eyring and M. O'Keefe), North-Holland, Amsterdam, 1970, p. 1.
2. J. S. Anderson, in *Natl. Bur. Standards Special Publ.* **364**, 'Solid State Chemistry', (ed. by R. S. Roth and J. S. Schneider), Nat. Bur. Stand., Washington, 1972, p. 295.
3. B. E. F. Fender, in M.T.P. Int. Rev. Sci., Series 1, *Inorg. Chem*, vol. 10, Butterworths, Oxford, 1972 p. 243.
4. L. A. Bursill and B. G. Hyde, in *Prog. in Solid State Chem.*, vol. 7 (ed. by H. A. Reiss and J. O. McCaldin), Pergamon, Oxford, 1972.
5. R. R. Merritt and B. G. Hyde, *Phil Trans.* **274** (1973), 627.
6. W. Schottky and C. Wagner, *Z. phys. Chem.* **B11** (1930), 163.
7. J. S. Anderson, *Proc. Roy. Soc. A* **185** (1945), 69.
8. F. A. Kröger, *The Chemistry of Imperfect Crystals*, North-Holland, Amsterdam, 1964.
9. A. D. Wadsley, in *Non-Stoichiometric Compounds* (ed. by L. Mandelcorn), Academic Press, New York, 1963, p. 98.
10. A. D. Wadsley and S. Andersson, in *Perspectives in Structural Chemistry*, vol. 3 (ed. by J. D. Dunitz and J. A. Ibers), Wiley, New York, 1970, p. 1.
11. B. T. M. Willis, in *The Chemistry of Extended Defects in Non-Metallic Solids* (ed. by L. Eyring and M. O'Keeffe), North-Holland, Amsterdam, 1970, p. 272.
12. B. T. M. Willis and J. Williams, in *Structural Characteristics of Materials*, (ed. by H. M. Finniston), Elsevier, Amsterdam, 1971, p. 290.
13. J. G. Allpress and J. V. Sanders, *J. Appl. Cryst.* **6** (1973), 165.
14. J. S. Anderson and R. J. D. Tilley, in *Surface and Defect Properties of Solids*, vol. 3 (ed. by M. W. Roberts and J. M. Thomas), The Chemical Society, London, 1974, p. 1.
15. J. S. Anderson, in *Problems in Non-Stoichiometry* (ed. by A. Rabenau), North-Holland, Amsterdam, 1970, p. 1.
16. F. B. Koch and M. E. Fine, *J. Appl. Phys.* **38** (1966), 1470.
17. B. T. M. Willis, *Proc. Roy. Soc. A* **274** (1963), 134; *J. Phys.* **25** (1964), 431.
18. P. Caro, *J. Solid State Chem.* **6** (1973), 396.
19. L. Kihlborg, *Arkiv Kemi* **21** (1963), 471.
20. J. W. Pierce, H. L. McKinzie, M. Vlasse, and A. Wold, *J. Solid State Chem.* **1** (1970), 332.
21. O. Glemser and G. Lutz, *Z. anorg. allgem. Chem.* **264** (1951), 17; O. Glemser, U. Hauschild and G. Lutz, *Z. anorg. allgem. Chem.* **269** (1952), 93; O. Glemser, G. Lutz and G. Meyer, *Z. anorg. allgem. Chem.* **285** (1956), 173.
22. K. A. Wilhelmi, *Acta Chem. Scand.* **23** (1969), 419.
23. L. A. Bursill, *Proc. Roy. Soc. A* **311** (1969), 267.
24. L. A. Bursill, *Acta Crystallogr.* **A28** (1972), 187.
25. B. M. Gatehouse and P. Leverett, *J. Solid State Chem.* **1** (1970), 484.
26. W. G. Mumme and J. A. Watts, *J. Solid State Chem.* **2** (1970), 16.
27. R. J. D. Tilley and B. G. Hyde, *J. Phys. Chem. Solids* **31**, (1970), 1613.
28. S. Andersson, J. Galy and K. A. Wilhelmi, *Acta Chem Scand.* **24** (1970), 1473.
29. K. A. Wilhelmi and K. Waltersson, *Acta Chem. Scand.* **24** (1970), 3409.
30. K. A. Wilhelmi, K. Waltersson and L. Kihlborg, *Acta Chem. Scand.* **25** (1971), 2675.
31. M. L. F. Bayard, T. G. Reynolds, M. Vlasse, H. L. McKinzie, R. J. Arnott, and A. Wold, *J. Solid State Chem.* **3** (1971), 484.
32. P. Hagenmuller, in *Prog. in Solid State Chem.* vol. 5 (ed. by H. Reiss), Pergamon, Oxford, 1971, p. 95.
33. J. Galy, A. Casalot, M. Pouchard and P. Hagen Muller, *Compt. Rend. Acad. Sci.* **262C** (1966), 1055.
34. J. Galy and A. Carpy, *Compt. Rend. Acad. Sci.* **268C** (1969), 2195.
35. A. Carpy, A. Casalot, M. Pouchard, J. Galy and P. Hagenmuller, *J. Solid State Chem.* **5** (1972), 229.
36. S. Andersson, *Acta Chem. Scand.* **19** (1965), 1371.
37. J. Galy, D. Lavaud, A. Casalot, and P. Hagenmuller, *J. Solid State Chem.* **2** (1970), 531.
38. J. S. Anderson, in *Surface and Defect Properties of Solids*, vol. 1, (ed. by M. W. Roberts and J. M. Thomas), The Chemical Society, London, 1972, p. 1; R. J. D. Tilley, in M.T.P. Int. Rev.

Sci., Series 1, *Inorg. Chem*, vol. 10 (ed. by L.E.J. Roberts), Butterworths, London, 1972, p. 279.

39. M. Poulain, M. Poulain, and J. Lucas, *J. Solid State Chem.* **8** (1973), 132.
40. A. F. Wells, *Structural Inorganic Chemistry*, Oxford, London, 3rd ed., 1962.
41. G. M. Clark, *The Structures of Non-Molecular Solids*, Applied Science, London, 1972, p. 146.
42. A. Tressaud, J. Portier, R. de Pape and P. Hagenmuller, *J. Solid State Chem.* **2** (1970), 269.
43. S. N. Ruddlesden and P. Popper, *Acta Crystallogr.* **10** (1957), 538; **11** (1958), 54.
44. E. T. Keve and C. Skapski, *J. Solid State Chem.* **8** (1973), 159.
45. S. Andersson and J. Galy, *Acta Crystallogr.* **B25** (1969), 847.
46. J. K. Brandon and H. D. Megaw, *Phil. Mag.* **21** (1970), 189.
47. A. Carpy, P. Amestoy, and J. Galy, *Compt. Rend. Acad. Sci.* **275C** (1972), 833.
48. B. Krebs, *Acta Crystallogr.* **B28** (1972), 2222.
49. J. R. Günter, *J. Solid State Chem.* **5** (1972), 354.
50. M. L. Freedman, *J. Amer. Chem. Soc.* **81** (1959), 3834.
51. J. T. Kummer, in *Prog. in Solid State Chem.*, vol. 7 (ed. by J. O. McCaldin and H. O. Reiss), Pergamon, p. 141.
52. R. C. DeVries and W. L. Roth, *J. Am. Ceram. Soc.* **52** (1969), 364.
53. W. L. Bragg, C. Gottfried, and J. West, *Z. Kristallogr.* **77** (1931) 255.
54. C. A. Beevers and S. Brohult, *Z. Kristallogr.* **95** (1936), 472; C. A. Beevers and M. A. S. Ross, *Z. Kristallogr.* **97** (1937), 59.
55. C. R. Peters, M. Bettman, J. W. Moore, and M. D. Glick, *Acta Crystallogr.* **B27** (1971), 1826.
56. M. Bettman and C. R. Peters, *J. Phys. Chem.* **73** (1969), 1774.
57. M. Bettman and L. L. Terner, *Inorg. Chem.* **10** (1971), 1442.
58. N. Weber and A. F. Venero, Ford Motor Co., publication preprint, 1969.
59. Y. Le Cars, J. Théry, and R. Collongues, *Compt. Rend. Acad. Sci. Paris* **274C** (1972), 4; *Rev. Int. Hautes Tempér. et Refract.* **9** (1972), 153.
60. M. Harata, *Mat. Res. Bull.* **6** (1971), 461.
61. J. P. Boilot, J. Théry and R. Collongues, *Mat. Res. Bull.* **8** (1973), 1143.
62. W. L. Roth, *J. Solid State Chem.* **4** (1972), 60.
63. W. van Gool and P. H. Bottelberghs, *J. Solid State Chem.* **7** (1973), 59.
64. Y. Le Cars, R. Comes, L. Deschamps, and J. Théry, *Acta Crystallogr.* **A30** (1974), 305.
65. P. B. Braun, *Phil. Res. Repts.* **12** (1957), 491.
66. J. A. Kohn and D. W. Eckart, *Z. Kristallogr.* **119** (1964), 454.
67. J. A. Kohn and D. W. Eckart, *Am. Mineral.* **50** (1965), 1371.
68. J. A. Kohn, D. W. Eckart, and C. F. Cook, *Science* **172** (1971), 519.
69. J. van Landuyt, S. Amelinckx, J. A. Kohn, and D. W. Eckart, *Mat. Res. Bull.* **8** (1973), 339, 1173; *J. Solid State Chem.* **9** (1974), 103.
70. F. Hulliger, *Structure and Bonding* **4** (1968), 83.
71. R. Huisman, R. De Jonge, C. Haas and F. Jellinek, *J. Solid State Chem.* **3** (1971), 56.
72. F. Jellinek, *Arkiv. Kemi* **20** (1962), 447.
73. F. Jellinek, in M.T.P. Int. Rev. Sci., *Inorg. Chem.*, Series 1 vol. 5, Butterworths, London, 1972, chapter 9.
74. F. Jellinek, in *National Bureau of Standards Special Publication* **364**, 'Solid State Chemistry', (ed. by R. S. Roth and S. J. Schneider), 1972, p. 625.
75. F. Jellinek, in *Inorganic Sulphur Chemistry*, (ed. by N. Nickless), Elsevier, Amsterdam, 1968, Chapter 19.
76. M. Chevreton, *Bull. Soc. Fr. Mineral Cristallogr.* **90** (1967) 592.
77. Y. Jeannin, *Ann. Chim.* **7** (1962), 57.
78. P. J. Silvester and R. J. D. Tilley, unpublished results.
79. J. Flahaut, in M.T.P. Int. Rev. Sci., *Inorg. Chem.*, Series 1, vol. 10, Butterworths, London, 1972, chapter 6.
80. E. Flick, G. A. Wiegers and F. Jellinek, *Rec. Trav. Chim.* **85** (1966), 869.
81. E. Tronc and M. Huber, *Compt. Rend. Acad. Sci. Paris* **272C** (1971), 1013.
82. R. J. D. Tilley, *J. Solid State Chem.* **7** (1973), 213.
83. E. Tronc and M. Huber, *J. Phys. Chem. Solids* **34** (1973), 2045.
84. J. S. Anderson, *J. Chem. Soc., Dalton Trans.* (1973), 1107.
85. R. S. Roth, J. L. Waring and H. S. Parker, *J. Solid State Chem.* **2** (1970), 445.
86. N. C. Stephenson and R. S. Roth, *Acta Crystallogr.* **B27** (1971), 1010, 1018, 1031, 1037.

87. R. F. Brebrick, in *Chemistry of Extended Defects in Non-Metallic Solids* (ed. by L. Eyring and M. O'Keeffe), North-Holland, Amsterdam, 1970, p. 183.
88. G. A. Wiegers and F. Jellinek, *J. Solid State Chem.* **1** (1970), 519.
89. J. P. Delmaire and H. Le Brusq, *Compt. Rend. Acad. Sci. Paris*, **276C** (1973), 779.
90. F. K. McTaggart and A. D. Wadsley, *Aust. J. Chem.* **11** (1958), 445.
91. J. A. Wilson and A. D. Yoffe, *Advan. Phys.* **18** (1969), 193.
92. J. E. Iglesias and H. Steinfink, *J. Solid State Chem.* **6** (1973), 119.
93. R. H. Plovnick, D. S. Perloff, and A. Wold, *Inorg. Chem.* **7** (1968), 127.
94. T. Takahashi and O. Yamada, *J. Solid State Chem.* **7** (1973), 25.
95. S. Muranaka, *Mat. Res. Bull,* **8** (1973), 679.
96. M. Danot, J. Bichon, and J. Rouxel, *Bull. Soc. Chim. Fr.* (1972), 3063.
97. J. Rouxel, M. Danot, and J. Bichon, *Bull. Soc. Chim. Fr.* (1971), 3930.
98. J. Bichon, M. Danot, and J. Rouxel, *Compt. Rend. Acad. Sci. Paris* **276C** (1973), 1283.
99. F. Jellinek, *Acta Crystallogr.* **10** (1957), 620.
100. K. Igaki, N. Ohashi, and M. Mikami, *J. Phys. Soc. Japan* **31** (1971), 1424.
101. D. J. Young, W. W. Smeltzer, and J. S. Kirkaldy, *J. Electrochem. Soc.* **120** (1973), 1221.
102. T. J. A. Popma and C. F. Van Bruggen, *J. Inorg. Nucl. Chem.* **31**, (1969), 73.
103. A. W. Sleight and T. A. Bither, *Inorg. Chem.* **8** (1969), 566.
104. M. Mikami, K. Igaki, and N. Ohashi, *J. Phys. Soc. Japan* **32** (1972), 1217.
105. J. C. Ward, *Rev. Pure Appl. Chem.* **20** (1970), 175.
106. H. T. Evans, *Science* **167** (1970), 621.
107. M. E. Fleet, *Acta Crystallogr.* **B27** (1971), 1864.
108. E. F. Bertaut, *Bull. Soc. Fr. Minéral. Cristallogr.* **79** (1956), 276.
109. E. F. Bertaut, *Acta Crystallogr.* **6** (1953), 557.
110. H. Nakazawa and N. Morimoto, *Mat. Res. Bull.* **6** (1971), 345.
111. M. Chevreton, B. Petit, S. Brunie, and J. M. Kauffmann, *Compt. Rend. Acad. Sci. Paris* **270C** (1970), 426.
112. N. Morimoto and H. Nakazawa, *Science* **161** (1968), 577.
113. N. Morimoto, H. Nakazawa, K. Nishiguichi and M. Tokonami, *Science* **168** (1970), 964.
114. H. Nakazawa and N. Morimoto, *Mat. Res. Bull.* **6** (1971), 345.
115. R. C. Erd, H. T. Evans and D. H. Richter, *Amer. Mineral.* **42** (1957), 309.

PHYSICAL AND CHEMICAL PROPERTIES OF
PHYLLOSILICATES

S. CAILLÈRE

National Museum of Natural History, Paris

and

S. HÉNIN

National Institute of Agronomic Research, Versailles

The study of these minerals has been the subject of a great deal of research within the last thirty years. Even when we limited the subject to physical and chemical aspects only, it did not appear possible to give a coherent presentation of their very varied properties and of the methods of investigation which were used. Since we are chemists, we have chosen the chemistry of crystals as a basis for this representation.

We have described the fundamental principles which govern these structures, and summarized the experimental evidence provided by physical and physico-chemical methods. Special attention has been paid to work carried out on minerals in clay form, which lend themselves best to a study of the relationship between structure and physico-chemical properties.

Considering the informative nature which this article should possess, we considered it useful to give briefly a number of references which readers might consult for descriptions of the methods or developments of the demonstrations which will only be outlined in this text.

Brown, G.: 1961. *The X-Ray Identification and Crystal Structure of Clay Minerals.* Published in collaboration. Mineralogical Society, London, 544 pp.

Devoted to research techniques by means of X-rays, and their application to the main types of minerals belonging to the phyllitous class.

Caillère, S and Hénin, S.: 1963, *Minéralogie des Argiles.* Masson, Paris, 355 pp.

This work may be regarded as a complement to the present article since it contains a list of mineral substances and their descriptions.

Fripiat, J. J., Chaussidon, J., and Jelly, A.: 1971. *Chimie Physique des phénomènes de surface. Application aux Oxydes et Silicates.* Masson, Paris.

Review of the theoretical principles governing surface phenomena and of techniques used.

Grim, R. E.: 1962, *Applied Clay Minerology.* Mc Graw-Hill, New York and London, 422 pp.

Description of properties of clays and of their applications in industrial use.

Mackenzie, R. C.: 1957, *The Differential Thermal Investigations of Clays.* Published in collaboration. Mineralogical Society of London, 456 pp.

F. Lévy (ed.), Crystallography and Crystal Chemistry of Materials with Layered Structures. 185–267. *All Rights Reserved.*
Copyright © 1976 by D. Reidel Publishing Company, Dordrecht-Holland.

Devoted to application of thermal methods to the minerals of sedimentary rocks. Several chapters deal with phyllitous minerals.

Mackenzie, R. C.: 1970, *Differential Thermal Analysis*. Academy Press, London and New York, 2 vols. I: 775 pp., II: 607 pp.

A more general treatment than the previous one; less space is given to phyllosilicates.

Magnan, C.: 1961, *Traité de microscopie électronique*. Published in collaboration. Hermann, 2 vols. I: 656 pp., II: 656 pp.

Principles of the electron microscope: conditions of operation and utilisation. Principal methods of electron diffraction, application to investigation of various materials, in particular to minerals.

Millot, G.: 1964, *Géologie des Argiles*, Masson, Paris, 499 pp.

Essentially concerned with formation, synthesis and conditions of deposition.

Strunz, H.: 1966, *Mineralogische Tabellen*, Leipzig, 560 pp.

Classification tables of minerals according to chemical composition and crystallographic nature, with an introduction giving definitions and laws of crystallography.

The illustrations to the present paper have been produced by Mr Jean Barrandon, Technical Collaborator at the National Centre for Scientific Research, to whom we express our thanks.

1. Introduction and Historical Background

Introduction

The study of phyllosilicates is linked with technical problems the origin of which goes back to the origin of civilisation itself. The potter's clay, the brick-maker's clay or the fuller's earth are all types of materials owing their properties to the presence of phyllitous minerals. For this reason there has been a progressive characterisation of their properties for technological purposes and later of their crystallo-chemical properties. But the two aspects have never been separate and we still find, in nomenclature problems, that crystallo-chemical criteria and certain aspects of physico-chemical behaviour are used.

In the first chapter of this article we present a short historical review of the investigations of clay minerals. The first experimental methods and the basic models are mentioned. In the second chapter, we describe the exact structures of the phyllosilicates in a more or less deductive way. The structural data are discussed in relation to the crystallochemical analysis. The problems relating to the structure of the clay minerals are exposed in the third chapter. We review the methods used to check the structural data, in particular infrared spectroscopy and differential thermal analysis. The actual difficulties concerning the structure are suggested. Chapter four is devoted to general aspects of chemical and physical properties of phyllosilicates. Some references are also given regarding their behaviour in various solutions. The last chapter gives a general classification of clay minerals.

Particle Sizes and Rheological Properties

Houghton [1] showed in 1681, by separating the fine particles by washing in the pan, their role in the rheological properties of these materials. In 1807 Brongniart [2] classified clay materials in a systematic manner according to their response to heat: they melt or do not melt.

In 1850 Way [3] showed that earth is able to exchange cations with a saline solution with which it is in contact.

In 1874 Schloesing [4] was the first to obtain a stable suspension of clay and laid the basis for a rational granulometric analysis. He reported the lamellar nature of elements when in suspension.

Le Châtelier [5] examined the conditions of dehydration of these materials and devised the first apparatus for differential thermal analysis in 1887. This method of investigation has now become traditional.

In 1912, Atterberg [6] starting from the idea that the clay fraction can originate in the crushing of material, initially in large crystals, proceeded to a systematic investigation of various minerals. He subjected micas, chlorites, feldspaths and quartz to mechanical treatments. He separated the fine fraction by dispersion and showed that only minerals which appeared macroscopically in the phyllitous state can be crushed so as to give substances with the rheological properties of clays. Further, he drew up tests to characterise clays, the results of which are known by the name of 'Atterberg's constants', and are still widely used by specialists in civil engineering, and in ceramics as well as by agronomists.

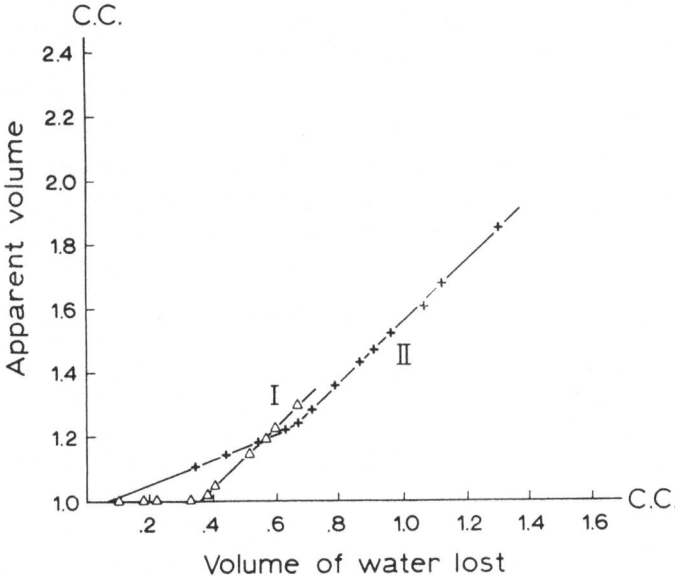

Fig. 1.   Graph showing the reduction in volume of a rod of clay as a function of the water loss in cc. Curve II is for an agricultural soil and curve I for a kaolinite.

These results are of great importance since it is the first major argument establishing the relationship between mainly crystalline minerals which can be subjected to the traditional methods of mineralogy, and materials which even recently were regarded as amorphous or colloidal and are now regarded as crypto-crystalline.

To complete this analysis it is necessary to mention Haines [7], who in 1923 investigated variations in volume of a rod of clay as a function of the loss of water. In the case of small particles extracted from a sample of soil, the curve of the variation in volume corresponds exactly to the weight of water lost by the sample and is linear up to a certain humidity. Then the volume decreases less rapidly than would be expected from the loss of water, and in certain cases the volume remains constant.

It is curious that, working on the only clay identified at that time, Haines established that the range over which the change in volume is equal to that of the water loss is very narrow. Beyond this range a loss of weight leads to no further volume change. This led the author to conclude that a particle of clay consists of a nucleus of kaolinite surrounded by a strongly hydrophile colloidal substance.

2. Chemical Investigation; Definition of Species

Over this period mineralogists had commenced their study of materials which crystallised in two dimensions and could be cleaved easily. These were micas and chlorites.

Mention must be made of minerals whose crystal dimensions sometimes permit examination under the optical microscope, but which are more often found in very small sizes, and cannot be usefully examined by this method. Kaolin is of this type. Due to its importance in ceramics, it was one of the first intensively investigated.

Thus in 1862 Delafosse [8] was able to isolate a white substance with the properties then associated with kaolin, which was found to have a chemical composition of $2 SiO_2, Al_2O_3, 2H_2O$, which is still accepted today.

Very fine materials appear to be formed by mechanical or chemical change of truly crystallised compounds, quartz, feldspaths, micas, chlorites.

Apart from the case of kaolin, analyses showed a wide variation in chemical composition, which led to the concept of type formulae, the constituents of which could be the subject of substitutions or associations in a different form.

In 1930 André [9] said

It is wrong to attach an exaggerated importance to these formulae; they are only a convenient means of explaining the mechanism of formation of soil dissolution.

But we cannot repeat too often that none of the constitutional formulae hitherto suggested for silicates is a perfect representation of the different modalities of this so complex group of chemical compounds.

By way of example, the formula suggested for kaolin by Glinka [10] is:

$$HO-Si=AlOH$$
$$|$$
$$O$$
$$|$$
$$HO-Si=AlOH$$

Here is another formula suggested for phlogopite in 1889 by Clarke [11]:

$$
\begin{array}{l}
\diagup SiO_4 \equiv MgK \\
Al - SiO_4 \equiv MgH \\
\diagdown SiO_4 \equiv MgH
\end{array}
$$

These ideas also led researchers to attempt to separate such molecules by controlled attack by acids (e.g. Roborg [12], 1935).

In any case, the chemical composition of the materials investigated could be explained in terms of two hypotheses: either there are associated impurities, such as iron or aluminium oxides, or there are substitutions such as Fe^{2+} H for MgH in the above formulae. However difficult the interpretation of chemical analysis and despite the very hypothetical nature of the formulae suggested, the guidelines and principles which emerged from this group of investigations are valid today.

In support of the first hypothesis, Dana [13] suggested in 1888 that within the crypto-crystalline minerals there was a distinction between magnesian and ferromagnesian, (named serpentine and talc) and minerals predominantly aluminious (classified in the kaolin division).

For the second explanation it is necessary to mention the concept of isomorphic series as stated in 1890 and 1891 by Tschermak [14]. This author started from molecular compositions as was then the practice (that is by rounding off the results of chemical analysis), allotting the coefficient 1 to the constituent present in the smallest quantity, and whole number coefficients increasing for the others. For chlorites he used:

1 type Am	2 MgO	1 Al_2O_3	1 SiO_2	2 H_2O
1 type Sp	3 MgO	0 Al_2O_3	2 SiO_2	2 H_2O

Intermediate substances were fitted into the scheme by imagining mixtures of the two types mentioned above, such as:

Am 8 Sp 2   or   Am 4 Sp 6

This leads to a simplification in nomenclature.

In 1925 Winchell [15] established that it was possible to set down the chemical composition of micas starting from 12 oxygens and compensating their charges by positive ions; 7 are needed in the case of muscovite: 3 Si 3 Al 1 K, which is typical of the heptaphyllite group; eight are required for the phlogopites and the biotites which therefore become the octophyllites: 3 Si Al 3 Mg K.

It must be stressed that the mineralogists of the time were impeded by the fact that it was almost impossible to ascertain the characteristic chemical compositions. Mauguin [16] was the first to undertake an investigation of the micas and he arrived at the following results:

All micas are composed of an elementary sheet about 10Å thick. All the particles contain 24 oxygens or fluorines and a number of cations (between 14 and 16), not including H ions. These latter amount to 4 when there is no fluorine.

Mauguin presented his results in the following manner:

```
        K
        6 O
    Si  Al  Mg  H
        6 O
    Si  Al  Mg  H
        6 O
    Si  Al  Mg  H
        6 O
    Si  Al  Mg  H
        6 O
        K
```

Since such a grouping must be electrically neutral, he concluded that the total positive charge must be 48, thus balancing the 48 negative charges due to the oxygen. He attributed changes in chemical composition to substitutions such as:

$$O H \rightarrow F$$
$$Mg \rightarrow 2 Li \quad etc.$$

This type of substitution, known as diadochic, clearly takes account of the diversity of compositions observed for the different minerals belonging to the same family.

This forms a turning point in the study of silicate materials.

Two years later, in 1930, Mauguin [18] approached the problem of chlorites by the same method, and also gave the principle of the crystallographic constitution of these minerals.

These results opened the way to modern research, but they required completion by specifying the arrangement of the oxygen ions and of the cations. Further, Mauguin had reached his results by use of single crystals, and we have seen that many silicates, those which form clays, exist only in the form of extremely fine particles submicroscopic in nature. It is now necessary to refer to work carried out on these materials by means of X-rays before reaching the key work of Pauling [19] in 1930.

3. Use of X-Rays

As soon as the method developed by Debye and Scherrer became known, numerous researchers applied it to materials consisting of submicroscopic particles such as rocks and mineral clays. In 1923, Hadding [20] suggested the application of this method to the identification of cryptocrystalline materials. He determined the diagrams for quartz, crushed muscovite and kaolinite and compared them with those obtained from products of different origins, thus providing direct proof of the presence in clays of minerals which could also exist in the macro-crystalline state.

In 1924 Rinne [21] described an arrangement for determination of such diagrams and investigated bauxite from Les Baux, showing that kaolinite, gibbsite and diaspore were present. Research continued on this line, in particular de Jong's work [22] in 1927 on the laterites and Boehm's [23] in 1928 on the hydroxides of iron. The publications of Hendricks and Fry [24] in 1930 showed in the fine fraction of cultivated soils, the minerals already identified by their Debye-Scherrer diagram: halloysite and montmorillonite.

II. Crystal Structure of Phyllosilicates

4. Determination of Structure

Credit is due to Pauling [19] for his suggestion of the types of arrangement characteristic of phyllitous silicates. Taking note of the different possible arrangements of the constituents, Pauling [19] sought in 1930 to define a certain number of elementary

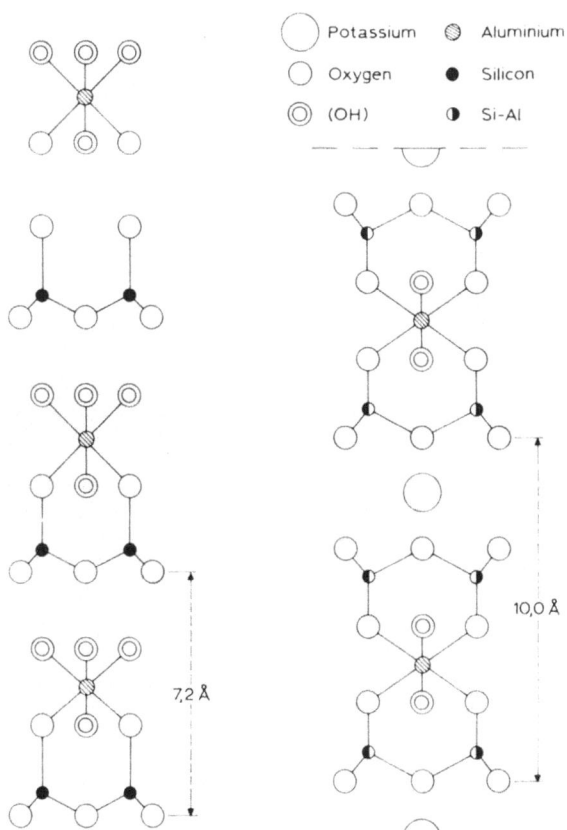

Fig. 2.   Diagram of the cation-anion association. Above left: Al surrounded by six oxygens and by OH, forming an octahedron. The figure below represents the Si tetrahedron where Si is surrounded by four oxygens. In the same column two figures show tetrahedral and octahedral grouping in a kaolinite. In the right hand column: grouping of two pairs of tetrahedra placed face to face, and an octahedron. The diagram is given twice so as to place the K ion between its two base elements: diagram of a mica.

groupings of anions and cations. These groupings were used by Bragg [25] in 1937 in his first work on silicates, and are tetrahedra, octahedra and more rarely, other polyhedra. By arrangement of these elementary groupings. Pauling was able to constitute the structures of the principal silicates, obtaining with these models the same structural data as those found by X-ray investigations of these minerals (Mauguin [16] in 1928).

Pauling established rules for grouping of these polyhedra:

(1) Around each cation a co-ordinated polyhedron of anions is formed, usually O, OH or sometimes F ions. The cation-anion distance is determined by the sum of the radii. The co-ordination number of the cation is determined by the ratio of anion radii to cation radii.

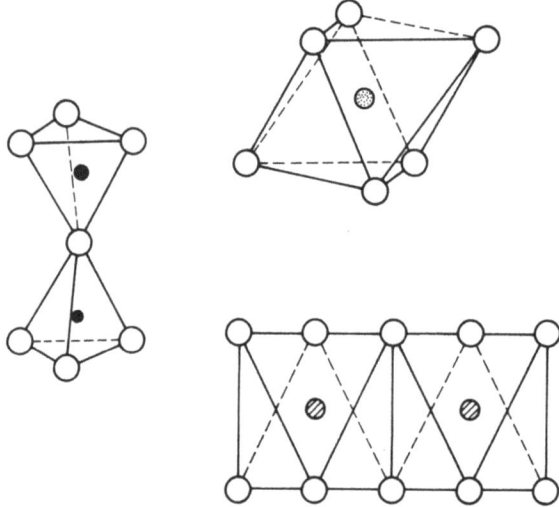

Fig. 3. Examples of associations of anion polyhedra. Left: two tetrahedra with a common oxygen, hence a peak. Right: top: an isolated octahedron. Below: association of two octahedra with two oxygens in common, hence a ridge (A).

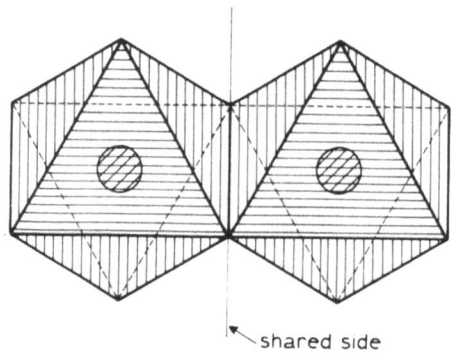

shared side

Fig. 4. Octahedra resting on a base. This view is perpendicular to image A of Figure 2. The dotted lines represent the ridges at the base. In the centre the hatched circle indicates a cation.

(2) In a stable ionic structure, the valence of each anion absolute value is exactly or approximately equal to the sum of the electrostatic forces of attraction arising from the adjacent cations which act on it.

(3) The presence of shared edges and even more of shared faces between the poly-hedra forming part of a co-ordinated structure, leads to an increase in stability. This effect is large for cations of high valence and low-co-ordination number.

(4) In a crystal containing different cations, those of high valence and low co-ordination number tend not to have common polyhedric elements. This fourth rule

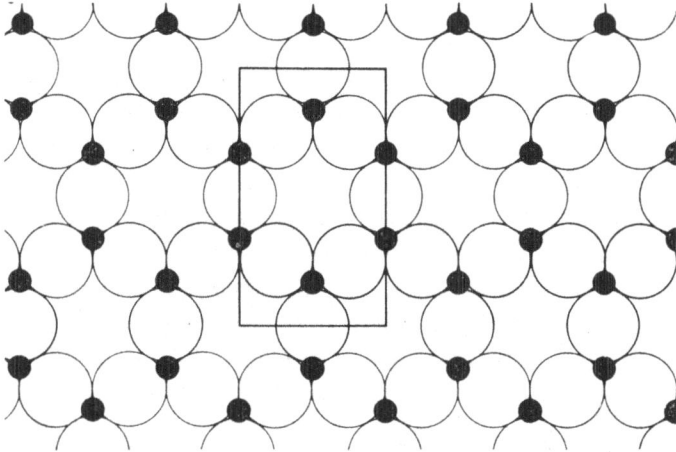

Fig. 5a.   Hexagonal layer seen perpendicularly to plane *ab*; the black circles represent the Si atoms. In the structure these silicons are applied to the succeeding layer (Figure 5b); they dominate the oxygens. The rectangle represents the base of a unit cell.

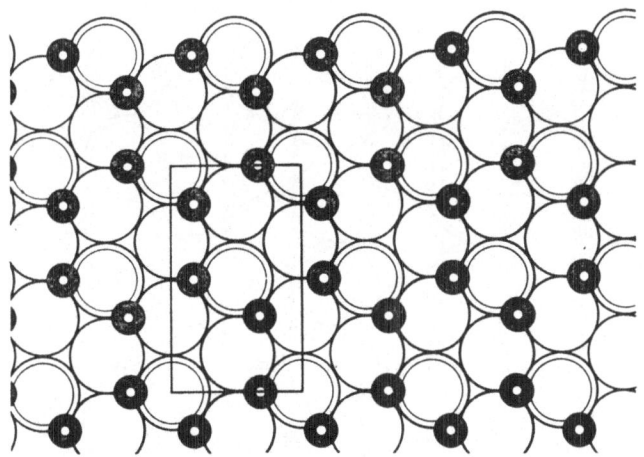

Fig. 5b.   Compact layer seen perpendicularly to plane *ab*; The double circles correspond to hydroxyls and the black circles with white centres to aluminium, which are also applied to the succeeding layer (Figure 6). Note their distribution at the apices of a hexagon. The rectangle represents the base of the unit cell. The hexagonal layer is situated behind the plane represented on this figure.

is a corollary of the third and as far as phyllitous silicates are concerned, rules (1) and (2) are the most important.

These polyhedra are associated in such a manner that a two-dimensional lattice of tetrahedra is formed, with their bases in the same plane. The groups of three oxygens forming the bases of the tetrahedra are associated so as to form groups of six oxygens delimiting a space within which a hexagon can be inscribed. This plane is sometimes known as the hexagonal layer and corresponds to Figure 5a.

The black mark representing silicon will be noted in the centre of the groups of three oxygens.

The fourth oxygen ions forming the apices of the tetrahedra cover up the silicons. The empty spaces still existing between the oxygens, and which correspond to the hexagonal cavities, are occupied by OH or F ions. The O or OH group are associate in a compact assemblage shown diagrammatically in Figure 5b.

It is obvious that the diameter of the oxygens in the hexagonal plane is smaller than that of the oxygens or OH groups in the lower lying compact plane.

The order of magnitude of difference in diameter is about 0.17 Å if the diameter of the oxygens in the hexagonal plane is 2.5 Å. The second plane of oxygen ions rests on a third with the same structure. It may be formed like the second, by juxtaposition of O, OH or F ions, or entirely of OH ions. The superposition of planes 2 and 3 is such that a group of three ions belonging to one of them with their centres situated at the points of an equilateral triangle, is superposed on a group of three ions belonging to the succeeding plane. The two equilateral triangles thus delimited are arranged in such a manner that the assembly forms an octahedron (see Figure 3).

In Figure 5 there are, at the junction of three spheres, black circles with white centres which represent the cations placed at the centres of the octahedra. This figure shows an arrangement in which the straight lines demonstrate that the alignment

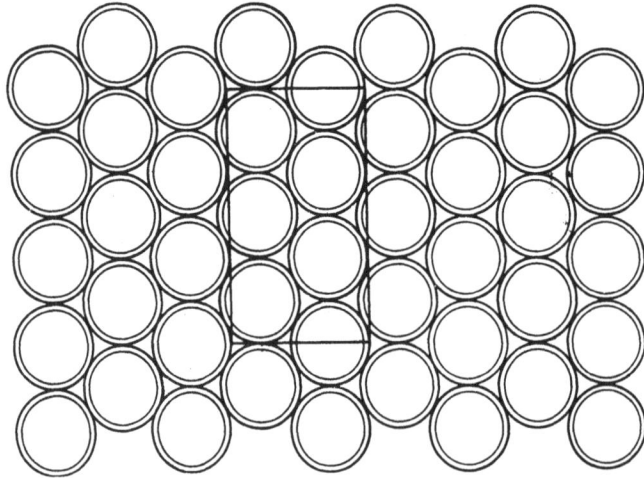

Fig. 6.   Arrangement of a compact layer of hydroxyls (third of kaolinite). Layers *b* is applied on layer 6, with the Al situated at the centre of curvi-linear triangles. The rectangle represents the base of the unit cell.

of the cations is continuous, instead of being punctuated by separate groups of two cations. Figure 5a corresponds to a dioctahedral type, whilst a distribution at regular intervals corresponds to a tri-octahedral type. In this latter case, all the octahedral cavities formed by superposition of two compact planes are occupied by a cation.

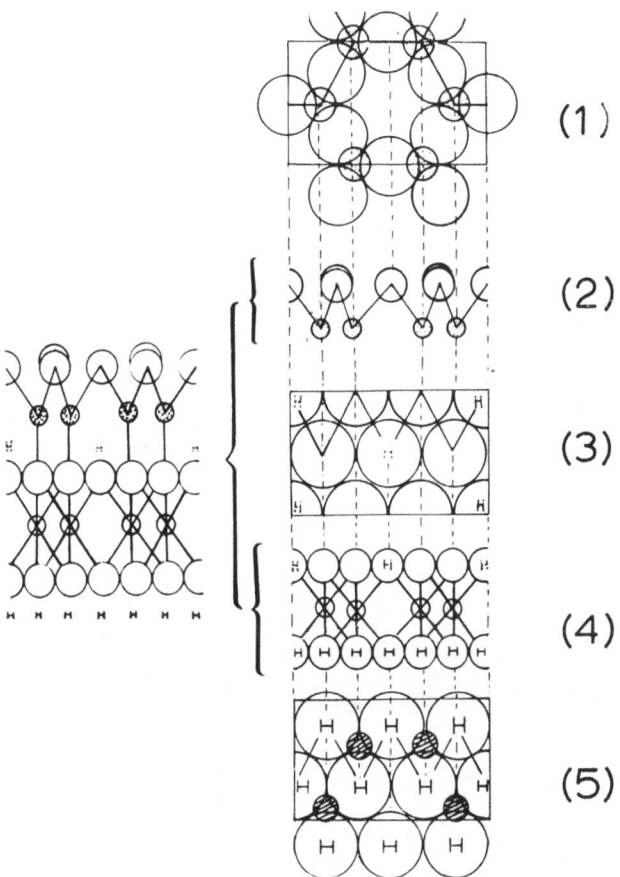

Fig. 7a. Correspondence of the various representations of the structure of kaolinite.
The circles represent the oxygens, the small dotted circles the silicons and the shaded circles the aluminiums.

At the top right: hexagonal layer (oxygen and silicon) seen perpendicular to plane *ab* (1).

The same elements projected onto plane *bc* are shown immediately below (2). The thick line shown on (1) connects the ions shown on projection *bc* (2). The dimensions of the elements have been reduced.

On the third figure starting from the top, a compact layer seen perpendicularly to *ab* (3).

On the last figure in the right hand column (5), a compact layer seen also in a plane with the distribution of the Al (view perpendicular to *ab*) intermediate layer (4) shows the association of layers 3 and 5 seen in projection on *bc*. The straight line segments shown on the two compact layers connect the oxygens and the OH groups appearing on projection *bc*. Left side: projection on the same plane *bc* of the constituents of a unit cell of kaolinite resulting from the association of the two projections on *bc* of the diagrams (2) and (4) of the right hand column.

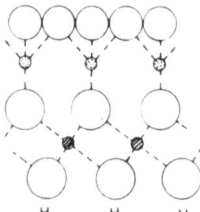

Fig. 7b.   Projection on *ac* of the structure of kaolinite. In order to demonstrate better the difference in arrangement of the elements, the scale (dimensions of ions and distances) has been modified.

The first layer corresponds to diagram (2) of Figure 7a.

The two superposed oxygen layers and the Al correspond to diagram (4) (Figure 7a).

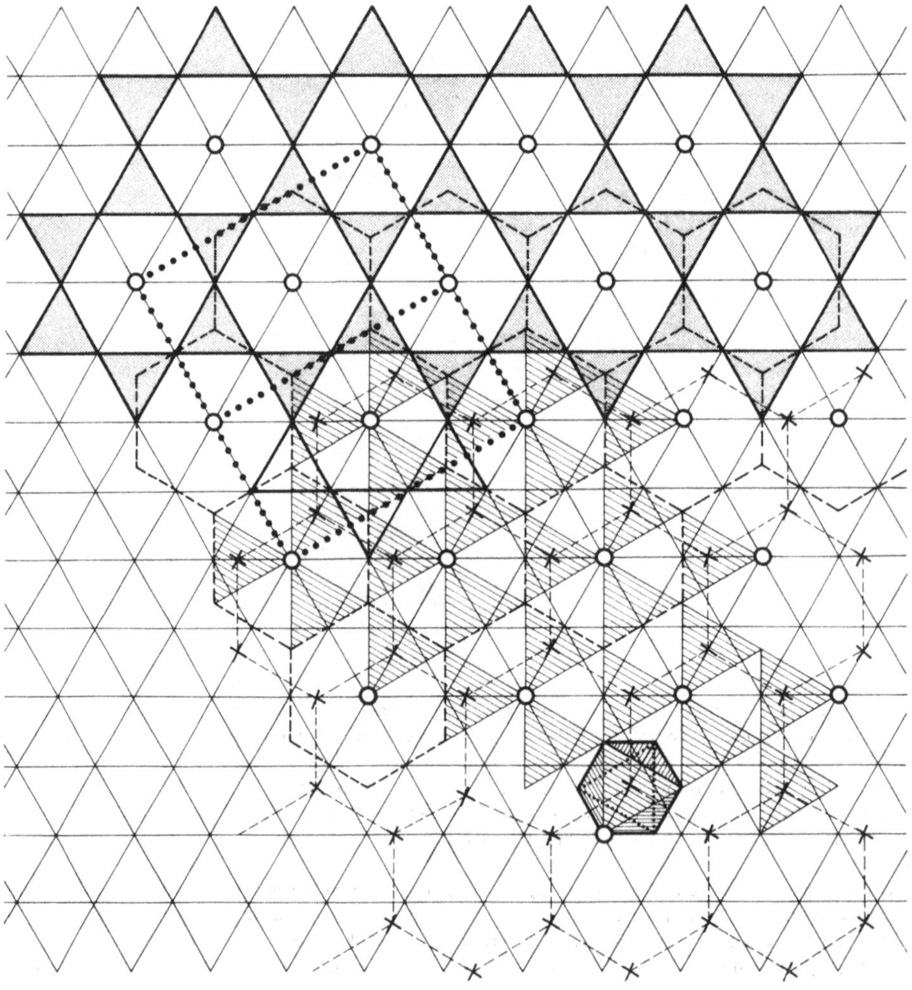

Fig. 7c.   *General presentation of the arrangement of the two first layers of oxygens, silicons and aluminiums, in a phyllitous material.* This representation is a projection on a plane *a-b*. The various

In Figure 4 a rectangle will be seen which delimite the base of the unit cell; which placed on Figures 5a and b coincides with its preceding one and so enables the two planes to be superposed.

Figure 6 represents a type of compact assembly composed of OH only, and also shows the base of the unit cell.

Structures are frequently represented diagrammatically by means of a section perpendicular to the planes which have just been studied; some elements of this representation have already been included in Figure 2. Figure 7 shows the correspondence between plane and section representations. On the right hand side at the top of the figure there is a hexagonal layer, six oxygens arranged along a broken line are connected by full lines forming four angles the peaks of which are situated at the centre of the silicons. These six oxygens connected by full lines are shown above the silicons corresponding to the second level of this figure. The following rectangle shows a first compact layer formed by oxygen and OH. A broken line joins the centers of the ions which are shown side by side on the first layer shown in the fourth diagram of the figure.

The fifth diagram of this figure also shows the six OH groups forming the second layer of the fourth diagram. The four shaded circles represent the octahedral cations which constitute the intermediate plane of Figure 4.

To sum up, representation by section corresponds to a diagrammatical projection where the elements disposed on two rows in the spatial diagram are here shown side by side. Finally, by associating diagrams 2 and 4 in the right hand column, we obtain

---

elements are separately represented at the extremities of the figure; as we approach the centre, they are superposed to a greater or a lesser extent.

The system of intersecting lines shown for example at the bottom right corner of the figure, represents the position of the centres of a certain number of O and Si constituents.

The first line at the top of the figure shows a series of triangles containing separate points, forming the base of the tetrahedra constituting the hexagonal plan.

In the next line we find, superposed on the previous system, a second series of hexagons shown by dotted lines; the apices of these latter locate the centre of the oxygen atoms forming the apices of the tetrahedra. The sides of this second series of hexagons make an angle of 30° with those of the first hexagons. The silicons are found at the center of the tetrahedra on the same vertical, not shown in the diagram at this plan. At this level a rectangle of large dots, on the left hand site, it represents the base of a unit cell.

In the lower part of the figure there are hexagons shown by lightly dotted lines. Their apices are marked by a cross showing the position of the Al ions in the system.

Moving back to the central part on the right, there are triangles shaded by continuous lines. Certain of the apices of these triangles are marked by a circle; this latter indicates an OH group. The points of intersection of these triangles correspond to the position of an oxygen. The triangles themselves represent the upper surfaces of the octahedra which exist as discrete units in the structure. It will be noticed that there is a displacement of the hexagons which have Al ions at the apices (crosses) with respect to the system of triangles. This is normal, the Al ions being at the centres of the octahedra.

In the lower part of the drawing, near the centre, the projection of an octahedron on the observation plane is drawn. The oxygen ions, the centers of which are located at the apices of the triangle shown on this picture, are dealing with a third plane of oxygen.

If we move upwards again towards the center of the diagram, we find all these elements superposed. It can be verified that they include all the constituents shown in diagrams 1 and 3, and the aluminiums of diagram 5 (Figure 7a). The length of the rectangles represents the b axis.

the one placed in the left hand column, which represents the complete structure of a unit cell of kaolinite. The interesting feature of this representation is that it shows the whole of the constituents of the unit cell and the linkages which exist between anions and cations. These diagrams are essentially intended to show the positions of the anions which are in general the most voluminous elements of the structure, and of which they form the framework.

It is now appropriate to fix the position of the cations in the different types of polyhedra; it is the ratio between the radii of the cation and of the anion which determines these positions. It must also be noted that the nature of the polyhedron corresponds to the coordination number of the cation. These relations as given by L. Pauling are as follows:

Ratio of radii $\dfrac{r_{cations}}{r_{anions}}$ R	Co-ordination number	Position of anions
$0.225 < R < 0.415$	4	tetrahedron
$0.415 < R < 0.732$	6	octahedron
$0.732 < R < 1$	8	cube
1	12	cubic octahedron

This table shows that for one anion arrangement there can be a large number of corresponding chemical compositions.

The only condition, to which we shall return, is that the variations in charge caused locally by substitution of an ion of valence $n+1$ or $n-1$ for an ion of valence n is to be compensated for at the level of the unit cell.

It must also be noted that the smallest value of the ratio R corresponds to the radius of the sphere which would be just tangential to the spheres representing the oxygens in the cavity situated in the centre of the assembly. There are therefore some problems of adjustment.

It will be seen that the rule giving the type of anion polyhedra formed round the cations as a function of R is only approximately obeyed. According to this rule, silicon with $R=0.3$ and aluminium with $R=0.360$ would always be tetrahedral; but we find Al in the octahedral position. As for Fe(III) with $R=0.456$ should not be in the tetrahedral position, but it certainly is found there in certain phyllitous minerals. We will see in fact that the choice of radius 1.40 Å for oxygen is itself open to discussion. We have indeed seen from simple geometrical considerations that the dimensions of the oxygens in the octahedral layer must be greater than those in the tetrahedral layer, which must therefore alter the value of R.

L. Pauling's original hypothesis is that the bond is essentially ionic. It can in fact be covalent and this same author has given a function enabling us to predict the extent of ionic or covalent bonds as a function of the differences in electronegativity between the different atoms.

	Radii as given by Arrhens (1952)	$R_a = 1.40$ $\dfrac{R_c}{R_a} = R$
Tetrahedral positions		
Si	0.42	0.300
Al	0.51	0.360
Fe^{+++}	0.64	0.456
Cr^{+++}	0.64	0.456
Octahedral positions		
Al	0.51	0.360
Fe^{+++}	0.64	0.456
Cr^{+++}	0.64	0.456
Mn^{+++}	0.66	0.470
Mg	0.66	0.470
Ti^{++++}	0.68	0.485
Li	0.68	0.485
Ni	0.69	0.492
Co^{++}	0.72	0.514
Cu^{++}	0.72	0.514
Zn	0.74	0.528
Fe^{++}	0.74	0.528
Ti^{+++}	0.76	0.542
Mn^{++}	0.80	0.570
Hexagonal positions		
Na	0.97	0.692
Ca	0.99	0.708
K	1.33	0.947

Above are a certain number of values corresponding to the cation-oxygen bonds for the main elements of phyllosilicates, calculated for binary compounds.

This table shows the general tendency, which may be substantially modified when complex compounds are concerned, when several cations can react on the same anion. It must be stressed that as the number of covalent bonds increases: (1) the distance between cation and anion tends to become smaller due to interpenetration of the electronic orbits. (2) In addition these bonds tend to become directed.

By comparing the above table with that of ionic radii on p. 198, the elements which can be in the tetrahedral position are characterized by a relatively small percentage of ionic bonds. Thus the fact that Fe(III) is sometimes found in the tetrahedral position, if it departs slightly from the standard behaviour as given by the ion ratio, is compensated for by the fact that the FeO bond is more covalent or less ionic than that of aluminium.

In addition, the cations situated in the hexagonal layer are those with strongly onic bonds.

Ions	Electronegativity as given by D. P. Grigorieff [26] in 1964	Δ O-electronegativity-electronegativity of ion	% Ionic bonds
0	3.5		
Tetrahedral positions			
Si	1.8	1.7	50
Al	1.5	2	58
Fe^{+++}	1.8	1.7	50
Cr^{+++}	1.5	2	58
Octahedral positions			
Al	1.5	2	58
Fe^{+++}	1.8	1.7	50
Cr^{+++}	1.5	2	58
Mn^{+++}	1.2	2.3	70
Mg	1.2	2.3	70
Ti^{++++}	1.6	1.99	59
Li	0.95	2.55	80
Ni	1.8	1.7	50
Co	1.7	1.8	55
Zn	1.5	2	58
Cu^{++}	2	1.5	56
Fe^{+++}	1.7	1.8	55
Mn^{++}	1.4	2.1	60
Hexagonal positons			
Na	0.9	2.6	80
Ca	1	2.5	76
K	0.8	2.7	82

Pauling had himself also indicated the corrections to be applied and had suggested means of carrying out these adjustments. In short, allowance must be made for the effect of the ratio between the radii, for the contact of the anions, and for the phenomenom of double repulsion. As a result the repulsive forces are larger than they would be separately for an anion-cation contact or for an anion-anion contact. Further, equilibrium with the Coulomb forces is reached through lattice characteristics, the distance from cation to anion being greater than the sum of the radii, and the distance between anion and anion greater than twice the radius of the anion. [27].

Experience has shown, however, that there is no general need for these complicated calculations, unreliable in nature, even for substances with asymmetric valences, and that interionic distances are sufficiently accurately given by the sum of the radii in the crystal.

The problem is then knowing to what extent the bond remains of the same nature. Taking account of the fact that the bonds are partly ionic and partly covalent,

Pauling introduced mesomeric data. Thus for the SiO_4 ion we can imagine a resonance between a formula of the type:

$$
\begin{array}{c}
O^- \\
| \\
{}^-O-Si-O^- \\
| \\
O_-
\end{array}
$$

and

$$
\left[
\begin{array}{c}
O \\
\| \\
O=Si=O \\
\| \\
O
\end{array}
\right]^{----}
$$

For the covalent bond, the distance Si—O is smaller than for the ionic bond, and the actual value of the cation-anion distance is a measure of covalent versus ionic bonding.

Thus in the Si—O bond the sum of the radii of the ions is $0.42 + 1.40 = 1.82$, whilst the observed distance is 1.60. The difference, 0.22 Å, is a measure of the covalent contribution to the bond.

It follows from these considerations that in an AlO_4 tetrahedron the bonds are strongly covalent, whilst they are more ionic in character in the $Al(OH)_6$ octahedron. The inter-ionic distances reflect this situation.

4.1. ELEMENTS OF STRUCTURES

In the descriptions which have just been given, we have considered the existence of polyhedra of anions as imagined by W. L. Bragg. We have described how they associate with each other. The anions are arranged in successive planes. It is characteristic of phyllitous materials that they consist of the systematic association of a number of these layers to form a sheet. Figure 7 shows for instance a cross-section of a sheet of kaolinite.

Thickness of sheets Mauguin [16–18] Hendricks [24]	Arrangement of polyhedra of anions in the sheets Pauling [19]	Names G. Marshall [28]	Brown [29]
7 Å	Te – Oc	1/1	Diphormic minerals
10 Å	Te – Oc – Te	2/1	Triphormic minerals
14 Å	Te Oc Te Oc	2/1/1	Tetraphormic minerals

Above is the nomenclature used to denote the different types of sheet. The work of Pauling and other crystallographers has shown that there are in practice three categories of assembly.

The symbols Te and Oc represent tetrahedral and octahedral layers.

It follows from this that by returning to the global representation of Mauguin, we can make a diagram of the composition of the archetypes of the different families in the following manner (Caillère and Hénin [30], 1963):

7 Å diphormic	10 Å triphormic	14 Å tetraphormic
6 O 4 Si 4 O, 2 OH } Te	6 O 4 Si 4 O, 2 OH } Te	6 O 4 Si 4 O, 2 OH } Te
4 Al ou 6 Mg 6 OH } Oc a	4 Al ou 6 Mg 4 O, 2 OH 4 Si 6 O } Te	4 Al ou 6 Mg Oc 4 O, 2 OH 4 Si 6 O } Te
		6 OH 4 Al ou 6 Mg } Oc 6 OH

The number of 6 O which is taken as a basis of reference corresponds to the dimensions of the unit cell. These latter are directly derived from X-ray measurements.

We will see later how, starting from these diagrams, it is possible to deduce formulae representative of these minerals.

5. Crystallo-Chemical Interpretation of Analyses of Minerals

Mauguin's approach [16] of 1928 can be summarized as follows: having determined the unit cell of the mineral, he multiplied this by Avogadro's number so as to obtain a 'molecular unit'. This volume, when multiplied by the density, gave the weight of the substance contained in one unit cell. Since chemical analysis was also available, he transformed the weights of the oxides into numbers of molecules and multiplied these by the ratio weight of a unit cell/100. He thus obtained the number of ions present in the unit cell. A numerical example showing this procedure in concrete form is the following. The theoretical dimensions of the cells are assumed to be:

$$a = 5 \text{ Å} \qquad b = 9 \text{ Å} \qquad c = 10 \text{ Å}$$

The measured density of the crystal is 3.05.

The chemical analysis is as follows:

		Number of molecules	Number of cations	Number of oxygens
SiO_2	43.30	0.722	0.722	1.444
Al_2O_3	12.25	0.120	0.240	0.360
MgO	28.80	0.720	0.720	0.720
K_2O	11.30	0.120	0.240	0.120
H_2O	4.35	0.242	0.480	0.240
	100.00			

The 'molecular' volume of the cell is:

$$10^{-24} \times 5 \times 9 \times 10 \times 6.06 \times 10^{23} = 272.7 \text{ cm}^3$$

Weight of molecular unit: $3.05 \times 272.7 = 832$ from which we find for the unit cell:

$$\text{Number of Si} = \frac{832 \times 0.722}{100} = 6.03, \quad \text{Number of O} = \frac{832 \times 2.884}{100} = 24$$

$$\text{Number of Al} = \frac{832 \times 0.240}{100} = 1.99, \quad \text{Number of H} = \frac{832 \times 480}{100} = 4$$

$$\text{Number of Mg} = \frac{832 \times 0.720}{100} = 6.03, \quad \text{Number of K} = \frac{832 \times 0.240}{100} = 1.99$$

In consequence, if the determinations of the cell dimensions and density are accurate and if the chemical analysis is correct, the unit cell is rigourously shown to contain (in rounded numbers):

$$\left.\begin{array}{l} 6 \text{ Si} \\ 2 \text{ Al} \\ 6 \text{ Mg} \\ 2 \text{ K} \\ 4 \text{ H} \end{array}\right\} \text{ associated with 24 oxygens.}$$

Thus we find the number of ions indicated for triphormic minerals. We must now distribute them.

For all the phyllosilicates considered up to now, we have per unit cell:

4 sites available for cations in a tetrahedral layer
6 sites available in the octahedral position
2 hexagonal cavities for the unit cell.

Examining the arrangement of the oxygens in the rectangle defining the unit cell at the level of the hexagonal layer, we find that there is one hexagonal site and four $\frac{1}{4}$ (see Figure 5).

We might therefore think that there are two hexagonal sites available. In reality, and especially in the case of micas, since the K ion penetrates partially into the layer to which it belongs as well as into the layer above, it is only possible to use for a given unit cell half of the available hexagonal sites, that is one, the other being reserved for the potassium of the neighbouring layer. Since this is the case for a mineral of mica type with two tetrahedral layers, there are:

8 tetrahedral sites

6 octahedral sites

2 hexagonal positions (one for each face of the sheet)

to be occupied.

As a result of the previous calculation, we place the Si 6 in the tetrahedral position and 2 Al which can also occupy these positions. 6 Mg occupy octahedral positions and 2 K occupy the free hexagonal sites.

The dimensions of the units are relatively well known, and the principles of their construction agreed; the interpretation of the chemical analysis is carried out as follows:

If x is the number of cations found by analysis, there are x/nO for one oxygen and for the m oxygens of the unit cell there are mx/nO. Since the value of m/nO is applicable to all the cations in the analysis, it is evaluated for the whole of the results. The following is an example of the interpretation of the chemical analysis of a phlogopite from Snake Creek, Utah [30].

		1/1000 of molecules	1/1000 of oxygene	1/1000 of cation
SiO_2	40.6	676	1352	676
Al_2O_3	14.6	143	429	286
Fe_2O_3	1.1	6	18	12
FeO	0.4	5	5	5
CaO	0.1	1	1	1
MgO	27.0	685	685	685
Na_2O	0.2	3	3	6
K_2O	10.7	113	113	226
H_2O	4.7			
	99.8		2600	

If we consider the dehydrated mineral, the analysis leads to 2.606 for the number of oxygens, in the dehydrated unit cell being 22.

$$\frac{m}{nO} = \frac{22}{2.606} = 8.46.$$

By applying this coefficient to the various cations found by analysis, we obtain the following values for the cations in the unit cell:

Si	5.72
Al	2.42
Fe^{3+}	0.10
Fe^{2+}	0.04
Ca	–
Mg	5.71
Na	0.05
K	1.91

In order to obtain the formula of the mineral, it is sufficient to distribute these cations among the various tetrahedral and octahedral layers in the hexagonal cavities, referring to the following table:

Number of sites	Distribution of ions in the analysis
Tetrahedral 8	5.72 (Si)+2.28 (Al)=8
Octahedral 6	0.14 (Al)+0.1 (Fe^{3+})+0.04 (Fe^2)+ 5.71 (Mg)=6
Hexagonal 2	0.06 (Mg)+0.05 (Na)+1.91 (K) =2.02

This calculation illustrates the procedure followed; we fill each layer with the cations having the diameter and co-ordinance desired in such a manner as to accommodate all the ions found by analysis. There is a slight excess of ions in the hexagonal layer since there are 2.02 ions for 2 places. This difference is negligible, considering the accuracy of the analysis on the one hand and to the inevitable approximate nature of the calculations on the other hand.

These results will now be examined in relation to the equilibrium of charges. The theoretical tetrahedral layer contains 8 Si therefore 32 positive charges. In the present case their charge is:

$$(5.72 \times 4) + (2.28 \times 3) = 29.72 \text{ charges.}$$

Now, given the number of hexagonal cavities available we can only compensate two charges, so there should be 30 positive tetrahedral charges, and the deficit of 0.28 can be compensated by substitution of divalent ions for monovalent ions in hexagonal positions, or more usually by excess charge in the octahedral position. This latter is equal to:

$$(0.14 \times 3) + (0.10 \times 3) + (0.04 \times 2) + (5.71 \times 2) = 12.22$$

so that there should normally be 12 charges. These 0.22 charges compensate almost the whole of the tetrahedral deficit of 0.28.

In the hexagonal position we have:

$$(0.06 \times 2) + (0.05 \times 1) + (1.91 \times 1) = 2.08,$$

i.e. 0.08 in excess of the normal charge. This balances to within 0.02 the deficit remaining after partial compensation of the tetrahedral deficit by the excess of octahedral charges. Here again, the difference of 0.02 is to be attributed to the approximate nature of the calculations.

We have given this example in its entirety, corresponding to the analysis of a mineral, so as to illustrate the procedure. We have assumed that a small quantity of magnesium is found in the hexagonal position. Such a hypothesis is not acceptable in the case of a mica, unless there is a very small quantity of this element.

The micas in which compensation of charge is effected by divalent ions contain calcium (margarite); in that case the charge deficit of the tetrahedral layer is nearly 2. However, in certain minerals such as vermiculites and the montmorillonites, the hexagonal cation, frequently called the compensating cation, may be any cation whatsoever. Under these conditions the sheets are not attached to one another; they are separated by layers of water. On the other hand we cannot consider that in a mica a cation such as calcium can compensate the two charges*. The distance between the cation and the sites which neutralize its charge would then be too great. Such a mineral would be unstable; the sheets would separate under the influence of the water molecules which would hydrate the cations.

5.1. CALCULATION OF WATER CONTENT

The above calculation has been made on the supposition that the mineral was anhydrous; there is in theory no reason against this practice. This has the advantage of not introducing the water content into the calculations, the determination of which is often very delicate. This is not caused by the measurement of the total quantity of water, but by the fact that more than one form of water is present: hygroscopic water and water of constitution, and that it is frequently very difficult to distinguish the two. The traditional view of regarding hygroscopic water as that escaping below $100\,°C$ is entirely arbitrary. Nevertheless, it is necessary to be able to make use of existing determinations. With this aim, we will calculate the theoretical quantity of water of constitution which should be contained in the mineral, and we will compare it with the quantity measured in the above example.

The 'molecular' weight of the unit is equal to:

Weight of Si	$=28 \times 5.72=$	160.16
Weight of Al	$=27 \times (2.28 + 0.14)=$	65.34
Weight of Fe^{3+}	$=56 \times 0.1$ $=$	5.60
Weight of Fe^{2+}	$=56 \times 0.04=$	2.24
Weight of Mg	$=24 \times (5.71 + 0.06)=$	138.96
Weight of Na	$=23 \times 0.05=$	1.15
Weight of K	$=30 \times 1.91=$	71.49
Weight of O	$=16 \times 22$ $=$	352.00
Weight of water	$=18 \times 2$ $=$	36.00
		832.94

* With a deficit of a charge of one.

The water content with respect to 100 is therefore equal to:

$$\frac{36 \times 100}{832.94} = 4.32$$

compared with 4.70 found by analysis.

The difference is therefore very small and justifies the acceptance of the formula set up. Comparison of this result with the quantities of water lost below the temperature at which the water of constitution escapes, provides further support for this agreement. Various techniques are now available which enable us to determine these data, since exact knowledge of the OH and of the states of the water is indispensable for understanding the properties of these substances. Amongst those most commonly used we will mention thermogravimetric analysis infrared spectroscopy which will be discussed later, and Hertzian absorption, which is specially suitable for investigation of absorbed water.

6. Structural Formulae

We have already seen that attempts to express the constitution of silicates starting from the notion of molecules have met with an obstacle. But the fact that by definition the unit cell must express the properties of the crystal and in particular must be electrically neutral, enables us to give a representation of the crystallo-chemical constitution. The formula will indicate the distribution of the elements according to their position: tetrahedral, octahedral or hexagonal. We can simplify and halve the number of constituents without changing the representation. Thus we write for a diphormic mineral without substitution, such as kaolinite:

$$Si_4O_{10}Al_4(OH)_8 \quad \text{or} \quad Si_2O_5Al_2(OH)_4$$

for a triphormic mineral without substitution, such as pyrophillite:

$$Si_8O_{20}Al_4(OH)_4 \quad \text{or} \quad Si_4O_{10}Al_2(OH)_2$$

A mineral exhibiting substitution such as a mica will be written:

$$(Si_{(8-x)}Al_x)\, O_{20}Al_4\, (OH)_4 K_x \quad \text{or} \quad (Si_{(4-\frac{1}{2}x)}Al_{\frac{1}{2}x})\, O_{10}Al_2\, (OH)_2 K_{\frac{1}{2}x}$$

We can stress the nature of the bonds between the different constituents by arrows for covalent bonds and $+$ and $-$ signs for ionic bonds. Here for example is the formula for muscovite corresponding to the above mineral where $x = 1$ (Grigorieff [26]),

$$\{[(Si_2^{\downarrow\downarrow+} + Si^{\downarrow\downarrow++}Al^{\downarrow\downarrow+})\, O_{10\uparrow}]^{-5}\, (Al_2^{+3})\, (OH)_2^{-1}\}\, K^{+1}$$

We find that there are ten covalent bonds due to Si by Al substitution compensated by ten covalent bonds with the oxygens, five ionic bonds which balance out half of the ten ionic bonds of the oxygens; the left hand bracket therefore has a charge of -5. These bonds will be compensated by four positive bonds due to the octahedral Al.

The two remaining positive charges are compensated by those of the OH groups. The ultimate charge of the sheet is -1 so it is free to fix the K ion. As Pauling [27] pointed out in 1949 in connection with another formula, these distributions are based on considerations similar to those of the chemists who had previously suggested structural formulae, but we will see later by explaining certain detailed considerations about structures that they can now be justified by arguments based on experience.

III. Investigations Related to the Structure

7. Verification of Structural Data

7.1. STUDY BY X-RAY FLUORESCENCE

Recent work had as its purpose the verification of various hypotheses concerning the location of the ions within the structure. Amongst the most advanced we must mention the location of tetra- and hexa-coordinated Al by means of X-ray fluorescence. In 1924, Lindh and Lundquist [31] showed that the wavelength of the fluorescent radiation from certain elements slightly varied with the valence. In 1958 White *et al.* [32] showed that the wavelength of the $K\alpha$ radiation of Al varied with its co-ordination. In 1961 Brindley and MacKinsky [33] appear to have been the first to investigate this on phyllitous minerals. The applications of this phenomenon were generalized above all by Fripiat *et al.* [34] in 1964, and later by Steinberg [35] in 1970. The angular difference in reflection between $K\alpha$ Al_{IV} and $K\alpha$ Al_{VI} is of the order of $0.05°$. Therefore it is necessary to work very accurately. Whilst it is almost certain that the presence of Al_{IV} can be proved by this procedure, the quantities found are smaller than those resulting from calculations based on chemical analysis. More work of this nature is required so that all the conclusions may be drawn and that the crystallo-chemical composition of minerals may be determined with greater precision. We will see later (p. 216) that a precise measurement of the parameter b also enables us to detect the presence of certain ions substituted in the lattice. The same applies – less precisely – for the relative intensity of rays of a given rational series.

7.2. USE OF VARIOUS OTHER PHYSICAL METHODS

The need for exact determination of certain details of structure has led research workers to have recourse to the most modern methods of investigation of the solid state.

Electron paramagnetic resonance and nuclear magnetic resonance have been fairly extensively used for some years. Ché *et al.* [36] have given an account in 1974 showing the conditions for application of these methods and mentioning some of the results obtained. These deal particularly with hydroxides and zeolites. As far as true clays are concerned, research has mainly been on the behaviour of ions such as manganese, copper and chromium in the exchangeable state.

The only case considered in detail is that of iron entering into the structure. The authors reach the conclusion that electronic paramagnetic resonance enables us to define the type of symmetry of its environment, and hence to locate this ion. Three types can be distinguished: those present in tetrahedral sites corresponding to a substitution $Si^{+4} \rightarrow Fe^{3+}$, those included in the lattice of an oxide mixed with the silicate phase of type Fe_2O_3 or Fe_3O_4, and those in octahedral or exchangeable positions. This has enabled us to detect the presence of octahedral iron in kaolinites. In addition, the signals corresponding to the presence of Fe^{3+} ions are accompanied by unusual features leading to hypotheses concerning the deformations which they introduce into the lattice. These latter may also be investigated by refined crystallographical methods. It is very useful to be able to compare results obtained by different methods, considering the difficulty of interpretation of the results.

Finally we must make special mention of Mössbauer spectroscopy amongst the most recent methods. Most of this work has been carried out in France in 1968, 1971 and 1973 by Janot et al. [37]. By this method they have confirmed that Al^{3+} is substituted for Fe^{3+} in haematite and in goethite, and the reverse in diaspore and kaolinite.

7.3. USE OF THE ELECTRON MICROSCOPE

The electron microscope provides images of elementary particles, and is of great potential value in the study of very small crystallites.

A review of these investigations will be found in the reference works quoted, in particular in those by Grim and Magnan.

We can say that there is a general tendency to find the characteristic hexagonal forms of these minerals in the macrocrystalline state.

Characteristic features are also found. Halloysite may be in the form of tubes, the sheets being rolled up. Hectorite is found in the form of lath. Allevardite is found in the form of very long ribbons, sometimes rolled like wood shavings. The elementary particles of sepiolite and palygorskite, two fibrous minerals, also exhibit this aspect, when examined by this technique.

These morphological data are valuable since it is clear that with such asymmetrical shapes the concept of mean diameter no longer has any meaning. A knowledge of shapes is therefore fundamental information for an understanding of the rheological behaviour of these materials.

Techniques are steadily improving and amongst the most remarkable pictures obtained recently we can quote those of sepiolite which show the small channels in the structure. (M. Rautureau et al., 1974*).

But the electron microscope has another advantage, that of producing microdiffraction pictures. Since it is possible to work on a very small surface of a mineral located by its shape, leads to determination of the very fine special features of its structure, by examination of the symmetry of the diagrams. A paper by J. Méring

---

* Rautureau, M. and Tchoubar, C.: 1974, *J. Appl. Cryst.* **7**, 293–294.

and his collaborators* should receive special mention in this connection. They have shown fairly directly that nontronite corresponds to the three-dimensional group: $c(2/m)$ and that the structure is derived from the ideal type by a slight rotation of the tetrahedra of the hexagonal layer and by a distortion of the octahedral layer.

8. The Use of Infrared Spectroscopy

Infrared spectroscopy is one of the most important methods used for investigation of details of structure. This technique has not only provided the means of determining the arrangement of the basic framework, but has also given essential information about the properties of hydroxyls.

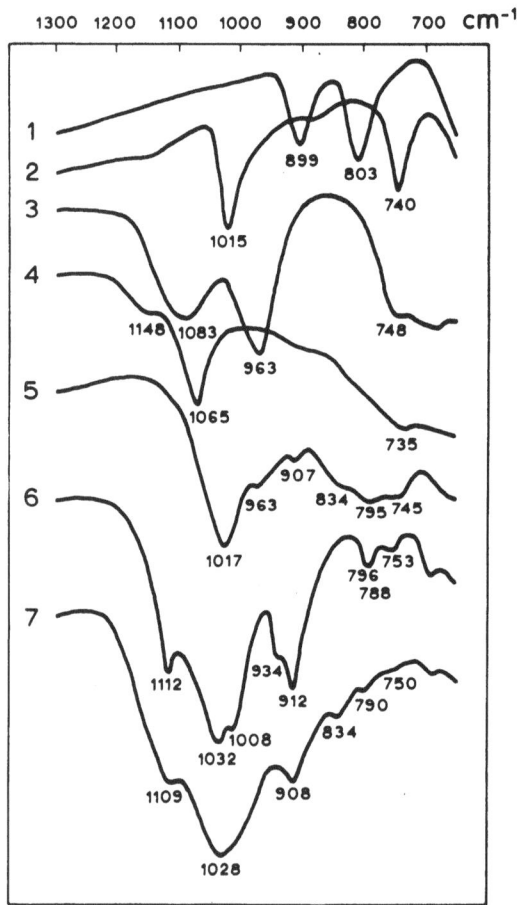

Fig. 8. Infra-red absorption curves for various minerals: clays and hydroxides. 1 – Goethite; 2 – Lepidocrocite; 3 – diaspore; 4 – synthetic boehmite; 5 – Gibbsite; 6 – Kaolinite; 7 – Montmorillonite, obtained with a double beam Parkin-Elmer 21 spectrograph of low resolving power enabling spectra to be exibited on a small scale.**

* Méring, J. and Oberlin, A.: 1964, *Bull. Gr. fr. des Argiles* **14**, 147.
** Caillère, S. and Pobeguin, Th.: 1965, 'Mineralogical composition and formation of bauxites', *Mémoire Muséum National d'Histoire Naturelles* **12**, part 4, p. 212.

8.1. Infrared Spectroscopy Applied to Phyllitous Materials

When incident radiation of intensity I_0 passes through a sample of thickness u, the transmitted intensity I is such that:

$$\frac{I}{I_0} = e^{cu} \frac{\cos^2 \theta}{\cos r},$$

where c is an absorption coefficient depending on the nature of the sample and the wavelength; and θ is the angle between the electric vector of the incident radiation field and the direction of the moment of transition.

The absorption coefficient depends on the frequency v of the vibration, which is determined by the mass of the atoms and by the coupling forces between them. This is given by the following formula:

$$v = \frac{1}{2\pi} \sqrt{\frac{k}{\mu}},$$

where μ is the reduced mass of the oscillating system, and k the restoring constant which is a measure of the coupling forces.

In place of the frequency v the wave number

$$v' = v/c$$

is generally used, where c is the speed of light. v' is therefore expressed in cm^{-1}.

The information obtained by use of infrared rays is characterized by a number of bands, by their frequencies or wave numbers, to some extent by their intensities, and by the fact that absorption may or may not depend on the orientation of the specimen with regard to the incident radiation.

Since absorption bands possess a definite width, they can overlap to a greater or lesser extent, forming a sort of group. Thus low intensity bands may be masked. It is possible to analyse these groups by separating them into elementary bands (Chaussidon and Prost*).

In the case of simple molecules it is possible to predict their absorption spectra, taking into account their symmetry. But application of these fundamental arguments to materials as complex as phyllitous silicates comes up against insurmountable difficulties. Researchers have therefore made use of analytical methods by comparing the spectra obtained with the crystallo-chemical structure of typical materials, in order to relate the absorption data to the composition of the mineral. This approach neglects the interactions. By comparing more complex materials, we find evidence of progressive modifications which are thus regarded as second order effects. The observed results may always be discussed by taking account of the point symmetry of groups of atoms.

Another method of verification uses the changes in the spectrum due to physical or physico-chemical changes of the lattice.

* Chaussidon, J. and Prost, R.: *Bull. Gr. f. Argiles* **19**, 25–38.

We can group the main types of vibration and the wave number of the absorbed frequencies in the following manner:

Vibrations	ν cm^{-1}
Valency vibrations M$-$O	400–550
Deformation vibrations Si$-$O$-$M$\}$	
Deformation vibrations M$-$OH$-\}$	550–960
Valency vibrations:	
$\quad$ Si$-$O, Al$-$O$-$Si, Si$-$O$-$Si	900–1100
Valency vibrations OH	3000–4000

In this table, M represents any cation of the octahedral layer. In the AL$-$O$-$Si system, the aluminium is tetrahedral.

The vibration of the OH bond corresponds to the change in the anion-proton distance. This is due to the mobility of the proton, because its weight is small compared with oxygen.

8.2. SOME RESULTS IN CONNECTION WITH THE BASIC FRAMEWORK OF MINERALS

These have been selected by way of examples in order to illustrate the principles which have just been stated. In this section we will consider the phenomena concerned with the set of bonds which ensure rigidity of the sheet.

For example, an analysis reported by V. C. Farmer [38] on the subject of talc, treats the Si$-$O vibrations independently of the vibrations linked to Mg. He distinguishes, according to their symmetry, five possible types of vibration:

– ν_1 is a pure frequency of elongation of the O with regard to the Si, which he associates with an absorption at 1045 cm^{-1}.

– ν_3 implies an elongation and a torsion corresponding to 1018 cm^{-1}.

– the two vibrations ν_2 and ν_4 combine an elongation and a torsion of the oxygens vibrating partially perpendicularly to the plane ab; their wave numbers are 690 and 670 cm^{-1}.

– the vibration ν_5 corresponds essentially to a torsional vibration and contributes a band centred on 450 cm^{-1}.

He locates the Mg vibrations by comparing absorption in talc with that in brucite. A vibration at 458 cm^{-1} appears to be linked with a motion in plane ab.

Examination of all the data and comparison with the spectra of two other minerals containing magnesium, saponite and hectorite, results in the following table (p. 213).

In 1961, Stubican and Roy [39] introduced certain interactions and based their reasoning on comparison of spectra of products obtained by synthesis in which the substitutions might have been modified in a fairly systematic manner. Thus certain Si$-$O vibrations are displaced towards the lower frequencies when Al replaces Si in the tetrahedral layer. The same is observed when Al(VI) is replaced by Mg(VI) in the octahedral position. The M$-$OH deformation vibrations as mentioned in 1960 by Serratosa [40] are situated at 820 cm^{-1} for Fe^{3+} and at 915 cm^{-1} for Al^{3+}.

Interpretation of the $I-R$ spectra of talc, saponite and hectorite after
Farmer (1958). B. Siffert (1962)[a]

	Talc (cm^{-1})	Saponite (cm^{-1})	Hectorite (cm^{-1})	Brucite (cm^{-1})
Si$-$O ν_1 $\perp$	1045 cm^{-1}	1056 cm^{-1}	1073 cm^{-1}	
Si$-$O ν_3 $\parallel$	1018 cm^{-1}	1005 cm^{-1}	1011 cm^{-1}	
Si$-$O ν_2 $\perp$	690 cm^{-1}	692 cm^{-1}	696 cm^{-1}	
Si$-$O ν_4 $\parallel$	670 cm^{-1}	655 cm^{-1}	655	
MgO ν_6 $\perp$	539 cm^{-1}	534 cm^{-1}	533 cm^{-1}	560 cm^{-1}
MgO ν_7 $\parallel$	467 cm^{-1}	464 cm^{-1}	465 cm^{-1}	458 cm^{-1}
SiO ν_5 $\parallel$	426 cm^{-1}	420 cm^{-1}		

[a] B. Siffert, Complementary Thesis, Strasbourg. New physical methods
in the study of clays: The infrared spectroscopy (1962).

In 1960 Lyon and Tuddenham [41] observed that the ν_3 band moves towards shorter wavelengths as the degree of substitution of Si by Al increases in the tetrahedral position in the case of chlorites.

The following experiment shows how data may be derived from differential attack of minerals: in an acid medium and subject to certain precautions, the octahedral layer of glauconite is affected much more rapidly than the tetrahedral layer. The ν_3 band is greatly attenuated and is displaced towards shorter wavelengths as the attack takes place. It disappears when the octahedral layer is sufficiently changed. On the other hand, the characteristic band of the tetrahedral layer at 9.2 μ approximately is reinforced and retains its frequency.

These different results show that, despite the complexity of the problem, it is possible, by associating the chemical data and certain fundamental principles about structure, to obtain confirmation of a crystal-chemical nature.

8.3. THE STUDY OF HYDROXYLS

This investigation is carried out by using the vibration numbers of OH, their frequency and the polychroism effect.

Following Pimentel and McClellan's work [42] of 1960 we can draw up the following table:

Valence vibrations (elongation of OH distance)	2000 to 3750 cm^{-1}
Deformation vibrations in the R OH plane	1000 to 1700 cm^{-1}
(Rotation of the proton around oxygen in this plane)	
Deformation vibrations out of the R OH plane	300 to 900 cm^{-1}
(Same deformation perpendicular to the above)	

The study of kaolinite enables us to illustrate the techniques used in this field. The spectrum is characterised by three bands situated at 3690, 3668, and 3620 cm^{-1}.

Attempts have sometimes been made, e.g. by Fripiat [43] in 1960, to bring the number of vibrations observed closer to the number of sites available for the proton, which itself depends on the occupation of the four tetrahedral orbitals of the oxygen. Figure 9 shows the existence of these sites. It will be seen that there is a unit cell of kaolinite, and for the elements of the octahedral layer there are 12 oxygens and hence 48 possible orbitals. Four are linked to the silicon, and because of the six co-ordinate bonds due to the four Al, 24 others are therefore occupied. There remain 20 available

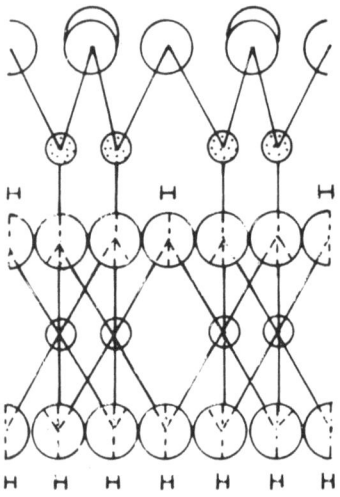

Fig. 9.   Projection onto a plane *bc* of the constituents of the unit cell of a kaolinite. The dotted arrows represent the oxygens of the two compact layers, and indicate the orbitals of the oxygens.

for the 8 hydrogens, i.e. 2.5 orbits per proton. We should observe between two and three bands for the vibrations linked to these oxygens.

For the equivalent mineral containing magnesium, antigorite, the six Mg ensure the occupation of 36 orbitals which, added to the four bonded to the silicon, leave only 8 orbitals free for the 8 hydrogens, and we should therefore observe only one single vibration.

Unfortunately this reasoning is not sufficiently convincing. In 1964 Ledoux and White [44] established for kaolinite that the band with the lowest frequency originates from the hydroxyl turned towards the tetrahedral layer, and the three bands with the highest frequencies from the hydroxyls turned towards the surface of the sheet. This leads us to think that there are three moments of vibration for surface hydroxyls: the one oriented perpendicular to the sheet gives the band at 3697 cm^{-1} and the two others more or less inclined account for the bands at 3669 and 3652 cm^{-1}.

This diagram is more satisfactory since it appears to explain the spectra of nacrite and of dickite, which differ from kaolinite only in the details of superposition of the sheets (See Farmer and Russell [45]).

It must be added that in addition to these symmetry considerations, the hypotheses can be confirmed by deuteration. This is effected by placing the mineral in contact with heavy water, but as not all the hydrogen atoms are accessible to this treatment, it is possible to differentiate certain OH groups by this method. The frequency changes predicted by the formula given at the beginning of this section are very appreciable. Thus we move from a zone of 3700 cm^{-1} for OH to 2600–2700 cm^{-1} for OD. In 1970 Russell *et al.* [46] used this method to attribute certain absorption bands for a series of triphormic minerals.

The orientation of the OH groups has also been connected with the crystallochemical composition, but in this case it is necessary to take into account all the particular features of the structure. The main feature is the difference between the di- and the tri-octahedrals clearly reported by Bradley and Serratosa [47] in 1958.

In 1964, Vedder [48] resumed the study of micas, and considered the interpretation of five bands of phlogopite with frequencies extending from 3700 to 3565 cm^{-1}. The first two at 3700 and 3665 cm^{-1} are called N and I respectively. They are greatly affected by the orientation of the sample with respect to the incident radiation. They are therefore due to vibrations of the OH groups approximately perpendicular to the plane of the sheet. The three other components at 3620, 3600 and 3565 cm^{-1} are almost independent of the orientation and are known as V bands. The interpretation is connected with the details of the structure in the following manner:

Band N   corresponds to OH ions dominating the octahedral sites occupied by divalent ions.

Band I   OH ions dominating octahedral sites occupied by two divalent and one trivalent ion.

Band V   OH ions situated in the neighbourhood of a vacant octahedral site.

In accordance with this scheme, trioctahedral minerals give essentially intense N and I bands, whilst dioctahedral minerals tend to give V bands.

This general scheme is then modified by a certain number of special features; for example the orientation of the OH groups forming the N and I bands will be affected by tetrahedral substitutions which create local charge deficits.

The nature of the cations of the octahedral layer causes a modification of the frequency of vibration of the OH groups. For example, an OH group above a group of cations $Fe_1^{++}Mg_2$ absorbs at 3685 cm^{-1} and for the $Fe_2^{++}Mg_1$ group absorption takes place at 3660 cm^{-1}, the line then being wider. If we consider phlogopite in place of biotite, we find a line at 3665 cm^{-1} linked with the Al_1Mg_2 group. A great amount of work has been carried out on these effects (see Farmer and Russell [45]).

The influence of compensatory cations and adsorbed molecules on the surface of the sheet should be mentioned. In the case of micas, the presence of potassium modifies this frequency. For example, Chaussidon showed in 1970 that natural phlogopite [49] exhibits two bands at 3712 and 3669 cm^{-1}. After extraction of the potassium, replacement by lithium and dessiccation of the sample (brought to 10 Å) these same bands are observed at 3661 and 3630 cm^{-1}. This latter sample hydrated exhibits bands at 3677 and 3645 cm^{-1}.

There is a considerable amount of literature devoted to the study of interactions between the sheets of mineral clays, the liquid molecules and the cations fixed in them (see Farmer and Russell [45]). When these latter are complex they sometimes undergo fundamental transformations; for example, an amine may liberate the carbon chain and leave the basic group on the site exchangeable in the form of ammonia. This was reported in 1965 by Chaussidon and Calvet [49b]. In addition the water molecule itself exhibits a much greater dissociation when it is fixed in the interlayer spaces, as reported in 1963 by Calvet et al. (*Bull. Gr. Fr. des Argiles* **15**, 59–98) The reader will have to refer to specialized journals to obtain more information about the great number of papers published on this subject.

9. Current Problems Concerning the Structure

9.1. DIMENSIONS OF UNIT CELL AND CHEMICAL COMPOSITION

We have already drawn attention to the fact that the anion-cation distances should vary with the type of bond.

In 1953 various authors including Brindley and McEwan [50] drew up an empirical formula connecting the composition of the unit cell with the dimensions of its parameters. Thus they reached the following two formulae:
Dioctahedral minerals:

$$b(\text{Å}) = 8.92 + 0.06x + 0.09q + 0.18r + 0.27s$$

Trioctahedral minerals:

$$b(\text{Å}) = 9.19 + 0.06x - 0.12p - 0.06q + 0.06s$$

where for each half unit cell:

p = number of Al ions ⎫
q = number of Fe^3 ions ⎪
r = number of Mg ions ⎬ in the octahedral layer
s = number of Fe^2 ions ⎭
x = number of Al ions in tetrahedral positions.

Use of this formula in a few simple cases shows good agreement between the calculated values and the experimental results.

If we consider the case of muscovite in relation to pyrophyllite where $b_0 = 8.92$ Å, we must add one tetrahedral Al, i.e. $8.92 + 0.06$, giving 8.98 compared with the measured 9.04 Å, whereas paragonite, having the same formula as muscovite with potassium replaced by sodium, has a value $b_0 = 8.88$ Å. The calculated value lies between the two experimental values, the deviation being attributed to the replacement of sodium by potassium. In the case of phlogopite, one Si is replaced by one Al in the tetrahedral position. From the formula we calculate: $9.19+0.06=9.25$ compared with the measured 9.23 Å. The value 9.19 corresponds to b for talc in which no substitution takes place.

The authors suggest similar formulae for chlorites based on the hypothesis that the octahedral cations are distributed equally between the two layers. The aluminium present will frequently be distributed equally between the octahedral and tetrahedral positions since the excess charge due to replacement of this trivalent ion by a divalent ion compensates the charge deficit created by the tetrahedral substitution.

Under these conditions, the two coefficients $+0.06$ and -0.06 in the formula multiplied by the same quantity cancel each other. The evaluation of the parameter b for the chlorites as a function of chemical composition is as follows:

$b\text{Å} = 9.19 + 0.03y^*$ compatible with those of various authors.
$b\text{Å} = 9.21 + 0.032y^*$ (von Engelhardt [51] and McMurchy [51b])
$b\text{Å} = 9.20 + (0.028\ Fe_{tetra}) + 0.047\ Mn$, Hey [51c]
$b\text{Å} = 9.21 + 0.037(Fe^2 + Mn)$, Shirozu [51d]
y^* represents in these cases the number of Fe^{2+} ions in the formula.

It is of interest to compare these results with the differences in diameter of the cations which are able to replace the cations of the most representative mineral of the series. These values are shown in the following tables:

Tetrahedral elements

(Radii of substitution cations) – [(radius of Si)=0.42]

$$Al\quad 0.51 - 0.42 = 0.09$$
$$Fe^3\quad 0.64 - 0.42 = 0.22$$
$$Cr^3\quad 0.64 - 0.42 = 0.22$$

Octahedral elements

(Radii of substitution cations) – [(radius of Al)=0.51]

$$Fe_3\quad 0.64 - 0.51 = 0.13$$
$$Mg\quad 0.66 - 0.51 = 0.15$$
$$Ni\quad 0.69 - 0.51 = 0.18$$
$$Co\quad 0.72 - 0.51 = 0.21$$
$$Fe^2\quad 0.74 - 0.51 = 0.23$$

(Radii of substitution cations) – [(radius of Mg)=0.66]

$$Al\quad 0.51 - 0.66 = -0.15$$
$$Fe_3\quad 0.64 - 0.66 = -0.02$$
$$Ni\quad 0.69 - 0.66 = \quad 0.03$$
$$Co\quad 0.72 - 0.66 = \quad 0.06$$
$$Fe^2\quad 0.74 - 0.66 = \quad 0.08$$

It is obvious that the sign of the differences corresponds with the signs of the coefficients used. For dioctahedrals, all the cations are larger than aluminium, therefore the difference is positive. On the other hand, for trioctahedrals, the differences are either positive or negative depending on the sign of the weighting factor in the formula.

These different results are in agreement, at least qualitatively with the hypotheses of additivity of ionic radii corresponding to the properties of bonds mainly ionic.

Since the dimensions of certain elements of the unit cell are changed, we may wonder how the others are affected. For example, how will the tetrahedral layer adjust itself to layers as different as:

(1) kaolinite, where the octahedral layer should correspond effectively to the parameters of gibbsite, i.e.

$$a = 5.07 \text{ Å} \qquad b = 8.64 \text{ Å}$$

or

(2) antigorite, where a tetrahedral layer which is identical in principle must adapt itself to an octahedral layer of dimensions:

$$a = 5.45 \text{ Å} \qquad b = 9.44 \text{ Å}$$

In 1954, Smith [52], investigating a considerable number of minerals, showed that the most probable $Si-O$ distances were 1.60 ± 0.01 Å. A tetrahedral layer formed in accordance with the model given in the first part of this paper, would, based on these values, have parameters corresponding to:

$$a = 5.22 \text{ Å} \pm 0.04 \text{ Å} \qquad b = 9.05 \pm 0.06 \text{ Å}$$

Various authors have attempted to explain this particular feature. It was necessary to modify the organisation of the hexagonal layer, whilst still preserving a certain number of features corresponding to the properties of the unit cell, that is the fact that its translation about the axes of reference occupies the whole plane.

In order to obtain the structure of dickite, Newnham and Brindley [53] imagined, a pivoting of the tetrahedra about the oxygens situated in the first compact plane. Their diagram is reproduced in Figures 10 and 11.

It will be seen that this pivoting effect leads to mutual approach of two oxygens, which leads in turn to a slight stress of the hexagon which loses its six-fold symmetry and becomes ditrigonal. On referring to Figure 10 above, it is seen that within the dotted rectangle forming the unit cell, the six oxygens forming the base are to be found, along with the four silicons. The elements of the original diagram are therefore accounted for.

A 1954 paper by Smith [52] showed that the $Al-O$ bond of tetrahedral aluminium was: 1.78 Å ± 0.02 Å, a slightly higher value than that deduced by the additivity rule, which would be 1.69 Å. This therefore indicates that the $Al-O$ bond is less covalent than $Si-O$.

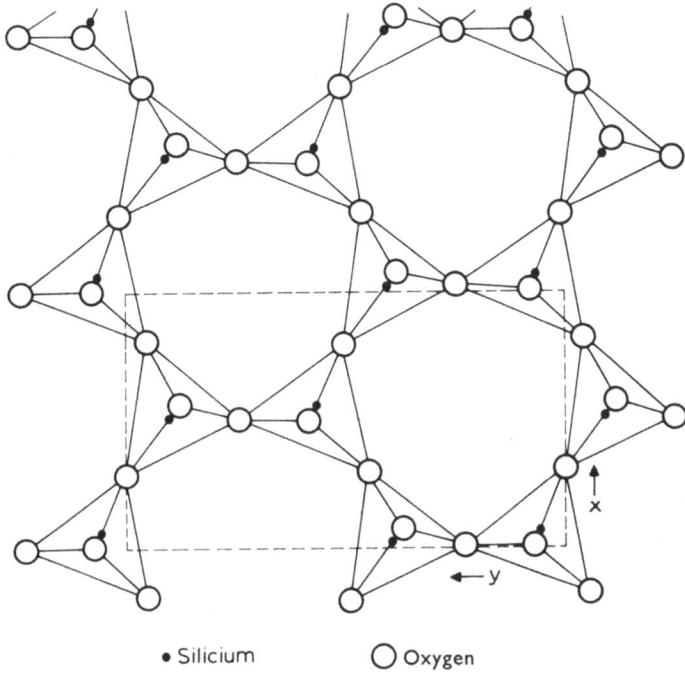

• Silicium   ○ Oxygen

Fig. 10. Rotation of silica tetrahedra according to Brindley and Newham. The elongation of the hexagon is oblique with respect to the b axis.

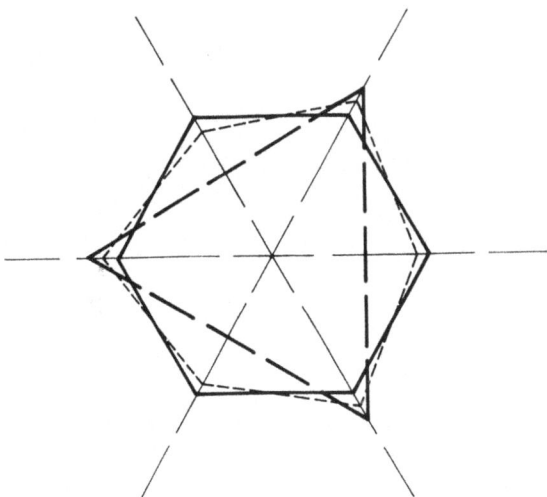

Fig. 11. Deformation of a hexagon as far as the extreme case of the equilateral triangle. The length of the side of the triangle is equal to the sum of the lengths of the two sides of the hexagon; these distances representing the Si−O distances are constant. The deformed hexagon shown by dotted lines represents an adaptation to a dilated octahedral layer.

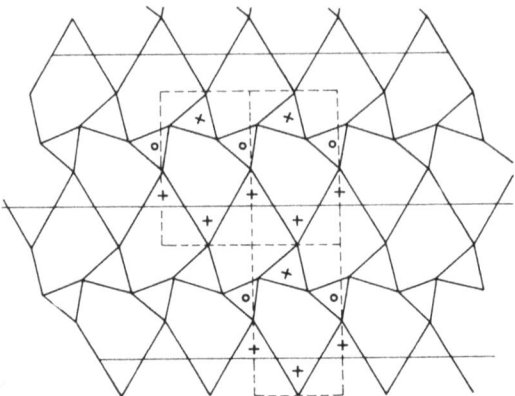

Fig. 12. Deformations of the structure due to Al—Si substitutions. These deformations imply localisation of the Al—Si substitutions. The dotted lines correspond to the bases of a unit cell (from E. W. Radoslovich).

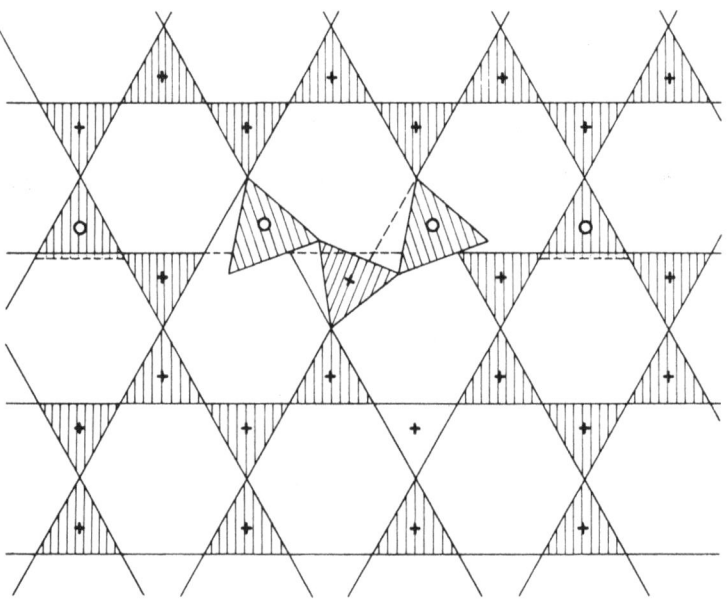

Fig. 13. Adaptation at the centre of the figure, to the structure of tetrahedra expanded by substitution of Al (white circles) for Si (black crosses) from H. S. Yoder. The dotted lines show the positions of the tetrahedra with Si at their centres. The other triangles to the right and to the left of the deviated triangles are expanded as shown by the dotted line. It is found that a rotation is necessary to ensure the assembly whilst maintaining the same dimensions *ab* for the unit cell.

The authors also sought to determine the form of the tetrahedral layer when it is perturbed by the introduction of tetrahedra of a size greater than that of the Si—O tetrahedra.

In 1956, Smith and Yoder Jr. [54] devised the diagram reproduced in Figure 12.

Here we are concerned with muscovite with the substitution of 1 Al for 1 Si/4. In order to maintain the regularity of the system, the substitutions were located on the (010) alignments. In order to 'hold' the largest sized elements, they were moved from their natural positions. It follows from this that the orientations of the bases of the equilateral triangles representing the positioning of the tetratedra on the plane, are inclined with respect to the straight line on which they are normally arranged in such a manner as to form a broken line. Due to this, the space occupied by these triangles is reduced to their projection onto the straight reference line; we thus return to the normal space occupied. However, the internal surface of the hexagon is found to be reduced. This will be more easily appreciated by consideration of Figure 13, above on which three triangles have been modified, two corresponding to tetrahedra containing one Al (white circle) and one to a tetrahedron containing one Si.

The position which these triangles would occupy in the regular theoretical diagram is shown by dotted lines. The expansion caused by introduction of the aluminium corresponds to the diagonally shaded area in the triangle to the right of the figure. This latter is left in its normal position so as to allow comparison; adjustment to it would result in its inclination towards the left; at the same time the triangle situated at its upper left apex would also pivot in the same direction. These modifications have the effect of changing the symmetry. As a result of this, reflections which were forbidden on the X-ray diagrams now become possible, which explains certain anomalies exhibited by the diagram whilst still confirming the form of the model suggested.

It has been possible to carry out very extensive studies of diffraction diagrams with the aid of Fourier projections, by the use of computers. This information enables us to determine the details of the structure by measurement of the anion-cation distances resulting from these new arrangements. In 1960 Radoslovich [55] found the following values:

	Parameters measured	Distances
Tetrahedral groups	Si_1-O	1.69
	Si_2-O	1.60
	Si_1-O apical	1.69
	Si_2-O apical	1.61
	$O-O$	2.76
Octahedral groups	$Al-O$ ⎱ $Al-OH$ ⎰	1.95
	$O-O$	2.76
Interfoliary cation	$K-O$	⎱ 2.81 ⎰ 3.39

These results lead us to assume two groups of different distances: Si_1-O and Si_2-O. This latter group corresponds to the $Al-Si$ substitution. On the other hand, the distance $O-O$ from Si and Al is always the same whilst $Al-O$ is greater than

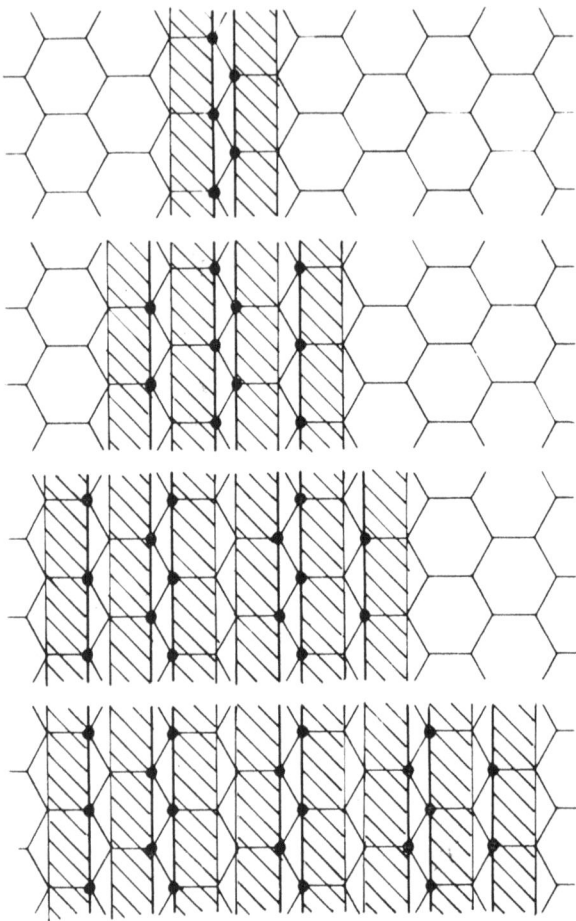

Fig. 14. Arrangement seen perpendicularly to plane *ab*, of the distribution of substitutions suggested by L. Gatineau.

Si—O. The K ion can be at two different distances from the oxygens of the base, corresponding to the deformation of the hexagon.

These results were queried in 1963 by Gatineau [56] who suggested the single value 1.64 Å for the distance Si 3/4 Al 1/4 O.

The relations between potassium and oxygen were studied by Gatineau [57] in 1965. The diffusion surrounding diffraction spots led him to the conclusion that the Si—Al substitutions are arranged regularly as can be from Figure 14.

As we pass from one sheet to the next, the bands where the substitutions are found alternate from one sheet to another as is shown by Figure 15.

This investigation has been extended to types of mica other than muscovite. In 1958 Méring and Gatineau [58] found that micas can be classified in the following manner: in margarite (2 Si 2 Al) there is a biperiodic organisation of the tetrahedral atoms. On the other hand, in muscovite, phlogopite and biotite (3 Si 1 Al) we find the arrange-

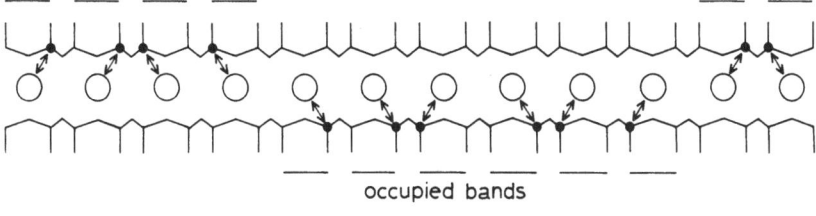

occupied bands

○ potassium

• tetrahedral aluminium

Fig. 15. Systematic distribution of charge deficits and potassium in a mica, projected on plane *bc*, as given by Gatineau.

ment shown in Figure 15. Substitution of compensating cations may modify the parameters of the unit cell for sheets of the same composition. This is apparent when we compare:

paragonite (sodium) $b = 8.88$ Å
muscovite (potassium) $b = 8.995$ Å *

These different considerations led Radoslovich and Norrish [59] to suggest the following rules for the adjustments of tetrahedral and octahedral layers:

The size of the unit cell is determined essentially by the octahedral layer and the interactions between the layers. The configuration of the surface of the unit cell (surface of tetrahedra) depends on the size of the tetrahedral layer with regard to the '*b*' axis.

(1) In all silicate layers, the tetrahedra may rotate freely to reduce the dimensions of these layers but the rigidity of groups of tetrahedra limits the extension as in serpentines.

(2) In all silicates the octahedral layer can be expanded or contracted with slightly more difficulty by changing the angle of the bonds rather than by altering the length of the bonds. A consequent modification to the thickness of the sheet occurs.

(3) For micas in particular, the oxygen tetrahedra rotate until part (probably half) of the distance between the compensating cation and the oxygen have a normal value, that is until half the oxygens are blocked on the interstrand cations.

The authors give a table of the rotation of the tetrahedra for a series of micas, with values of the parameter *b* between 8.67 Å and 9.32 Å. The tetrahedra rotate through angles between 0 and 23°.

These adjustments in the orientation of the tetrahedra lead to a certain displacement of their apices, and in consequence we can no longer regard the oxygens of the hexagonal layer as arranged in the same plane. This results in a modification of the apparent thickness of the sheets; this, however, is of little importance; see work published in 1960 by Gatineau and Méring [60].

Considering all the deformations, we are led to seek empirical relationships between the dimensions of the unit cell and the composition, by means of multiple regressions, not by simple regressions as in the above formulae. In 1962 Radoslovich determined the following relationships [61]:

* Certain authors give 9.04 Å for this parameter; see p. 216.

Diphormic minerals

$$b = (8.923 + 0.125 \text{ Mg} + 0.229 \text{ Fe}^{2+} + 0.079 \text{ Fe}^{3+} + 0.28 \text{ Mn}^2$$
$$+ 0.17 \text{ Ti}) \pm 0.014 \text{ Å}$$

Triphormic minerals

$$b = (8.944 + 0.096 \text{ Mg} + 0.096 \text{ Fe}^{3+} + 0.037 \text{ Al}_{\text{tetra}}) \pm 0.012 \text{ Å}$$

Tetraphormic minerals

$$b = (9.23 + 0.03 \text{ Fe}^{2+}) \pm 0.03 \text{ Å}$$

The different models suggested, in particular at the level of the octahedral layer, lead us to represent the grouping of the oxygens as compact. It becomes difficult to account for a singularity known as the Hofmann-Klemen effect [62]: if we fix lithium, as an exchangeable cation, on a montmorillonite, and then heat it to between 250 and 300°, the mineral exhibits a greatly reduced capacity for exchange of bases, together with a modification of the associated properties of hydration, solvation, and expansion. The hypothesis is that the Li ion has passed from the exchangeable state into the state of a component of the octahedral layer by passing through the compact oxygen layer. This transition takes place because the exchange capacity of the mineral is due to a charge deficit with regard to this layer. It appears that the Li is attracted to this position.

Much controversy exists on the subject of whether the Li reaches the octahedral layer, or whether it is only at the bottom of the hexagonal cavity.

There are some arguments in favour of the passage into the octahedral layer. For example, in the case of dioctahedral minerals such as beidellite, where the charge deficit is purely tetrahedral, no Hofmann-Klemen effect takes place, as shown in 1952 by Green-Kelley [63]. If, however, beidellite exhibits octahedral deficits, it was shown in 1967 by Glaeser *et al.* [64] that the Hofmann-Klemen effect did take place. Also it is not observed in the case of trioctahedral minerals since there are no sites available at this level. But the conclusive proof was given in 1968 by Calvet and Prost [65] who showed that after reduction of the exchange capacity by fixation of Li and heating, there was a modification in the orientation of the OH groups. In fact, whilst the direction of the OH vector is at a large angle to the *ab* plane in montmorillonite and in general in dioctahedral minerals, some of the OH groups tend to straighten; after fixation of the lithium the mineral tends to become trioctahedral.

It also appears that exchangeable Al when these montmorillonites with H ions as their exchangeable cations are dessicated, may correspond to a departure of the Al from the octahedral layer by passage through the compact oxygen plane, H ions taking their place in the hexagonal layer.

The Hofmann-Klemen effect which has just been described, and the hypothesis which has just been formulated in connection with the appearance of exchangeable Al, imply that cations pass through the compact oxygen layer; there are therefore further details of the structure which should be determined in order to interpret this phenomenon.

IV. Chemical and Physical Properties

10. Surface and Exchange Capacity

One of the characteristics of phyllitous materials, of silicates in particular, is that of existing in the form of crystals of very different sizes. This applies to minerals belonging to the same family although the macrocrystalline appearances differ in certain features from their homologues found in the microcrystalline state; certain families exist only in the latter form. It follows that the distinguishing features of these minerals are their size and their surface.

Measurements of the size of the particles may be carried out by means of an electron microscope. The conditions of preparation must be stated precisely since there may be, between the crystallites, clusters with a greater or lesser degree of stability which resist the treatments aimed at separating them. This is particularly apparent when there is a mixture, such as oxide-sesquioxide, crystallized or not, silica gels or complex gels of silica and various hydroxides.

The separation of a mixture of clay particles of different sizes is carried out by sedimentation for the finer particles, where the initial material is subjected to physico-chemical treatments to form a stabilised suspension; the larger particles are separated by sieves. We will not enter into details of the techniques used, but refer the reader to specialized publications.

We recall that there are several classifications of particles according to their size, and we will mention one of these for which the principle is due to Atterberg, since it is in current use by specialists.

Names of class of particles		Dimensions
Coarse sand		$\begin{cases} 2 & \text{mm to } 0.5 & \text{mm} \\ 0.5 & \text{mm to } 0.2 & \text{mm} \end{cases}$
Fine sand $\begin{cases} \\ \end{cases}$	True sand	0.2 mm to 0.05 mm
	Very fine or coarse loam	0.05 mm to 0.02 mm
Laom		0.02 mm to 0.002 mm
Clay $\begin{cases} \\ \end{cases}$	Coarse	0.002 mm to 0.0005 mm
	Fine	<0.0005 mm

Depending on different authors, there are slight modifications to the limits of the intermediate classes; it can also happen that only three classes are allowed for:

> sands from 0.2 to 0.05 mm
> loams from 0.05 mm to 0.002 mm
> clays <0.002 mm

When such a separation is carried out on a complex material it is generally found that the mineralogical composition varies appreciably with the class of particle. The finest class, the clay fraction, especially the fine clay fraction, contains almost systematically minerals belonging to the phyllitous silicates whether associated with hydrox-ides or not. The minerals in this fraction are considered as clay minerals, or briefly clays.

10.1. SURFACE AND FORM

The surface of a set of particles is defined by the specific surface, Ss, expressed as surface area per unit weight; we can therefore write:

$$Ss = \frac{S}{P}$$

where S is the surface and P the weight of the sample.

There are two means of determining the surface of a sample: the size of the particles from which it is composed, or measurement of the surface directly by absorption methods.

In order to evaluate the size from the surface the unitary surface:

$$Su = \frac{S}{V}.$$

We pass from S_s to S_u by putting:

$$\frac{S_u}{\Delta} = S_s,$$

where Δ is the density of the mineral.

In order to evaluate S_u we must initially suppose that we are concerned with a sample consisting of n particles, of surface s and volume v. We have therefore:

$$\frac{S}{V} = \frac{ns}{nv} = \frac{s}{v}$$

It becomes necessary to make an assumption as to the shape of the particles. Supposing they are cubic:

$$S_u = \frac{6a^2}{a^3} = \frac{6}{a},$$

where a is the side of the cube.

For spherical particles we find $S_u = 6/d$ where d is the diameter.

These hypotheses regarding shape are implied in the methods of separation which classify particles according to their diameters, assuming the particles are effectively isodimensional.

Assuming that the particles are assymmetrical shaped as right prisms with rectan-gular bases, of sides x and y, and thickness z, we then have:

$$S_u = \frac{2xy + 2xz + 2yz}{xyz}.$$

Some conclusions can be drawn from this general equation. If we assume a constant shape with $x=a$, $y=b$ and $z=c$, the equation becomes:

$$S_u = \frac{2ab + 2bc + 2ac}{abc}.$$

It is sufficient to fix the absolute value of two of these coefficients, in order to calculate the surface the third being variable.

For very asymmetrical particles, those where one dimension is much smaller than the other two, that one determines the surface; and it does so to such an extent that the other dimensions may vary fairly considerably without changing the order of magnitude of the final result.

If we now put $a=b=x$ being variable and $c=z$, then

$$S_u = \frac{4x^2 + 2xz}{x^2 z} = \frac{4}{z} + \frac{2}{x}$$

If we assume x constant and z variable, $Su = A + 4/z$ which tends to A when z tends to infinity, corresponding to the case of fibrous particles. If on the other hand we assume z to be constant, then $S_u = B + 2/x$ tends to B for very large x, that is for very thin plates. This is in effect what takes place in montmorillonite where the sheets are 10 Å thick. This calculation assumes that we can vary the dimensions and shape as if the particles were progressively deformed. However, the variation in thickness of the minerals that we are investigating can only be achieved in jumps, that is by a certain number of elementary sheets. For this reason in 1969 Hénin et al. [66] suggested that calculation of these surfaces should be based on the characteristics of the unit cell, but on the assumption that they are associated either in two or three dimensions, giving the following formulae:

(1) Case of platelets, assembled in two dimensions:
where

$$s = (2N_1 s_b + 2\sqrt{N_1} sl_1 + 2\sqrt{N_1} sl_2),$$

$$S_u = \frac{2N_v}{N_1} (N_1 s_b + \sqrt{N_1} sl_1 + \sqrt{N_1} sl_2),$$

N_1 is the number of unit cells forming the platelet of an elementary particle;
N_v is the number of elementary cells;
s_b is the basic surface of the elementary cell;
sl_1 and sl_2 are the lateral surfaces of the elementary cell;
s_u is the surface relative to the unitary volume.

(2) In the case of a prism we have:

$$S_u = \frac{2N_v}{N_1 N_2} (N_1 s_b + N_2 \sqrt{N_1} (sl_1 + sl_2),$$

where the symbols in this formula have the same meaning as in the previous formula and N_2 is the number of superposed platelets.

In order to make these concepts precise, the number n_v of particles present in unit volume is

$$n_v = \frac{N_v}{N_1}$$

in the case of platelets, and

$$n_v = \frac{N_v}{N_1 N_2}$$

in the case of prisms.

S_u can be related to the specific surface, that is, it can be calculated per unit weight:

$$S_s = \frac{S_u}{\text{density}}$$

10.2. MEASUREMENT OF SURFACES

Starting from theories on the mechanism of adsorption, it is possible to deduce methods enabling us to measure the surface of the particles constituting a given sample. Comparison of the results obtained enables us to test the validity of the hypotheses on which these methods are based. For this reason we will describe some of them.

One of the most extensively used, known as the B.E.T. method (S. Brunauer, S. P. H. Emmett, and E. Teller) [67], published in 1958, depends on the adsorption of a non-polar gas. This is most often nitrogen but gases with heavier molecules may be used, such as argon or propane. Equivalent results are obtained if the surfaces are really plane.

In such gases the molecules do not tend to take up any particular orientation, as would be the case for polar molecules, water in particular. Measurements are carried out in the region of the liquefaction temperature of the gas used.

The fundamental equation is put in the form:

$$x = ac(S - y),$$

where
x is the quantity of adsorbed molecules,
c is the concentration of molecules in the external medium,
S is the surface of the adsorber,
y is the surface of the adsorber already covered by the adsorbed molecules,
a is the equilibrium constant of the reaction.

The authors used such an equation to express the equilibrium of the successive layers of gas fixed by the adsorber. They were able to establish a general equation relating the quantity of gas adsorbed at the surface of a solid with its relative pressure.

The equation, limited to adsorption of the first layer, is as follows:

$$\frac{X}{V(1-X)} = \frac{1}{V_m C} + \frac{C-1}{V_m C} X,$$

where

$X = p/p_0$ (ratio of actual vapour pressure to saturated vapour pressure);

V_m is the volume of molecules of adsorbed substance necessary to provide the adsorber with a monomolecular layer;

V is the total volume adsorbed;

C is the equilibrium constant.

The straight line obtained by plotting the first term of this equation as a function of X has as ordinate $1/V_m C$ for $X=0$, and the slope is $(C-1)/V_m C$.

The sum of these two values is:

$$\frac{1}{V_m C} + \frac{C-1}{V_m C} = \frac{1}{V_m}.$$

Therefore the surface covered by a monomolecular layer can be calculated directly by this very simple equation, if we know the surface occupied by a volume V_m of molecules. This expression is valid for $X < 0.35$.

Other methods may be used, such as that of Harkins and Jura, which depends on the following equation:

$$\log(p/p_0) = b - kA^2/V^2,$$

where

A is the surface of the adsorber;

V is the volume adsorbed;

b and k are constants;

p and p_0 are the actual and saturated vapour pressures respectively of the adsorbed substance;

A is obtained by graphing $\log(p/p_0)$ against $1/V^2$.

This relation is generally found true for $0.5 < p/p_0 < 0.9$. A complete account of these theories has been published in 1971 by Fripiat et al. [68].

Here is some of the data published in 1952 by Escard [69] for the surfaces presented per gram of certain clay minerals:

Mineral	m² g⁻¹	Form of surface
Attapulgite	140	small channels
Illite	113	compact sheets
Kaolinite	22	compact sheets
Dehydrated montmorillonite	82	external surface
Dehydrated nontronite	75	external surface
Sepiolite	394	small channels

Adsorption of a non-polar molecule implies that we are dealing with dehydrated materials, this being incomplete in presence of humidity. Certain minerals such as montmorillonites and vermiculites have their elementary sheets separated by layers of water molecules, in such a manner that measurements of the surface give only the envelope of the sets of sheets and not the surfaces developed in contact with a liquid phase composed of polar molecules such as water.

In 1950, Hendricks and Dyal [70] devised another technique consisting of the use of ethylene glycol. The molecules of this substance, due to their behaviour as an alcohol, can take the place of water molecules on the adsorbing substance, and their needle-like shape allows their regular arrangement on the adsorbing surface. Knowing the molecular volume of this substance, we can calculate a priori the weight adsorbed; it is 1 mg for 3.22 m^2.

The technique consists to imbibe the mineral being investigated with glycol. It is then placed in a container which is evacuated. Using a quartz spring balance, the change in weight of the system is measured as a function of time, and the existence of a desorption plateau is shown. X-ray examination shows that such a plateau corresponds to the adsorption of a monomolecular layer on the surfaces of the minerals.

Below are some values obtained by this method:

Mineral	Origin	$\text{m}^2 \text{ g}^{-1}$
Kaolinite		45
Illite		89.6
Montmorillonite	Wyoming	761
Montmorillonite	Tehachapi	719
Montmorillonite	Cheto	780
Halloysite		44.4

A third type of technique is based on negative adsorption. We will see later that when clays are in suspension, the adsorbed cations become dissociated and that the agglomerations behave as if they were negatively charged particles. This leads to repulsion of the anions present in the solution close to these particles. This phenomenon is known by the name of negative adsorption. It is only exhibited by certain anions. Theories put forward in 1910 by Gouy [71] and in 1913 by Chapman [72] justify relating this repulsion effect with the charge carried by the particle. Knowing the exchange capacity, and hence the potential charge after dissociation of the cations from the unit weight of clay, we need only divide this exchange capacity by the charge density to find the contact surface of the unit weight of the material concerned. Below are some results showing the agreement between measurements carried out by various techniques.

There is therefore a certain degree of agreement between these different values. Measurements of negative adsorption relate to the surface of the mineral only if the

Mineral	External surface measured with N	Internal and external surface	Surface measured by negative adsorption	Authors
Kaolinite	15.5 m² g⁻¹	–	16 m² g⁻¹	Laudelout [73]
Montmorillonite Camp Berteau	–	800 m² g⁻¹	750 m² g⁻¹	Laudelout [73]
Complex clay extracted from a sol	184 m² g⁻¹	444 m² g⁻¹	saturated Na: 670 m² g⁻¹ saturated Sr: 180 m² g⁻¹	Chaussidon]74]

concentration of ions in the solution in contact is sufficiently high. In addition, for divalent cations, we find systematically lower values for the surfaces, due to the lower ionic concentration, which is, however, compensated to some extent by the presence of ions of the type $(M\ OH)^+$.

10.3. EXCHANGE PHENOMENA AND EXCHANGE CAPACITY

it has been known since the time of Thomas Way [3] in 1850 that if a clay material is placed in contact with a saline solution, cations may be exchanged between the solution and the solid phase. For example we can put:

$$Clay_{Li} + NaCl \rightleftharpoons Clay_{Na} + LiCl$$

in the same way:

$$Clay_{Ca} + 2NaCl \rightleftharpoons Clay_{Na} + CaCl_2$$

Equilibrium is governed by Gapon's equation which can be written as:

$$\frac{adsorbed\ Li}{adsorbed\ Na} = \frac{sol\ Li}{sol\ Na} \times c$$

in the same way:

$$\frac{adsorbed\ Ca}{adsorbed\ Na} = \frac{sol\ Ca}{sol\ Na^2} \times c_1$$

where c and c_1 are constants.

We can in fact write the second term of the second equation as: $Na/\sqrt{Ca}$. In practice we use a ratio $Na/\sqrt{Ca + (Mg/2)}$ which expresses the probability of observing a soil becoming enriched with sodium. Such an expression shows that the equilibrium of solutions containing monovalent and divalent cations in contact with the clay is a function of the dilution. In 1949 Schofield [75] systematically verified the validity of the expression not only for ratios between monovalent and divalent cations, but also for trivalent cations. He thus established that if H ions are introduced in the exchanges, maintenance of equilibrium in dilution requires the introduction of Al ions. This unexpected result means that, in acid media, minerals tend to liberate Al ions from their octahedral layer.

Even allowing for these special features, it would be possible to predict the trend of desorption of one cation by another, by considering a series of contacts; for example, between a fresh solution of NaCl and clay containing calcium, if the constant k were independent of the relative concentration of the cations, on the clay phase. Unfortunately this is not the case and there is no general theory of exchange phenomena. It has been shown by direct measurement that the activity of the adsorbed cations varies in a fairly unpredictable manner as a function of their ratio, compared with the exchange capacity. These measurements demonstrate the preferential adsorption of certain ions, potassium in particular. In a general way, ions are more adsorbed as their valence increases and as their degree of hydration decreases. In 1935 Jenny and Reitmeier [76] suggested the concept of symmetrical value so as to characterize the adsorption tendency. If we place a monoionic clay in contact with a solution containing a quantity of cations equivalent to that adsorbed, an exchange equilibrium is established. If we always use the same solution to make the exchange, we obtain a relative value for the energy with which the different cations are fixed on this clay. The quantity of desorbed ions is called 'the symmetrical value'. The following table gives a set of results for a number of cations.

Data on ionic size, adsorption, charge and flocculation (After Jenny and Reitmeier [76])

Ions saturating the clay	Size of hydrated ions in Å	Symmetrical values (%) [a]
Li^+	10.03	68
Na^+	7.90	66.5
K^+	5.32	48.7
NH^+	5.37	50
Rb^+	5.09	37.4
Cs^+	5.05	31.2
H^+	–	14.5
Mg^{++}	–	31.32
Ca^{++}	–	28.80
Sr^{++}	–	25.76
Ba^{++}	–	26.75
La^{+++}	–	13.96
Th^{++++}	–	1.85

[a] Symmetry value = percent release of an adsorbed ion when replacing ion is added in equivalent amounts to the ion adsorbed. NH_4Cl used with monovalent ions; KCl used for polyvalent ions and NH_4^+.

Certain authors have suggested general formulae to express the exchange between a mineral clay and a solution. They are all based on classical equations, although they are purely empirical. As an example, in 1928 Rothmund and Kornfeld [77] suggested:

$$\frac{C_{1\,ad}}{C_{2\,ad}} = a\left(\frac{C_{1\,sol}}{C_{2\,sol}}\right)^b$$

where $C_{1\,ad}$ and $C_{2\,ad}$ are the cations fixed by the clay $C_{1\,sol}$ and $C_{2\,sol}$ are the concentrations in the solutions. a and b are constants where $b < 1$.

This formula shows that the extraction of a cation becomes more difficult as the quantity fixed becomes smaller.

If we start from a monoionic clay, we can, by a series of successive exchanges, replace the cations which it carries by another cation. It therefore appears that for a given type of clay there is a constant capacity of fixation; this is the exchange capacity.

10.4. EXCHANGE CAPACITY, ITS MEASUREMENT AND ITS ORIGIN

The most typical experiment demonstrating this property is due to Hissink [78] published in 1925. This author replaced the exchangeable bases in a sample of clay by successive contacts with an acid solution, and so obtained an H clay. Dividing this material into fractions of equal weight, he added a different quantity of base to each. After centrifuging the sample he titrated the quantity of base remaining in the solution. We can in fact put:

$$\text{H clay} + \text{NaOH} \rightleftharpoons \text{Na clay} + \text{H}_2\text{O}$$

so long as the quantity of NaOH added is less than the quantity of H ions available.

By drawing a graph with the added quantities as abscissae and the basicity of the solution after contact as ordinate, he obtained a diagram which showed that up to a certain amount of added base, the basicity remains zero, whilst above that value the basicity is equal to the surplus amount of added base. The value x corresponding to this change is a measure of the exchange capacity of the material.

In 1946 Glaeser [79] repeated these experiments with montmorillonite, using several monovalent bases: NaOH, LiOH, NH_4OH or divalent bases: $Ca(OH)_2$ and $Ba(OH)_2$. It was established that the points obtained with monovalent bases lay along a straight line which was not exactly at 45°. For divalent bases the slopes vary to some extent and the intersection with the abscissa is by an intermediate straight line segment. These divergences from the theoretical aspect are due to various causes:

It has been too easily assumed that the agglomeration surface is inert; in fact in a very acid or a very alkaline medium there is an attack on the octahedral layer. For example, it is impossible to obtain a mineral containing magnesium where the exchangeable bases are H ions; the structure changes and Mg ions are liberated. This is equally true for clays containing aluminium, but the change is slower.

Nevertheless, during desiccation, the Al ions are rapidly released, showing that in a

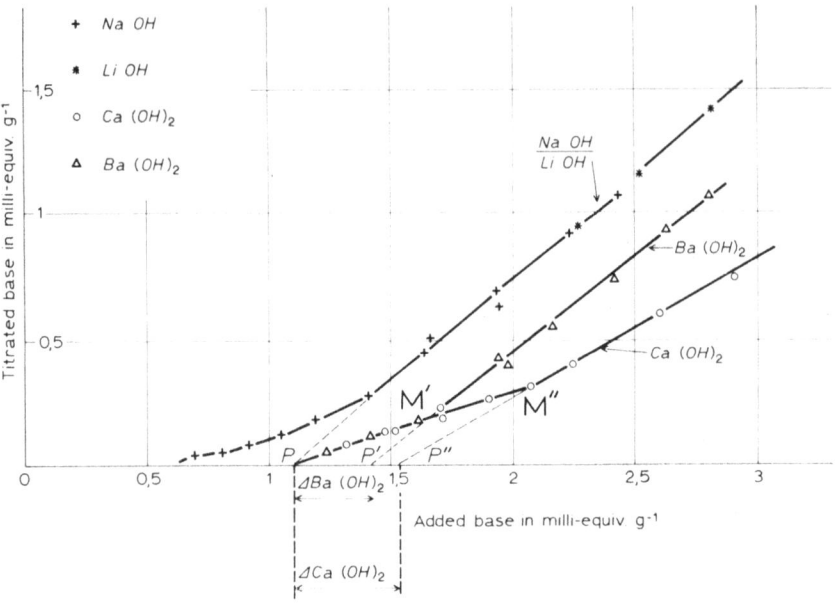

Fig. 16.   Determination of the exchange capacity of cations using the principle of Hissink's method
(R. Glaeser's results).

medium with low hydration the material behaves like a strong acid. When the particles
are formed from mica crystallites, the captive ions between the sheets are slowly transformed into the exchangeable state, which modifies the exchange capacity. This
concept is therefore valid only in a medium where the crystallites remain stable.

Use of divalent ions presents another problem: the exchange sites are sometimes
some distance apart on the same surface, so that if attempts are made to fix a divalent
cation, it will tend to be blocked in the form of a hydroxyl cation, for example
$CaOH^+$. Thus the quantity of remaining ions would be double the exchange capacity.
The true position is in between, because when the exchange sites are closer together,
the calcium can be fixed in the form of Ca^{++} ions, and because an equilibrium is
established between sites which have fixed an H ion and those which have fixed a
$CaOH^+$.

Whatever the circumstances, measurements of exchange capacity enable us to
characterise the different types of mineral (see the table on p. 236).

It should be possible in principle to evaluate the exchange capacities by establishing
saturation curves. Such curves have simple or complex sygmoidal trends, which differ
to some extent depending on, the nature of the cations. Thus, for different cations, the
proportion fixed by clay for the same pH is not the same. Hissink's method has a
general application because it is used at saturation and therefore at high pH.

We are indebted to Schofield [75] for a very illuminating experiment published in
1949. He placed in contact a clay, and solutions of hydrochloric acid and ammonia
in an alcoholic medium, and after equilibrium had been reached he determined the

number of Cl^- and NH_4^+ ions in solution. He established a fixation of Cl^- ions which decreased with pH and a fixation of NH_4^+ ions which varied only slightly with pH at first, then increased considerably above a pH of about 6.5, as is shown by the following table:

	pH 2.5–pH 6	pH 7	Fixation due to dissociation of hydroxyls at pH of 7
Kaolinite	4	10	6
Montmorillonite	95	100	5

He concluded from this that some of the exchange sites behave as strong acids and the remainder as weak acids. This led to a search for the origin of the exchange capacity.

These considerations show the importance of the meaning measurements carried out in accordance with the methods of determining the exchange capacity. Measurements made up to pH values of 6 to 6.5 appear to be related to strong acidities arising from substitutions in the structure. They are therefore measured by methods of direct adsorption as will be seen in the paragraph on this subject.

It appears in addition that clay materials, even having lost the hydroxides which may be associated with them, also have some capacity for fixation of anions. Nevertheless, the behaviour of these latter differs appreciably according to their chemical properties. In 1932, Demolon and Bastisse [80] had suggested a classification of anions into two types corresponding to the following table:

Active anions	Inactive anions
ortophosphoric	hydrochloric
silicilic	nitric
hydrofluoric	sulphuric
citric	acetic
tartaric	
oxalic	

Active anions are strongly adsorbed in presence of a mineral clay. The type of reaction is:

$$\text{clay OH} + \text{anion Na} = [\text{clay} - \text{anion}] + \text{NaOH}$$

Furthermore, they form slightly dissociated chemical bonds with the constituents of an octahedral layer. For example this is the case for phosphoric and hydrofluoric anions.

Inactive anions are not only not adsorbed, but the solution extracted after contact with a clay appears more concentrated than it was at the start; we therefore have negative adsorption. These different behaviours depend mainly on the pH values. It seems that in a sufficiently acid medium, although it is not at present possible to give specific values to the term 'sufficiently', all anions are capable of being adsorbed.

It is possible to fix an exchange capacity for active anions and for a given clay. The following table gives a certain number of values of this anion exchange capacity.

Mineral	Cation exchange capacity Grimm [81]	Anion exchange capacity Hofmann et al. [82]
Kaolinite	3–15	6.6–20.2
Halloysite $2H_2O$	5–10	– [a]
Halloysite $4H_2O$	40–50	–
Montmorillonite	80–150	23–31
Illite	10–40	–
Vermiculite	100–150	4 – –
Chlorite	10–40	–
Sepiolite	20–30	–
Palygorskite		
Saphonite	–	21
Nontronite	–	12–20
Beidellite	–	21

[a] I.e. not determined

10.5. ORIGIN OF EXCHANGE CAPACITIES

The problem is that of seeking sites where there can be charge deficits leading to fixation of ions. For this purpose we will refer to Figure 7, showing a diagram of the structure.

At the level of the upper tetrahedral layer there is one Al ion in the tetrahedral position. This is the origin of a charge deficit as found in micas which is compensated for in the classical model by potassium, but which becomes an interchangeable ion when the hexagonal layer comes into contact with a solution.

A second case corresponds to substitution of an Al by an Mg, in the octahedral position. It is also manifested as a charge deficit which will be compensated at the exterior of the sheet by an exchangeable ion. This is the case for montmorillonite, for instance.

But other exchange sites can be predicted.

On examining Figure 17, we see that after rupture free valences appear. Water molecules or OH groups will saturate the free valences.

Equilibrium of charges in the case of silicon presents no problem since the co-ordination value of 4 is equal to the valence of four, and one water molecule will form two $\equiv Si-O-H$ radicals with the free valences.

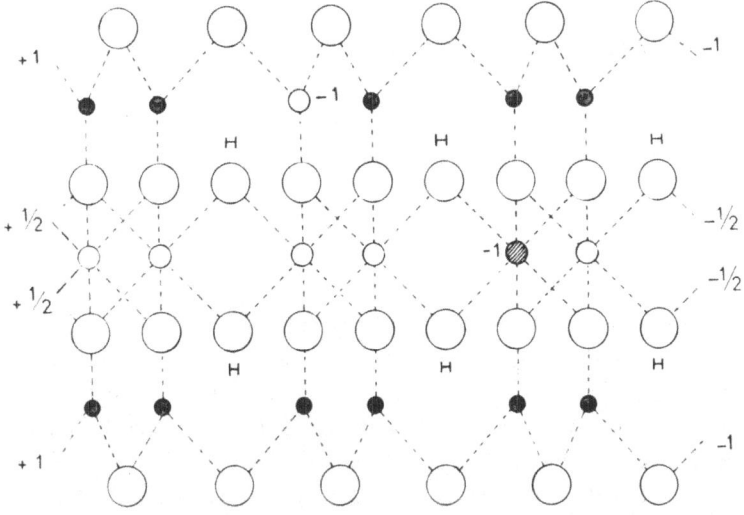

Fig. 17.   *Origin of charge deficits on a triphormic sheet.*

I(a) In the structure, at the level of the Si layer, the white circles represent a substitution of Si by Al (a −1 charge). At the level of the octahedral layer, substitution of an Al ion (white circle) by an Mg (shaded circle): charge deficit −1.

(b) On the edges of the sheet, rupture is supposed to take place at the level of the OH groups.

II. At the level of the oxygens of the hexagonal layers, charge −1, oxygen although bivalent has only one of its valences occupied in the direction of an Si.

On the left: at the same level, the +1 charges correspond to the free valencies of Si.

III. At the level of the octahedral oxygens on the right, the coordination bonds represent only half a valence; the OH groups have only one half valence which is not saturated. On the left: two half valences, both positive, linked to Al are also not saturated.

On the other hand, at the level of the octahedral layer, the situation is different. It is difficult to re-establish equilibrium of charges since the co-ordination bonds correspond only to fractions of valence. Let us examine in more detail the different relative arrangements of cations and oxygens. In 1954 Fripiat *et al.* [83] gave the following formula for a kaolinite:

$$
\begin{array}{lll}
 & \text{H} & \\
\text{HO} & \text{O} & \text{O}{-}0.5 \rightarrow \\
\text{(I)} \quad \text{HO}{-}\text{Al}{-}1.5 \rightarrow \quad \text{HO}(-) & & \text{(IV)} \\
\text{HO} & \text{HO}{-}\text{Al}{-}\text{O}{-}0.5 \rightarrow & \\
 & \text{O} \quad \text{H} & \\
\end{array}
$$

$$
\begin{array}{lll}
\text{O} & \text{O} & \\
\text{HO} & & \\
\text{(II)} \quad \text{HO}{-}\text{Al}{-}1 \rightarrow \quad \text{O}{-}\text{Si}{-}1 \rightarrow & & \text{(V)} \\
\text{HO} & \text{O} & \\
\end{array}
$$

$$\text{(III)} \quad \begin{array}{c} O_{\diagdown} \\ HO-Al-1.5 \rightarrow \\ HO^{\diagup} \end{array} \quad \begin{array}{c} O_{\diagdown} \\ O-Si-0-1 \rightarrow \\ O^{\diagup} \end{array} \quad \text{(VI)}$$

$$\text{(VII)} \quad \begin{array}{c} \qquad Al \\ O_{\diagdown} \quad | \\ O-Si-O-0.5 \rightarrow \\ O^{\diagup} \end{array}$$

In order to compare these situations with the previous diagram it would be necessary to replace all the Si—O bonds of the lower tetrahedral layer by OH bonds in order to re-establish kaolinite. We then find easily the different cases mentioned by J. J. Fripiat. The schemes given represent the electrostatic bonding force suggested by L. Pauling. If we allow for the fact that the electronegativity of Al is 1.5, that of Si is 1.8, and that the strength of acids and bases is a function of the electronegativity of the central atom and of the presence of non-hydroxyl oxygen bound to this atom, we can draw up the scale of acidity of the OH groups fixed on the free valences of the radical found in the previous diagrams.

This scale is drawn up in the following order:

$$\text{V,} \quad \text{VI} > \text{VII} > \text{IV} > \text{II,} \quad \text{III} > \text{I}$$

It is possible to measure the different types of OH appearing on the surface of a kaolinite. In 1954 Fripiat et al. [83] attempted to characterise them by an appropriate chemical reaction.

Investigation of the crystal shows that for each gram of sample there is:

	SiO bonds (V) (VI)	Acidoid OH groups (VII, IV)	Al bonds (I, II, III)
Theoretical values	94	80	54
Experimental values	75 to 112 cation equivalents mean 87	58 to 104 OH equivalents	17.9 to 56 anion equivalents

Nature of lateral surfaces

The exchange capacity of the cations was determined by the help of Ca, H, Li and Na. The fixation capacity of the anions can be determined by fixation of the Cl anion in an acid medium and the acidoid OH groups by the reaction published in 1947 by Berger [84]. The principle of this reaction is that a Si—OH group in contact with diazomethane (N_2CH_2) forms a CH_3 group and a molecule of N_2 is liberated.

A large number of papers have been published on the reactions between organic substances and mineral clays; see Deuel's paper of 1952 [85]. They have led to the conclusion, that reactions of this type were possible, but depended on the reactivity of the OH groups on the surfaces. Estimation of the $O-CH_3$ groups by Zeissel's method enables us to make determinations of the type of those given in this table.

The authors have also attempted to determine the number of OH groups present on these surfaces. For this purpose they made use of the hydrogen-deuterium exchange. The clay is carefully dehydrated in vacuum and then placed in contact with heavy water. The sample is then again degassed. Then a further exchange with ordinary water makes it possible to determine the number of D ions fixed on the clay in place of the H ions.

The results are much more surprising in this case, since the number is very variable (between 830 and 1170) depending on the exchangeable cations.

A critical study of these results has been made in the following manner: from the theoretical number of OH groups, calculated to be present on the surface of a gram of clay, subtracted from the number of OH groups found by experiment, the excess OH groups can be attributed to hydration of exchangeable cations. The values obtained lie between 3 and 6, which agrees well with determinations made under other conditions.

It therefore appears that the given picture is close to the truth.

In addition, it has been found possible to give direct evidence for the existence of fixation sites, by causing the clays to adsorb clusters of hydrosols, charged positively or negatively. Provided that soils of heavy metals are used, these particles are clearly distinguished from the clay constituted light elements, in an electron-microscope analysis of Follet [86] showed in 1965 that iron hydroxide was adsorbed on the base surfaces of kaolinite. It appears, however, that only one of these faces is able to fix the colloid. At low pH, a negative gold hydrosol is generally fixed on the circumference of the particles, thus demonstrating the anion-fixing sites. This effect disappears almost completely at a pH of 5.5.

Since 1942, when Thiessen appears to have been the first to make use of this method [87], the results obtained have sometimes been divergent. In fact under certain conditions the negative hydrosol appears to have been fixed on the clay by the intermediary of a bridge of divalent cations of the type:

clay – Ca – negative hydrosol

(Méring et al. [88]).

The existence of polarity on the clusters may be proved by other means, the tendency of dispersed clays, on flocculation to form edge-face associations, bringing together the negatively and positively charged zones, gives the whole assembly the structure of a house of cards.

Further, if these considerations developed are true, it must be possible to observe a relationship between the dimensions of a mineral and, for example, its exchange capacity.

Let us imagine a mica; only the base faces in contact with water will be able to exchange K which can serve as a compensating ion in the structure for the ions of the solution. In the same manner, the number of Si—OH functions in the minerals will depend on the lateral surfaces. But in a montmorillonite, the sheets will separate spontaneously, and the exchange capacity will be almost independent of the dimensions of the crystallites.

The following results, due to Grimm [81] in 1963 illustrate these considerations:

Exchange capacity in meq/100 g

Mineral	Initial state	After 72 h grinding
Muscovite	10.5	76
Biotite	3	72.5
Kaolinite	8	70.4
Montmorillonite	126	238

It is easily seen that in minerals with low exchange capacities this value is eightfold or more, corresponding to the increase in surface, while for montmorillonite it is only doubled. This is certainly what would be expected. A more detailed interpretation of tests of this type becomes very difficult, since some of the mineral passes into the amorphous state and form permutites with very high exchange capacities.

10.6. Measurement of the Exchange Capacity

We have already had occasion to mention Hissink's method, which illustrates the concept of exchange capacity. It operates in a very alkaline medium, which tends towards the use of all available fixation sites. The method commonly used consists of washing the clay with a solution containing an easily titrated cation; most generally ammonium acetate. The solutions are renewed until after contact the only cation is NH_4. The salt used is buffered to some extent, which facilitates $H \rightarrow NH_4$ exchanges at $pH = 7.5$. After saturation of the clay by NH_4, the saline solution in contact with the clay is treated with a mixture of water and alcohol, in order to avoid dispersion of the material. It is also possible to wash with water, while centrifuging violently. The clay saturated with NH_4 and separated from the solution is placed in a flask containing a solution of $MgCl_2$ and magnesia, and NH_4 is displaced by magnesium. The mixture is boiled and NH_4 is removed and the quantity measured. It is thus possible to measure an exchange capacity corresponding to a given pH. Since this procedure is fairly lengthy, attempts have been made to measure the exchange capacity using a cation which is energetically adsorbed, in such a manner that the clay is practically saturated by one single contact with a solution containing an excess of these cations. In 1957 Morel [89] used trivalent cobaltihexamine chloride for this purpose. Unfortunately this cation is unstable when the medium is not sufficiently acid, and certain precautions must be taken. In 1969 Mantine [90] used ions rendered complex by EDTA for this measurement. In this method increasing quantities of complex ions ($CuEn_2$, $NiEn_3$) are placed in contact with the same weight of clay. The clay is shaken in the solution. It is allowed to settle, the clay becomes flocculent and the excess cations are estimated as bases were estimated in Hissink's method.

We must also mention the use of methylene blue.

These last methods attempt to measure the exchange capacity due to substitutions. It can be calculated the formula of the mineral, as for example:

$$(Si_{2.8}Al_{1.2})\, O_{10}\, (Mg_{2.8}Fe^{3+}_{0.2})\, (OH)_2\, \text{C.E.}, \quad \text{where C.E.} = 1.$$

The molecular weight of this mineral is 383.

Since the exchange capacity is expressed in milliequivalents per 100 g, this value is:

$$\frac{1000 \times 100}{383} = 261 \text{ meq}/100 \text{ g}.$$

This calculation is a means of verification of the formula established, starting from chemical analysis by the method given on p. 203. If the sheets of the mineral were not separated it would be possible to carry out the same calculation, but it would then be necessary to know the true surface of the sample so as to apply a correction coefficients.

10.7. COLORATION PHENOMENA

As a result of the properties of clay surfaces, the adsorbed substances undergo changes of state. These latter may lead to the appearance of colours or to changes in colour.

There has been little fundamental research in this region, but it can be said that certain modifications have been linked with oxido-reduction phenomena. A fair number of authors have suggested the use of colour reactions for identification of mineral clays, both on site and in course of extraction procedures. Maillard* published a summary of these techniques and of the results obtained.

11. Thermal Behaviour

Study of thermal behaviour is based on the observation of three types of phenomenon: loss of material, i.e. thermogravimetric analysis, release of heat, i.e. differential thermal analysis, and changes in dimensions, i.e. dilatometric analysis.

11.1. THERMOGRAVIMETRIC ANALYSIS

The principle of this method is simple: a sample of the material under test is introduced into a crucible which is placed at the end of an arrangement for detecting changes in weight. The crucible is surrounded by a heater, where the temperature rise can be regulated and recorded at the same time as the changes in weight. Under these conditions, silicated materials, hydroxides which behave as monovariant systems, exhibit

---

* Maillard, P.: *Bull. Gr. fr. des Argiles* **16**, 17–23, 1965.

a loss in weight which increases with temperature until the volatile body has completely escaped.

The curve as recorded has a characteristic S shape. Whilst such a procedure may appear satisfactory, its correct operation implies that attention is given to a number of conditions. In particular, the rate of heating must be sufficiently slow for the assumption that at every temperature the substance is in equilibrium with the medium. A rate of heating of $100° h^{-1}$ should not be exceeded, and even at this rate the material is not in equilibrium in the neighbourhood of the point of inflexion of the curve. A further loss in weight is observed even when the heating device is turned off and the temperature maintained constant.

The mass of product used in the experiment, the shape of the crucibles, the composition of the atmosphere, are all factors influencing the equilibrium. In fact, the decomposition process is slowed down to such an extent that the volatile product requires a certain time to diffuse.

The equipments available on the market provide two types of data: the loss of weight and the temperature intervals over which this loss is produced. We have seen the factors which could modify the thermal data; the greatest difficulty lies in knowing to what the loss of weight actually corresponds. We do not know the nature of the substance lost; impurities such as carbonates or organic matter could also be the cause of a further loss in weight.

The process most difficult to account for is the change in oxidation state. The following example illustrates this difficulty:

A hydroxide $Fe(OH)_2$ ought to decompose giving:

FeO and H_2O

In fact the reaction is:

$$2Fe(OH)_2 \rightarrow Fe_2O_3 + H_2O + H_2$$

The thermogravimetric curve will therefore not give information as to the amount of this hydroxide present.

This hydroxide does not exist in the natural state but the Fe^{+2} ions present in the structure of a mineral can give rise to processes of this nature.

Ions other than iron are also capable of reoxidation when they become dehydrated; this is the case with manganese Mn^{2+}. Oxides of manganese can pass to states with progressively lower degrees of oxidation as the temperature rises. Thus MnO_2 changes into Mn_2O_3 and then into Mn_3O_4.

But there are further effects; if a silicate such as a biotite or a chloride containing iron is maintained at a temperature a little below its dehydration temperature, the mineral undergoes a loss of hydrogen compensated by an increase in the valence of the cation contained in the structure. This process is too slow to have any appreciable effect on a thermogravimetric curve, which is measured over a time which is relatively short compared with that taken by the phenomenon in question; this possibility is

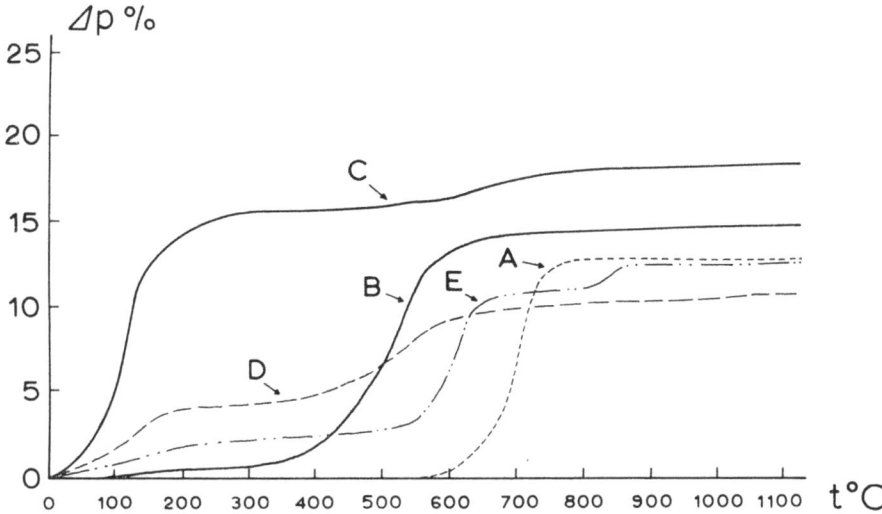

Fig. 18. Thermogravimetric curves for various phyllitous materials: A antigorite; B kaolinite; C montmorillonite; D illite; E chlorite.

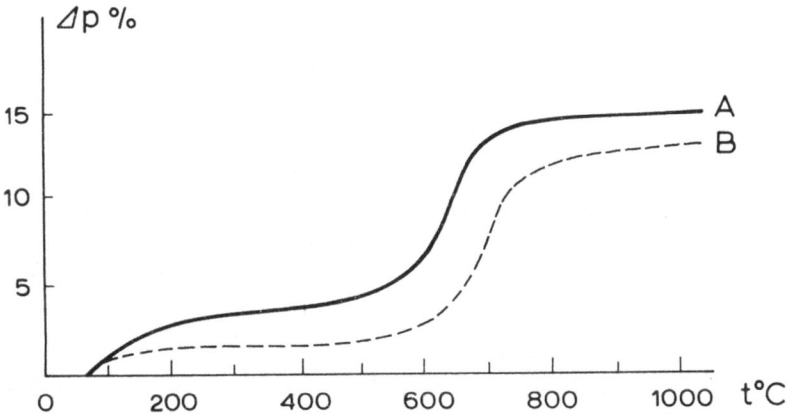

Fig. 19. Thermogravimetric curves for an antigorite with different particle sizes. A: grains $<2\mu$. B: normal grains. Note in case A the increased loss of hygroscopic water between 0 and 300°C.

none the less disturbing since sometimes these minerals are found in the natural state and it is then very difficult to interpret the chemical analysis in mineralogical terms. For this reason it is useful to verify the nature of the chemical products produced during heating; see the work published in 1953 by Brindley and Youell [92].

Figure 18 shows some thermogravimetric curves obtained with different types of minerals. When the temperatures of the water losses are clearly distinguishable, it is possible to deduce the nature of the constituents, this latter being fixed by the proportions of the mixture. Let us stress however that there is no rule which enables us to predict all the anomalies which can be encountered, especially in the case of mixtures of silicates with other minerals, since reactions take place between the compounds present. Interpretation of the results must therefore be carried out with prudence.

Finally the effect of the size of the crystals, is mentioned which is clearly seen by an examination of Figure 19 in which are plotted two thermogravimetric curves, one obtained with an antigorite from Ambindavato, Madagascar, with crystals between 0.2 and 0.3 mm in size, and the other with the same mineral, ground down to particles $<2\mu$. See Caillère *et al.* [91].

Fortunately there are certain tendencies which are sufficiently well marked in general to overcome the special features just mentioned. These are the crystallo-chemical constitution and the nature of the constituents. We will return to this at the end of this section.

11.2. DIFFERENTIAL THERMAL ANALYSIS

This method is due to Le Châtelier [93]; he applied it to kaolinite. Later, in 1927, Orcel [94] made use of it to distinguish chlorites, and it is to Caillère [95] that we owe its systematic application in 1955 to phyllitous silicates and to the minerals with which they are generally associated. It is now very widely used and is applied to a wide range of materials.

The principle is as follows: we place two small cups in an oven with a constant rate of heating; these cups contain small quantities of substance which originally were about 1 g, but with modern equipment can be only a few centigrams.

One of the recipients contains the material under investigation, the other an inert substance, generally calcined alumina. It is also possible to use another material as control, such as another variety of the species to be identified. The essential feature is that the control substance has a thermal conductivity and specific heat close to those of the sample under test.

Two thermocouples connected in opposition are placed in position, one in the reference and the other in the sample. A third thermocouple can be placed in the reference cup or alternatively in a third container also containing the reference sub-stance. The purpose of this second couple is to determine the temperature of the system when the three containers are placed in cavities made in a block of metal or a ceramic holder. The whole assembly is then put into a closed oven so as to avoid turbulence of gases as the temperature rises progressively. When this temperature reaches one of the critical values corresponding to a transformation of the mineral under test, e.g. loss of water, recombination, change of state or fusion, these phenom-ena are accompanied by absorption on release of heat, leading to a relative cooling or heating of the substance. The two thermocouples connected in opposition will then show a potential difference, and the current passing is recorded on a galvanometer.

Differential thermal analysis curves for various minerals are shown in the Figure 20.

We must stress the rather special nature of this technique which acts as a calori-meter when the temperature of the system is increased, assuming that it is not thermally insulated. The differential effects are produced by adjusting the rate of heating. Heating must be relatively fast, about $600°$ h^{-1} in order to bring out the discontinuities caused by monovariant reactions.

This method shows, of course, larger or smaller endothermal changes when a loss of water or of substances is recorded on the thermobalance. However, there is a systematic shift of temperature due to the high rate of heating in the case of differential thermal analysis, since the temperature of the system and that of the reaction are not in equilibrium. The shift is about $100°$ for monovariant phenomena. On the other hand, invariant phenomena (change of state) are only slightly affected.

Impurities or exchangeable cations exert much more influence on differential thermal analysis than on thermogravimetric measurements. For example, in the case of a mineral such as montmorillonite, with a high exchange capacity, a vitreous fusion occurs when it is saturated by Na ions, and there is a transformation into

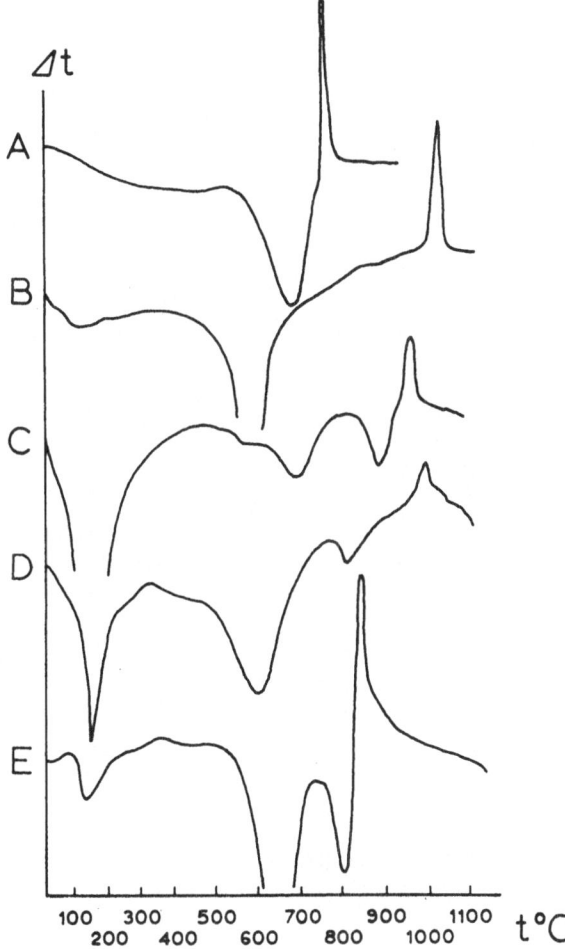

Fig. 20. Differential thermal analysis curves showing a succession of endothermal and exothermal peaks. A: antigorite; B: kaolinite; C: montmorillonite; D: illite; E: chlorite.

other crystalline substances, when it is saturated by H or Al ions. The exothermal phenomenon caused by this transformation does not appear when the vitreous fusion takes place.

The presence of iron hydroxide closely linked with a mineral also causes a reduction of the transformation temperature and a flattening of the corresponding exothermic peaks in the curves.

It follows that the identification of a mineral may be facilitated by certain treatments bringing out characteristic features; see Caillère and Hénin [30].

By taking certain precautions, information can be obtained about the proportion of the constituents present in a mixture.

This technique was also used to determine the order of magnitude of heats of reaction, by Sabatier [96] in 1949.

To sum up, even if differential thermal analysis is neither very precise nor very reliable, its convenience and speed enable us to obtain very useful information.

Amongst the different possibilities of this technique, we will mention the use of a controlled atmosphere, such as nitrogen or carbon dioxide which, by modifying certain reactions, enables us to determine their nature.

In the study of phyllosilicates, the transformations are irreversible and the differential thermal curve during cooling gives no further information. However, in the case of other minerals, it is useful to record the cooling curve, since it may exhibit certain special features corresponding, for example, to a reoxidation. A typical case is that of quartz where the $\alpha \rightarrow \beta$ transformation taking place at 565° can be screened by a dehydration of the kaolinite. During cooling, the calcined kaolinite shows no change, so that the change of the curve marking the $\beta \rightarrow \alpha$ transition of the quartz can be clearly seen in the region of 565°.

The following table summarizes various treatments capable of altering the effect of certain phenomena and of simplifying identification of minerals by this technique. See Caillère and Hénin [97].

Mineral	Nature of changes in gradient of temperature curve, variations	Treatments capable of accentuating or reducing the changes in gradient
Kaolinite	100° endothermal, variable with alteration 500–550° endothermal, well marked, relatively constant. 900–1050° exothermal, very variable	10% cold AlCl₃ solution for 48 h tends to accentuate change at 950°. Removal of iron increases change at 950° and sometimes splits it, if its small size is due to sesquioxides.
Halloysite	100° endothermal, usually substantial; do not dry the mineral above 50°. 280° endothermal, may be confused with the effect of sesquioxides. 500–550°, 900–1050° as for kaolinite.	

Table (continued)

Mineral	Nature of changes in gradient of temperature curve, variations	Treatments capable of accentuating or reducing the changes in gradient
Anauxite	140° as for kaolinite 500° medium connected with the following 580° larger.	In 10% Al (NO₃)₃ solution in cold for 3 days, prevents splitting into two peaks at 500 and 580°, which combine to form one single at 500°, very slight exothermal effect at 950°.
Montmorillonite Beidellite	100° endothermal, very substantial, sometimes splitting near 175–250° 600–650° endothermal, of very variable size. 700–800° endothermal, very variable size, frequently absent, usually smaller than the previous. 900–1000° rare exothermal in natural samples, sometimes preceded by a slight endothermal peak.	in 10% Al (NO₃)₃ solution in cold for 3 days, frequently accentuates peak at 700–800°, which is very slight to start with: frequently gives rise to exothermal phenomenon. Same treatment as before followed by flocculation with NH₄OH, considerable increase of peak at 550–650°. Same result on treatment with 5% solution of potassium silicate precipitated by HCl q.s.
Nontronite	100° endothermal, very large. 400° endothermal, variable size. 850–900° exothermal, very weak and rounded.	3 days in cold solution of Al(NO₃)₃, reduction and elimination of peak at 400. 3 days in cold 5% (NH₄)₂CO₃ causes peak at 400° to grow as large as that at 100°.
Palygorskite	100° endothermal, not well marked. 320° slightly endothermal variable in size; may be absent. 500–550° medium sized endothermal 800–1050° weak exothermal, may be absent; in this case weak endothermal.	3 days in cold 5% (NH₄)₂CO₃ causes the appearance or increase of peak at 320°. 4 days in cold 10% Al(NO₃)₃ solution reduces peak at 500–550° to vanishing point, produces and develops exothermal peak.
Sepiolite	100° endothermal, not well marked. 420° weak endothermal. 780–800° very weak endothermal 800–850° small exothermal.	Boiling for 4 periods of 20 min each in N/4 H₂SO₄ destroys the inflexions except for the exothermal peak which is strengthened.
Bravaisite Illite	100° moderate endothermal not well marked. 550° small, joining with the next; may be absent. 620° medium; when this occurs alone it starts near to 550°. In certain samples it is formed at a higher temperature, and is then separate from the previous peak.	3 to 15 days in cold 10% Al(NO₃)₃ leads to splitting of peak at 550°. Effect fairly constant. Boiling for 10 periods of 20 min each in N/4 H₂SO₄ prevents splitting of peaks.

11.3. INVESTIGATION BY THERMAL EXPANSION

This method is less used than the two previous methods although it brings out a special aspect of the phenomena and so is valuable as a control.

In principle it is the heating of a sample of material with a bar of a non-oxidable metallic alloy (Chevenard [98]). The two samples are placed side by side and two silica rods are placed on them, their lengths being such that they extend beyond the end of the oven. Their external ends are applied to the two sides of a right triangle with its apex supported on a fixed point. During heating, if the two substances expand in the same manner, they push the side of the triangle on which they are supported and the triangle pivots about the fixed point. A mirror fixed to the triangle reflects a ray of light and its deviation is recorded. In the case just described, the beam traces out a straight line on a flat photographic plate. The length of this line is a function of the oven temperature, which can be read from a control scale. If the substance tested expands or contracts with respect to the control, the triangle rotates about the side determined by the fixed point and the point of contact with the silica rod resting on the control substance.

The spot of light will move above or below the horizontal line which corresponds to equal expansion of the two substances compared.

This method enables us to demonstrate certain consequences of the loss of water. Mica crystals, for example, piled on top of each other expand when a dehydration temperature is reached. This illustrates the mechanical resistance which the mineral exerts against the circulation of water. This is confirmed by the fact that the same material when crushed no longer exhibits this phenomenon. Loss of hygroscopic water is frequently accompanied by a contraction, for instance in montmorillonites.

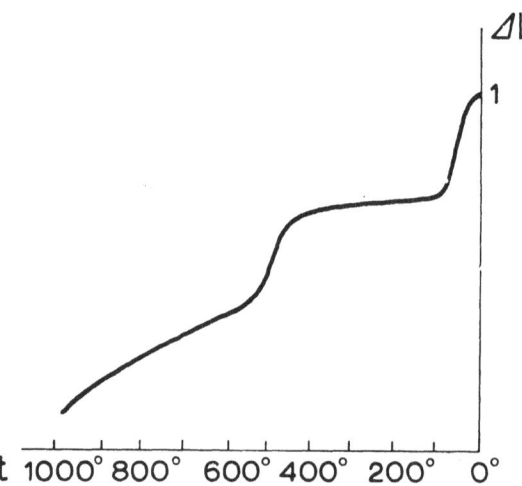

Fig. 21.   Expansion curve of a halloysite showing two contractions, one due to loss of hygroscopic water and the other due to release of water of constitution and to the perturbation of the structure at about 500°.

However, for chlorites, at first loss of water of constitution is not accompanied by a volume change. This is confirmed by X-ray examinations, which show that there is no appreciable change in dimension of the unit cell. Changes of state, of course, fusion and recrystallisation, are frequently accompanied by changes in dimension, which are demonstrated by this type of equipment.

This method is used to a relatively small extent except by ceramic specialists, but is worthy of more widespread use in the investigation of phyllitous minerals.

11.4. RELATIONSHIP BETWEEN CRYSTALLOCHEMICAL PROPERTIES OF HYDROXIDES AND SILICATES AND THEIR THERMAL BEHAVIOUR. (S. Caillère, S. Hénin [30])

We may ask how the relationship between crystallographic and thermal properties are established. Techniques based on the latter are frequently regarded as diagnostic methods.

In general, when a substance passes from hydroxide to diphormic or triphormic

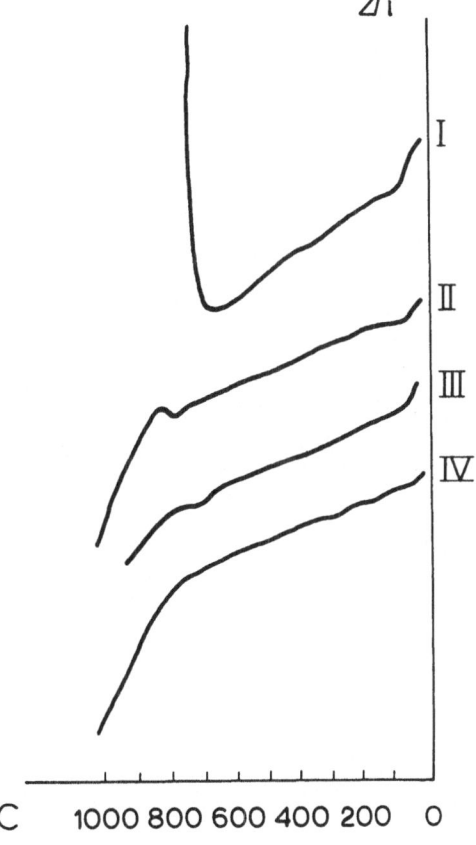

Fig. 22. Expansion curves for muscovite, with sheets perpendicular to the rods showing expansion and contraction. I – curve for a crystal; II – curve obtained for particles <0.2 mm; III – curve obtained for particles of the order of 100 μ; IV – curve obtained with particles $\leqslant 2\mu$.

silicates, the water content is reduced since some of the OH groups are replaced by
O—Si bonds. We can therefore draw up the following table:

| | Hydroxides | Oxyhydroxides | | |
		Diphormic silicates	Tetraphormic silicates	Triphormic silicates
Water content % (order of magnitude)	30	15	15 in two lots 5 10	5
Dehydration temperature	200–350°	350–450°	450–650°	650–900°

These values do not account for hygroscopic water.

The number of peaks in the differential thermal analysis curve is also characteristic.

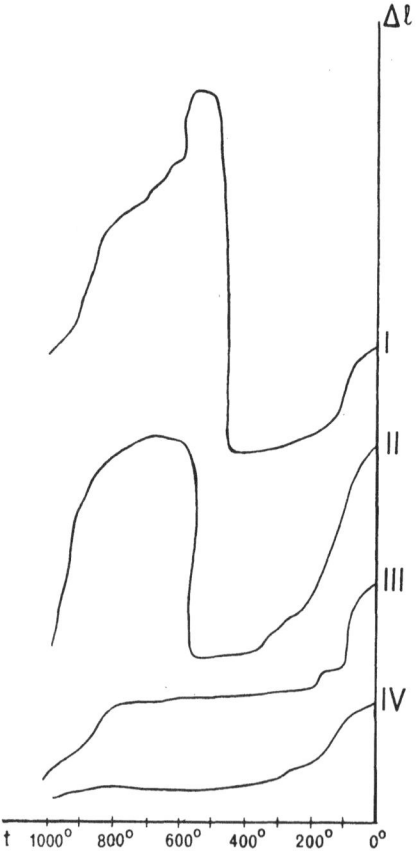

Fig. 23. Expansion curves for vermiculite. Sheets are perpendicular to the rods showing expansion
and contraction. The expansion, due to loss of water of constitution appears in the region of 600° in
curves I and II (crystals of the order of one cm). Curves III and IV are for the crushed mineral, and
expansion is no longer observed, since water vapour escapes easily from small particles.

	Hydroxides oxyhydroxides	Diphormic silicates	Tetraphormic silicates	Triphormic silicates
Phenomenon	1 endo + sometimes 1 exo rather flat	1 endo + usually 1 exo	2 endo + 1 exo	1 endo + sometimes 1 exo

In this case the temperatures at which the endothermal phenomena take place correspond to those measured in thermal equilibrium augmented by about 80°.

Exothermal phenomena appear at much more variable temperatures. Sometimes they follow immediately the endothermal phenomenon, which is considerably larger and may screen it. In other cases, they occur at considerably higher temperatures (kaolinite, for example).

Finally, for one given type of structure, the nature of the cation forming the octahedral layer acts in a characteristic manner. For example, for the same structure, magnesium minerals become dehydrated at temperatures about 100° above those for aluminium minerals. Nickel and zinc minerals decompose at temperatures below those for aluminous silicates. The temperature also tends to be lower when the minerals contain iron. There are even smaller differences depending on the type of association of the sheets or on the nature of the compensating cations when these are present in appreciable quantities.

When several cations are present in the octahedral layer of the same material, we can determine approximately the dehydration temperature by calculating the weighted mean of the decomposition temperatures of each cation present.

12. Relations between Chemical Composition and Physical Properties

The ions in the composition of a mineral modify its properties. Those most in evidence are refractive indices, polychroism, colour and density.

12.1. DENSITY

The density is affected in a complex manner since replacement of one ion by another can lead to expansion of the unit cell. The observable effect is the result of opposite causes, which may be an increase in weight of the ion or a variation in volume of the unit cell. This is of interest only when comparison between chemical and crystallographic data is attempted.

We must stress one difficulty peculiar to minerals existing in the form of very fine particles. When determination is carried out in water, the structure of the liquid is modified on contact with the solid phase and the density measured differs from the true density. These measurements are made in non-polar liquids which are only slightly affected by the charges on their surface.

12.2. OPTICAL PROPERTIES

Optical properties are very sensitive to the presence of certain constituents: transition elements such as Fe, Co, Ni, Ti, V, Cr, and Mn. Due to their variable valences there is a variation in the colours of the compounds in which they are found. Apart from the characteristics exhibited by minerals depending on these elements and on their states, we can also deduce from this the valence state of their constituents. We have assembled, in the table below, the colours exhibited by the cations along with their corresponding valence states.

Elements	Violet	Indigo	Blue	Green	Yellow	Orange	Red	Uncoloured
Ti								4
V	2	4		3				
Cr	3	2		3	6			
Mn	7			6			2.3	
Fe				2	3			
Co			2				3	
Ni				2				
Cu		2						1
Zn								2

The figures in the different columns show the valences of the elements.

The following table gives the colours exhibited by a certain number of minerals, considering the chromophoric elements which they contain.

Minerals	Ti	V	Cr	Mn	Fe	Co	Ni	Zn
Fuschite (mica)			green					
Glauconite					Fe^{++} & Fe^{+++} green			
Biotite					dark green			
Antigorite (nepouite)							green	
Chlorite (kamererite)			violet					
Montmorillonite (sauconite)								colourless
Montmorillonite (volchonskoite)			green					
Muscovite (roscoelite)		brown green						
Montmorillonite (nontronite)					yellow			
Montmorillonite (delanouite)			pink					

These tables show a certain similarity between the colour of the ions and that of the minerals. Few investigations have been made of the dependence of the overall colour on the size of the sample. Haematite is black in bulk and red when pulverized.

Certain effects of coloration may be due to internal reflections. The conditions of observation must therefore be defined by preliminary investigation.

A lot of research has recently been carried out on this subject and has led to certain conclusions. For example, titanium is always octahedral and of the form Ti^{3+}, the same is true of Cr^{3+}. These data enable us to confirm or determine the ion balance found by simple analysis.

In addition, the structure modifies the optical properties. Light passes unchanged through cubic crystals; in cases of non-cubic crystals double refraction occurs. Under these conditions, it is necessary to distinguish uniaxial crystals, quadratic, hexagonal and rhombohedral from biaxial crystals belonging to the orthorhombic, monoclinic and triclinic systems.

Phyllosilicates are usually biaxial, but the angle between the axis is sometimes very small. They belong to the monoclinic and triclinic systems and are sometimes pseudo-hexagonal.

Refractive indices are greatly influenced by chemical composition, and systematic research has been carried out to link these two sets of values for certain families of minerals. Here, for example, are the results obtained in 1927 by Orcel [94].

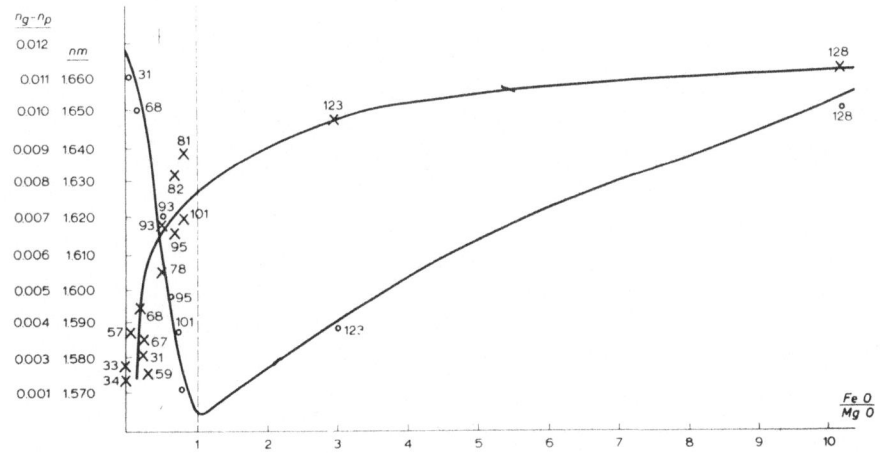

Fig. 24.   Variation in refractive indices of chlorites as a function of the ratio of the constituents FeO and MgO (from J. Orcel).

V. Classification

13. Classification of Mineral Clays

One of the aims of the naturalist is to group individuals into types and to introduce them into a system of classification.

These are two complementary procedures and it is clear that different classifications will be reached, according to the criteria used for grouping of individuals.

Grouping of minerals according to colour or density would lead to very different classifications from those based on shape or structure.

From the time of R. J. Haüy's discovery, the concept of structure has dominated crystallography and mineralogy. Nonetheless most minerogical classifications are based on the nature of the anions.

The minerals that we discuss in this chapter are all silicates, and belong to the same mineralogic class.

More detailed criteria are found at the level of crystallochemistry; they can nevertheless be applied in different ways. For example, these minerals may be subdivided into di- and tri-octahedrals, and each of these groups may be subdivided according to the number of layers of silica. We can also start from the number of layers of silica and then distinguish the di- and tri-octahedral types.

Let us recall that the terminology to be applied to the description of minerals, as set by the latest recommendations of the nomenclature committee [99] is as follows:

The term 'plane' will be used for the surface defined by the positions of the centres of the atoms or ions constituting the mineral, for example the plane defined by the cations of the octahedral layer or of the tetrahedral layer, the plane of the OH groups in the structure of kaolinite.

The term 'layer' will be used for a combination of planes characterized by an arrangement of anions: tetrahedral layer, octahedral layer.

A 'sheet' is a combination of associated layers linked by ionic or covalent bonds.

The space between two sheets is the 'interfoliary space'.

The association of a sheet and an interfoliary space defines a 'unit of structure'.

In dealing with piling up of sheets, we speak of 'polytypism' and 'polymorphism'.

'Polytypism' occurs when sheets of the same nature and the same structure are piled together in different ways.

'Polymorphism' occurs through association of sheets in which the arrangement of the layers is different.

13.1. INTERNAL CRITERIA – CONSTITUTION OF THE SHEET

Let us recall the characteristics of the constitution of a sheet with reference to the work by Pédro [100] published in 1965.

Two types of octahedral layer can be found, characterised by the fact that all the available sites are occupied, 6 per unit cell or 3 per half unit cell; these are trioctahedral minerals. On the other hand, we can predict 4 occupied sites out of the 6 in the unit cell, or 2 per half unit cell; these minerals are known as dioctahedral.

In trioctahedral minerals, when the cations are divalent and the three holes occupied, the arrangement is centro-symmetrical and symmetry is maximum (holohedral).

In the case of dioctahedral minerals, with only two holes out of three occupied, there are several possibilities:

(a) the two trivalent ions may be distributed at random, in complete disorder.

(b) the two ions may be distributed in a determinate manner, but with a greater or lesser degree of symmetry, that is:

– either centrally symmetrical (holohedral)

– or not centrally symmetrical (hemihedral)

With regard to tetrahedral substitutions, we have already had the opportunity of seeing that they can exhibit a certain degree of regularity (see p. 222, Figure 14).

13.2. SUPERPOSITION OF SHEETS

The different sheets are piled up in space in such a manner as to constitute the crystal. It is characteristic for these minerals, that they thus have two types of arrangement.

One is the sheet which may be regarded as a two-dimensional crystal.

The other is the piling up of these sheets leading to a three-dimensional crystal.

Since sheets may be superposed without any given relationship, there is complete disorder in the piling up. Such a structure is known as turbostratic. It has no hkl reflections in the X-ray diagram. The only reflections observed are the 001, determined by the thickness of the sheet or by the unit of structure and the hk0 lines determined by the structure in the plane of the sheet. These are asymmetrical.

On the other hand, we can have a perfect order, but the periodicity along the c axis can correspond to a constant number of sheets, varying with the material in question. For example, monoclinic micas will be denoted by 1 M or 2 M, where M represents monoclinic, and 1 or 2 shows that the arrangement of a sheet chosen as base is found in the nearest neighbour sheet or in the next nearest.

This notation was suggested in 1956 by Smith and Yoder [54] and can be generalised. Thus we find types 3 T, 2 O, 6 H etc. for micas.

This notation is applicable to other forms of phyllitous silicates. However, the crystallographic data necessary to determine these characteristics cannot always be obtained from powder photographs.

Between both extreme cases there are partial disorders due to translations along the b axis, with values corresponding to $nb_0/3$ where b_0 is the length of a unit cell, n is 1 or 2; n cannot equal 3 since this would return to perfect superposition. Displacements can also be observed along the a axis.

Partial disorder arises when n takes the values 0, 1, or 2 at random from one sheet to the next. As a result, part of the hkl reflection disappears, or at least the corresponding lines are widened and diffuse. Such piles are frequent in certain phyllites, such as kaolinites, talcs, etc.

13.3. BONDS BETWEEN THE SHEETS

The bonding of the sheets may be due to electrical attractions or to dispersive forces.

13.4. BONDS DUE TO THE CHARGE ON THE SHEET

The simplest case of micas has already been mentioned. The sheets are in this case highly charged as a result of the substitutions, varying from 1 to 2 Al for 4 Si. The sheets are held in position because of the presence of monovalent ions, usually K, sometimes Na, or divalent Ca ions.

Such minerals may be placed in water or in polar liquids without any separation of the sheets. By placing them in solutions of electrolytes, it is possible to extract the potassium with some difficulty. Under these conditions the sheets separate, and one or two layers of water are intercalated. There exists a set of minerals with approximately the same structure as mica but with either a lower amount of tetrahedral substitutions, less than 1 Al for 4 Si, or with octahedral substitutions. The charge deficit is still compensated for by an intercalated cation, but in this case the system is unstable since when placed in water or in certain polar liquids, the sheets separate spontaneously, allowing one or more layers of water molecules to penetrate, depending on the nature of the material.

Whilst minerals with octahedral substitution always behave in this manner, the number of tetrahedral substitutions for which this behaviour appears is not precise. There are two opposing schools of thought: one considers that the determining factor is the charge on the sheet (Brindley [101]), and the other supposes that both the overall charge and its position (Pédro [102]) are involved.

Recent work by a number of authors in 1973, e.g. Tchoubar and Méring [103] has led to the conclusion that if the isomorphic substitutions are made in a totally disordered manner, the whole set of sheets is unstable in contact with water. Starting at a charge deficit of 0.6, the sheets start to order themselves within the structure. Depending on the charge distribution with such a deficit, there may or may not be spontaneous penetration of water into the structure. This limit provides a distinguishing feature between smectites (monmorillonite, beidellite, hectorite, saponite) and micas.

Another family of minerals where the sheets are held in position by electrical charges are the chlorites. They are constituted from di- or tri-octahedral sheets of mica type, but these sheets are separated by an octahedral layer (which can also be di- or tri-octahedral) with a positive charge. This excess charge is due to the substitution of a trivalent ion for a divalent ion in the trioctahedral layer, or to an excess of ions of any sort but usually trivalent when the layer is dioctahedral. Such systems can be plunged into water without causing any separation of the sheets. Further, the bond between the two types of sheet is so stable that it has not yet been possible to dissolve selectively the interfollary octahedral layer. This is a phenomenon which can take place in a natural medium. Orcel et al. [104]. There are still some intermediate cases, that is minerals giving X-ray photographs not very different from chlorites, but with sheets which separate on contact with water and with certain polar liquids; they are expanding chlorites. Our present knowledge suggests that the interfoliary layer is neutral and more or less complete.

13.5. Bonds due to forces other than those connected with charge deficits

There are quite a number of minerals with neutral sheets, where no substitution takes place, or where the charge deficit created in the tetrahedral position is compensated by an excess of charge in the octahedral position. When these minerals are placed in water, the sheets remain attached to each other.

For triphormic minerals such as talc and pyrophillite, no detailed investigation has led to an explanation for the stability of the assembly of sheets. The reason is perhaps that they have no affinity for the liquid placed in contact.

The case of diphormic materials is different. For these the superposition of the sheets gives a layer of hydroxyl groups with the hexagonal layer of oxygen of the following sheet. It has been thought possible to explain the stability of the assembly in terms of hydrogen bonds (p. 214). There is always a very distinct difference between di- and tri-octahedrals. If the former are placed in a certain variety of liquids such as formamide, dimethyl sulphoxide, or concentrated solutions of potassium acetate, the sheets separate. This group of minerals includes one variety, halloysite, which has a layer of water between the sheets. There is a reversibility in the sense that this layer of water when removed by heating to 100°, can be reformed after various treatments. The relative arrangement of the sheets of halloysite is disordered. These facts show that the association of sheets can be stable only if certain conditions are satisfied. A recent, more detailed investigation of the problem by Cruz et al. [105] in 1972 led them to consider the three types of bond which can exist between the sheets:

(a) the hydrogen bond; 4(OH) can act as bonds, corresponding to 4 cal per unit cell.

(b) the van der Waals forces; calculated energy corresponds to 3.75 cal per unit cell.

(c) the electrostatic energy due to the location of the charge distribution, which can vary between 28.8 and 48.8 cal per unit cell, according to the hypothesis on which this distribution is based.

To summarize, the total bonding energy resulting from all the elementary effects varies from 36.55 to 56.55 cal per unit cell. R. F. Giese Jr [106] has found an order of magnitude close to this by another calculation.

Calculations for trioctahedral minerals give higher values since for the electrostatic effect alone the bonding energy is 53.7 cal per unit cell. Therefore they do not react to the action of the various active substances on dioctahedral minerals. The bonding energy, calculated from the variation in vibration frequencies of the OH groups after fixation of the active substance, is lower than the bonding energy of the sheets even for dioctahedral minerals. In order to understand their action, it is necessary to refer to the reduction in electrostatic attraction due to the high dielectric constant of the substances used. In any case, these attempts at explanation require confirmation.

13.6. THE PROBLEM OF INTERSTRATIFICATION

We have up to now been considering the piling up of sheets of the same type. These are known as monophyllites, but especially for materials in the clay state, there are crystallites composed of sheets of several types: these are polyphyllites (see Pédro [100]). Between these two extremes, it is possible to observe intermediate types where similar sheets can be separated by a variable number of layers of water. Sabatier [107] explained this in 1961 by the idea that this is a straightforward interstatification. The 001 reflections observed with X-rays, showing an average equidistance between the sheets. Indices of higher orders do not form a rational series and in addition, the first order line is often diffuse. The typical case of polyphillites corresponds to a fundamental interstratification; according to Sabatier [107] we must regard this as the juxtaposition of two or more sheets of different nature.

Some polyphillites are arranged regularly, that is two sheets A and B are arranged systematically in the order AB, AB, for instance allevardite and rectorite.

Under these conditions, the 001 series is rational, and the first order corresponds to the equidistance $A + B$. Polyphillites are irregular most of the time, that is the sheets may be arranged at random, in a series of type: $A B A A B A B A B B$ etc, for a mixture of equal numbers of constituents A and B, but these proportions may vary. Under these conditions, X-ray photographs will exhibit the features described in the intermediate case for straightforward interstratifications.

It is assumed that polyphillites with regular interstratification can be given a name. When the interstratification is irregular, the names of the two constituent sheets are used, mentioning the fact that the mixture is interstratified.

When the interstratification is straightforward, the mineral is given a name characterising the nature of the constituent sheet as in the case of ordinary minerals.

13.7. THE DIFFERENT TYPES OF SHEET

If we make a short summary of the different types of minerals mentioned as examples in this paper, we can define the following groups:

(a) *diphormic minerals*: a tetrahedral layer, an octahedral layer, either di- or tri-octahedral. In the first case we are concerned essentially with aluminium; this is part of the kaolinite family. The different members are distinguished depending on the piling up of the sheets.

The trioctahedral group is much more varied: different divalent cations can replace each other; the differences are related to chemical composition. Some tetrahedral substitutions can also be observed.

(b) *triphormic minerals*: the sheet comprises an octahedral layer sandwiched between two tetrahedral layers. The archetypes are pyrophyllite for the dioctahedrals and talc for the trioctahedrals. In each of these cases we distinguish the families characterized by tetrahedral or octahedral substitutions and by different behaviours. The differences are chemical, related to substitution of a given ion by an ion of the same valence.

(c) *tetraphormic minerals*: Constituted from a triphormic sheet and an octahedral sheet, the first having a charge deficit and the second an excess, as for instance chlorites. We can distinguish four families depending on whether the two octahedral layers are one trioctahedral and the other dioctahedral, or whether they are one di-, the other tri-, or inversely. The different possibilities are chemical in nature.

(d) *minerals with fibrous needles*: Two types of these have not yet been mentioned:

palygorskites and sepiolites. They occur macroscopically in the form of packets of fibres, of cardboard or of chiffon. But even when they form a compact needle, it is found that after milling examined in the electron microscope they occur in greatly elongated units. The structure has the following special features:

A continuous plane of oxygen with a hexagonal arrangement is found as in all minerals considered up to now, and it is this feature which leads the authors to consider them as truly phyllitous. While in general all the silicate tetrahedra are orientated in the same way, they are arranged pointing sometimes upwards and sometimes downwards. The oxygens at the apices are arranged in a compact plane, as in the case of other phyllites, but the octahedral layer formed is a narrow ribbon containing 10 oxygens in the case of sepiolite and 7 in that of palygorskite. There are 9 octahedral cations in the first case and 5 in the second for each half unit cell. These ribbons are sometimes found on the lower part of the reference plane and sometimes on the upper part. We could also describe this structure by saying that it consists of ribbons of talc displaced with respect to each other in an octahedral layer on a vertical plane and fixed by the plane of hexagonal organisation. Such an arrangement therefore resembles a hollow brick. Hygroscopic water is found lodged in the small channels. When we consider the constitution of these minerals, we find that while sepiolite is clearly trioctahedral, the palygorskites are rather dioctahedral. The first group have an equidistance of 12 Å, and the second of 10 Å. The differences are chemical in nature; the only name given is that of xylotile, to the iron-bearing variant of sepiolite.

(e) *Minerals resembling phyllitous silicates*: We could mention silico-aluminous gels which are frequently associated with the microcrystalline forms of these minerals, but if they exhibit properties similar to those of phyllosilicates, such as exchange capacity, their amorphous nature leads us to separate them.

A new variant of phyllosilicates has recently been reported: imogolite, which is a rarity still awaiting classification within the classes which have just been described. It is found in the form of a tubular structure, the details of which require further examination.

(f) *Classification tables*: All these considerations are summarized in tables which enable us to compare the different characteristics of phyllitous minerals. Although they are based on microcrystalline types, it is possible to find space in them for the main macrocrystalline forms, in particular chlorites and micas.

This classification is shown in Tables I–VI.

(g) *Identification tests using X-rays*: a number of criteria are used for identification of minerals by means of X-rays. For classic monophyllites, micas and chlorites, the lines on the photograph are used. For other minerals, tests of properties are used, such as measurement of the change in 001 under the influence of various treatments. These latter must be supplemented by various chemical or thermal characteristics, leading to a regrouping of mineral families. For example, the smectite type includes montmorillonite, beidellite, stevensite, hectorite, and saponite. The aim of the tests is to distinguish these from vermiculites.

The table on page 263 lists these different tests. It also has the advantage of recalling some of the essential properties of these minerals.

TABLE I

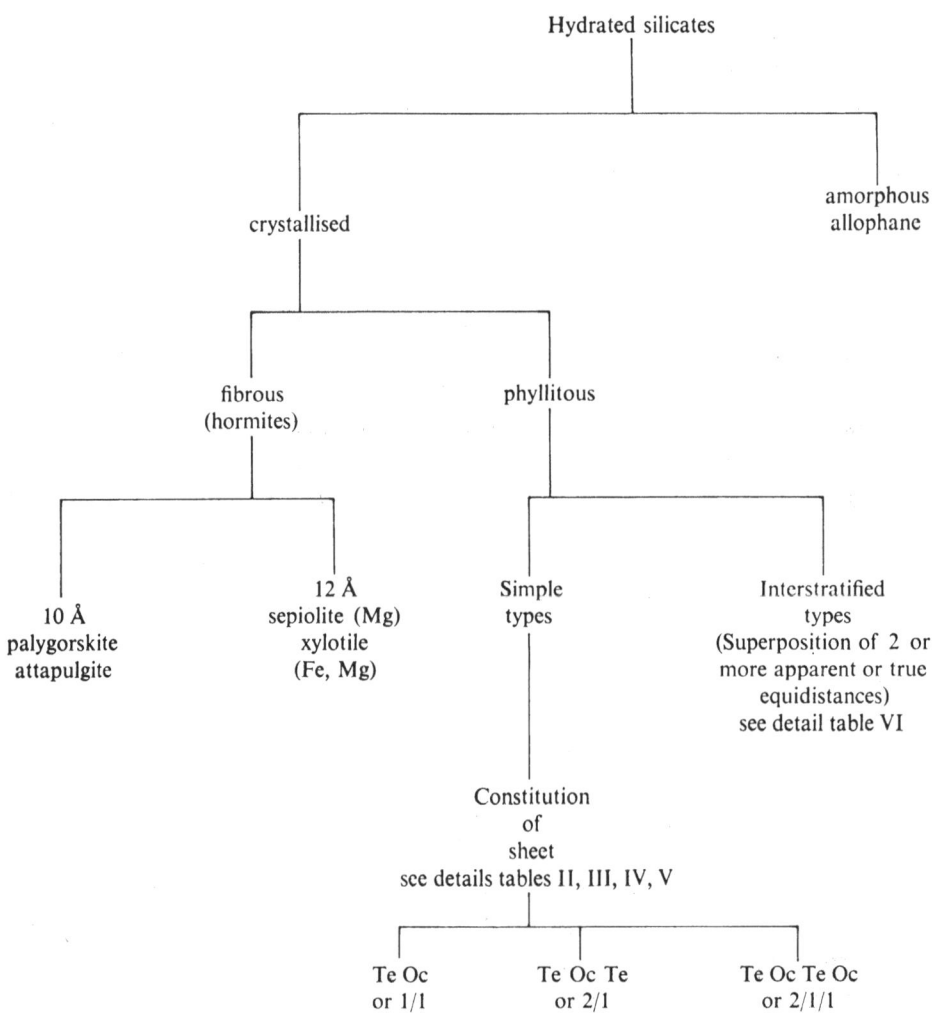

TABLE II

Sheet of type Te Oc or 1/1
True thickness 7 Å

Dioctahedral Oc=4/6		Trioctahedral Oc=6/6	
Te=4 Si	Te<4 Si	Te=4 Si	Te<4 Si
Oc=12/12	Oc>12/12	Oc=12/12	Oc>12/12
Apparent equidistance stable	Apparent equidistance stable	Apparent equidistance stable	Apparent equidistance stable

Dioctahedral, Te=4 Si, Apparent equidistance stable:

variable — halloysite Al

stable — kaolinite, nacrite, dickite, fireclay

Dioctahedral, Te<4 Si: dombassite

Trioctahedral, Te=4 Si:

Ni nouméite

Mg $\}$ antigorite chrysotile

Fe greenalite

$\left.\begin{array}{l}Fe^2 \\ Mg\end{array}\right\}$ jenkinsite

Trioctahedral, Te<4 Si:

$\left.\begin{array}{l}Fe^3 \\ Fe^3\end{array}\right\}$ cronstedite

Al, Fe — normal berthierine

Al, Mg — amesite

Al, Mg Fe² — orthoantigorite

Al, Mn — grovesite

TABLE III

Sheet of type Te Oc Te 2/1
True thickness 10 Å
Dioctahedral
Oc=4/6

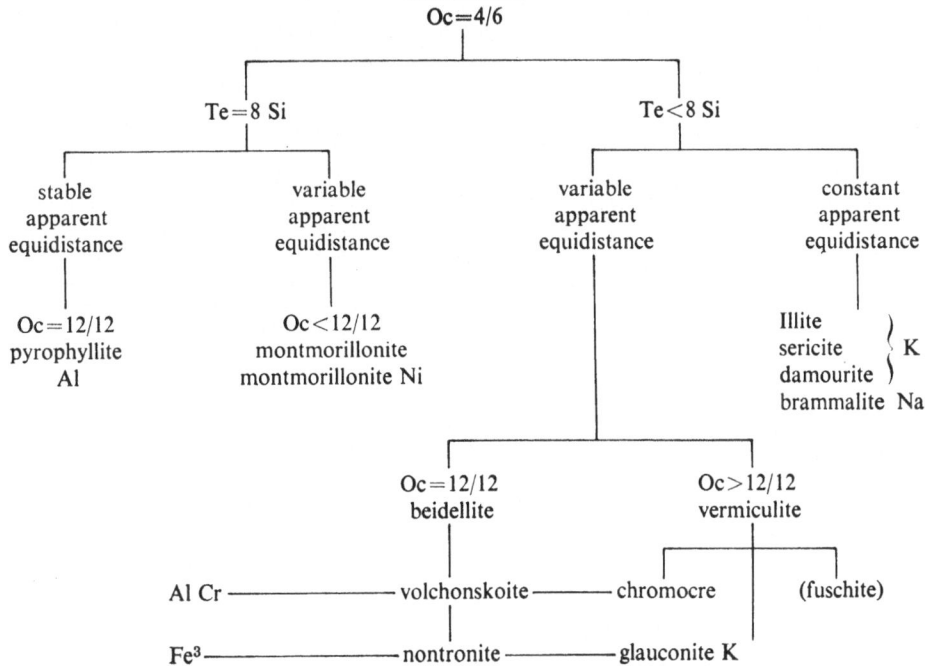

Te=8 Si		Te<8 Si	
stable apparent equidistance	variable apparent equidistance	variable apparent equidistance	constant apparent equidistance
Oc=12/12 pyrophyllite Al	Oc<12/12 montmorillonite montmorillonite Ni		Illite sericite damourite $\}$ K brammalite Na

Te<8 Si, variable apparent equidistance:

Oc=12/12 beidellite   Oc>12/12 vermiculite

Al Cr ——————— volchonskoïte ——————— chromocre   (fuschite)

Fe³ ——————— nontronite ——————— glauconite K

TABLE IV

Sheet of type
Te Oc Te 2/1 true thickness of sheet 10 Å
Trioctahedral
Oc=6/6

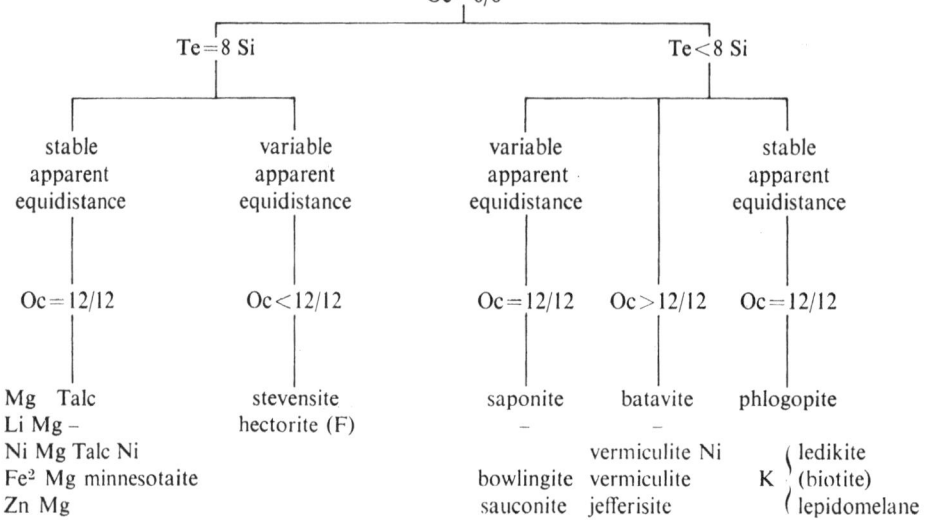

Te=8 Si		Te<8 Si	
stable apparent equidistance	variable apparent equidistance	variable apparent equidistance	stable apparent equidistance
Oc=12/12	Oc<12/12	Oc=12/12 Oc>12/12	Oc=12/12
Mg Talc	stevensite	saponite batavite	phlogopite
Li Mg –	hectorite (F)	–	
Ni Mg Talc Ni		vermiculite Ni	⎧ ledikite
Fe² Mg minnesotaite		bowlingite vermiculite	K ⎨ (biotite)
Zn Mg		sauconite jefferisite	⎩ lepidomelane

TABLE V

Te Oc Te Oc 2/1/1
true thickness 14 Å

Pseudochlorites variable equidistance more or less complete brucitic layer	true chlorites (clays, leptochlorite) stable equidistance normal brucitic layer

	mica sheet		
	dioctahedral hydroxyde sheet		trioctahedral hydroxyde sheet
dioctahedral	trioctahedral	dioctahedral	trioctahedral
sudoite	manandonite some substitutions Si→B	cookeite	subdivided with regard to the extent of the Si—Al substitution and the octahedral cations. By far the commonest type.

TABLE VI

Interstratified types

10 Å types	10 to 14 Å types	14 Å types
10 Å stable	10 Å V. 10 Å S. 14 Å.S.	14 Å variable
10 Å variable	14 Å V 14 Å S 10 Å S.	14 Å stable

Piling up:
irregular: hydromicas
regular: rectorite
allevardite

Practical identification of the main mineral clays (simple varieties) by tests of their properties.

N=normal RH=relative humidity G=glycerol EG=ethylene glycol CH=heated DMSO=dimethyl sulphoxide K$^+$=ion fixed

I – 2/1 Minerals

Type of clay	Type of saturating cation and test carried out	d (001) Å	Supplementary observations or treatments
	Mg RH>99%	18 to 20	
	Mg N	14.5 to 15.5	Hofmann Klemen test
	Mg G	17.5 to 18.0	Li 300° EG→10 Å montmorillonite.
	Mg EG	≃17	Li 300° EG→17 Å beidellite
Smectites	vacuum	10 to 12	
	CH 400°C	10	
	K$^+$	12 to 14	
	K 110°C	10 to 12	
	K 110°C EG	17	
	Na G	17 to 18	
	Na (RH>99%)	>20	
Transformation smectites	Mg G	17.5 to 18	Minerals which exhibit properties intermediate between smectites and vermiculites
	K 110°C	≠10	
	K 110°C EG	≠10	
	Na G	14 or 18	
	Na (RH>99%)	<18	
Vermiculites	Mg N	13.8 to 14.5	* for certain vermiculites of low charge (0.6–0.7) expansion between 16 and 17 Å with EG. (060)=1.53 trioctahedral vermiculite (060)=1.49 dioctahedral vermiculite
	Mg G	14	
	Mg EG	14*	
	CH 400°C	10	
	vacuum	10 to 12	
	K	≃10 to 11	
	K 110°C EG	10	

Table (continued)

Type of clay	Type of saturating cation and test carried out	d (001) Å	Supplementary observations or treatments
Illites and micas	N	10	dioctahedral micas — muscovite
	CH	10	illites (060)=1.49 glauconite
	G or EG	10	(060)=1.53 trioctahedral mica (060)=1.53 (biotite)
Intermediate minerals (smectites and vermiculites with hydroxide layer or chlorites in process of degradation.	N	14	The presence of the interfoliary hydroxide layer in a smectite doesn't necessarily limit expansion (G=17 to 18 Å) Behaviour on heating can be very variable (it depends on the quantity of fixed Al. The essential feature is to bring the (001) line towards 10 Å for degraded chlorites; d (002) is intense in the natural state.
	vacuum	14	
	CH 200 to 450 °C	10 to 14	
	G	14	
	K citrate, vacuum	10 to 12	
Chlorite	N	14	The essential test is to check the stability of the peak at 14 Å; the peak must never move to 10 Å (060) from 1.53 to 1.56: (d (002)>d (001) no expansion with DMSO)
	vacuum	14	
	CH 450 °C	14	
	CH 550 °C	14 or vanishes d (001)	
	G	14	
1/1 and interstratified minerals			
Kaolinite	Mg N	7	(060)=1.485
	Mg 400 °C	7	Expansion test with DMSO
	Mg G	7	
	CH 500 °C	vanishes	→11 Å
Halloysite	N	10	
	CH 110 °C or vacuum	7	
Interstratified	N (Mg²)	16 or between 10 and 14	Regularity checked by presence of the different reflections (example: 24–12–8 Å)

Tables drawn up by Mr Robert, CNRA Versailles, not yet published, and kindly provided by the author.

References

1. J. J. Hougton, in B. A. Keen (ed.), *Physical Properties of Soils*, London, 1931.
2. A. Brongniart, *Minéralogie* **1**, 512 (1807).
3. Th. Way, *J. Roy. Agricultural Soc.* **11**, 373–379 (1850).
4. Th. Schloesing, *Compt. Rend. Acad. Sci. Paris*, **79**, 475 (1874).
5. H. Le Châtelier, *Bull. Soc. Fr. Minéral.* **10**, 204–211 (1887).
5. H. Le Châtelier, *La Silice et les Silicates*, Paris, 476–482 (1914).
6. A. Atterberg, *Int. Mitteil. f. Bodenkunde* **1**, 29–33 (1911) **2** 185 (1912).
6. A. Atterberg, *Int. Mitteil. f. Bodenkunde* **3**, 291–330 (1913).
7. W. B. Haines, *J. Agricult. Sci.* **13**, 296–310 (1923).

8. G. Delafosse, *Cours de Minéralogie*, **3**, 356–362 (1862).
9. G. André, *Chimie du sol.* vol. 1, Encyclopédie Wery J. B. Baillère et fils, 2 vols., Paris, 1930.
10. K. Glinka, in G. André.
11. F. W. Clark, *Ann. J. Sci.* **38** (1889), 391.
12. R. H. J. Roborg, *A Study on the Nature of Clay*, 1 vol., Veerman & Zonen, Wageningen, 1935.
13. J. D. Dana, *Manual of Mineralogy and Petrography*, Trubner, London, 1888.
14. G. Tschermak, *Akad. Wiss. Wien* **99** (1890), 174–278.
14. G. Tschermak, *Akad. Wiss. Wien* **100** (1891), 29–107.
15. A. N. Winchell, *Amer. J. Sci.* **9** (1925), 309–327 and 415–430.
16. Ch. Mauguin, *Bull. Soc. Fr. Mineral* **51** (1928), 285–332.
17. R. J. Haüy, *Traité de Minéralogie*, 2nd edn., vol. III, 1822, 132.
18. Ch. Mauguin, *Bull. Soc. Fr. Mineral.* **53** (1930), 279–300.
19. L. Pauling, *Proc. Natl. Acad. Sci. Washington* **16** (1930), 123.
20. A. Adding, *Z. Kristallogr.* **58** (1923), 108–112.
21. F. Rinne, *Z. Kristallogr.* **60** (1924) 55–69.
22. W. F. de Jong, *Z. Kristallogr.* **66** (1928), 303–308.
23. J. Boehm, *Z. Kristallogr.* **68**, (1928), 567–585.
24. S. B. Hendricks and W. H. Fry, *Soil. Sci.* **29** (1930), 457–479.
25. W. L. Bragg, *Atomic Structure of Minerals*, Oxford Univ. Press, (1937) 292 pp.
26. D. P. Grigorieff, 'Fundamentals of the Constitution of Minerals' (transl. from the Russian) Israel Program for Scientific Translations 1964, 50 pp.
27. L. Pauling, in *La nature de la liaison chimique et la structure des molécules et des cristaux*, Presses Universitaires de France, Paris, 1949, 352–53.
28. C. E. Marshall, *J. Phys. Chem.* **41** (1937), 935–942.
29. G. Brown, 'Report on the Clay Minerals', Group Subcommittee on nomenclature of clay minerals (M. H. Hey, D. M. C. Mac Ewan, and R. C. MacKenzie), *Clay Minerals Bull.* **2**, (1955) 294–302.
30. S. Caillère and S. Hénin, *Minéralogie des Argiles*, Masson Edit., Paris, 1963, p. 43.
31. A. E. Lindh and O. Lundquist, *Ark. Mat. Astron. Fys.* **18** (1924), 14–35.
32. E. W. White, H. A. MacKinsky, and T. F. Bates, *Proc. VIIth Ann. Conf. Ind. Appl. K Ray Anal. Denver* **2** (1958) 239–245.
33. G. W. Brindley and H. A. MacKinsky, *J. Amer. Ceram. Soc.* **44**, (1961), 506–507.
34. A. J. Leonard, Sho Suziki, J. J. Fripiat, C. de Kimpe, *J. Phys. Chem.* **68**, (1964), 2608–2617.
35. M. Steinberg, *Bull. Gr. Fr. Argiles.* **22**, (1970), 115–126.
36. M. Ché, J. Fraissard, and J. C. Vedrine, *Bull. Gr. Fr. Argiles* **16** (1974), 1–5.
37. C. Janot, M. Chabanel, and E. Herzog, *Bull. Soc. Fr. Minéral. Cristallogr.* **91** (1968), 166.
37. C. Janot, H. Gibert, X. De Gramont, and R. Biais, *Bull. Soc. Fr. Minéral Cristallogr.* **94** (1971), 367–380 and **96**, (1973).
38. V. C. Farmer, *Mineral. Mag.* **21**, (1958), 829–845.
39. V. Stubican and R. Roy, *Amer. Mineral.* **46** (1961), 32–51.
40. J. M. Serratosa, *Amer. Mineral.* **45** (1960), 1101–1104.
41. R. P. J. Lyon and W. H. Tuddenham, *Nature* **185** (1960), 374.
42. C. Pimentel and A. L. MacClellan, *The Hydrogen Bond*, Freeman, London 1960, 475 pp.
43. J. J. Fripiat, *Bull. Gr. Fr. Argiles* **12** (1960), 25–41.
44. R. L. Ledoux and J. L. White, *Science* **145**, (1964), 47–49.
45. V. C. Farmer and J. D. Russell, *Clays and Clay Minerals* 15th Conf. on Clays and Clay Mineral 1967, p. 121–142.
46. J. D. Russell, V. C. Farmer, and B. De Velde, *Mineral Mag.* **37** (1970), 869–879.
47. W. F. Bradley and J. M. Serratosa, *J. Phys. Chem.* **62** (1958), 1164–1167.
48. W. Vedder, *Amer. Mineral.* **49** (1964), 736–768.
49. J. Chaussidon, *Clays and Clay Minerals* **18** (1970), 139–149.
49b. J. Chaussidon and R. Calvet, *J. Phys. Chem.* **69** (1965), 2265–2268.
50. G. W. Brindley and D. M. C. MacEwan, *Ceramics Symposium* (Brit. Cer. Soc.), (1953), 15–59.
51. W. Von Engelhardt, *Z. Kristallogr.* **104** (1942), 142–159.
51b. R. C. MacMurchy, *Z. Kristallogr.* **88** (1934), 420–432.
51c. M. H. Hey, *Mineral. Mag.* **30** (1934), 277–292.
51d. H. Shirozu, *Mineral. J.* **2** (1958), 209–223.

52. J. V. Smith, *Acta Cryst. Camb.* **7** (1954), 478–481.

53. R. E. Newnham and G. W. Brindley, *Acta. Cryst. Camb.* **9** (1956), 759–764.

54. J. V. Smith and H. S. Yoder Jr., *Mineral. Mag.* **31** (1956), 209–235.

55. E. W. Radoslovich, *Acta Cryst.* **13** (1960), 919.

56. L. Gatineau, *Compt. Rend. Acad. Sci. Paris* **256** (1963), 4648–4649.

57. L. Gatineau, *Bull. Gr. Fr. Argiles* **16** (1965), 3–10.

58. J. Méring and L. Gatineau, *Compt. Rend. Acad. Sci. Paris* **246** (1958), 960–963.

59. E. W. Radoslovich and K. Norrish, *Amer. Mineral.* **47** (1962), 600–615.

60. L. Gatineau and J. Méring. *Clay Mineral. Bull.* **3** (1958), 238–243.

61. E. W. Radoslovich, *Amer. Mineral* **47** (1962), 617.

62. U. Hofmann and R. Klemen, *Z. Anorg. Chemie* **262** (1950), 95.

63. R. Green Kelly, *Nature, London* **2** (1952), 1131.

64. R. Glaeser, I. Mantine. and J. Méring, *Bull. Gr. Fr. Argiles* **19** (1967), 125.

65. R. Calvet and R. Prost, *Compt. Rend. Acad. Sci. Paris* **269** (1969), 539–541.

66. S. Hénin, J. Chaussidon, and R. Calvet. *Bull. Gr. Fr. Argiles* **21** (1969), 31–45.

67. S. Brunauer, S. P. H. Emmett, and E. Teller, *J. Amer. Chem. Soc.* **60** (1958), 309–319.

68. J. J. Fripiat, J. Chaussidon, and A. Jelly, *Chimie Physique des phénomènes de surface – application aux oxydes et aux silicates*, Masson Ed., Paris, 1971.

69. J. Escard, *Bull. Gr. Fr. Argiles* **3** (1952), 83.

70. S. B. Hendricks and R. S. Dyal. *Transt. 4th Inter. Cong. of Soil Sci.* **2** (1950), 71–72.

71. G. Gouy, *J. Phys.* **9** (1919), 457.

72. D. L. Chapman, *Phil. Mag.* **25** (1913), 475.

73. H. Laudelout, *Bull. gr. Fr. Argiles* **9** (1957), 61–65.

74. J. Chaussidon, *Bull. gr. Fr. Argiles* **10** (1958), 27–35.

75. R. K. Schofield, *J. Soil Sci.* **1** (1949), 1–8.

76. Jenny and H. R. F. Reitmeier, *J. Phys. Chim.* **39** (1935), 593–604.

77. V. Rothmund and G. Kornfeld, *Anorg. alg. Chem.* **103**, (1918), 129.

78. H. Hissink, *Trans. Faraday Soc.* **20** (1925), 560–562.

79. R. Glaeser, *Compt. Rend. Acad. Sci. Paris* **222** (1946), 1179–1181.

80. A. Demolon and E. M. Bastisse, *Compt. Rend. Acad. Sci. Paris* **195** (1932), 790.

81. R. E. Grim, *Clay Mineralogy*, MacGraw Hill, New York, 1963, p. 33.

82. U. Hofmann, A. Weiss, A. Mehler, and A. Scholz, *Nat. Acad. Sci. Publ.* **456** (1956), 273–287.

83. J. J. Fripiat, M. C. Gastuche, and G. Vancompernolle, *Compt. Rend. 5° Congrès Int. Sci. Sol. Léopoldville* **2** (1954), 401–422.

84. G. Berger, *Compt. Rend. Conf. Pédologie Méditerranée Alger-Montpellier* (1947), 119.

85. H. Deuel, *Clay Min. Bull.* **1** (1952), 205–214.

86. E. A. C. Follett, *J. Soil Sci.* **16**, (1965), 334–341.

87. P. A. Thiessen, *Z. Elek. Angew. Phys. Chem.* **48**, (1942), 675.

88. J. Méring, A. Mathieu-Sicaud, and I. Perrin-Bonnet, *Compt. Rend. 19° Congrès Géol. Alger.* **18** (1953), 103–107.

89. R. Morel, Thèse Paris, *Ann. Agron.*, 1957, 97 pp.

90. I. Mantine, *Compt. Rend. Acad. Sci. Paris* **269** (1969), 815–818.

91. S. Caillère and S. Hénin, in R. C. MacKenzie (ed.), *The Differential Thermal Investigation of Clays*, Mineral Soc. London, 1957, p. 227.

92. G. W. Brindley and R. F. Youell, *Mineral Mag.* **30** (1953), 57–70.

93. H. Le Châtelier, *Bull. Soc. Fr. Minéral.* **10**, (1887), 204–215.

94. J. Orcel, Thèse Paris, 'Recherches sur la composition chimique des chlorites', *Bull. Soc. Fr. Mineral* **50** (1927), 361 pp.

95. S. Caillère, *Public. Pédologie Agronon. Gouvern. Génér. Algérie Série: I* (1955) 27.

96. G. Sabatier, 'Recherches sur la glauconie', Thèse Paris, *Bull. Soc. Fr. Minéral.* **72** (1949), 475–541.

97. S. Caillère, and S. Hénin, *Ann. Agron.* (1949), 69–70.

98. P. Chevenard, *Compt. Rend. Acad. Sci. Paris*, **164** (1917), 516–518.

99. G. W. Brindley and G. Pedro (Comité Nomenclature AIPEA 1972. Congrès Madrid), *Bull, Gr. Fr. Argiles* **25** (1973), 37–42.

100. G. Pedro, *Ann. Agron.* **16**, (1965), 108 pp.

101. G. W. Brindley, in G. Pedro 100 Clay Mineral. Soc. U.S.A.

102. G. Pedro, *Bull. Gr. Fr. Argiles* **19** (1967), 69–85.

103. C. Tchoubar, and J. Méring, *Bull. Gr. Fr. Argiles* **25** (1973), 155–160.
104. J. Orcel, S. Caillère, and S. Hénin, *Compt. Rend. Congrès Soc. Sav. Grenoble* (1952) 203–207.
105. M. Cruz, H. Jacobs, and J. J. Fripiat, *Compt. Rend. Int. Clay Conf. Madrid*, vol. I., (1972) p. 59–70.
106. R. F. Giese Jr., *Geol. Soc. America* (abstracts) **7** (1972), 579.
107. G. Sabatier, *Colloque intern. CNRS Genèse et synthèse des argiles Paris* (intervention), (1962).

POLYTYPISM AND STACKING FAULTS IN
CRYSTALS WITH LAYER STRUCTURE

G. C. TRIGUNAYAT

Department of Physics and Astrophysics, University of Delhi, Delhi-110007

and

AJIT RAM VERMA

National Physical Laboratory, Hillside Road, New Delhi-110012

1. Polytypism

The phenomenon of polytypism stems from the capacity of a solid to crystallize into more than one modifications, which have essentially the same chemical composition but which differ in the number and the manner of stacking of layers in the unit cell. The layers may have a composite character and are all identical. They are stacked one over the other in the close-packed planes of the solid. The modifications are called 'polytypes', or simply 'types', of the particular solid. Because of the structural identity of the constituent layers, the polytypes of a solid have the same magnitude of the unit cell dimensions lying in the plane of stacking of the layers. They vary in the cell dimension which is directed normal to the plane of stacking of layers and hence is commonly referred to as the cell height of the polytype. The variation may be very considerable, ranging from a few ångström units to several thousands of ångström units in certain substances. Another consequence of the identity of the layers is that all the polytypes of a substance necessarily have their cell heights equal to an integral multiple of the fundamental repeat period of a single layer. A polytypic substance usually crystallizes into a small period 'basic' or 'ideal' structure which is referred to as its most common polytype. Only occasionally it crystallizes into other polytypes which, barring a few exceptions, have higher periodicities than the common type. The frequency of occurrence of these higher types determines the degree of polytypism, which varies from one substance to another. Sometimes, besides the most common polytype, one or more of the other polytypes of a substance also occur more frequently than the others. They are usually referred to as second commonest polytype, third commonest polytype, etc.

1.1. HISTORICAL

The predominant occurrence of a common polytype in all polytypic substances tends to obscure the existence of other polytypes. This factor seems to have been largely responsible for holding up the discovery of the phenomenon of polytypism for a long time as also for impeding the progress of early explorations in the field. The phenomenon remained unknown until 1912 when Baumhauer [1], through optical studies, discovered two new polytypes of silicon carbide, besides the hexagonal 6-layered

F. Lévy (ed.), Crystallography and Crystal Chemistry of Materials with Layered Structures. 269–340. All Rights Reserved.

most common polytype. He later confirmed their existence through X-ray diffraction studies [2]. He also suggested that the occurrence of the polytypes was related to their colours. Until about mid 1940's, Baumhauer's findings had not been followed by any major or consistent efforts in the field for unearthing the nature of polytype growth. However, during this period a small number of intermittent investigations were carried out which led to the discovery of a few more polytypes of silicon carbide, as also of a few polytypes of cadmium iodide, lead iodide, graphite, etc., through X-ray and electron diffraction studies [3, 4]*. Through spectrochemical and X-ray diffraction evidence, correlations of polytype growth with impurity content of polytypes and their rate of crystallization were also sought.

Different names, viz. 'plurimodism' [5], 'superperiodicity' [6], and the general term polymorphism, were employed in the literature to designate the phenomenon of polytypism. The first two names have not been popularly used but the use of the third one still continues, presumably on account of polymorphism being a well known phenomenon in crystallography for more than two centuries. Even some standard texts like those by Wyckoff [7]** and Evans [8] † have employed the terms polymorphism and polymorphs, for describing polytypism and polytypes, respectively. This usage is not very appropriate because extensive theoretical and experimental studies made in the last two decades have clearly revealed that polytypism is physically different from polymorphism. A substance is said to be polymorphous when it can exist in two or more forms with different crystal structures. A familiar case is that of zinc sulphide, which occurs in a cubic form (β-ZnS; sphalerite) and a hexagonal form (α-ZnS; wurtzite). The former reversibly converts into the latter around 1020 °C. Many substances have more than two polymorphs, e.g. quartz exists in as many as six different forms, which are stable or metastable in different temperature ranges. The polymorphs of various substances are not always obtained by a direct process of crystallization. In many cases they result from phase transformations in the solid state. Consequently, polymorphism should be regarded to include every possible difference encountered in the crystal structure of a substance, excepting homogeneous deformations [9]. Structurally, therefore, polytypism can be justifiably regarded as a special case of polymorphism and called 'one-dimensional polymorphism'. The structural similarity is supported by energy considerations, too. To take a specific example, in the two polymorphic forms of zinc sulphide, the cubic β-ZnS and the hexagonal α-ZnS, the immediate environment of zinc and sulphur atoms is identical and the difference between the structures appears only when next nearest neighbours are considered. The two forms, therefore, negligibly differ in their potential energies, the energy difference being hardly amenable to calculation. The same environmental situation exists in the polytypes of zinc sulphide (the α-form is polytypic, with more than 150 polytypes being known so far), which consist of mixtures of cubic and hexagonal packings in various proportions. However, unlike polytypism, the phenom-

---

* vide Chapter 5 of Reference [3] and Section 3 of Reference [4].
** vide p. 108 and 120.
† vide p. 144.

enon of polymorphism can be fairly well understood, at least qualitatively, on thermo-dynamic grounds. The physical factors most commonly involved in polymorphic transitions are changes of temperature and pressure. Consequently, the polymorphic transitions are in many ways analogous to changes of state from solid to liquid or liquid to vapour; both are first order phase transformations, accompanied by a discontinuous change in volume and the evolution or absorption of latent heat, and obey Gibb's phase rule and Clausius-Clapeyron equation (they do, however, differ in one important way: while the changes of state are almost instantaneous, the velocity of polymorphic transitions varies nearly from zero to infinity, because such transi-tions involve disruption or rearrangement of structural bonds). Considering the effect of pressure alone, it is generally found, as expected, that increasingly higher pressures produce structures of increasingly higher density and coordination. The effects of temperature changes are more complicated, but for transitions that involve a change in nearest-neighbour coordination, it can be said that higher temperatures promote structures of lower coordination and higher symmetry. No such physical factors have been found to govern the formation of polytypes. The polytypes of a substance appear to form under the same conditions of temperature and pressure. In fact, more than one polytype are frequently observed to coexist within the same crystal piece. Only for a few small period polytypes of silicon carbide some vague temperature-structure relationship has been suspected to exist. Therefore, in spite of their structural resemblance, polymorphism and polytypism should be regarded as two physically distinct phenomena.

1.2. RECENT ADVANCES

Between the mid 1940's and early 1950's detailed morphological and structural studies on several new polytypes of silicon carbide were carried out [10–12]. Also, towards the early 1950's the ideas concerning the kinetics of crystal growth had undergone a radical change. It had been realized that the growth of real crystals should necessarily involve imperfections [13] and Frank et al., had advanced their celebrated screw dislocation theory of crystal growth, according to which even at low supersaturations a crystal could grow through the presence of a step on the surface provided by a screw dislocation in the crystal [14]. The crystal thus grown would not have a molecularly flat face, but a spirally terraced surface. The advent of the theory was followed by an intensive search for growth spirals on crystal surfaces, as predicted by the theory. Remarkable success was achieved by Verma [15, 16] and Amelinckx [17, 18] who employed phase-contrast microscopy to observe and photograph a variety of beautiful growth spirals on the basal faces of SiC crystals. Verma also carried out precise measurement of the step heights of the spirals by multiple-beam interferometry. He found that they equalled the X-ray unit cell heights of the crystals, thus indicating that the spirals actually originated from screw dislocations. A detailed account of these findings, as also of the growth spirals on other crystals, can be found in a book by Verma [19]. From the experimental results, a spectacular and simple explanation of the phenomenon of polytypism, based on the screw dislocation model of crystal growth was advanced by Frank [20], which heralded the beginning of extensive investigations in the field. The ensuing years have witnessed a spurt in the study of both theoretical and experimental aspects of the phenomenon. A host of substances have been found to be polytypic. Their list contains a rich variety and includes compounds with different types of interatomic binding, e.g. predominantly ionic, predominantly covalent, etc. Both artificially grown and naturally occurring crystals have been found to exhibit the phenomenon. A multitude of polytypes of certain substances have been discovered. From a theoretical standpoint, the main problem has been to account for the propensity of a compound to crystallize into a vast number of modifications with memories extending upto thousands of ångström units, and at least seven different theories of polytypism have been propounded so far. The origin of the phenomenon has been attributed to a variety of causes, like thermodynamical factors, lattice vibrations, second order phase transitions, etc. While a final solution is still awaited, the picture has been becoming increasingly clearer and the very effort of solving the problem has been rewarding. The experimental findings have indicated that stacking faults and dislocations play a decisive role in the growth of polytypes. The investigations hold promise of useful applications in solid state physics, solid state chemistry, geology and mineralogy, apart from the theoretical interest involved in the study of the phenomenon [4]. Recent X-ray topographic studies in polytypic ZnS crystals have revealed the existence of crystallographically perfect regions, which hold immense interest for basic dislocation theory.

Judging from the list of known polytypic substances, it appears that the polytypism is restricted to compounds with close-packed and layer structures. In these structures, if one attempts to lay down a layer over another, the first coordination of some atom can be achieved in more than one way, as is the case if the structure of an element is built up by (i) a cubic close packing of layers and (ii) a hexagonal close packing of layers. Consequently, the polytypes of a substance, consisting of packings of identical layers in various manners, have the same first nearest-neighbour relationship. They differ only in the second or higher coordinations and, therefore, differ inappreciably in their potential energies. In this article, we shall restrict ourselves to a detailed description of the polytypism of layer structures alone. The other polytypic structures will be described only briefly and only when found necessary and relevant. A book [3] and four review articles [4, 21–23] dealing with all the known polytypic substances together, has been published earlier. Another article, dealing with the polytypism of silicon carbide alone, has also been published before [24].

2. Description of Polytypes and Their Identification

The existence of numerous and closely related polytypes of certain compounds has led to the evolution of a number of schemes of notation. As polytypism is exhibited by close-packed and layer structures and all the layers of one kind are structurally identical, it is usually sufficient to specify the number and relative positions of the layers in the unit cell for obtaining a representation. The various important forms of representation, suggested and used in literature, are described in the following. An account of the X-ray methods employed for identifying the polytypes has also been provided.

2.1. THE FUNDAMENTAL ABC NOTATION

It is a well known classical device to describe three-dimensional close-packed crystal structures through arrangements of two-dimensional layers. A close-packing of spheres can be dissected into identical layers, which can occupy three possible positions A, B and C, which are mutually related through translations of $\pm(\frac{1}{3}, \frac{2}{3})$ (equivalent to rotations of $\pm 60°$). The layers themselves are close-packed, with each sphere in a layer in contact with six others around it, and no two successive layers can be alike. A polytype is specified by the stacking sequence of the layers in its unit cell. Although a polytypic compound comprises more than one kind of atoms, it is usually enough to specify the sequence for one atomic species alone, since the various polytypes of the compound are built up of identical units of structure in different numbers and arrangements; the position of the other atomic layers follows from crystal symmetry. Thus the four-layered polytype of cadmium iodide is represented by the stacking sequence of the iodine atoms alone as $ABCB$. If the position of the cadmium atoms, which lie in the octahedral voids between successive close-packed layers of the larger I atoms, is also intended to be shown, the sequence is written as $(A\gamma B)(C\alpha B)$, where the Greek letters denote the positions of the cations. The spatial location of the atoms in the unit cell can be easily worked out from the knowledge of the ABC sequence, which is characteristic of a given polytype. In this way, the ABC notation affords a complete and unambiguous representation of the various polytypic structures. However, it does not reveal the crystal symmetry immediately and also it becomes inconveniently large for high polytypes.

2.2. OTHER SCHEMES OF NOTATION

The X-ray diffraction methods enable a quick determination of the unit cell height, and hence of the number of layers in the unit cell of a polytype. Moreover, usually the crystal symmetry of the polytype can also be ascertained in the same process. Accordingly, a simple and convenient way of representing a polytype thus identified is to express the number of layers followed by an appropriate letter denoting the lattice type, e.g. the four-layered CdI_2 polytype possessing hexagonal symmetry is denoted as $4H$. In general, an n-layered hexagonal type is represented as nH. This system of notation was suggested by Ramsdell and Kohn [12] and is popularly known as Ramsdell notation. They employed letter subscripts, e.g. nH_a, nH_b, etc., for distinguishing between two polytypes having the same cell size and symmetry but different stacking sequences of layers. However, as discussed in the next subsection, it is preferable to use number subscripts, e.g. nH_1, nH_2, etc.

While the Ramsdell notation is very concise and expresses the symmetry of the lattice, too, it does not disclose, unlike the ABC notation, the relationship between the layers. The other schemes of notation aim at fulfilling the latter purpose in various compact ways. The first attempt was made by Ott [25] who expressed the polytype structure by the sequence of intervals between layers of one sort, A, B or C, in the unit cell. Thus the rhombohedral CdI_2 polytype $12R$, with the ABC sequence as $ABACBCBACACB$, is represented by the interval sequence (2523). However, the notation is unsuitable for representing hexagonal polytypes, because their interval sequence differs along the three vertical axes passing through $A(0, 0, 0)$, $B(\frac{1}{3}, \frac{2}{3}, 0)$ and $C(\frac{2}{3}, \frac{1}{3}, 0)$. Moreover, it becomes progressively more unwieldy for larger cell sizes, although it is shorter than the ABC notation.

The remaining systems of notation make use of the relative position of neighbouring atomic layers. As mentioned earlier, the three orientations, A, B and C, of the layers are interrelated, since one can be converted into the other either through a translation of $\pm(\frac{1}{3}, \frac{2}{3})$ or a rotation of $\pm 60°$ in its own plane. Accordingly, the conversions, which can be arranged in two general categories as cyclic $(A \rightarrow B \rightarrow C)$ and anticyclic $(A \rightarrow C \rightarrow B)$ can be visualized in two different ways, viz. through translation, as done by Hägg [26], or through rotation, as adopted by Frank [20]. Hägg used the symbols $+$ and $-$, and Frank $\triangle$ and $\triangledown$, for expressing the cyclic and anticyclic changes, respectively, e.g. a structure $ABAB,\ldots$ is denoted as $+ - + - \cdots$. The sequence of $\triangle$'s and $\triangledown$'s (or $+$'s and $-$'s) in the unit cell is laid down to represent a polytype. This provides some compactness of notation for rhombohedral polytypes in which the same sequence repeats thrice in the unit cell, e.g. $(\triangle\triangledown\triangledown\triangledown)_3$ for the CdI_2 polytype $12R$, but no advantage results for the hexagonal polytypes for which the number of symbols remains the same as the number of layers in the ABC sequence. However, great compactness is obtained when successive symbols of one kind are added up and expressed as a succession of numeral figures, e.g. as (22) for $4H$ and $(13)_3$ for $12R$. This useful scheme of notation was evolved by Zhdanov [27]. The succession of symbols also represents the geometrical zigzag sequence of atoms in the $(11\bar{2}0)$ plane, which comprises all the three atomic positions A, B, C upon it, as first pointed out by Ramsdell and Kohn [12].

A system of notation taking account of the relative position of layers in the immediate neighbourhood of a particular layer on both sides, has been independently employed by Pauling [28]*, Wyckoff [7]** and Jagodzinski [29]. If a layer has similar layers situated on its two sides, e.g. BCB, it is designated as 'h'. If the orientations of the surrounding layers are different, e.g. ACB, it is denoted as 'c'. The symbols derive from the well known forms of hexagonal close-packing ($ABAB\ldots$) and cubic close-packing ($ABCABC\ldots$) of layers. Thus the CdI_2 polytype $4H$, having the ABC sequence as $ABCB$, is denoted as $(hc)_2$. Instead of using the symbols h, c, Wyckoff and Jagodzinski have employed the letters H, C and h, k, respectively.

A monoclinic single layer of mica possesses a characteristic structural symmetry, which has led to the formulation of schemes of notation that are specifically contrived for describing the polytypes of mica. The symmetry permits the single monoclinic mica layer to be superimposed on an adjacent layer in six possible orientations,

* vide p. 408.
** vide p. 115.

involving relative layer rotations of 0, ± 60, ± 120, or 180°, about an axis normal to the layers. An n-layered mica polytype can, therefore, be represented in terms of the rotational sequence of its constituent layers. Zvyagin [30] designated the six possible orientations as A, B, C, $\bar{A}$, $\bar{B}$, $\bar{C}$, while the rotations were envisaged with respect to a 'standard' layer in the C orientation, which was assumed to have the single layer monoclinic axial setting. A polytype with n layers in its unit cell is denoted by a series of n letters, with the jth letter of the series expressing the Zvyagin orientation of the jth mica layer. For instance, a 4-layered polytype of mica with triclinic a lattice, represented as $4T_c$ in Ramsdell's notation, is denoted as $\bar{C}\bar{C}\bar{A}A$ in this system of notation. The succession of letters, expressing the stacking sequence of layers, is referred to as Zvyagin oriented stacking symbol. Recently, a modified scheme of notation has been suggested by Ross *et al.* [31], which visualizes relative rotations between adjacent layers, instead of rotations with reference to a 'standard' layer. The six possible rotations of 0, ± 60, ± 120 and 180° are represented by the numeral figures of 0, ± 1, ± 2, and 3, respectively. In this notation the 4-layered triclinic polytype above, is represented as $4T_c$ [0132], referred to as the vector stacking symbol of the polytype. Here T_c symbolizes the triclinic symmetry of the polytype.

2.3. Suitability of the various notations

Unfortunately, variations exist in the use of the same scheme of notation by different writers, leading to unnecessary confusion in the literature. In view of the recent rapid increase in the volume of work on polytypism, it is highly desirable that a uniform pattern is followed by all workers to avoid confusion and to obtain the most complete representation in specific problems. The ensuing brief account includes suggestion towards this end.

For representing the polytypes having known cell sizes but undetermined crystal structures, the Ramsdell notation is most convenient. The lattice symmetries, viz. cubic, tetragonal, orthorhombic, hexagonal, rhombohedral, monoclinic, and triclinic, may be denoted by the natural abbreviations, C T_l, O, H, R, M, and T_c, respectively. For ensuring uniformity of notation, the use of any other letter symbols, e.g. of the letter T for representing hexagonal polytypes with trigonal symmetry [32], should be discontinued. However, in the cases where it is not possible to identify the symmetry of the lattice, a general symbol L may be used to signify 'layers', e.g. $72L$ for describing a 72-layered polytype. Earlier, several workers, e.g. Brafman *et al.* [33], have employed this symbol for designating all polytypes of zinc sulphide, having both hexagonal and rhombohedral symmetry. To discriminate between two polytypes of a substance having the same cell size and symmetry but different layer arrangements, numeral subscripts, e.g. nH_1, nH_2, etc. should be used, instead of the letter subscripts, viz. nH_a, nH_b, etc., originally suggested by Ramsdell and Kohn [12]. This has become very necessary in view of the fact that an unusually large number of polytypes of the same cell size have been discovered in certain substances, e.g. 24 CdI_2 polytypes of 20 layers each have been reported. In this connection, it is further suggested that the use of the subscripts may be limited to polytypes of known crystal structure, since only these have an established and permanent identity. The polytypes of unknown structure may be represented without attaching any subscripts, e.g. nH.

Zhdanov notation is compact and also elegantly preserves the interrelationship of the layers. As such, it is eminently suited for representing the polytypes with known crystal structure. But a word of caution is necessary here. Rarely, one comes across two or more Zdhanov symbols which prima facie appear to represent the same polytype, but really prove to be representing different polytypes when extended into their respective ABC sequences. For instance, the symbols (2321) and ($\overline{1}$232) initially appear to denote the same CdI_2 type, but when transformed into their ABC sequences, viz. $(A\gamma B)(C\alpha B)(A\beta C)(A\gamma B)\dots$ and $(A\gamma B)(A\beta C)(A\gamma B)(C\alpha B)\dots$, they are clearly seen as representing two different crystal structures. This also emphasizes the basic significance of the ABC notation, which is wholly unambiguous and should, therefore, be always employed in doubtful situations, particularly when conjecturing the probable atomic models for determining the crystal structure of a polytype.

The hc notation is valuable in problems involving considerations of the interaction between layers. The vector stacking symbol is particularly suited to describe mica polytypes, although it becomes clumsy for higher cell sizes.

2.4. IDENTIFICATION OF POLYTYPES

Usually X-ray diffraction methods are employed for the identification of polytypes. Since most polytypic substances have hexagonal or rhombohedral lattice symmetries, we shall restrict the discussion of the methods to identifying only such symmetries. The prominent polytypic layer structures like CdI_2 and PbI_2 possess these symmetries, which essentially derive from the close-packed character of the constituent layers. Even a non close-packed layer-structured polytypic substance like mica has a pseudo-hexagonal symmetry of structure. Electron diffraction methods have also been employed for the identification of a few polytypes, but since they are based on the same premises as the X-ray diffraction methods, they will not be treated separately. The early morphological and optical methods, which now have only a historical interest, will also be left out.

The identification necessarily involves the determination of (i) the lattice type and (ii) the cell size and hence the number of layers in the unit cell, both of which can be inferred from the position of reflections on suitable X-ray diffraction photographs. Both hexagonal and rhombohedral lattices can be referred to either hexagonal or rhombohedral axes. The former are more convenient to use and, therefore are universally employed. When a hexagonal lattice is referred to them, the unit cell is primitive, as shown in Figure 1. When a rhombohedral lattice is so referred to, the hexagonal cell, defined by the dimensions a, b, c, is nonprimitive, although the rhombohedral cell (a_1, a_2, a_3) is primitive (Figure 2). The indices of lattice planes in this lattice, referred to the hexagonal cell, are governed by the condition

$$-h + k + 1 = 3n \, (n = 0, \pm 1, \pm 2, \dots) \tag{2.1}$$

For a rhombohedral polytype two kinds of setting, obverse and reverse, for the

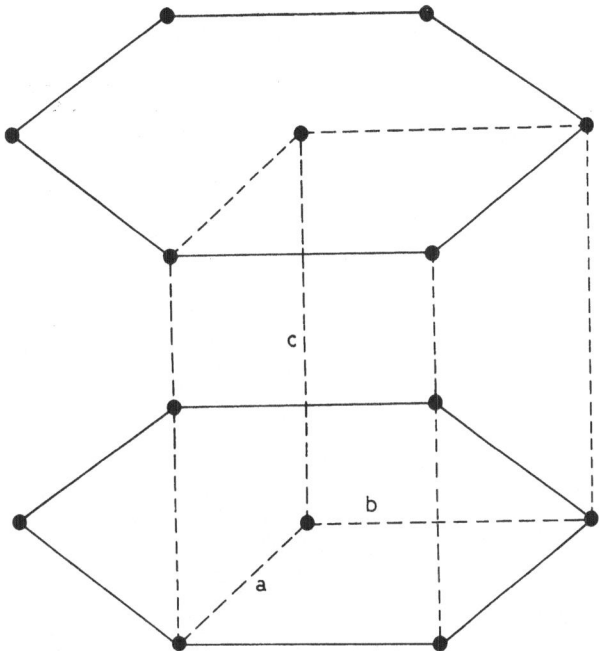

Fig. 1.   A hexagonal lattice (lattice points shown by black circles) referred to hexagonal axes (a, b, c).

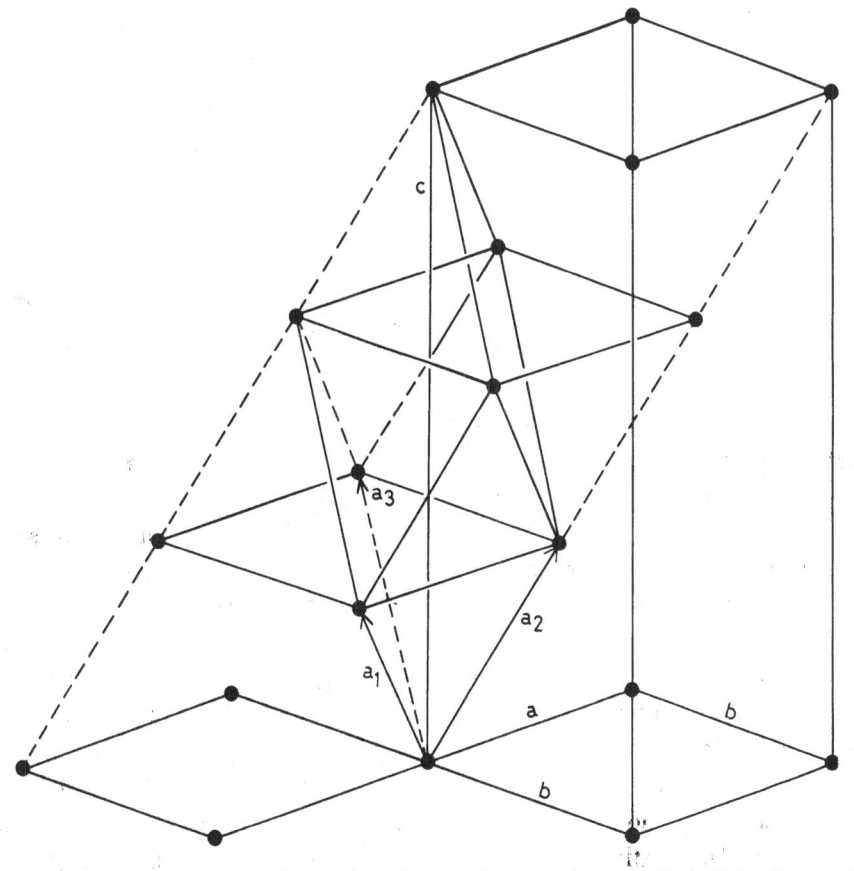

Fig. 2.   A rhombohedral lattice (a_1, a_2, a_3) referred to hexagonal axes (a, b, c). (After Buerger [34]).

unit cell are possible. These are totally equivalent, merely differring by a rotation of 60° about the c axis, but by international convention the former has been adopted as the standard [35].

For the purpose of obtaining X-ray diffraction photographs, the crystals are mounted either about the c or the a axis. Therefore, it is useful to examine the orientation of the reciprocal lattice for these positions. Since c is perpendicular to the plane of a^* and b^*, the a^*b^* reciprocal lattice nets lie in parallel horizontal planes when the c axis is vertical. One such net is shown in Figure 3. According to the well known

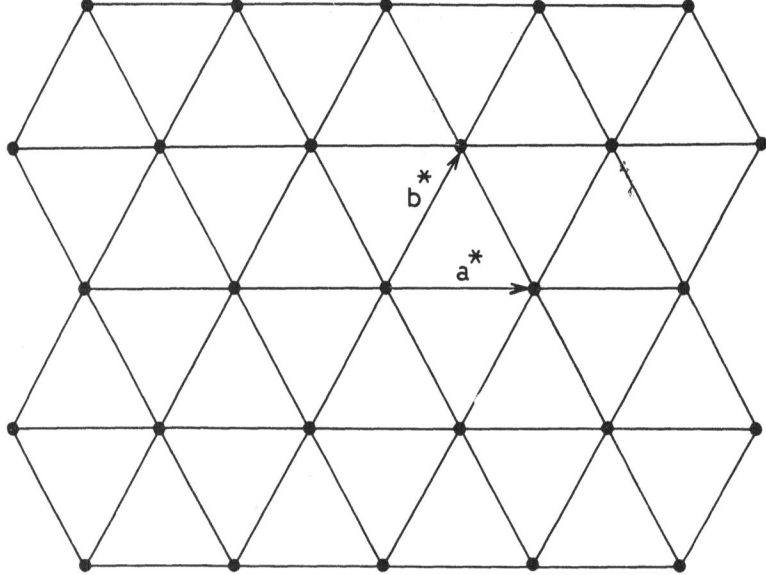

Fig. 3.   An $a^*\ b^*$ reciprocal lattice net for a hexagonal or rhombohedral polytype. c axis is perpendicular to the plane of paper.

relation between the direct and reciprocal lattice vectors, $\mathbf{a} \times \mathbf{a}^* = \mathbf{b} \times \mathbf{b}^* = \mathbf{c} \times \mathbf{c}^* = \lambda$, the various a^*b^* nets for all the polytypes of a substance are identical, because the magnitudes of the a and b dimensions are equal for all of them. If the a axis is held vertical, the reciprocal lattice nets b^*c^* lie in horizontal planes, and now these are not the same for the different polytypes on account of the variation in the magnitude of c, and consequently of c^*. One such net is depicted in Figure 4. Here the density of the lattice rows along b^* remains constant for all polytypes of a substance, but varies along c^*. The higher the polytype, the greater is the density of rows along c^*. Two oscillation photographs of the most common polytype $4H$ of cadmium iodide, taken about the c and a axes, respectively, are shown in Figure 5. The respective ranges of oscillation have been chosen to record a large succession of 10.l reflections, which are found to be most suitable for polytype identification as well as for structure determination. The continuous chain of spots on the zero layer line in Figure 5b

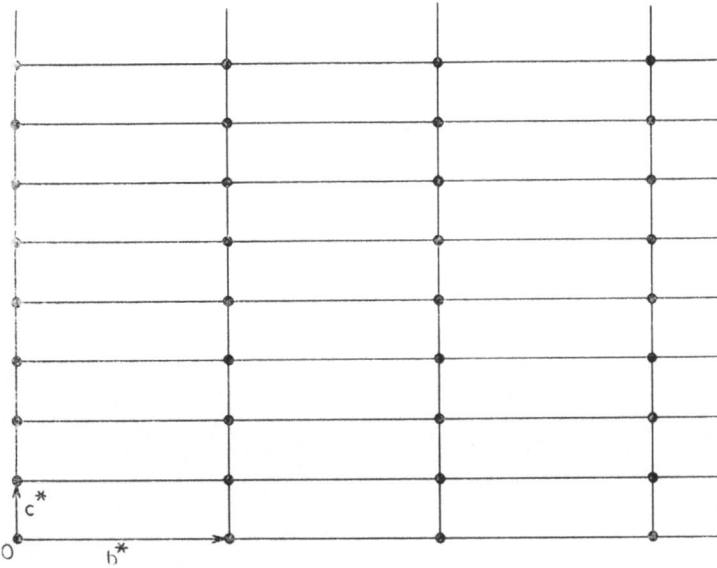

Fig. 4. A b^* c^* reciprocal lattice net for the most common type 4H of cadmium iodide ($b^* = 0.27$, $c^* = 0.07$).

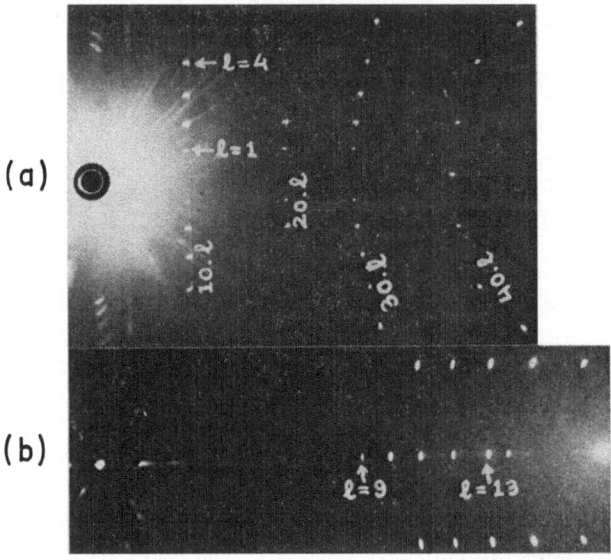

Fig. 5. 15° oscillation photograph of the most common CdI_2 polytype 4H, taken about (a) the c axis and (b) the a axis, recording a large succession of 10.l reflections. Camera radius 3 cm; CuKα radiation.

really consists of 01.*l* reflections, but since *a* and *b* axes are indistinguishable in a hexagonal lattice, the same can as well be regarded as representing 10.*l* reflections.

A *c* axis oscillation photograph is most suited for determining the lattice type. For polytypes with a hexagonal lattice, the reflections are symmetrically distributed about the zero layer line, as e.g. in Figure 5a. This mirror symmetry in the photograph results from the existence of a six-fold axis plus the centre of symmetry invariably added in the usual X-ray photographs. It does not arise from the existence of a mirror plane perpendicular to *c* in the crystal structure, as one may initially be led to conclude.

Fig. 6.   *c* axis 15° oscillation photograph of CdI₂ polytype 24R_1, showing asymmetrical distribution of spots about the Laue streak along zero layer line. Conditions as in Figure 5.

For a rhombohedral lattice, the distribution of the 10.*l* reflections in a *c* axis photograph is essentially unsymmetrical. In this case, according to Equation (2.1), the possible reflections are limited to the values $l = 3n + 1$, so that on the 10.*l* row of diffraction spots, the permitted reflections above and below the zero layer line are 10.1, 10.4, 10.7,... and 10.$\bar{2}$, 10.$\bar{5}$, 10.$\bar{8}$, ..., respectively. The asymmetry is illustrated in Figure 6, which is the oscillation photograph of CdI₂ type 24R_1. It will be seen that the Laue streak along the zero layer line divides the distance between the nearest spots (10.1 and 10.$\bar{2}$) on its upper and lower side in the ratio 1:2, which is the characteristic of all rhombohedral polytypes. The inequality arises because the intervening spots, 10.0 and 10.$\bar{1}$, are missing and hence the reciprocal lattice ζ-values of the two spots bear this ratio, viz. $\zeta_{10.\bar{2}} = 2\zeta_{10.1}$. The upper part of the 10.*l* row also includes some close reflections of a higher unidentified polytype other than 24*R*. The reflections of 24*R* are well separated from one another and uniformly spaced and hence they are easily distinguished. The existence of additional reflections on the upper side indicates that the upper half of the crystal contains an admixture of another polytype. This phenomenon, in which more than one polytype coexist one on top of the other within the same crystal, is called syntactic coalescence, and occurs very frequently in polytypic crystals.

The determination of the number of layers in the unit cell of a polytype necessitates the measurement of its c parameter, which lies perpendicular to the plane of stacking of the layers in the hexagonal and rhombohedral polytypes. The successive unit layers of structure are spaced at regular intervals along c. For relatively small polytypes, c can be readily determined from c axis oscillation or rotation photographs, by measuring the reciprocal lattice ζ-values for the layer lines with the help of a Bernal chart. But the method is not satisfactory for high polytypes which give extremely close diffraction spots along the various row lines and therefore render the accurate measurement of the ζ-values almost impossible. Their identification is more conveniently carried out by comparing their X-ray photographs with the photograph of a small type which gives widely spaced spots along the row lines. Since the reciprocal lattices of different polytypes of a substance are all similar, except for a change in c^*, their X-ray photographs are also very similar. The chain of $10.l$ spots coincides for all the polytypes, such that the range of spots from 10.0 to $10.X_1$ for a polytype X_1H coincides exactly with the range 10.0 to $10.X_2$ for a second polytype X_2H. Consequently, the number of layers in the unit cell of a high polytype can be known by mere inspection, e.g. if m spacings of the polytype coincide with n spacings of a small type, say $4H$, the number of layers is given by $4(m/n)$, where m is not necessarily integral. The comparison is conveniently made by matching the X-ray films of the two polytypes against an illuminated background. The same information can also be obtanied from a-axis oscillation photographs in which the only difference is that the $10.l$ spots lie along

18 H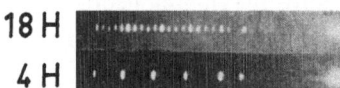

4 H

Fig. 7. a axis $15°$ oscillation photographs of types $4H$ and $18H$ showing comparison of corresponding $10.l$ reflections. Conditions as in Figure 5.

layer lines instead of the row lines. Two chains of these spots, occurring along the zero layer line and corresponding to CdI_2 polytypes $18H$ and $4H$, respectively, are shown juxtapositioned in Figure 7, which clearly illustrates the method of identification; here 9 consecutive spacings of the higher polytype coincide with 2 corresponding spacings of the lower polytype. The c^* value, and hence the c-value, can also be computed from a-axis photographs [36].

Moving film Weissenberg photographs also afford polytype identification in the same way as the oscillation photographs, but need much longer exposures and are therefore not usually employed unless complete structure determination is aimed at. However, for crystals of low X-ray absorption, like silicon carbide, they are valuable also for identification purposes [3]*. Laue photographs can also be used similarly as the oscillation photographs and provide high resolution of diffraction spots [37]. Many giant polytypes of silicon carbide, comprising upto as many as 1080 layers in the unit cell, have been identified by this method [38, 39]; two still bigger polytypes of 2400 and 4680 layers, respectively, have been approximately identified [39]. However, their identification has been possible because they occurred syntactically intergrown with a small polytype, so that the number of reflections from the unknown modifications, which lay between reflections of the known type, could be estimated with certainty. In the absence of such an intergrowth the method becomes uncertain, because then separate photographs of the two polytypes have to be matched against one another, which cannot be done accurately enough. A photographic enlargement for enhancing the separation between the sports is sometimes useful. The resolution of the spots can be improved by increasing the crystal to film distance but it needs a fine collimation of the incident X-ray beam and a proportionately much higher exposure time. A better alternative is to employ a microfocus X-ray source, which provides a highly intense and narrow beam of X-rays. These considerations also apply to oscillation and Weissenberg methods for polytype identification.

The lattice type can also be ascertained from a-axis oscillation photographs by looking for systematically absent reflections according to the condition $l = 3n + 1$ for a rhombohedral polytype $3nR$, around the positions of the corresponding hexagonal type nH, and also from a-axis Weissenberg photographs, in which the nearest reflections on the two sides of the central Laue streak are asymmetrically distributed in the ratio $1:2$ for a rhombohedral lattice. This is specially valuable for the CdI_2 crystals in which the existence of 'arcing effect' (vide Section 5.2) makes the reflections to be undesirably extended in an oblique direction on the c axis photographs. Rotation photographs about c axis can also be employed for lattice identification, in which the reflections appear in pairs for the rhombohedral lattice [3]**. However, this calls for a very accurate setting of the crystal, since a similar situation can also result from a slight misadjustment of the crystal.

The complete identification of a polytype necessarily requires the determination of its crystal structure, besides the determination of its cell size and the lattice type. Several different methods have been employed for this purpose, for a description of which the reader is referred to a recent review [4].

3. Polytypic Substances and Stacking Faults in Their Structures

While now polytypism has been recognised as a widely occurring property of solids, not all substances display it to the same extent. Two layer-structured substances, viz. the iodides of cadmium and lead, have been found to be predominantly polytypic. A short description of them, including their mode of growth, salient structural features, growth form, list of polytypes, etc., follows. Other polytypic substances

* vide p. 164.
** vide p. 149.

with layer structures will also be described but in lesser detail. Certain non-layered prominently polytypic substances will also be briefly referred to. In the end, a short description of stacking faults in the polytypic structures will be provided.

3.1. CADMIUM IODIDE

The cadmium iodide structures are formed of piles of parallel composite layers, each of which consists of a sheet of cadmium atoms sandwiched between two close-packed sheets of much larger iodine atoms. The various polytypes contain different numbers and arrangements of these extended molecular sheets in their unit cells. The smallest CdI_2 polytype, $2H$, consists of just one molecular sheet, which has also been referred to as a 'minimal sandwich' in literature. In the ABC notation, it is represented as $(A\gamma B)$, $(B\alpha C)$, or $(C\beta A)$, where the Roman letters stand for the I atoms and the Greek ones for the Cd atoms. The thickness of an isolated sandwich $(I-Cd-I)$ is 3.42 Å. Two such sandwiches in a 2H structure have been outlined in Figure 8. The Cd atoms

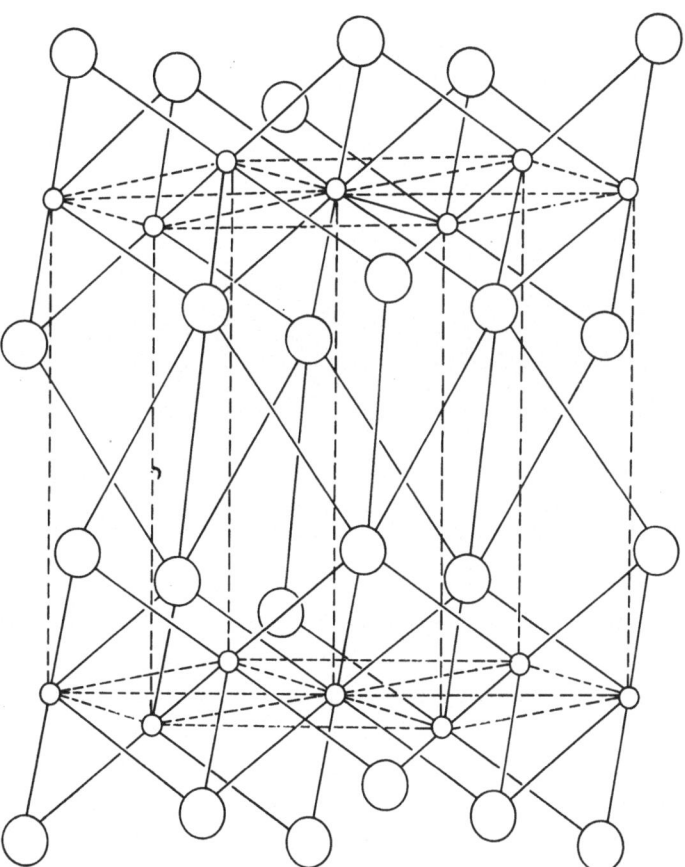

Fig. 8. Two minimal sandwiches of the 2-layered hexagonal polytype, $2H$ of cadmium iodide. The Cd and I atoms are represented by small and large circles, respectively (After Wells [40]).

occupy alternate octahedral voids between the I layers and if one disregards their lack
of compactness, they also form a close-packed layer, with each atom symmetrically
surrounded by six others. In this sense, the structure can be regarded as a packing
of layers of linked CdI_6 octahedra, as depicted in Figure 9. The octahedra are fairly
regular, three of the I−I distances 4.24 Å being, the other three 4.21 Å. Since
these values are nearly equal to the sums of the ionic radii of iodine, a hexagonal
close-packing of the iodine atoms is obtained $(r_I=2.16$ Å, $r_{Cd}=0.97$ Å [28]*). The
cell dimensions of the 2H structure (see Figure 9) are $a=4.24$ Å, $c=6.84$ Å [7]**.
This gives the axial ratio c/a as 1.61, which is slightly less than the ideal value of 1.63
for a close-packed structure.

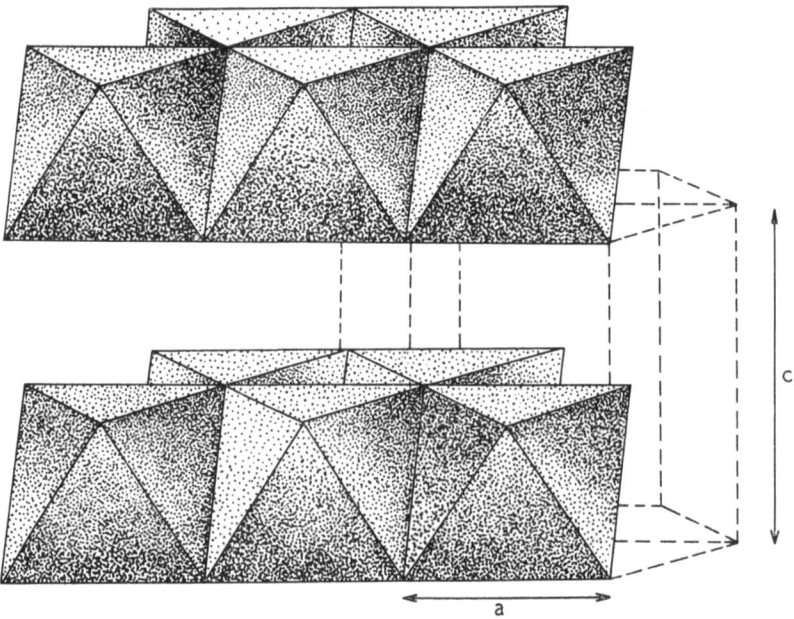

Fig. 9.   Packing of layers of linked CdI_6 octahedra in the CdI_2 2H structure. The I atoms occupy the
corners and the Cd atoms the centres of the octahedra (After Verma and Krishna [3], p. 79).

    The bonding within a sandwich is generally regarded as ionic in character. However,
according to a later view, there exists a resonance between covalent and ionic bonding
in which the influence of the former predominates; the electronic configurations of the
Cd and I atoms permit the formation of sp^3d^2 hybrid covalent bonds [8]†. In any
case, the binding forces within the sandwiches are far stronger than the forces between
the sandwiches, which are held together by van der Waals bonds between like
I atoms. This heterogenous nature of the structure imparts to the CdI_2 crystals the

* vide p. 514.
** vide p. 268.
† vide p. 152.

characteristic physical properties of a layer structure, e.g. an excellent cleavage parallel to the layers and a markedly anisotropic thermal expansion.

The CdI_2 crystals have been grown from solution, vapour, melt and gel. They are extremely soft and usually occur as flat hexagonal platelets, but whisker growth has also been reported [41]. The platelets are colourless and transparent. Growth spirals are frequently found on their basal surfaces. The most common polytype is $4H$. The next but much less common type is $2H$. More than one polytype are commonly found to be syntactically coalesced within the same crystal. The X-ray photographs of the crystals often show features of one-dimensional disorder, manifesting as 'streaking' and arising from the existence of random stacking faults in the structure (Section 5.1).

The crystals grown from solution and vapour have been generally found to be richly polytypic, but the melt-grown crystals have been reported to be of the type $4H$ alone [41]. Nearly 240 polytypes have been discovered so far, the complete crystal structures of 60 of them have been worked out. An uptodate list of the known polytypes, including their crystal structure, if determined, is presented in Table I. The polytype notation adopted here follows the suggestions made in Section 2.3. The discoverer's reference has been given in the last column. In many cases, more than one reference is provided for the same polytype, signifying that different investigators have come across a polytype with the same number of layers in the unit cell. These differently reported polytypes are likely to be different in their atomic structures by virtue of differences in the stacking sequence of their layers, since for an n-layered polytype 2^{n-1} sequences are theoretically possible. Sometimes several different polytypes of a particular periodicity have been encountered by the same investigator, which has been indicated in the table by attaching a numeral subscript to the corresponding reference; the subscript denotes the number of polytypes of the given periodicity observed by the investigator. These considerations should be borne in mind while counting the total number of polytypes of a given cell size.

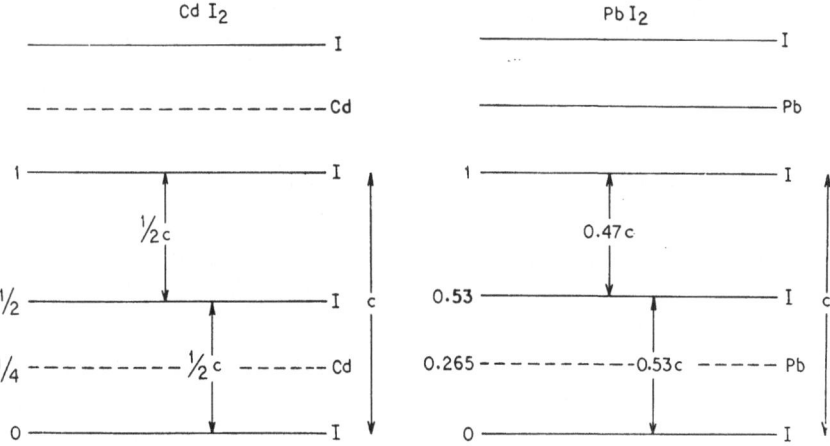

Fig. 10. Comparative layer spacings in the smallest polytype $2H$ of cadmium iodide and lead iodide, respectively.

TABLE I

Known polytypes of cadmium iodide

Serial No.	Polytype	Structure (Zhdanov symbol)	Reference	Serial No.	Polytype	Structure (Zhdanov symbol)	Reference
1	$2H$	11	[42]	30	$16H_6$	$(22)_2112211$	[58]
2	$4H$	22	[43]	31	$16H$	–	[43], [47],
3	$6H_1$	2211	[43]				[52], [53],
4	$6H_2$	33	[44]				[54], [57]
5	$8H_1$	$22(11)_2$	[43]	32	$18H_1$	$(22)_411$	[60]
6	$8H_2$	$(121)_2$	[45]	33	$18H_2$	$22(11)_7$	[61]
7	$8H_3$	1232	[46]	34	$18H_3$	1222322211	[62]
8	$8H$	–	[47]	35	$18H_4$	$22(212)_2(11)_2$	[58]
9	$10H_1$	$(22)_211$	[43]	36	$18H_5$	$(22)_31221$	[48]
10	$10H_2$	$(221)_2$	[48]	37	$18H$	–	[43], [52],
11	$10H$	–	[49]				[53], [54],
12	$12H_1$	222123	[43]				$[56]_3$
13	$12H_2$	$(21)_2(12)_2$	[43]	38	$20H_1$	$(22)_4(11)_2$	[53]
14	$12H_3$	$(22)_2(11)_2$	[43]	39	$20H_2$	$22(11)_2(2112)_2$	[53]
15	$12H_4$	$22(211)_2$	[50]	40	$20H_3$	$22(11)_8$	[46]
16	$12H_5$	11123211	[50]	41	$20H_4$	$(11)_72112$	[63]
17	$12H_6$	22111221	[51]	42	$20H_5$	$(22)_221122211$	[63]
18	$12H$	–	[43], [49],	43	$20H_6$	$(22)_3(211)_2$	[63]
			[52], [54],	44	$20H$	–	[43], [47]6,
			[47]				[49], [52],
19	$12R_1$	$(13)_3$	[55]				[53], [54],
20	$14H_1$	$(22)_311$	[43]				$[56]_3$,
21	$14H_2$	$(1122)_211$	[46]				$[64]_4$
22	$14H_3$	11212322	[51]	45	$22H_1$	$(11)_5(2211)_2$	[65]
23	$14H_4$	$(22)_2(11)_3$	[48]	46	$22H$	–	$[47]_3$,
24	$14H$	–	[47], [56],				$[49]_2$, [53],
			[57]				[56]
25	$16H_1$	$(22)_2(211)_2$	[46]	47	$24H_1$	$(2222211)_2$	[45]
26	$16H_2$	$(22)_2(11)_4$	[46]	48	$24H$	–	$[43]_4$, [46],
27	$16H_3$	$22(212)_211$	[46]				$[52]_2$, [53],
28	$16H_4$	$(22)_3(11)_2$	[58]				[56], [57],
29	$16H_5$	12223222	[51]				$[64]_2$

(continued on next page)

3.2. LEAD IODIDE

The crystals of lead iodide are isostructural with cadmium iodide. The thickness of an isolated minimal sandwich $(I-Pb-I)$ is slightly greater, viz. 3.49 Å. Also unlike the cadmium iodide crystals, in which the successive iodine layers are equispaced, the $I-I$ separations are not the same but vary alternately. As illustrated in Figure 10 for the smallest type $2H$, the separation within the sandwich $(=0.53c)$ is a little larger than the separation outside the sandwich $(=0.47c)$. The cell dimensions for the type $2H$ are $a=b=4.56$ Å, $c=6.98$ Å [7]*. The c/a ratio $=1.54$ is considerably further away from the ideal value of 1.63 for perfect hexagonal close-packing,

* vide p. 269.

Table I (Continued)

Serial No.	Polytype	Structure (Zhdanov symbol)	Reference	Serial No.	Polytype	Structure (Zhdanov symbol)	Reference
49	$24R_1$	$(2213)_3$	[61]	73	$38H$	–	[47], [49]$_3$, [70]$_5$
50	$26H_1$	$(21111)_411$	[66]				
51	$26H_2$	$(22)_611$	[66]	74	$40H_1$	$(22)_721122211$	[58]
52	$26H_3$	$(222211)_22112$	[59]	75	$40H$	–	[43]$_3$,
53	$26H$	–	[43]$_2$,				[47]$_2$, [54],
			[47]$_2$, [54]				[56]$_2$,
54	$28H_1$	$(22)_6(11)_2$	[67]				[70]$_4$
55	$28H_2$	$(22)_411(22)_211$	[58]	76	$42H$	–	[47]$_4$, [54],
56	$28H_3$	$(22)_5112211$	[58]				[56]$_2$
57	$28H$	–	[47], [54]$_2$,	77	$42R_1$	$(22221212)_3$	[68]
			[56]$_3$	78	$42R$	–	[47]
58	$30H_1$	$(2211)_41122$	[61]	79	$44H$	–	[43], [47]$_4$
59	$30H_2$	$(22_2(211)_2(22)_311$	[63]	80	$46H$	–	[47], [54]
60	$30H_3$	$(22)_711$	[63]	81	$48H$	–	[47]$_2$,
61	$30H_4$	$(22)_4211222(11)_2$	[63]				[54]$_2$, [56],
62	$30H$	–	[43]$_2$, [47],				[57]
			[56]	82	$48R$	–	[47], [53],
63	$30R_1$	$(221212)_3$	[68]				[57]
64	$32H_1$	$(22)_5321123$	[69]	83	$50H$	–	[43], [52],
65	$32H$	–	[43]$_2$,				[54]$_2$, [64]
			[54]$_3$, [57]	84	$52H$	–	[48], [70]
66	$34H_1$	$(222211)_322$	[59]	85	$54H$	–	[47], [52]
67	$34H$	–	[47]$_2$,	86	$56H$	–	[43], [54]
			[49]$_2$,	87	$60R_1$	$[(22)_31223]_3$	[72]
			[54]$_3$, [57]	88	$60R$	–	[47], [54]
68	$36H_1$	$(22221111)_2(11)_2$	[58]	89	$62H$	–	[43], [47]
		$(22)_2$		90	$64H$	–	[43], [54]
69	$36H$	–	[47]$_5$, [48],	91	$66H$	–	[47]
			[52], [53]$_5$,	92	$72R_1$	$[(22)_41223]_3$	[72]
			[56], [70]$_2$	93	$84R$	–	[53], [54]$_2$
70	$36R_1$	$(22112121)_3$	[61]	94	$90R$	–	[54]
71	$36R_2$	$(22212111)_3$	[71]	95	$108R$	–	[47], [57]$_2$
72	$36R$	–	[47]				

than in cadmium iodide ($c/a = 1.61$). Hence the PbI$_2$ structures are relatively less close-packed.

The PbI$_2$ crystals have been grown from solution, vapour, and gel. They occur as yellow platelets which are flat and have triangular or hexagonal profiles. Growth spirals are often seen on their basal faces. Solution and vapour growths generally yield very small and thin platelets, looking like little wafers. Gel growth, on the other hand, can yield appreciably larger and thick crystals and hence it has been widely employed in recent years. The crystals are soft but harder than cadmium iodide. Unlike the latter, they usually have well-ordered structures, free from the features of one-dimensional disorder on their X-ray photographs. They have been found to be considerably less polytypic than cadmium iodide and, possibly for the same reason, the incidence of syntactic coalescence is not so common in them. Nearly 45 polytypes have been reported so far, the complete crystal structures of 12 of which have been determined. They are listed in Table II, in the same manner as for the CdI_2 polytypes in Table I.

TABLE II

Known polytypes of lead iodide

Serial No.	Polytype	Structure (Zhdanov symbol)	Reference	Serial No.	Polytype	Structure (Zhdanov symbol)	Reference
1	$2H$	11	[73]	16	$18H$	–	[76]$_2$
2	$4H$	22	[74]	17	$18R_1$	(1311)$_3$	[80]
3	$6H$	2211	[74]	18	$18R_2$	[(21)$_2$]$_3$	[80]
4	$6R$	∞	[75]	19	$18R$	–	[76]$_2$
5	$8H$	–	[76], [77]$_3$	20	$20H_1$	(11)$_7$2112	[78]
6	$10H_1$	(11)$_3$22	[78]	21	$20H$	–	[76]
7	$10H$	–	[76]$_2$	22	$24H$	–	[76], [77]
8	$12H_1$	(11)$_3$2112	[79]	23	$24R$	–	[76]$_2$, [81]
9	$12H$	–	[76]$_2$	24	$28H$	–	[76]
10	$12R_1$	(31)$_3$	[74]	25	$30H$	–	[76]
11	$12R_2$	(13)$_3$	[80]	26	$30R$	–	[76]
12	$12R$	–	[76], [77]	27	$36H$	–	[76]
13	$14H_1$	(11)$_5$22	[78]	28	$36R$	–	[76]
14	$14H$	–	[76]	29	$42R$	–	[76]
15	$16H$	–	76]$_3$	30	$48R$	–	[76], [77]

3.3. OTHER SUBSTANCES WITH LAYER STRUCTURES

The mineralogically important compounds, micas, occur in several varieties. Certain common forms are muscovite $[KAl_2 (OH)_2 (Si_3Al) O_{10}]$, phlogopite $[KMg_3 (OH)_2 (Si_3Al) O_{10}]$, margarite $[CaAl_2 (OH)_2 (Si_2Al_2) O_{10}]$, etc. Their structural features, particularly with reference to the understanding of polytypism, have been described in literature [30, 82, 83]*. A single mica layer is monoclinic and consists of three parts, viz. a sheet of octahedrally coordinated cations sandwiched between two identical sheets of linked (Si, Al) O_4 tetrahedra. The other cations, e.g. the interlayer cations K^+ in muscovite, are located between these composite mica layers. A schematic representation of the structure of this variety of mica has been illustrated in Figure 11. The other varieties have a similar structure. The surface oxygen atoms of the tetrahedral layer possess pseudo-hexagonal or pseudo-trigonal symmetry, which permits individual composite layers to occupy one of the six possible orientations in the structure, as discussed earlier in Section 2.2. Since the orientations are all equivalent, it is believed that this factor is basically responsible for the incidence of polytypism in micas [84]. Until recently, only four mica polytypes of small repeat periods, viz. $1M$, $2M_1$, $2M_2$, and $3H$, had been known to exist [3, 83]. But of late, several new modifications have been discovered, particularly in the biotites and lithium-rich

* (See Section 1)

micas [31, 85]. The total number of known mica polytypes now stands at 19, of the crystal structures of all but two which have been worked out. They are listed in Table III. They frequently display features of one-dimensional disorder on their X-ray photographs. Syntactic coalescence is also commonly observed.

Besides the micas, more layer silicates have been found to display polytypism. These include chlorites, the ferro-silicate cronstedtite, Mn-silicates of the pyrosmalite group, and certain clay minerals. The chlorites are a group of layer lattice silicates of very variable composition that can be approximately expressed as $X_m Y_4 O_{10}(OH)_8$, where m lies between 4 and 6 and X, Y represent positive ions in octahedral and

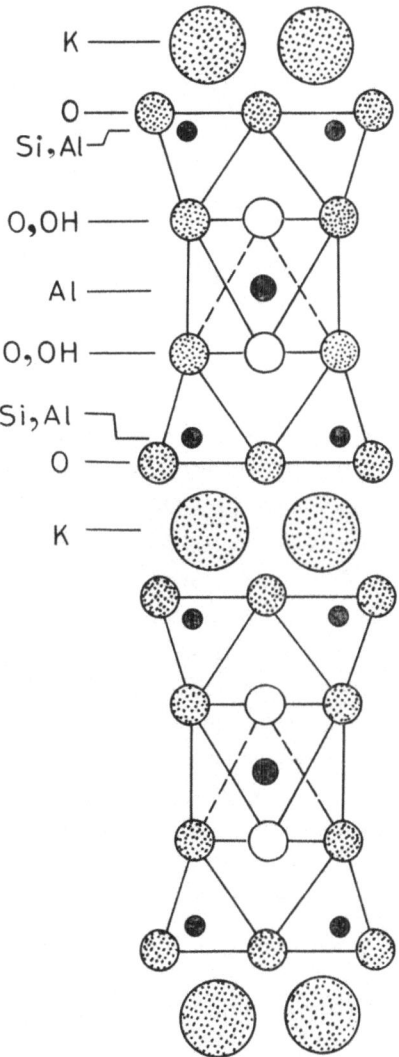

Fig. 11. A diagrammatic outline of the crystal structure of muscovite mica. Two composite layers have been drawn (After Evans [8], p. 251).

TABLE III

Known polytypes of mica

Serial No.	Polytype	Structure (Vector stacking symbol)	Reference
1	$1M$	0	[4]
2	$2M_1$	$2\bar{2}$	[4]
3	$2M_2$	$1\bar{1}$	[4]
4	$3H$	222	[4]
5	$3T_c$	$02\bar{2}$	[31]
6	$4M$	2220	[31]
7	$4T_c$	0132	[31]
8	$5M$	$(222)2\bar{2}$	[85]
9	$8M$	$(222)_2 2\bar{2}$	[31]
10	$8T_{c_1}$	$(0)_6 2\bar{2}$	[31]
11	$8T_{c_2}$	$0002\bar{2}\bar{2}02$	[31]
12	$9T_c$	$(0)_7 2\bar{2}$	[85]
13	$10T_c$	$(22)_2 2\bar{2}\bar{2}200$	[31]
14	$11M$	$(222)_3 2\bar{2}$	[31]
15	$12M$	–	[85]
16	$14T_c$	$(0)_{12} 2\bar{2}$	[31]
17	$14M$	$(222)_4 2\bar{2}$	[85]
18	$20M$	–	[31]
19	$23T_c$	$(0)_{21} 2\bar{2}$	[31]

tetrahedral positions, respectively; usually X stands for Mg with substitution by Al, Fe, Cr, or Mn, and Y for Si and Al [86]. It has been noted that in most chlorites adjacent layers are stacked with a considerable degree of randomness, as manifested by the existence of heavy streaking on the X-ray photographs [86]. Regular layer sequences are less common. Bailey and Brown [87] have proposed that theoretically four different types of single chlorite layers may be formed, leading to 12 different 1-layer polytypes with monoclinic or triclinic symmetry, 3 of which have been actually discovered in natural specimens. Further, 2- and 3-layer polytypes, having monoclinic symmetry, have been reported in one specimen each [86]. Their crystal structures, and that of the three 1-layered types, have been determined [86, 87]. 13 polytypes of the hydrous ferro-silicate mineral cronstedtite have been reported. Assuming that the structures are composed of trioctahedral kaoline-type layers stacked together and adopting the chemical formula $(Fe^{2+})_3(SiFe_3^+) O_4(OH)_5$, Steadman and Nuttall [88] have succeeded in determining the crystal structures of 8 of these polytypes, viz. $1H, 2H_1, 2H_2, 2H_3, 3H, 6R, 1M$ and $2M$. The remaining 5 polytypes with unknown structures include a $9R$ structure [89] and two each of 2-layer and 6-layer structures [88]. Like the chlorites, the crystals of cronstedtite frequently display the features of one-dimensional disorder on their X-ray photographs. Syntactic coalescence has also been observed. The unit cell parameters of the rare Mn-silicates of the pyrosmalite group, viz. manganpyrosmalite, schallerite and friedelite, suggest that they can be regarded as a group of polytypes $1H$, $2H$, and $3H$, respectively. The parameters are $a =$ nearly 13.4 Å and $c =$ nearly 7.1, 14.2 and 21.6 Å, respectively [90]. The three silicates have the same type of crystallochemical formula $(Mn, Fe, ...)_8$ $(OH, Cl)_{10} Si_6O_{15} . nH_2O$. Similarly, the three kaoline (china clay) minerals kaolinite, nacrite, and dickite, have been shown to consist of different stacking sequences of complex layers of $Al_2O_3 . 2SiO_2 . 2H_2O$, perpendicular to c [91]. They have the same a and b dimensions and thus they can be regarded as polytypically related.

From electron microscopic studies of etched fracture surfaces of the precious gem opal, the existence of a $9R$ polytype, derived from the close-packed $3C$ structure, has been inferred [92]. The electron micrographs clearly reveal that the gem is composed of packings of spheres of amorphous silica (SiO_2), having a uniform diameter of about 0.2 μm, which frequently show stacking faults and twinning.

The crystals of $CdBr_2$, which is isostructural with CdI_2 and PbI_2, have also been found to be polytypic but to a far lesser extent. The most common polytype is $6R$ (Zhdanov symbol: ∞). In the crystals grown from aqueous solution as well as by sublimation, Pinsker [91] reported the existence of three polytypes, $2H(11)$, $4H(22)$ and $6H(33)$. Subsequently, Mitchell [93] has observed that the sublimation-grown crystals are all of the $6R$ type, while those grown from solution show polytypism. He reported two additional types of longer periods, viz. $\approx 40H$ (or $\approx 120R$) and $12H$ (or $12R$) but failed to obtain any $2H$ or $6H$ types. The $CdBr_2$ crystals grow as platelets or needles. The platelets often display growth spirals on their basal surfaces and show syntactic coalescence of polytypes. They are generally found to be heavily disordered.

The group of structures known as hexagonal ferrites comprises six different phases, containing Ba, Mg, Fe and O in varying proportions. They are hexagonal and rhombohedral layer structures arising from variable stacking of three kinds of basic structural units [94]. The various possible arrangements of these units involve stacking of chemically different yet structurally compatible layers. Thus, strictly speaking, they may not be called polytypes. Nevertheless, they resemble them in all possible ways. 4 such 'polytypes', $84R$, $102R$, $40H$, and $138H$, have been discovered in syntactically grown crystals of the 'Z' phase $(Ba_3Mg_2^{2+} Fe_{24}O_{41})$ and the possible existence of a vast number of others has been suggested [94]. Two polytypes of the rare beryllium mineral taffeite, which is a Be–Mg–Al oxide, have been reported [95]. These are $4H$ and $9R$. The crystals had a platy habit and showed a perfect basal cleavage parallel to (0001). Therefore, presumably, they have a layered structure. A new polytype of the heterogenite group of minerals, which are hydrated cobalt oxides with formula Co 0.0H and normally occur in the rhombohedral $3R$ form, has been discovered. This new 2H polytype occurs in specimens having about the same chemical composition as the normal variety [96]. Its crystal structure has also been worked out [96].

The hydrated sulphate minerals coquimbite and paracoquimbite, both having the formula $Fe_2(SO_4)_3 . 9H_2O$ (often Fe is partially substituted by Al [97]), have been described as polytypic, the modifications of the former being hexagonal and of the latter rhombohedral [98, 99]. Similarly, three polytypic varieties of the carbonate minerals parasite and synchisite have been established [100].

Molybdenite (MoS_2), which has a crystal structure resembling that of CdI_2, ordinarily crystallizes in a hexagonal $2H$ modification*. Following the discovery of a rhombohedral $3R$ modification [101], several extensive theoretical and experimental investigations in the search of other possible polytypes have been carried out, notably by Wickman and his coworkers [102, 103]. More than 110 possible polytypic structures consisting of upto 6 layers were postulated and their calculated powder pattern intensities were listed [102]. However, an examination of over 100 natural specimens, picked up from over 80 different localities all over the globe, did not reveal any new polytypes [103]. About 80% of the specimens were found to be of the type $2H$, three of the type $3R$ and the rest as mixtures of $2H$ and $3R$.

Polytypism has been reported in the selenides of niobium and tantalum, too. 3 polytypes of $NbSe_2$, viz. $2H$, $3R$ and $4H$, and 2 of $TaSe_2$, viz. $2H$ and $4H$, have been discovered and their crystal structures worked out [104]. The 3-layer form of $NbSe_2$ is isostructural with the $3R$ form of MoS_2.

Graphite in its commonest form has a hexagonal 2-layered structure. The layers are not close-packed, but consist of sheets of linked hexagons of carbon atoms, with the $C-C$ separations substantially the same as in the benzene ring. Thus the 'hexagonal' graphite can be loosely described by the stacking sequence $|AB| AB...$ of the constituent layers. Some graphites show weak X-ray diffraction lines of a 3-layered rhombohedral structure, too, which, by comparison with the 2-layered structure, can be loosely represented by the sequence $|ABC| ABC...$; others have a structure in which the stacking sequence is not regular, being partly hexagonal and partly rhombohedral [8]**.

3.4. NON-LAYER STRUCTURED SUBSTANCES

Amongst the non-layer structured polytypic substances, the silicon carbide is the foremost. It is one of the most widely investigated polytypic compound and happens to be the prototype of all polytypic substances. Grown at high temperatures ranging from nearly 1000–2700 °C, it occurs in two crystalline forms, the cubic β-SiC and the hexagonal (or rhombohedral) α-SiC, of which the latter displays polytypism. A total number of 128 polytypes have been reported, the crystal structures of 38 of which have been determined. An uptodate list of these, excepting 3 polytypes [105, 106, 107], has been published elsewhere [4]. The crystals have been grown from vapour phase as well as from solution and have been found to occur as flat plates, needles, whiskers and dendrites. The polytypism has usually been discovered in plates, which often show growth spirals on their flat faces. The crystals frequently exhibit one-dimensional disorder and syntactic coalescence is quite common. The most common polytype is $6H$ and the next common is $15R$, which also occurs quite frequently. In the SiC crystals, each carbon and silicon atom is tetrahedrally coordinated

---

* The notations $2H$ and $3R$ here actually refer to the number of $S-Mo-S$ sandwiches in the unit cell, rather than to S layers, in line with the convention followed in the literature.
** vide p. 127.

by four atoms of the other kind. The silicon-carbon bond is known to be strongly covalent. The crystals are extremely hard.

The zinc sulphide is another prominent polytypic compound. Its crystal structure is similar to SiC, but the $Zn-S$ bond is known to be partially covalent and partially ionic; the crystals are comparatively quite soft. It is grown at high temperatures, i.e. around 1000 °C and above and like SiC, it also occurs in two crystalline forms, viz. the cubic form β-ZnS (called sphalerite or zincblende) and the hexagonal form α-ZnS (wurtzite). The cubic form reversibly transforms into the hexagonal around 1020 °C and 1240 °C [108]. The α-variety is polytypic. A total number of 154 polytypes have been reported, the crystal structures of 153 of which have been worked out. A list of 137 polytypes has been published earlier [4], to which now the following have been added: $14H_4$, $42R_7$, $42R_8$ [109]; $22H_3$ [110]; $36R_7$, $36R_8$, $60R_{16}$, $60R_{17}$ [111]; $24H_{12}$, $24H_{13}$, $24H_{14}$, $24H_{15}$, $72R_{17}$, $72R_{18}$, $72R_{19}$ [112]; $90R_1$, $90R_2$ [113]. The crystals are usually grown from the vapour phase and usually occur as needles or elongated plates. The work on polytypism has been mainly confined to the latter. One-dimensional disorder is frequent and syntactic coalescence is a common feature. The c axis, along which the crystal structure varies, lies along the length of the crystal. The basal planes often show growth spirals. The most common polytype is $2H$.

Apart from zinc sulphide, more chalcogenides have been found to be polytypic [114]. These include zinc selenide, zinc telluride and cadmium telluride. Besides the cubic $3C$ form, one hexagonal or rhombohedral form of each of them has been discovered. The X-ray photographs of cadmium sulphide and cadmium selenide also indicate the possible existence of polytypism. All these chalcogenides have a ZnS type of structure.

Polytypism has been reported in an extensive range of ABX_3 perovskites and perovskite-related structures, either synthesized at high pressures or obtained by high pressure transformations of compositions prepared at atmospheric pressure [115]. The list includes oxides (ruthenates, e.g. $BaRuO_3$; manganates, e.g. $SrMnO_3$; iridates, e.g. $BaIrO_3$; the titanate $BaTiO_3$; the chromate $BaCrO_3$, the ferrite $BaFeO_3$; the complex compound Ba_2FeSbO_6), halides (fluorides, e.g. $RbNiF_3$ and chlorides, e.g. $CsCdCl_3$), and sulphides, e.g. $BaTaS_3$. The compounds exhibit one or more of the following polytypic modifications: $2H$, $9R$, $4H$, $6H$, and the cubic $3C$. The pressures employed range from 1 atm to nearly 120 atm in the transformations and from a few kbars to 90 kbar in the syntheses. The temperature of formation differs for the various compounds, ranging from 700 to 1400 °C. Higher pressures favour the cubic perovskite phase. The $4H$ phase is relatively rare, being formed only in a few oxides.

Polytypism has been observed in non-stoichiometric compounds in the titanium-sulphur system [116]. The polytype formation occurs in the range of composition $TiS_{1.6}$–$TiS_{1.8}$. 8 polytypes have been discovered: $8H_1$, $10H_1$, $12H_1$, $48R_1$, $12H$, $24H$, $40H$, $696R$. The crystal structures of the first four have been worked out. Several unidentified types exhibiting one-dimensional disorder have also been encountered.

Six polytypic modifications of the compound potassium cobalticyanide,

$K_3Co(CN)_6$, [3]* and four of the acid hydrogen ferricyanide, $H_3Fe(CN)_6$, [117], have been reported. One polytype of sodium 2-oxocaproate, $CH_3(CH_2)_3 COCOONa$, has been reported [118].

One polytype, $21H$, of the interesting trimorphous compound AgI, which has a ZnS type of structure with primarily covalent binding, has been reported [119]. The so-called γ-form of AgI, which is stable at room temperature, has the cubic $3C$ structure, which transforms at 137 °C to the β-form having the hexagonal $2H$ structure. The α-form, stable above 146 °C, has a less tightly packed cubic body-centered arrangement of iodine atoms, with the silver atoms having no fixed positions and wandering freely in a fluid state throughout the structure, presenting an example of a defect structure [8]**.

3.5. STACKING FAULTS IN CdI_2 TYPE STRUCTURES

3.5.1. *Geometry of Partial Dislocations and Stacking Faults*

The stacking sequence of the CdI_2 type $2H$ can be expressed as (vide Section 3.1),

$$(A\gamma B) (A\gamma B) (A\gamma B) (A\gamma B). \tag{3.1a}$$

Any deviation from the above ideal stacking sequence represents a stacking fault. For example, the sequence given below contains a stacking fault:

$$(A\gamma B) (A\gamma B) (C\beta A) (C\beta A). \tag{3.1b}$$

Because of the weak bonding between adjacent minimal sandwiches and the consequent easy cleavage along the basal planes, the usual dislocation configurations in the CdI_2 structures are expected to be confined to the basal planes. The observations of dislocations confirms this [120 to 123]. Thus in the CdI_2 structures one is usually concerned with the basal dislocation configurations embodying both perfect and imperfect (partial) dislocations. Since the stacking faults are associated with the imperfect dislocations, we shall be particularly concerned with them in the ensuing discussion. In passing we mention that although the present discussion refers to the cadmium iodide structures, it holds good for any layer structure having the cadmium iodide type structure, such as tin disulphide, nickel bromide, etc.

Figure 12 shows the structure of the minimal sandwich of cadmium iodide in projection on (0001). Figure 13 shows the possible x, y positions of the atoms (A, B and C positions). The comparison of the two figures brings out the x, y positions of atoms of the $(A\gamma B)$... sequence. In this structure, the most stable configuration of perfect dislocations would have the Burgers vectors corresponding to the smallest lattice translation (least strain energy). Evidently these Burgers vectors are of the type **PR**, **PQ**, **QR** and their negatives. (Figure 14). These can be called for brevity **PR** type dislocations with their Burgers vectors having the magnitude $|\mathbf{PR}| = a$, where a is the

---

* vide p. 119.
** vide p. 137.

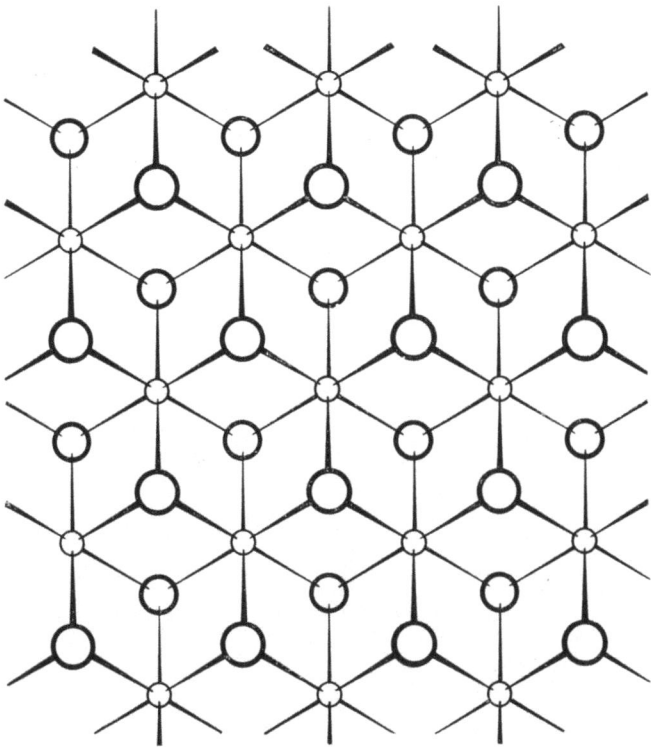

Fig. 12.   The structure of a 'minimal sandwich' ($A\gamma B$) of CdI$_2$ in projection on (0001). The small circles represent Cd atoms in the plane of the paper and the larger circles the I atoms above and below; the Cd atoms lie in the ocatahedral voids between the close-packed I layers.

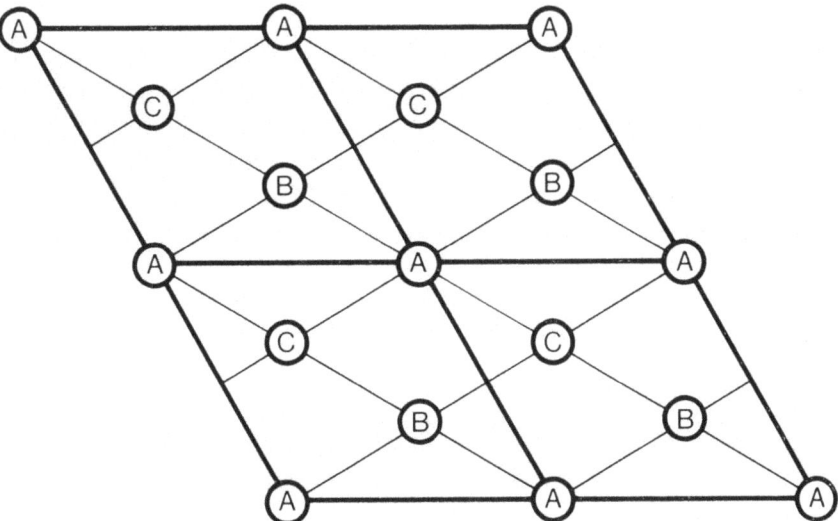

Fig. 13. x, y positions of the atoms (A, B and C positions).

basal lattice translation of the unit cell. As seen in Figure 14, since the projection of the possible layers other than A (or α) layers correspond to either σ or δ, the perfect dislocation of the type **PR** may clearly dissociate according to the scheme **PR** = **Pσ** + + **σR**. Other similar dissociation schemes easily follow, e.g. **QR** = **Qσ** + **σR**, **PQ** = = **Pσ** + **σQ**, etc. All the Burgers vectors of the type **Pσ** represent non-lattice vectors, since the positions σ or δ do not represent lattice points. Hence they correspond to burgers vectors of partial dislocations and as is well known, the two (or more) partials

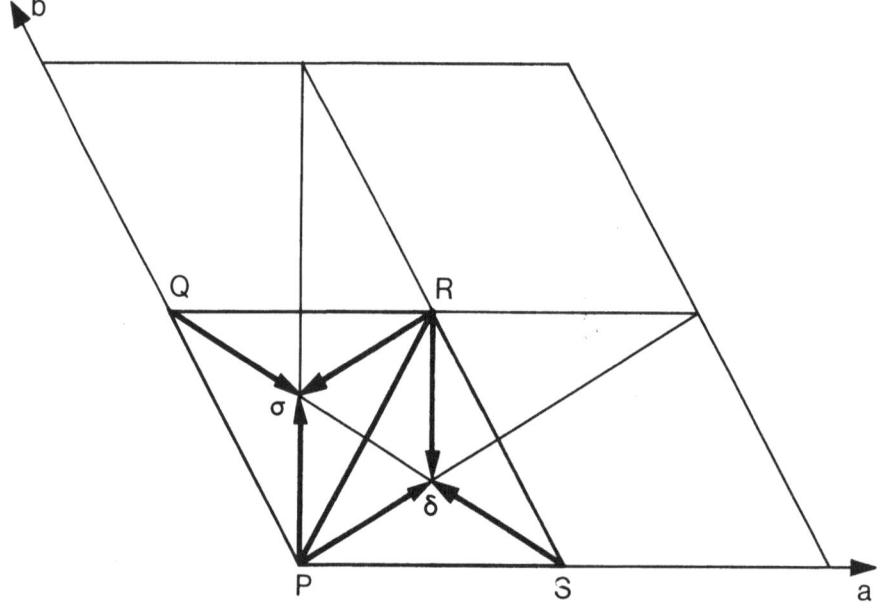

Fig. 14.   Burgers vectors for the basal unit and partial dislocations. Dissociation of unit dislocations takes place according to the reaction of the type **PR** = **Pσ** + **σR**.

ensuing from a perfect dislocation bound a stacking fault region. In cadmium iodide type structures the glide plane is the (0001) plane. There are however, two different types of glide planes – one corresponding to glide between neighbouring iodine layers (I–I glide) and the other between a cadmium and an iodine layer (Cd–I glide). The nature of the stacking faults depends on the type of glide and therefore we need to consider separately the stacking faults produced by the two types of glide processes.

(a) *Stacking faults associated with I–I glide*

The following sequence represents the original CdI$_2$ 2H structure

$$(A\gamma B)\,(A\gamma B)\,\big|\,(A\gamma B)\,(A\gamma B)\,(A\gamma B)\ldots \tag{3.2a}$$

The vertical line indicates the position of a glide plane, which has been chosen to lie between two iodine layers. The passage of a partial dislocation in this plane changes the above sequence to

$$(A\gamma B)\,(A\gamma B)\,(C\beta A)\,(C\beta A)\,(C\beta A)\ldots \tag{3.2b}$$

resulting in a stacking fault. The glide vectors are of the type $P\sigma$ and $Q\sigma$ (σ is the projection of C or γ atoms; the partial associated with $P\sigma$ changes the ideal sequence $...(A\gamma B)...$ to $...(C\beta A)...$, i.e. $A \rightarrow C, B \rightarrow A, C \rightarrow B$ and the partial σR throws back the changed sequence to the ideal sequence $...(A\gamma B)...$). Figure 15 shows schematically the change in the stacking sequence due to the creation of a fault of the above type ('a' type fault).

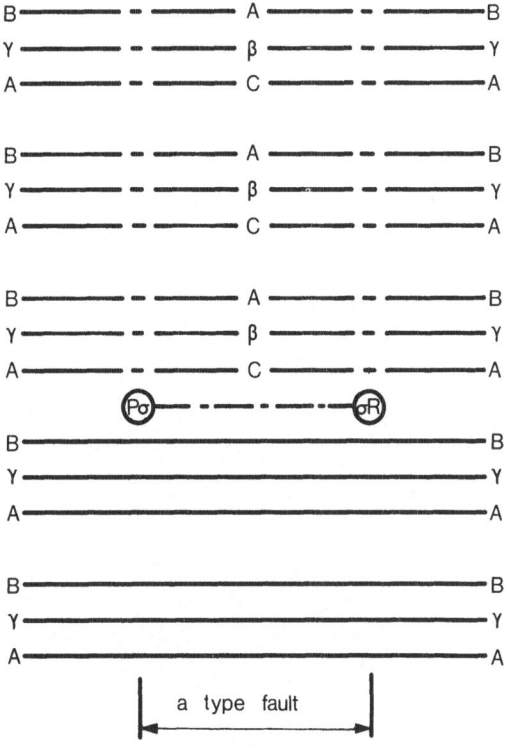

Fig. 15.   Schematic change in the stacking sequence due to the creation of a stacking fault ('a' type fault) with the glide vectors $P\sigma + \sigma R$.

It should be noted that for the above type of glide process a dislocation of the type $PR = P\delta + \delta R$ will be very improbable since $P\delta$ would change the ideal sequence according to $A \rightarrow B, B \rightarrow C, C \rightarrow A$, and thus produce a region of stacking fault in the ideal sequence $...(A\gamma B)(A\gamma B)...$ with a B over B configuration, which is energetically unfavourable.

(b) *Stacking faults associated with Cd−I glide*

(i) A glide process in which the glide takes place between cadmium and iodine layers leads to another type of stacking fault:

$$(A\gamma B)(A\gamma \mid B)(A\gamma B)(A\gamma B)... \tag{3.3a}$$

$$(A\gamma B)(A\gamma C)(B\alpha C)(B\alpha C)(B\alpha C)... \tag{3.3b}$$

The above stacking change would be effected by partials of the $P\delta + R\delta$ type and would bring one layer of Cd atoms, viz. those in the sandwich $(A\gamma C)$, into tetrahedral interstices, while the usual positions for the Cd atoms are the octahedral ones. In order to overcome this irregularity it is reasonable to expect that a synchro-shear motion should take place for bringing these Cd atoms back into the octahedral sites. Consequent to such a motion the stacking sequence (3.3b) would change to

$$(A\gamma B)\,(A\beta C)\,(B\alpha C)\,(B\alpha C)\ldots \qquad (3.3c)$$

The glide motions are indicated in Figure 16. The aforesaid type of motion and the term 'synchro-shear' were originally suggested in relation with dislocation structures in aluminium oxide [124]. (We shall call the resulting fault a 'b-(i)' type fault).

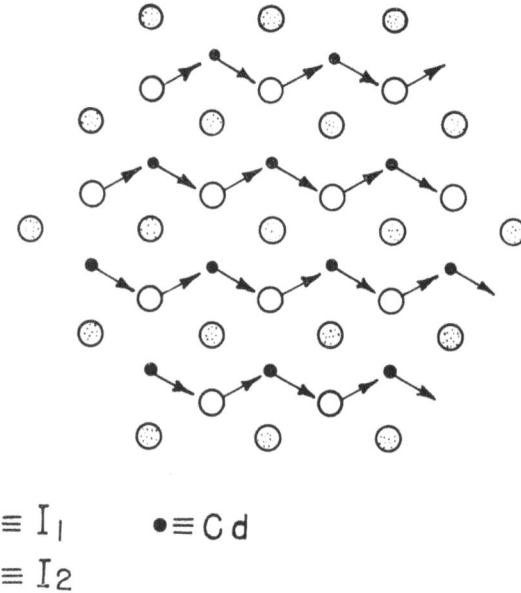

$$O \equiv I_1 \qquad \bullet \equiv Cd$$
$$\odot \equiv I_2$$

Fig. 16.   Synchro-shear movement consequent upon glide along Cd–I glide planes in the cadmium iodide $2H$ structure. The movement occurs because the Cd atoms have a tendency to stay in an octahedral environment. (Projection in the basal plane.)

(ii) In addition to the above, there is yet another possibility for the production of a stacking fault associated with the Cd – I glide. Taking the glide to occur between A and σ without synchro-shear, we have the following type of stacking fault:

$$(A\gamma B)\,(A \mid \gamma B)\,(A\gamma B)\,(A\gamma B)\ldots \qquad (3.4a)$$

The passage of a partial dislocation changes the above sequence to

$$(A\gamma B)\,(A\beta A)\,(C\beta A)\,(C\beta A)\ldots \qquad (3.4b)$$

Here the Burgers vector of the fault will be of the type $(\mathbf{P\sigma} + \mathbf{\sigma R})$. The above sequence contains a minimal sandwich $(A\beta A)$, which is found in the MoS_2 structure, for which the ideal sequence is $(A\beta A)(B\gamma B)(A\beta A)$.... It is evident that the presence of this type of sandwich will make the sequence $(A\gamma B)(A\beta A)(C\beta A)$... energetically unfavourable and therefore this type of fault (to be called 'b-(ii)' type) will be less probable.

Besides the above dissociated dislocations, which occur in the form of a pair of partials and will therefore form two-fold ribbons, a combination between two separate two-fold ribbons is also possible. For example, a ribbon with 'a' type-fault and total Burgers vector $\mathbf{PR}$ can combine with a ribbon of type 'b' on an adjacent basal plane, through the reaction $\mathbf{PR} + \mathbf{PQ} = \mathbf{P\sigma} + \mathbf{\sigma R} + \mathbf{\sigma Q} + \mathbf{P\sigma}$ (applying Frank's rule for dislocation reactions it can be seen from Figure 14 that $\mathbf{P\sigma} = \mathbf{\sigma R} + \mathbf{\sigma Q}$). This reaction leads to reduction in energy and produces a three-fold ribbon (consisting of three partials with the same burgers vector, resulting from the fusion of two single ribbons). The formation of the three-fold ribbon and the associated stacking faults are shown in Figure 17. It may be mentioned that the above description can be similarly extended to the CdI_2 structure $4H$.

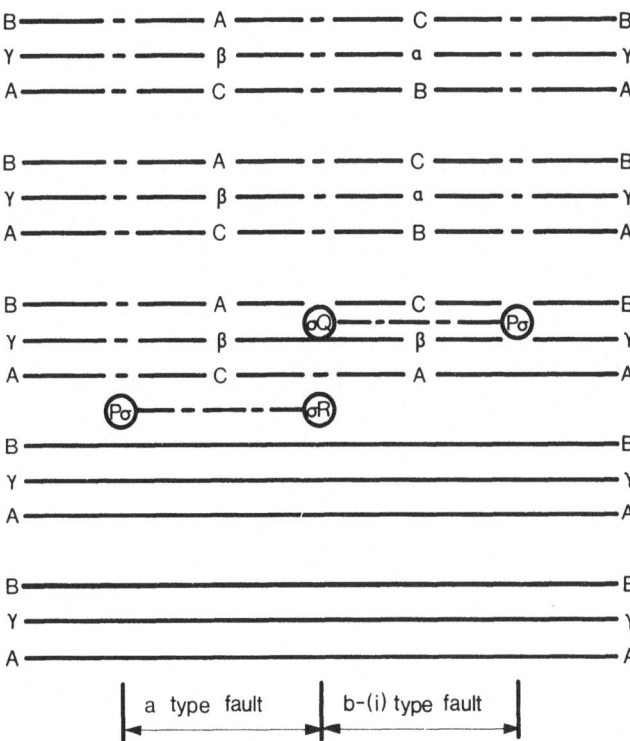

Fig. 17.   Formation of a three fold ribbon in cadmium iodide by the fusion of two separate two-fold ribbons, as for example one with 'a' type fault and total burgers vector $\mathbf{PR}$ and the other with 'b' type fault and total burgers vector $\mathbf{PQ}$.

It may be pointed out that the above stacking faults are produced through glide processes and hence the bounding partials are of the Shockley type – their Burgers vectors lie in the basal planes and the partials are therefore glissile. There is yet another way in which stacking faults can be created, viz. through the precipitation of vacancies. The existence of vacancies in layer structures (molecular vacancies) has been demonstrated [125–127]. Their clustering may result in the nucleation of stacking faults. These faults, in keeping with the other similar cases [128], would be bounded by Frank partials. Their Burgers vectors do not lie in the basal planes. The stacking faults of this type can be understood by taking the ideal structure as ...$(A\gamma B)\,(C\alpha B)$.... The following depicts the genesis of the stacking fault through vacancy precipitation:

$$(A\gamma B)\,(C\alpha B)\ \boxed{A\gamma B}\ (C\alpha B)\,(A\gamma B)\,(C\alpha B)\ldots \qquad (3.5a)$$

If the enclosed layer represents a possible vacancy then

$$(A\gamma B)\,(C\alpha B)\,(C\alpha B)\,(A\gamma B)\,(C\alpha B)\ldots(A\gamma B)\,(C\alpha B) \qquad (3.5b)$$

is the resulting faulted sequence.

It should, however, be mentioned that almost all the stacking faults in CdI_2 and similar layer structures correspond to the types produced through glide processes.

3.5.2. *Observation of the Various Partials and the Associated Stacking Faults*

Out of the various possible methods for the observation of imperfections, the transmission electron microscopy is well suited to provide direct information with regard to the partials and stacking faults in various materials, including the CdI_2-type layer structures. As is well known, under the two-beam kinematical approximation, the diffraction contrast at imperfections, in the transmission electron microscopy of crystals, is produced by the incorporation of an additional phase factor in the electron waves diffracted from the local region containing the faults. This additional phase factor α is given by $2\pi\mathbf{g}.\mathbf{R}$ where $\mathbf{g}$ is the reciprocal lattice vector corresponding to the operating reflection and $\mathbf{R}$ is the direct lattice displacement vector associated with the imperfection [128, 129].

The cadmium iodide crystals (thin crystals transparent to 60–100 kV electron beam are produced either by careful cleaving or by vapour growth under controlled conditions) observed under the electron microscope are always in the form of basal oriented platelets, as a result of which the diffraction pattern always corresponds to a $(00.1)^*$ oriented reciprocal lattice net (cadmium iodide is a hexagonal layer structure). The spots on the net are arranged on a hexagonal grid. Figure 18 shows a representative diffraction pattern obtained from a crystal flake. The diffraction spots arranged on the hexagonal grid next to it are of 11.0 type. As discussed previously for the observations leading to interpretations of dislocation configurations, a two-beam condition is necessary. This is obtained by fitting the crystal flake in such a way that in the observed diffraction pattern besides the direct beam only one diffraction spot is strong. The contrast observed at the dislocations depends on the factor $\mathbf{g}.\mathbf{b}$.

In cadmium iodide, as described previously, the Burgers vectors **b** of the partial dislocations are of the type **Pσ**, **Qσ**, **Rσ**, etc. (Figure 14). Figure 19 shows the Burgers vector in correct orientation with the two sets of low order diffraction vectors. The values of the factor **g.b** for different diffraction vectors have been presented in the Table IV, which shows that the expected partials in cadmium iodide structures have **g.b** terms as integers $(0, \pm 1)$ for the six 11.0 type diffraction vectors. For these diffraction vectors, therefore, the partials behave like perfect dislocations; the partials with parallel Burgers vector form ribbons, which can be best observed when a 11.0 type vector is the operating diffraction vector. In 11.0 type reflection, since **g.b** is integral, no stacking fault contrast appears. For a 10.0 type diffraction vector, the **g.b** term is always non-integral and therefore the stacking fault contrast appears for these reflections.

Fig. 18.   A representative diffraction pattern observed in cadmium iodide crystals used for electron microscopic investigations. The spots on the hexagonal grid nearest to the direct beam are 10.0 type and those on the hexagonal grid next to it, are of the 11.0 type.

The most frequently observed dislocation configurations in a cadmium iodide crystal are those of partials grouped together in the form of ribbons. The dissociation and the consequent observations of partials implies that the stacking fault energy of cadmium iodide is very small. Almost invariably all the dislocation configurations

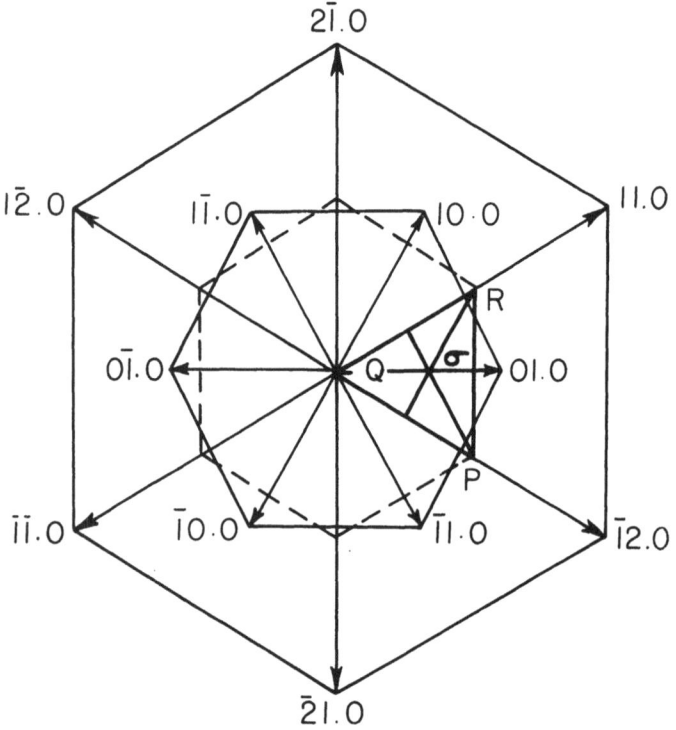

Fig. 19.   Burgers vector in correct orientation with two sets of low order diffraction vectors, $g_{10.0}$ and $g_{11.0}$.

TABLE IV

Values of **g.b** for different diffraction vectors

Burgers vector (b)	Diffraction vector (g)	g.b	Burgers vector (b)	Diffraction vector (g)	g.b	Burgers vector (b)	Diffraction vector (g)	g.b
Pσ	01.0	1/3	Qσ	01.0	2/3	Rσ	01.0	−1/3
Pσ	10.0	−1/3	Qσ	10.0	1/3	Rσ	10.0	2/3
Pσ	1Ī.0	2/3	Qσ	1Ī.0	−1/3	Rσ	1Ī.0	1/3
Pσ	0Ī.0	−1/3	Qσ	0Ī.0	−2/3	Rσ	0Ī.0	1/3
Pσ	Ī0.0	1/3	Qσ	Ī0.0	−1/3	Rσ	Ī0.0	−2/3
Pσ	Ī1.0	−2/3	Qσ	Ī1.0	1/3	Rσ	Ī1.0	−1/3
Pσ	11.0	0	Qσ	11.0	1	Rσ	11.0	1
Pσ	2Ī.0	−1	Qσ	2Ī.0	0	Rσ	2Ī.0	−1
Pσ	1Ẑ.0	1	Qσ	1Ẑ.0	−1	Rσ	1Ẑ.0	0
Pσ	ĪĪ.0	0	Qσ	ĪĪ.0	−1	Rσ	ĪĪ.0	−1
Pσ	Ẑ1.0	1	Qσ	Ẑ1.0	0	Rσ	Ẑ1.0	1
Pσ	Ī2.0	−1	Qσ	Ī2.0	1	Rσ	Ī2.0	0

consist of partials and the associated stacking faults. Figure 20 shows a dislocation network in cadmium iodide, seen in bright field and taken with a diffraction vector corresponding to a 11.0 type reflection. This figure, which is representative of the usually observed network, clearly shows several dissociated dislocations, which are visible as two-fold ribbons (some of the ribbons have been marked with arrows). The contrast shown by the partials is in keeping with the expected contrast [71, 128, 129]. Figure 20 also shows that the occurrence of the partials and the enclosed stacking

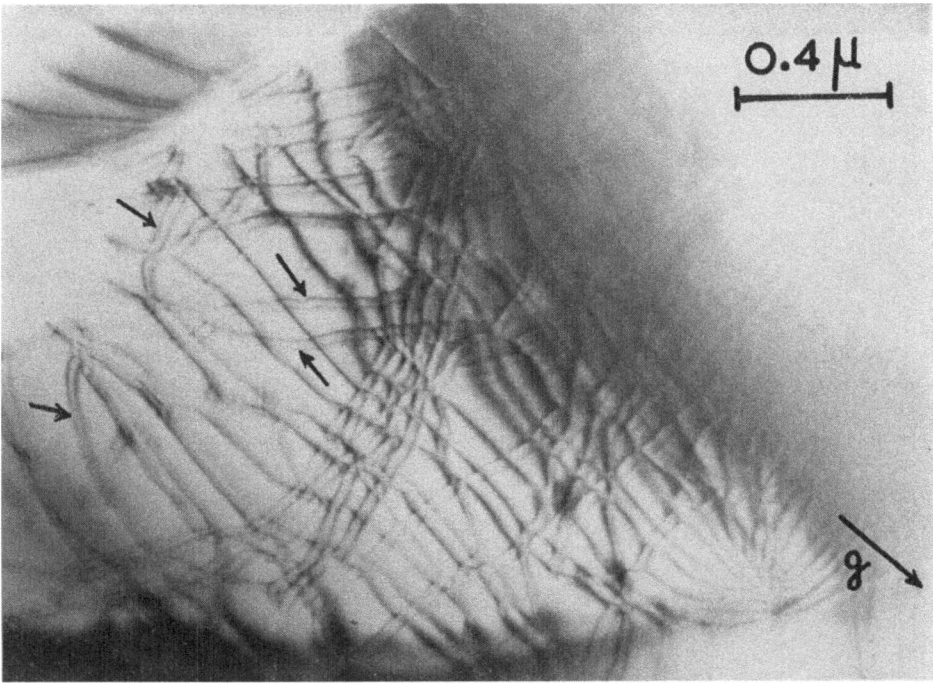

Fig. 20.   Transmission electron micrograph showing dissociated dislocations in a basal oriented thin platelet of cadmium iodide. Some of the two-fold ribbons are marked with arrows. The direction of 11.0 type operating diffraction vector is shown by the arrow in the right hand corner of the figure. Notice that nearly all the visible dislocations are dissociated.

faults is a common feature; nearly all the visible dislocations are in the form of two-fold ribbons. Figure 21 shows yet another dislocation configuration taken under two-beam conditions with a 11.0 operating reflection. As is evident, the dislocations are dissociated; some two-fold ribbons can be seen at places marked A, B, C. The three-fold ribbons are formed as a result of the fusion of the two-fold ribbons. It can be noted that the ribbons T_3 and C are jogged, which seems to be the result of elastic interactions with the ribbons T_1 and T_2. Figure 22 shows another dislocation configuration. Here several jogged ribbons are visible. The ribbon shown by ABC and DEF can be seen to undergo a transition from a plane of low energy stacking fault ($I-I$ type) to a plane of higher energy stacking fault ($Cd-I$ type).

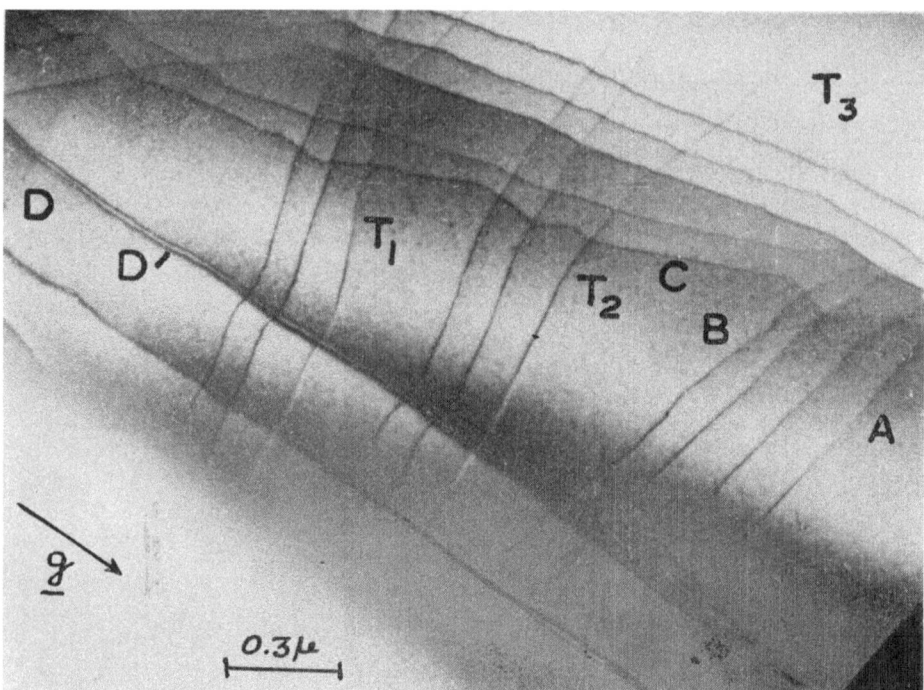

Fig. 21. Transmission electron micrograph showing two-fold and three-fold ribbons observed in a cadmium iodide crystal. Notice the discontinuous change in contrast across the middle partial of the three-fold ribbon T_3. The direction of the 11.0 type operating diffraction vector is shown in the left hand corner of the figure.

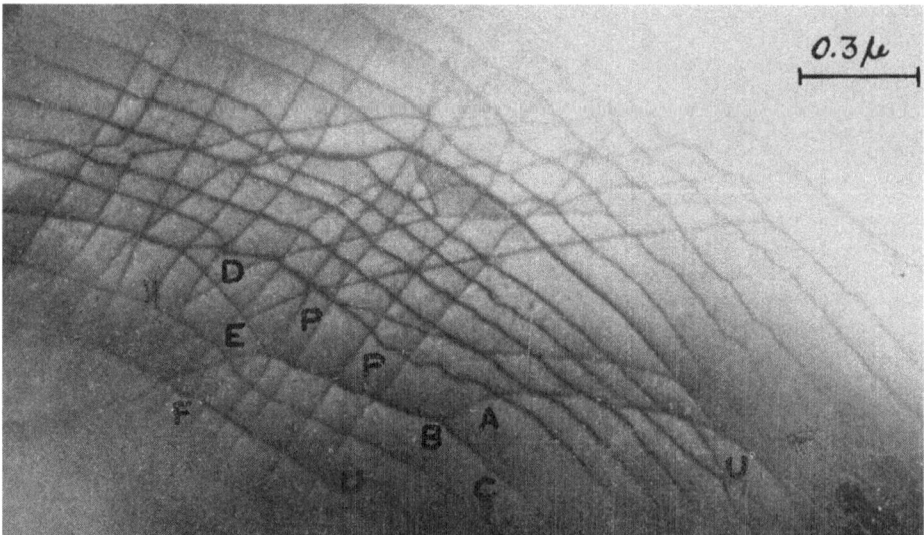

Fig. 22. Transmission electron micrograph showing partial dislocation ribbons in a cadmium iodide crystal. At several places jogged ribbons are formed by crossing of two extended dislocations in planes close to each other. The jogged regions around E are between ribbons of equal type and around U between ribbons of unequal type. The two-fold ribbons can be easily seen at several places, two of which have been marked with P. The ribbons shown by ABC and DEF can be seen to undergo a transition from a plane of low energy stacking fault (I-I type) to a plane of higher energy stacking fault (Cd-I type).

4. Theories of Polytypism

So many physical factors appear to be associated with the growth of polytypes that a larger number of explanations have been advanced from time to time to account for the origin of the phenomenon. These have been based on diverse considerations like the presence of impurities, presence of dislocations, influence of lattice vibrations, thermodynamical factors, internal rotations, electron energy, etc. A short account of the important ones will be provided in the following, along with the experimental evidence.

4.1. THE EARLY THEORIES

The possibility of a correlation between polytypism and impurity content has been expressed by several different investigators. Baumhauer [1] first suggested that the occurrence of SiC polytypes was related with their colour, which was supported by Espig [131] but negated by Thibault [10] in a comprehensive study; however, the latter did notice some influence of impurities on polytype formation. More definite investigations were made by Lundqvist [132], through spectrochemical and X-ray powder analyses, who concluded that the occurrence of the common SiC types $6H$, $15R$ and $4H$ was related to their aluminium content, which, of late, has been confirmed by Hayashi [133] in a chemical and X-ray powder analysis. But Knippenberg [134] failed to observe such a correlation in a detailed examination of SiC crystals grown with different proportions of aluminium at a constant temperature, which led him to conclude that the suspected impurity-structure correlation might be really a temperature-structure relationship. However, more recently, through their own investigations as well as from similar results of other workers, Knippenberg and his coworkers [135] have inferred that the nature and amount of impurity content, temperature and pressure have a combined influence on the formation of the common SiC types $6H$, $15R$ and $4H$; for higher polytypes, defects in the initial nuclei are also responsible. Employing powder method, Jagodzinski and Arnold [136] concluded that the impurities had no important effect on the growth of SiC polytypes, but very recently Jagodzinski [137] has opined that a connection between polytypism and impurities surely exists. The effect of Cu as impurity has been suspected in the formation of polytypes of MoS_2 [103]. The effect of impurities on polytype growth in ZnS crystals has also been visualized by Strock and Brophy [138], Kremheller [139] and others [140–142]. To sum up, the impurities certainly appear to wield influence on polytype formation, but as yet, their exact role remains largely unknown. More about the effect of impurities, temperature, pressure and dislocation content on polytype formation will be described in Section 6.1.

Employing X-ray powder analysis, Hägg [143] pointed out that the relative abundance of the solution-grown CdI_2 types $4H$ and $2H$ and of the disordered types was related to the speed of crystallization of the material. Recently, this influence has been confirmed in the gel-grown PbI_2 polytypes [144].

Ramsdell and Kohn [12] propounded a 'polymer' theory of polytypism in silicon carbide, according to which the polytypes could grow by accumulation of certain hypothetical clusters of atoms. However, the theory has several serious shortcomings [4, 22].

4.2. FRANK'S SCREW DISLOCATION THEORY

In Section 1.2 it was described that Frank [20] extended his screw dislocation model of crystal growth to the creation of polytypes. Assuming the step height to be the same for the different spirals on a crystal's surface, the growth spirals would essentially generate the same structure as the substrate if the Burgers vector happened to be an integral multiple of the height of the unit cell; the substrate, in which the dislocations occurred, was assumed to be the most common polytype of the substance under consideration. But the resulting structure would be different if the Burgers vector was a non-integral multiple of the cell height. That is how the screw dislocations could lead to the generation of several different polytypes of a substance.

According to the theory, the polytypes should necessarily display growth spirals on their faces, but in reality that is not always found to be true. Also the measured spiral step height should be equal to an integral multiple of the unit cell height. Such a correlation was observed by Verma for 7 SiC polytypes [146]. Making use of optical data of Forty [147] on step height measurement, Mitchell [43, 148] sought to establish a similar correlation for a large number of solution-grown CdI_2 polytypes examined by him through X-ray diffraction methods, which, subsequently, has been negated by Trigunayat and Verma [49], and later by Chadha and Trigunayat [149] in vapour-grown crystals, who performed the two kinds of measurement on the same crystal. Such a verification often demands a high degree of accuracy in step height measurement, which is permitted in CdI_2 crystals because they often present a multitude of internal interference fringes contouring the spiral steps, e.g. a growth spiral has been shown in Figure 23, where the accuracy attained has been ± 0.4 Å. Recently, using the experimental data of Trigunayat and Verma but employing a revised value for the refractive index of cadmium iodide, Tubbs [150] has re-estimated the magnitude of the step height and then purported to establish its correlation with the cell height. But subsequently, Mazumdar and Trigunayat [151] have pointed out the flaws in Tubbs' attempt and have shown that the reestimated values of the step heights leave the previously concluded non-correlation by Trigunayat and Verma unchanged.

Analysing detailed atomic positions inside SiC crystals, Mitchell [152] showed that certain dislocations left unstable gaps in the lattice, which rendered their generation improbable. Barring them, the formation of all of the known SiC polytypes could be explained on the basis of screw dislocations. The Mitchell scheme was further perfected by Krishna and Verma [153]. Extending this scheme of derivation to CdI_2 polytypes, Srivastava and Verma [67] ruled out the existence of rhombohedral polytypes of cadmium iodide. However, a formidable number of such polytypes have been reported of late (vide Table I, Section 3).

The existence of nearly 50 polytypes of lead iodide (vide Table II, Section 3) is in obvious disagreement with Frank's theory. This compound has $2H(11)$ as its basic structure, in which a screw dislocation of any given magnitude of the Burgers vector should always result in the generation of the original structure, thus forbidding the creation of any polytypes according to the screw dislocation mechanism. On the contrary, cadmium bromide, which has $6R$ as its basic structure, is expected to display rich polytypism but really possesses it in a very limited measure (vide Sec-

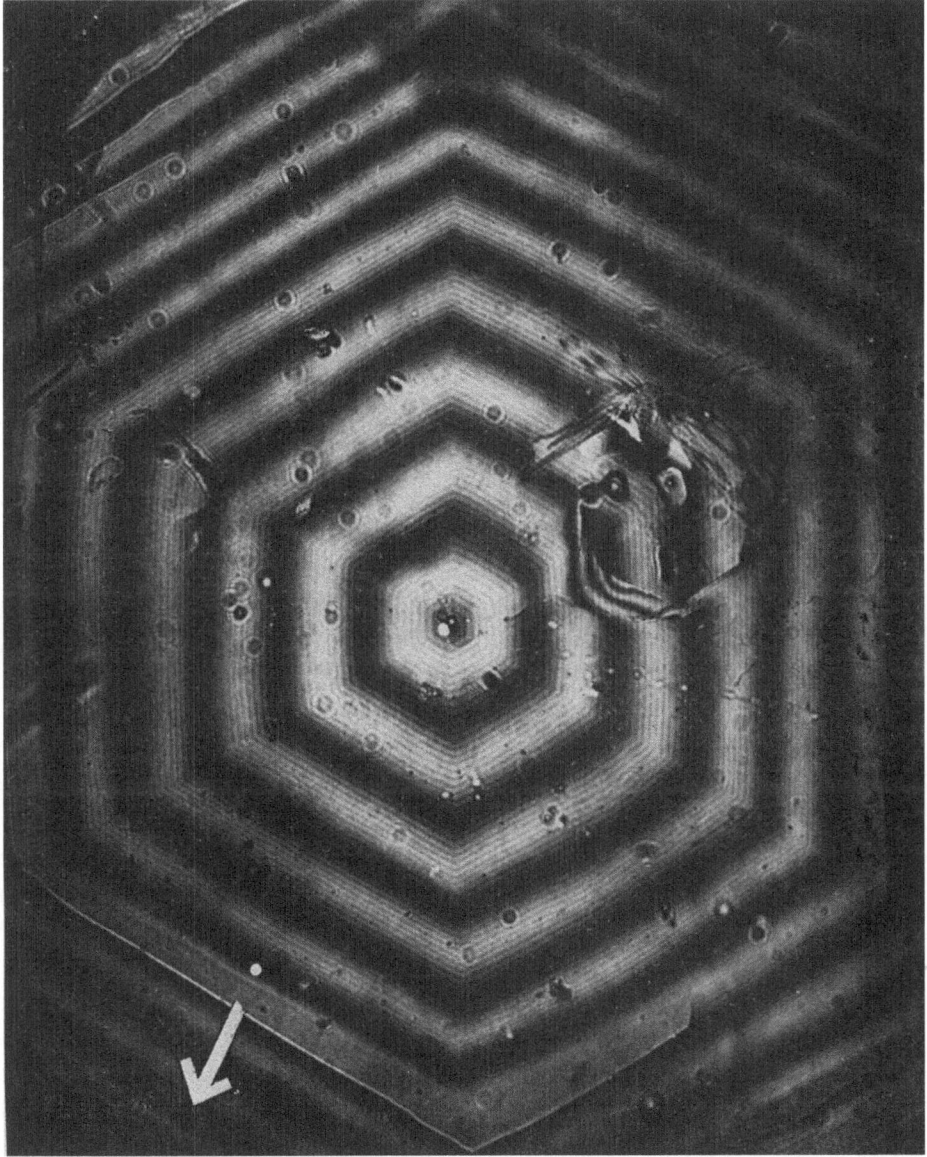

Fig. 23. Two-beam internal interference fringes contouring the spiral steps on a CdI$_2$ crystal. The step height has been measured to be equal to (116.3 ± 0.4) Å; (0001) face; Hg-green light.

tion 3.3). More objections against the validity of Frank's theory have been raised by Verma and Krishna [3]* and by Trigunayat and his coworkers [45, 53, 61, 78] after taking into consideration the detailed atomic structures of several SiC and CdI_2 polytypes. These have been summarised earlier by Trigunayat and Chadha [4]. Buckley [154, 155] has expressed doubts if the observed spirals really contribute towards the growth of crystals. Jagodzinski [156] has raised similar objections on energy considerations.

4.3. JAGODZINSKI'S DISORDER THEORY

As an alternative to Frank's simple geometrical model of the growth of polytypes, Jagodzinski [156, 157, 137] propounded a theory of polytypism based on thermo-dynamical considerations. He argued that edge dislocations could be created with great ease during crystal growth. Partial dislocations could produce stacking faults, consisting of displacement of layers from their normal positions in the structure (vide Section 3.5). Considering the formation of polytypes in silicon carbide, growing from vapour, Jagodzinski [157] suggested that these could result from layer transpositions in the cubic modification $3C$, which he regarded as the basic structure of SiC. The polytype $6H$ e.g. could be obtained as follows:

$$A\ B\ C\ A\ \boxed{B\ C}\ A\ B\ C\ A\ \boxed{B\ C}\ ...,\ 3C,$$
$$A\ B\ C\ A\ \boxed{C\ B}\ A\ B\ C\ A\ \boxed{C\ B}\ ...,\ 6H.$$

Similarly, all other polytypes could be derived. Normally, the faults should be randomly distributed, imparting one-dimensional disorder to the crystals, which is manifest on the oscillation photographs as 'streaking', in which reflections of the same h and k but different l values are connected by diffuse streaks of varying intensities (vide Figure 27, Section 5.1). All prominent polytypic substances have been found to exhibit this phenomenon. Under suitable conditions, several stacking faults might cooperate in a crystal to lower the energy of formation of the two-dimensional nucleus, necessary for further growth. In this manner, a multitude of 'superstructures' or 'superperiods', representing polytypes of a substance, might be produced. Jagodzinski [137, 156] has proposed that the cooperation of faults is brought about by vibrational entropy, which has a reverse dependence on the state of order of the system. This, in turn, renders the total entropy to have two maxima as a function of the degree of disorder, α, of the system, one for $\alpha_1 \simeq 0$ (completely ordered system) and another for $\alpha_2 \simeq 0.10$ (partially ordered system). The total entropy consists of vibrational and configurational components, but the former plays a dominant role in one-dimensionally disordered systems, since in a compound like SiC with two types of atoms there are altogether $2.3N_1N_2N_3$ eigen vibrations of the lattice but only N_3 composite layers subjected to configuration statistics. The variation in entropies as a function of α has been depicted in Figure 24. α is simply defined as the ratio of the number of layers displaced from their normal positions to the total

* vide p. 259.

number of layers. To sum up, the vibrational entropy helps to stabilize disorder in the system, which is manifested by the existence of a second maximum in the entropy curve of Figure 24.

A necessary implication of the theory is that the frequency distribution of the polytypes of a substance should be a function of their state of order. Jagodzinski himself verified it to be so in an investigation of 150 SiC crystals, nearly 60 of which were found to be disordered [156]. Jain and Trigunayat [47] confirmed the same

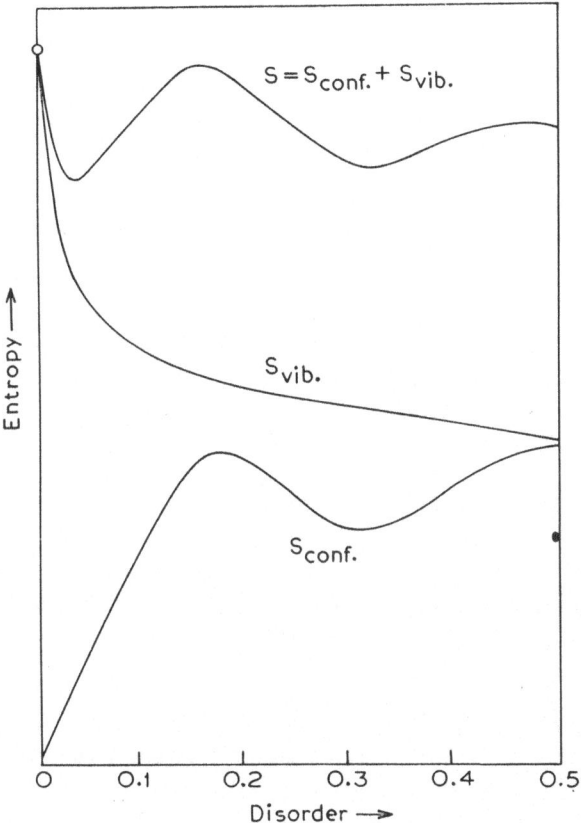

Fig. 24.   The variation of configurational entropy S_{conf}, vibrational entropy S_{vib}, and total entropy $S = S_{conf} + S_{vib}$, as functions of the state of order in a two-atom system like SiC, according to Jagodzinski [156].

result in their analysis of nearly 200 polytypes of CdI_2, of which nearly 80 were found to be disordered, but they found the two maxima to occur at somewhat different α-values than in SiC, viz. at $\alpha_1 = 0.06$ and $\alpha_2 = 0.26$. These are shown in Figure 25, where the ordinate represents the relative abundance of polytypes having a degree of disorder within a small range $\Delta \alpha$. The discrepancy in the location of maxima was attributed to the structural differences between SiC and CdI_2 [47]. In both the investigations the crystals with a high degree of disorder invariably had a tendency to form mixed polytypes of large periodicities, thus showing that random stacking faults did attempt to cooperate during crystal growth to produce ordered structures. If the attempt succeeded in producing a single long period polytype, the degree of disorder was expected to be essentially small, which was observed to be so in both the SiC and CdI_2 crystals.

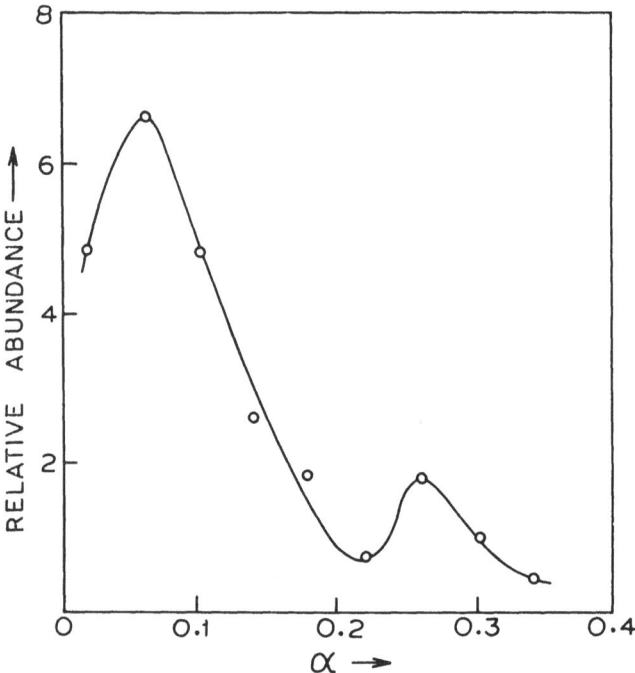

Fig. 25.   The frequency-distribution curve of CdI₂ polytypes as a function of the state of order, as obtained by Jain and Trigunayat [47].

Müller [158] observed the formation of several polytypes of ZnS when synthetic ZnS crystals were heated at temperatures between 870 and 900 °C. Since the polytypes consist of various mixtures of cubic (3C) and hexagonal (2H) packings of layers, they could be regarded as disordered states (with respect to the cubic or hexagonal phase, either of which might be taken as a completely ordered state) in thermal equilibrium, possibly stabilized through the agency of vibration entropy. Thus the thermodynamic basis of the Jagodzinski's disorder theory was upheld [156]. Similar results in ZnS have been obtained by Smith [159], Gobrecht *et al.* [160], and Baars and Brandt [161]; the latter two groups of workers have observed the polytype formation only while cooling down the 2H phase from a high temperature. The thermodynamic basis of the theory has been further supported in the observation of a broad structure-temperature relationship for small period SiC polytypes, 2H, 4H, 15R and 6H, by Knippenberg [134]. Inomata and his coworkers [162] have reported similar results in SiC, but their temperature ranges of formation of polytypes differ from Knippenberg's. Jagodzinski and Arnold [136] have held similar views, but their proposed temperature ranges are different, too. The difference possibly arises from the dependence of polytypism on the supersaturation of vapour during crystal growth [162] and on the stoichiometry and impurity content of the starting material. Also, solid state transitions in polytypes of SiC, ZnS, CdI$_2$, and PbI$_2$ have been recently observed by several investigations in heating experiments (vide Section 6.1.1) which further strengthen the contention that thermodynamic considerations may be relevant for the phenomenon of polytypism.

According to the Jagodzinski's theory, long period polytypes are generally expected to be disordered. But in disagreement with it, a large number of SiC, ZnS and CdI$_2$ polytypes of long periodicities have been reported, which show sharp reflections on their X-ray photographs, without any superposed features of one-dimensional disorder. Another snag in the theory is that the proposed variation of vibrational entropy has been arrived at only in a qualitative manner. In fact, adopting the simplification that the vibrational modes of SiC and ZnS polytypes can be considered as those of a linear diatomic lattice, recently Weltner [163] has estimated that the vibrational entropy differences between the various polytypes are too small to contribute significantly towards their stabilization.

4.4. SCHNEER'S PHASE TRANSFORMATION THEORY

For explaining the formation of SiC polytypes, Schneer [164] noted that they have very nearly the same internal energies and therefore he proposed that they might be the products of a second order phase transformation from cubic to hexagonal phase. His treatment is analogous to that of order-disorder in alloys by Bragg and Williams [165] and Bethe [166]. The polytypes are regarded as intermediate states in the phase transition, proceeding by infinitesimal steps over a given temperature range. Suitably defining the distribution of the layers and their contacts by two functions D and ϕ, respectively, and assigning the number of like contacts between the layers in hexagonal and cubic surroundings by n_{hh} and n_{cc}, respectively, the theory predicts a linear relationship between D and ϕ, which was found by Schneer for the 14 polytypes of SiC known till then and, barring a few cases, was later confirmed by Rai and Krishna [167] for more SiC polytypes as well as for the known polytypes of ZnS, CdI$_2$, and PbI$_2$. However, subsequently Alexander et al. [168] have exposed the fallacy that the proposed linear relationship really follows from the definitions of D and ϕ for polytypes having $n_{hh}=0$, and they found that for nearly 150 ZnS polytypes known till then n_{hh} was indeed zero, except for the polytype $12R(31)_3$, but this one did not fit on the D-ϕ curve. Hence they concluded that the observed D-ϕ linearity is physically meaningless. Trigunayat [22] has further pointed out that the linearity follows from the definition of D and ϕ also for $n_{cc}=0$, and that n_{hh} or $n_{cc}=0$ for all polytypes with known structure, excepting two in ZnS, eight in CdI$_2$, and one in PbI$_2$, none of which fits on the D-ϕ curve. In the meantime three more polytypes of ZnS, $14H_4, 42R_7$ and $42R_8$, and two of CdI$_2$, $14H_3$ and $16H_5$, have been reported, for which n_{cc} or $n_{hh}=0$, but these also do not lie on the curve. More shortcomings of the theory have been pointed out by Trigunayat and Chadha [4].

Envisaging the role of 'Internal rotation' forces in SiC and ZnS structures, lately Weltner [163] has sought to support Schneer's theory of the formation of ZnS polytypes. He has considered the lattices of either compound as mixtures of eclipsed (E) and staggered (S) puckered planes. This S-E notation is the same as the hc notation of Jagodzinski (vide Section 2.2), but it emphasizes the analogy with molecules, in which the terms 'staggered' and 'eclipsed' are freely used in the discussion of their internal rotation, e.g. in the ethane molecule (C_2H_6). The internal rotation forces, believed to extend over relatively short distances as compared to the ionic forces, persumably arise from repulsion between eclipsed layers, but their origin is not fully clear. The repulsion is manifested by the observed stability of the staggered form, viz. cubic zincblende structure, at low temperatures. As the temperature is raised, an increasing number of eclipsed layers become thermodynamically stable. Since a Zn atom opposes the S atom across an E layer, the partially ionic character of the ZnS structure helps the stabilization of the E layers by coulombic forces. Thus one can get mixtures of S and E layers, until at sufficiently high temperatures the wurtzite lattice, wholly consisting of the E layers, is obtained. The increasing long range coulombic forces, as the proportion of E layers increases, are taken as suggestive of a cooperative phenomenon, thus lending support to Schneer's theory that this transition is of second order. The consequent repulsion between the E layers should lead to distortions in the lattice, for which experimental evidence has been obtained. For a compound with a strongly covalent lattice, like SiC, one could at best obtain a mixture of equal numbers of E and S layers by heating the cubic phase to the highest possible temperature, but never a conversion to the completely eclipsed wurtzite form, which is also substantiated by observation. However, for the formation of higher polytypes of SiC, Weltner has proposed that Frank's screw dislocation mechanism may be operative. Similarly for the SiC type $2H$, some unspecified crystal growth mechanism has been proposed. Weltner's theory predicts polytypism in the elemental solids diamond, silicon, and germanium, which has not yet been observed, although lately the wurtzite forms of diamond and silicon have been synthesized at high pressures. Also, like the Schneer theory, Weltner's proposal fails to account for the ordering of layers in the polytypes, although through a simplified calculation it has been estimated that the vibration entropy contribution and hence the disordering effect, is small.

Very recently, from their experimental results on structural phase transitions in ZnS between 20 and 1200 °C, Baars and Brandt [161] have concluded that transitions between the $2H$, $4H$, and the $6H$ (33) hexagonal polytypes are second order transitions. They have employed the criteria developed by Landau and Lifschitz [169], based on the description of the crystal by a density function having symmetry properties of a particular space group.

4.5. PEIBST'S LATTICE VIBRATION THEORY

Peibst [170] has invoked the influence of thermal vibrations on crystal structure for explaining the growth of polytypes. The mathematical and other details of his treatment can be found elsewhere [3]*. The theory is very qualitative in nature. There are no experimental findings in its favour or disfavour. In fact, the theory provides little scope for them. Earlier, without going into mathematical details, Zhdanov and Minervina [171, 172] had mooted a somewhat similar suggestion that the presence of impurities on the face of a growing crystal might cause a periodic disturbance, which might affect the growth of the crystal and bring about the generation of polytypes. According to Peibst's theory, the order in a growing crystal is determined by the mutual action of lattice vibrations.

4.6. Vand and Hanoka's epitaxial theory

In formulating his screw dislocation theory of polytypism, Frank [20] had contended that the internal strain produced by the inhomogeneous distribution of impurities in an otherwise perfect platelet of a basic structure was responsible for the creation of screw dislocations. The strain could be severe enough to cause the crystal to warp and eventually crack, thus raising one or more terminated steps on its surface. Each point of termination is the seat of a screw dislocation. Vand and Hanoka [173] have alternatively suggested that the screw dislocation may be produced by epitaxial growth on a foreign particle, which may be a polycrystalline aggregate. Then the crystal protoplate need not be unduly strained as visualised in Frank's mechanism. At the same time the epitaxial conditions of growth may introduce an arbitrary number of faults in the initial stratum of layers, thus giving rise to polytypes with different stacking sequences during subsequent growth by the spiral mechanism. Thus the primary role of screw dislocations is retained as in Frank's model, but the epitaxial model is especially advantageous in explaining polytypism in substances like PbI_2 where the former model does not permit any polytype formation. Much earlier, both Frank [20] and Vand [174] had independently conceived of the possibility of insertion of stacking faults in the initial ledge exposed by the screw dislocation.

From a statistical analysis of their results from an X-ray study of nearly 650 PbI_2 crystals, Hanoka and Vand [76] have sought to substantiate their theory. They found that the occurrence of rhombohedral polytypes, of which they encountered nearly 180 specimens of eight varieties, viz $6R$, $12R$, $18R$, $24R$, $30R$, $36R$, $42R$, and $48R$, followed a Poisson's distribution of the form

$$p_s = e^{-m} m^s / s!$$

where $s = 6n$ ($n = 1, 2, 3, ...$), p_s is the probability of s to occur, and m is a constant, equal to 12 here. By implication this distribution has been taken to represent the distribution of the size of foreign nuclei, around which the epitaxial growth is proposed to take place [76, 175]. Since the nuclei can have various random sizes, it has been inferred that the different polytypes really grow by epitaxy. Hanoka and Vand have further asserted that their proposed model can account for the syntactic coalescence of a nH polytype with a $3nR$ polytype, which is very frequently observed

* vide p. 283.

in PbI$_2$ crystals, if the nH structure happens to include a single Shockley partial dislocation at some stage of its growth. However, several glaring exceptions to their proposal have been found to exist [4, 22]. Therefore, although the observed Poisson's distribution for the rhombohedral types of PbI$_2$ is a remarkable piece of information, it apparently needs a different interpretation than Hanoka and Vand's. In fact, there is little novelty in the proposed mode of creation of a $3nR$ polytype from a nH type by the introduction of a Shockley partial. Much earlier, Bhide and Verma [176] had proposed an almost identical way of achieving the same end, through inclined dislocations composed of edge and screw components.

4.7. MARDIX AND STEINBERGER'S THEORY OF EXPANSION OF STACKING FAULTS

To account for the formation of ZnS polytypes, Mardix and Steinberger and their coworkers [177, 178] have proposed another modification of Frank's theory, in which the screw dislocations have been envisaged to cause a periodic slip of the close-packed basal atomic planes. They have suggested a two-step process: (a) growth of ZnS crystals with the basic $2H$ structure from the vapour phase around a large screw dislocation and (b) introduction of stacking faults in the structure during the cooling period of furnace, when large thermal stresses exist, and their combined rotation and climb over the atomic planes, guided by the helical topology imparted to the crystal by the presence of the screw. If c_0 be the distance between successive (00.1) basal planes and mc_0 the burgers vector of the dislocation, a fault created in the mth layer will spread along the appropriate helical surface to the $2m$th, $3m$th,... layers, as sketched in Figure 26, thus transforming the original $2H$ structure into a new polytype

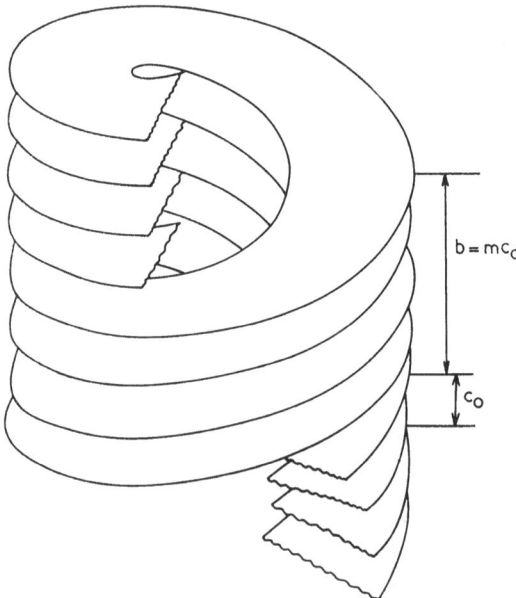

Fig. 26. Helical topology of atomic planes around a screw dislocation of burgers vector mc_0 ($m = 4$). The core has been removed and the faulted surface has been striped (after Alexander *et al.* [168]).

of cell height $c = mc_0$. Mardix and Steinberger's theory is based on an extensive optical and X-ray investigation of ZnS crystals, grown around 1250 °C. The low temperature form of ZnS is the cubic zincblende $(3C)$, which transforms above 1000 °C into the high temperature hexagonal form, wurtzite $(2H)$ (vide Section 3.4). Therefore, during the cooling period (when the temperature is well below the transition value) the stacking faults, which are bounded by Shockley partials, have negative free energies and hence they naturally tend to expand, pushing the partials ahead. Thus the necessary driving force for the propagation of the faults is provided.

The theory is well substantiated by experimental facts. Alexander *et al.* [168] have observed that needle-shaped crystals grow first, presumably by the screw dislocation mechanism, which later expand laterally into platelets. The direction of the needle is along the *c* axis. Several polytypes jointly exist in a crystal as parallel strips of varying widths, normal to *c*, and have different birefringences. The neighbouring strips are not coplanar but mutually tilted through small angles. Numerous zigzag lines, referred to as linear markings, run all through the length of the crystal nearly parallel to *c*. These structural features very well conform with theory. After a polytypic region has been formed by the expansion of faults in the $2H$ structure, it may develop new faults and similarly transform into a different polytype. Thus the existence of several parallel polytype strips in the same crystal becomes comprehensible. The periodic slip of atomic planes should manifest macroscopically as tilting of the various polytypic regions and of the linear markings, as actually observed. The angles of tilt will depend on the polytype structure and can be calculated. The calculated values agree with the measured ones [178]. X-ray diffraction topographic studies have revealed the existence of giant screw dislocations running across the length of the crystal platelets containing polytypes [168, 179]. Also, the dislocation line is found to be tilted like the linear markings. Direct evidence for the expansion of stacking faults in the wurtzite → zincblende transformation has been obtained in electron microscopic studies [180]. Since the various polytypic regions in a given crystal are generated from a common screw dislocation, they should be structurally related to each other, which agrees with observation and supports the concept of the observed 'polytype families' in the ZnS crystals [181]. If too many faults be introduced simultaneously, they may interact mutually or with other crystal defects to prevent their expansion and thus produce disordered $2H$ structure, which is very commonly observed. An uninterrupted free expansion of the faults should necessarily produce well-ordered polytypic regions, which is well substatiated by the general observation that the single crystal X-ray photographs of the ZnS polytypes do show sharp and well-defined reflections. Besides, the X-ray diffraction topographs and etching of these regions also show them as practically perfect crystals free from defects, except for the generating screw dislocation [182]. Since the polytypes originate from the basic $2H$ structure, they should have only even periodicities, e.g. $4H$, $6H$, $8H$, etc., which also agrees with observation. There are ample reasons to believe that the ZnS polytypes are not formed during crystal growth but afterwards. In a recent study of phase transitions in ZnS, Gobrecht *et al.* [160] and Baars and Brandt [161] have found that the polytypes did not form during the transition $3C \rightarrow 2H$, when the temperature was increasing, but during the reverse transition $2H \rightarrow 3C$, while the temperature was coming down. Direct evidence has been obtained that the tilt between polytypic strips occurs while the crystals are cooling down [183]. Lendvay and Kovacs [184] have reported that the X-ray patterns of thick ZnS rods show a strongly mixed cubic-hexagonal character, presumably owing to strong stresses arising from inhomogeneous cooling. But in the case of ZnS needles, cooling more rapidly and homogeneously than the rods, such mixed character is generally lacking. Also it has been observed that rapid cooling of ZnS crystals from 1200 °C down to room temperature freezes the $2H$ phase [160].

From a statistical analysis of disordered layers (called D layers) in n-type SiC crystals, studied by optical and X-ray diffraction techniques, Golightly and Beaudin [185] have concluded that the SiC polytypes form by screw dislocation-controlled expansion of stacking faults during crystal growth. More recently, in a study of solid state phase transitions in $2H$ SiC crystals by similar techniques, Powell and Will [186] have observed that the transition temperature is dependent on the dislocations in the crystals, whence they have surmised that the transition occurs through the periodic propagation of stacking faults around a screw dislocation, as suggested in the Mardix and Steinberger theory. They have also observed the appearance of ridges on the crystal faces, as a consequence of periodic slip.

Almost simultaneously with Mardix and Steinberger, Daniels [187] advanced a very similar model of polytype formation in ZnS, but replaced the long c-oriented screw dislocation by an axial dislocation having a component in the c direction. The axial dislocation was not visualized to play a role in the growth of the parent $2H$ crystals, thus implying that polytypes of odd periodicities could be generated equally well, which is grossly at variance with observation. Moreover, different axials could operate in different parts of the crystal, leading to structurally unrelated polytypes, which also does not agree with experiment. These difficulties are obviated in Mardix and Steinberger's model.

Rai [188] has criticized that should the Mardix and Steinberger mechanism be operative in the generation of ZnS polytypes from $2H$, which has the crystal structure (11) in the Zhdanov symbol, the polytypes should mostly contain the units 1 in their Zhdanov symbols. But in reality they are usually found to include much larger values of Zhdanov units in their stacking sequences. Hence Rai has conjectured yet another model for the genesis of ZnS polytypes, viz. by the ordering of f.c.c. micro-twins through screw dislocations, during crystal growth. The microtwins have been proposed to be produced by insertion of stacking faults at alternate layers of a $2H$ crystal containing widely separated growth faults, during the $2H \rightarrow 3C$ transformation. However, as mentioned earlier, the experimental evidence favours the formation of ZnS polytypes after the completion of crystal growth, rather than during the growth as envisaged by Rai. Also Rai's model assumes the existence of growth faults alone in the parent $2H$ crystals. The justification provided for ruling out the existence of deformation faults is not convincing. Evidence for the creation of f.c.c. microtwins has been obtained in electron microscopic studies of polycrystalline ZnS films during the $3C \to 2H$ transition (not the $2H \to 3C$ transition, as postulated in Rai's model) [189], but since properties in the thin film state can be appreciably different from the bulk state, the relevance of this observation to the proposed model is doubtful. Thus, Rai's model does not appear satisfactory.

4.8. KNIPPENBERG'S ELECTRON ENERGY THEORY

In a comprehensive investigation of the growth of SiC crystals, Knippenberg [134] has observed that the acceptor and donor impurities show a difference in behaviour regarding their preference for cubic and hexagonal forms. The acceptors and donors appear to stabilize the hexagonal and cubic structure, respectively, which has led him to conclude that the energy of electrons may be the primary controlling factor for the choice of a structure, particularly as other possible free energy differences between the various structures appear inappreciably small. He has reported large variations in the measured band-gaps for several SiC structures, viz. 3, 3.1, 2.9, 2.86, 2.77, 2.72, 2.56, 2.2 eV for the types $2H$, $4H$, $15R$, $6H$, $21R$, $24R$, $8H$, $3C$, respectively, which imply that the corresponding electron energy differences should also be appreciable.

Knippenberg has argued that the above band-gap variation qualitatively indicates a progressive increase in the lattice energy from $3C$ to $2H$, thus implying that the polytype $2H$ should be the most stable of all SiC modifications. However, this is contrary to observation. Structural changes in $2H$ have been observed to occur at temperatures as low as $400\,°C$ [186]. In fact, from the recent experimental evidence (vide Section 6.1.1) $2H$ appears to be the most unstable modification of SiC and $6H$ the most stable one. Knippenberg has sought to resolve this anomaly by proposing a suitable variation of the polytype band-gap with the temperature of polytype formation. However, no measurements to provide the free energy of polytypes as a function of temperature are yet available.

Variations in energy gap have been reported for 11 ZnS structures, too, by Baars [190] and Brafman and Steinberger [191], who have noted that the band-gap linearly reduces with increasing content of cubic packing. A similar inference has been drawn by Choyke et al. [192] for 7 SiC polytypes. This correlation demonstrates the influence of stacking faults on the magnitudes of the band-gap, which is well expected since the configuration of the electron energy bands depends on the periodicity and is also sensitively influenced by the long-range order in the crystal. It has been noticed that starting with the cubic structure, the band-gap steadily goes up as progressively more layers are displaced to acquire a hexagonal configuration. A scrutiny of the experimental data of Knippenberg reveals the existence of a similar correlation, viz. the percentages of cubic packing for the structures $2H$, $4H$, ..., $3C$, with band-gaps as 3, 3.1, ..., 2.2 eV, come out to be 0, 50, 60, 67, 71, 75, 75, 100, respectively. This raises the question whether it is the demand for a particular energy gap which requires the faults to arrange themselves into an ordered structure, or that the faults first arrange through some other device so that the observed band-gap is merely a consequential effect. Granting that the former alternative holds, the polytypes should show a definite correlation with their temperatures of growth, which, however, does not agree with observation. Therefore, most likely the observed energy-gap variation is a secondary effect, which follows the creation and stabilisation of the stacking faults into regular arrangements through a suitable mechanism.

5. Role of Stacking Faults in Polytype Formation

5.1. Evidence of Faults in Polytypic Crystals

While describing Jagodzinski's disorder theory of polytypism (Section 4.3), it was mentioned that all prominent polytypic substances exhibit features of one-dimensional disorder. The disorder manifests on an oscillation photograph as 'streaking', in which reflections of the same h and k but different l values are found to be linked through diffuse streaks of varying intensities. An example of a one-dimensionally disordered CdI_2 crystal is provided in Figure 27, which may be usefully compared with the cor-

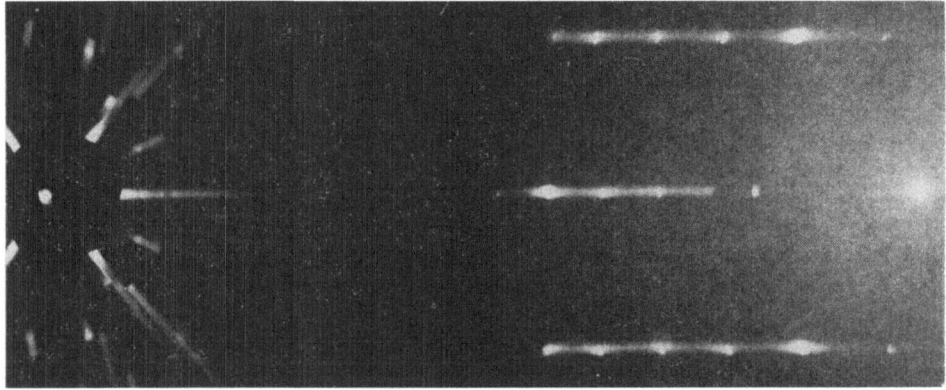

Fig. 27. Oscillation photograph of a disordered CdI_2 polytype $4H$, showing streaking on all the layer lines; compare with Figure 5b (Section 2.4), showing the oscillation photograph of a well-ordered $4H$ crystal, obtained under identical conditions.

responding photograph of a well-ordered crystal of the same cell size, depicted before in Figure 5b (Section 2.4). The disorder arises from the existence of random stacking faults, which disturb the regularity of the structure in the direction perpendicular to the layers [193]. The theory of diffraction from one-dimensionally disordered crystals is fairly well understood and has been developed, amongst several workers, by Jagodzinski [29, 156, 194]. The streaking occurs for the reciprocal lattice rows with $h-k=3n\pm1$; the rows with $h-k=3n$ are sharp. The role of stacking faults in polytype formation has been visualised in more than one way, as discussed earlier in Sections 4.3, 4.6, 4.7 and 4.8. The stacking faults are related to phase transformations and polytypism since they essentially represent a transition between well defined phases, with the degree of transition depending on the content and distribution of the faults. Hence the knowledge of the nature of the faults, viz. their mode of creation, distribution, kinetics, etc., is of immense interest in the study of polytypic crystals. Some fundamental information in this regard has been gained by the X-ray diffraction and optical studies of the 'arcing' phenomenon in CdI_2, PbI_2 and $CdBr_2$ crystals [195 to 197, 78] which will be presented in this chapter.

5.2. PHENOMENON OF ARCING

Sometimes diffraction spots on single crystal oscillation or Laue photographs are found to be extended in the shape of small arcs of a circle. Each arc consists of a succession of two or more spots. An example is shown in Figure 28. The phenomenon, termed as 'arcing' [193], is found to be widely prevalent in CdI_2 and $CdBr_2$ crystals and, to a lesser degree, in PbI_2 crystals. Detailed theoretical and experimental studies of the phenomenon in these crystals have been made, which have provided a satisfactory explanation of the formation of arcs, as also a convenient method for estimating the density of dislocations in the crystals. A brief account of the same follows.

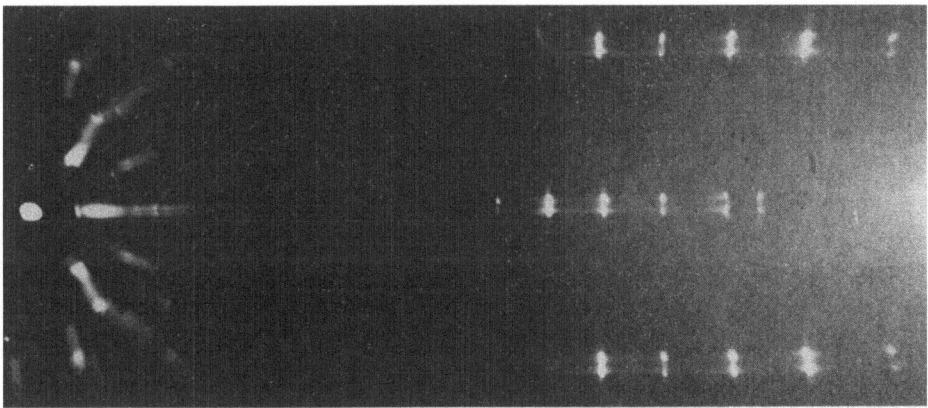

Fig. 28.   Oscillation photograph of a CdI_2 $4H$ polytype, showing arcing of the reflections; each arc consists of two spots; compare with Figure 5b (Section 2.4) obtained under identical conditions.

5.2.1. CdI_2 crystals

Investigations have been carried out on solution-grown crystals, which occur as hexagonal platelets, nearly 50 to 250 μm thick and $\frac{1}{2}$ to 1 mm in size. On account of the heterogeneous character of atomic forces in the CdI_2 structures (Section 3.1), two adjacent iodine layers can mutually slip with ease under a slight stress and give rise to an edge dislocation. The most likely slip planes and slip directions in CdI_2 crystals are the $\{0001\}$ basal planes and the $\langle 11\bar{2}0 \rangle$ directions, respectively, along which the atoms have closest packing. A slip of unit atom spacing along these produces an edge dislocation lying in the basal plane and having the Burgers vector $a/3\langle 11\bar{2}0 \rangle$. From energy considerations, the formation of such unit dislocations is most probable. A unit dislocation may also decompose into two Shockley partial dislocations, separated by a region of fault, as follows,

$$\frac{a}{3}\langle 11\bar{2}0\rangle = \frac{a}{3}\langle 10\bar{1}0\rangle + \frac{a}{3}\langle 01\bar{1}0\rangle.$$

The two partials repel and move away from each other, thus enlarging the faulted region between them, but their separation is limited by the stacking fault energy which has an inverse relationship with the magnitude of separation. The partials may also originate independently by slip along $\langle 10\bar{1}0\rangle$, which are not as close-packed as the $\langle 11\bar{2}0\rangle$ directions, yet enough close-packed for the slip to occur along them. Such a slip displaces the atoms at A-sites to B-sites over a part of the slip plane if the atoms in the preceding layer are at C-sites. The slipped part forms a region of stacking fault bound by a partial dislocation. Simple, as well as complex, sequences of partial dislocations and accompanying stacking faults have been observed as a common feature in the electron microscopic studies of CdI_2 crystals [123] (also see Section 3.5).

The dislocations thus created should be able to move freely during the growth of crystals, because for such motion they need stresses which are orders of magnitude lower than those responsible for creating them. Then the asymmetric nature of interaction between the dislocations of the same sign induces them to be vertically aligned one above another, producing the well-known macroscopically vertical tilt boundaries. The angle of tilt, θ, is determined by the dislocation spacing, h, i.e. $\theta = b/h$, where $\mathbf{b}$ represents the Burgers vector of a dislocation. A boundary effectively divides the crystal into two distinct blocks which are mutually tilted about an axis lying parallel to the dislocations, inside the contact plane of the boundary. Consequently, an X-ray reflection from the crystal consists of a pair of close spots. Since the spots are symmetrically situated on either side of the position of the usual single diffraction spot, now missing, the pair presents the appearance of a small arc of a circle. There are three equivalent directions of closest packing, viz. $[11\bar{2}0]$, $[\bar{2}110]$ and $[1\bar{2}10]$, along one, two or all three of which slip can take place, producing corresponding tilt boundaries. The partials created by the relatively less probable slip along the $\langle 10\bar{1}0\rangle$ directions may similarly produce their own tilt boundaries. The angle of tilt can be conveniently found out by measurement of the length of an arc on oscillation photographs, for which relevant relationships can be developed from the appropriate reciprocal lattice constructions. Depending upon the mutual orientation of the axis of tilt and the axis of oscillation, six cases may arise if the latter lies along the direction of an axis. These have been illustrated in Figure 29. Three cases correspond to the axis of tilt lying along $[1\bar{1}00]$ and the a axis along $[11\bar{2}0]$, $[2\bar{1}\bar{1}0]$, or $[\bar{1}2\bar{1}0]$ and the other three for the tilt axis along $[11\bar{2}0]$ and the a axis along one of the same three directions. Relevant mathematical relationships have been developed for each of these cases, from which general conclusions have been arrived at for the six respective cases, and fully substantiated by experimental evidence [195]. An example is provided in Figure 28, in which each arc is seen to be composed of twin spots on the oscillation

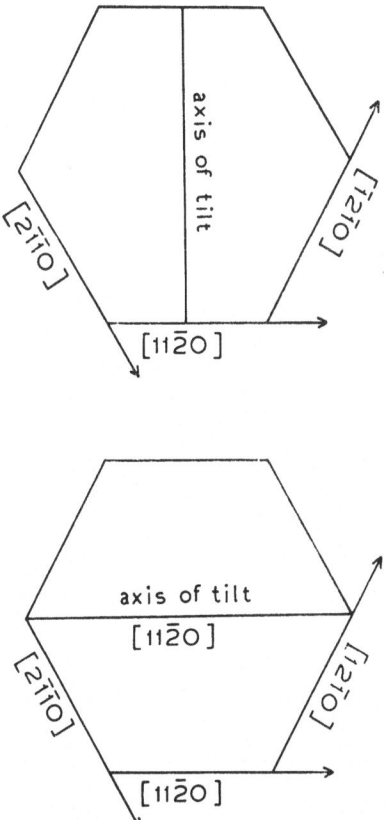

Fig. 29. Possible relative orientations of the axis of oscillation (the *a* axis) with respect to two possible orientations of the axis of tilt. The *c* axis is perpendicular to the plane of the paper.

photograph. The arc length progressively increases with the increase in *l*-value. It relates to the first of the six possible cases.

If *h* is the dislocation spacing in a boundary, the density of dislocations is given by,

$$\varrho = \frac{1}{h} = \frac{\theta}{b} \text{ dislocations cm}^{-1},$$

which yields a value of the order 10^5–10^6 dislocations per cm in the CdI_2 crystals. The dislocations residing within a crystal are located in the boundaries and inside the blocks, but the number inside a block is much less than in the boundaries because the tangential breadth of a spot is generally observed to be much smaller than the arc length. Hence the value of the dislocation density as determined above may very nearly be taken to represent the density in the crystal as a whole.

Simultaneous slip along two close-packed directions produces two tilt boundaries which may divide the crystal into 3 blocks, leading to composite reflections made up of 3 spots each, on the Laue photograph [195]. More interesting are the cases in which the slip takes place along three or more close-packed directions, leading to the formation of closed trigonal or hexagonal rings of diffraction spots on the Laue photographs. An example is shown in Figure 30. The preceding considerations of unidirectional slip are readily extended to understand these cases of multiple slip. Of the six equivalent directions of close-packing, viz. $[11\bar{2}0]$, $[\bar{2}110]$, $[1\bar{2}10]$, and their negatives, along which a unit slip can take place, simultaneous multiple slip along the first three will give rise to three identical edge dislocations $120°$ apart from each

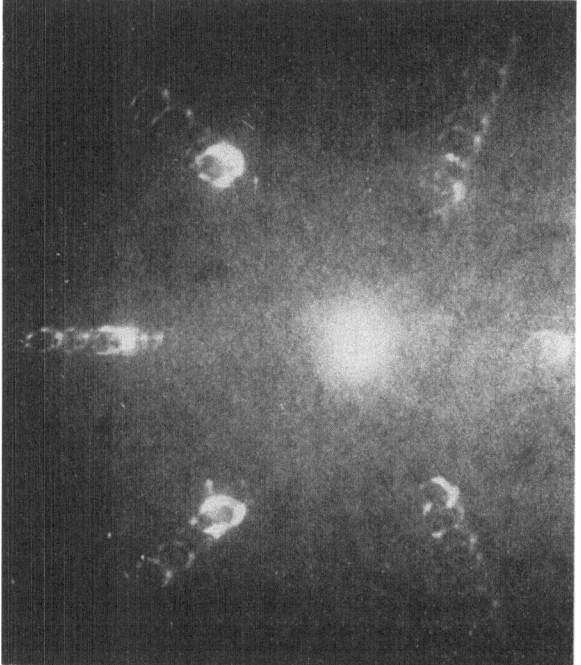

Fig. 30.   Laue photograph of a CdI$_2$ crystal showing closed hexagonal rings of diffraction spots; a axis vertical; angle between the X-ray beam and c axis is $25°40'$. Cu unfiltered radiation; camera radius 3 cm.

other, forming a triple node on the slip plane. Similarly another identical triple node of dislocations, rotated through $180°$ with respect to the first, will be produced by slip along the rest of the three directions, as shown in Figure 31. The various formations of the triple nodes lead to various corresponding features on the X-ray photographs, viz. hexagonal rings with alternate sides equal, trigonal rings, rings with elongated corners, irregular rings, etc., which find full agreement with the experimental results [196]. An example has been shown in Figure 30. The corresponding oscillation photograph (Figure 32) shows each reflection as composed of six spots. The tilting of the blocks across the boundaries is manifested on the basal faces of the crystals as well. In the crystals exhibiting hexagonal rings on the Laue photograph, optical observations have revealed that the basal surfaces are no longer flat, but composed of six mutually tilted regions. The angles of tilt measured with an optical goniometer agree with the calculated values obtained from the X-ray photograph [81].

To sum up, the existence of the phenomenon of arcing convincingly demonstrates that edge dislocations are profusely created during crystal growth. They also move freely during the growth and can sometimes arrange into regular formations. In Section 5.3 it will be examined how these features are directly linked with the generation and alignment of stacking faults and hence with the phenomenon of polytypism.

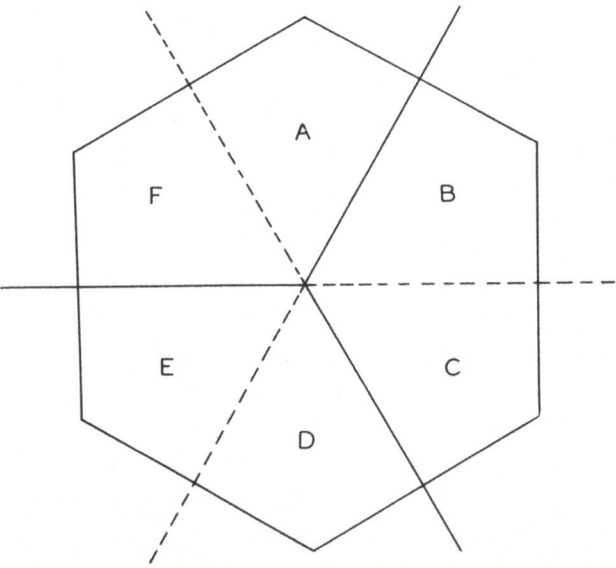

Fig. 31. Two triple nodes of dislocations, represented by the solid and dotted lines, respectively. The tilt boundaries along them divide the crystal into six blocks, *A*, *B*, *C*, *D*, *E* and *F*.

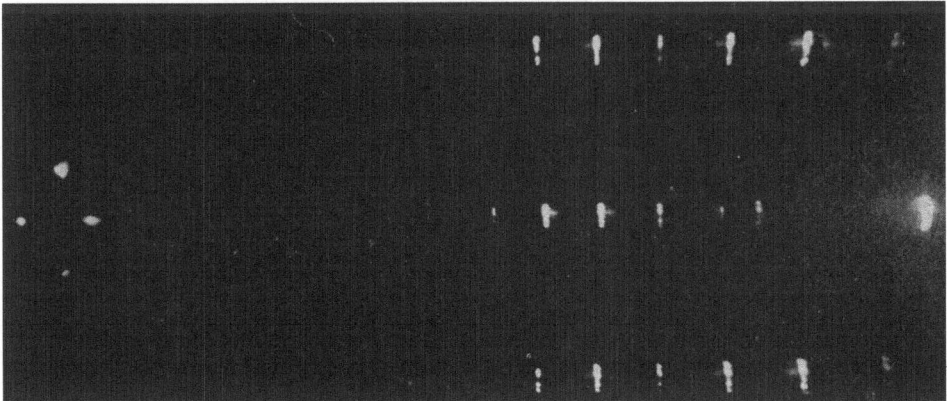

Fig. 32. Oscillation photograph corresponding to Figure 30, showing each reflection as composed of four spots; conditions as in Figure 5b.

5.2.2. PbI₂ Crystals

The PbI$_2$ crystals have the same MX$_2$-type of structure as CdI$_2$. Therefore, similar considerations of arc formation apply to them. Recent investigations in this compound have been carried out on the crystals growth as hexagonal platelets by the gel method and it has been found that the arcing phenomenon rarely occurs in them, although the nature of arcing is the same as in the CdI$_2$ crystals [78]. The comparatively far lesser incidence of arcing can be attributed to structural differences on points of fine details, which will be discussed in the next Section (5.3) in relation to the incidence of polytypism. Electron microscopic studies have furnished evidence for the existence of stacking fault sequences in the PbI$_2$ crystals [122].

5.2.3. CdBr₂ Crystals

The CdBr$_2$ crystals are also isostructural with the CdI$_2$ and PbI$_2$ crystals, but in sharp contrast to the latter, practically all the crystals display the arcing phenomenon on their X-ray photographs [197]. Investigations have been made on hexagonal plate-shaped crystals grown from aqueous solution. The formation of arcs follows the same principles as in CdI$_2$ and PbI$_2$, but the nature of arcing has been found to be markedly different and considerably complicated. It is principally to be attributed to the fact that unlike the CdI$_2$ and PbI$_2$ crystals, in which the slip is mainly confined to the close-packed basal planes, the slip can take place with equal ease on the non-basal $\{01\bar{1}2\}$ planes. The latter are equivalent to the basal planes in regard to the close-packing of atoms. The equivalence stems from the sequence of layers, viz. $(A\gamma B)\,(C\beta A)\,(B\alpha C)$, in the basic structure, $6R$, of cadmium bromide. The dislocations resulting from slip on the non-basal planes can interact with other parallel dislocations generated by slip on the basal planes to form asymmetrical tilt boundaries, having their planes inclined to the c axis. The shape of the reflections on the X-ray photographs is determined by the combined effect of the symmetrical and asymmetrical boundaries and a large variety of complicated patterns can be produced. Many such patterns have been observed [197], of which one is shown in Figure 33.

5.3. ARCING AND POLYTYPISM

Several factors indicate that the phenomena of arcing and polytypism are closely interrelated [78, 198]. We have seen that the arcing results from the interaction between edge dislocations, rendering them to be vertically aligned at a constant spacing. On the other hand, the various polytypes of a substance can be generated by introducing suitable stacking faults at a constant interval into some given structure of the substance, for which certain possible mechanisms have been suggested (vide Sections 4.3 and 4.7). But the introduction of a stacking fault follows the creation of a partial dislocation, and its subsequent movement over an atomic layer. As argued earlier in Section 5.2.1, the edge dislocations once created should be able to move freely during crystal growth. Thus the edge dislocations which give rise to the arcing phenomenon are also responsible for the formation of polytypes. This correlation between

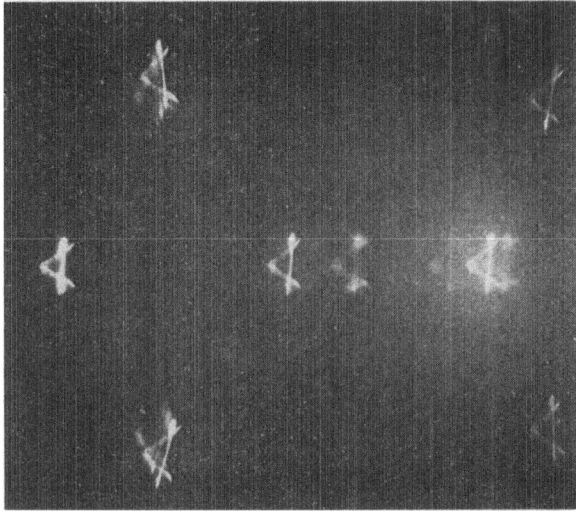

Fig. 33.   Laue photograph of a CdBr₂ crystal showing reflections of complex shapes. *a* axis vertical; angle between the X-ray beam and the *c* axis is 42°38′; other conditions as in Figure 30.

arcing and polytypism is manifested in the relative frequencies of occurrence of the two phenomena in the isostructural crystals of CdI_2 and PbI_2. The incidents of both polytypism and arcing in the PbI_2 crystals have been found to be far less than in the CdI_2 crystals. This is explained by taking a close look into the atomic structures of these crystals, which reveals that the probability of creation of edge dislocations, which, presumably, are basically responsible for the two phenomena, is considerably lower in the PbI_2 crystals [78, 195, 196]. The PbI_2 crystals show lesser incidence of streaking, too, as compared to the CdI_2 crystals. Since the streaking results from the existence of random stacking faults in the structure, this observation also suggests that relatively smaller number of edge dislocations are created in the PbI_2 crystals. These considerations are not tenable for the $CdBr_2$ crystals, which are also isostructural with the CdI_2 and the PbI_2 crystals, because the movement of dislocations on the basal planes in them are hampered by the dislocations on the equivalent but inclined non-basal planes.

The correlation between arcing and polytypism manifests in other ways, too. The upper and lower faces of a crystal, as picked up from a crystallizing dish, exhibit different degrees of arcing, with the lower face usually displaying more arcing. At the same time it has been found that generally the lower face is either a higher polytype than the upper one or it is relatively more disordered, as judged from the intensity of streaks on the oscillation photographs. Both these effects are to be attributed to a non-uniform rate of generation of the edge dislocations during crystal growth, the rate being comparatively much faster during the initial stages of growth when the supersaturation is high. Hence the dislocations themselves, as well as the resulting stacking faults, are initially closely spaced, giving rise to the observed increase in arcing and in the size of the polytype. Towards the concluding stages of growth, the growth conditions are fairly stabilized reducing the generation of the dislocations in turn. Hence the upper face of the crystal is relatively free from arcing and is also a smaller polytype. Another observation supporting the proposed correlation between arcing and polytypism is that the arcing is usually found to be displayed by the small-period common polytypes. The higher polytypes, which are believed to originate from the common types, display small arcs and that only rarely. Since the formation of the higher types requires the perpetuation of stacking faults by complete layer displacements at a constant interval in a smaller common type, the explanation for the observed reduced incidence of arcing in the higher types is that the energy available for aligning the dislocations into tilt boundaries is diverted into producing the necessary layer displacements in them.

The substances displaying arcing may not be proportionately polytypic, e.g. we have seen in $CdBr_2$ that despite nearly all the crystals showing arc formation, the interaction between the dislocations generated on the basal and non-basal planes, respectively, seriously hampers the formation of polytypes of this substance. The reverse may also not always be true, because the arcing results from a strong inteaction between dislocations of the same sign and, depending on the nature of dislocation systems and the density of dislocations in the substance, such an interacton may not always be permissible.

Of late, two possible alternative causes for arcing, and also for streaking in particular positions of the crystals, have been proposed. It has been conjectured that the effects of crystal shape [72] and paracrystalline distortions [199] may be responsible for these phenomena. However, it has been pointed out that the premises on which these proposals are based, and the deductions made therefrom, suffer from many obvious flaws and do not agree with the observed facts [200, 201].

6. Recent Experiments and Present Position

More information regarding the part played by stacking faults in the formation of polytypes has been obtained in some recent experiments, which also shed light on the kinetics of the faults. The experiments, to be described in the following, involve heating of the crystals to high temperatures, subjecting them to mechanical stresses and successively cleaving them. The present position regarding the progress in the study of polytypism has been summed up at the end.

6.1. EFFECT OF TEMPERATURE, PRESSURE AND GROWTH CONDITIONS ON PHASE TRANSITION IN POLYTYPES

6.1.1. *Effect of Temperature*

Marked structural changes with temperature were first observed in CdI_2 crystals that had been stored at room temperature for several years [22]. The observation stimulated a detailed investigation of the effect of heating on the CdI_2 crystals, grown from solution [202, 203, 54]. The temperature of heating was kept at about 260 °C, or somewhat higher, which was well below the melting point, viz. about 390 °C, of cadmium iodide. The X-ray examination of the crystals was conducted before heating and after each heating run, which usually consisted of nearly $1\frac{1}{2}$ h. The heating was repeated as many times as found necessary. Nearly 160 polytypes were investigated. The observed structural changes are manifest in four different ways on the X-ray photographs:

(i) Change in the intensity distribution of diffraction spots (indicating a transformation of polytype without a change in the cell dimensions).

(ii) Change in the spacing of spots (indicating a polytype transformation with change in the cell height).

(iii) Change in streaking (indicating a change in the distribution of random stacking faults).

(iv) Change in arcing (indicating a movement and redistribution of edge dislocations).

A common observation was that in one or more heating runs all polytypes eventually transformed to the most common CdI_2 polytype $4H$, thus establishing the dominating thermodynamic stability of the latter. Even the lower polytype $2H$, which corresponds to a single 'minimal sandwich' of CdI_2 and therefore prima facie appears to be very stable owing to the existence of strong ionic bonds within the sandwich, was found to transform into $4H$. In fact, it was invariably found to transform in just one heating run. These observations indicate that the CdI_2 polytypes, except $4H$, represent metastable crystalline states of cadmium iodide. In other polytypes, the final transformation was not always observed to be as quick as in $2H$ and often required several heating runs, particularly for the higher polytypes. Usually, intermediate types having the same cell height as the original polytype but different intensity sequences of X-ray reflections were formed in successive heating runs, i.e. a polytype nH transformed as $nH_1 \rightarrow nH_2 \rightarrow nH_3 \rightarrow \cdots \rightarrow 4H$. As the transformation proceeded, the intensity sequence of the intermediate types increasingly simulated that of the $4H$, until finally the $4H$ structure itself was obtained. An example has been presented in Figure 34 ($84R_1 \rightarrow 84R_2 \rightarrow \cdots \rightarrow 84R_{16} \rightarrow$ almost $4H$). The nature of the transformation clearly showed that the stacking faults (with respect to the stable $4H$ structure) were eliminated gradually and in a systematic way. Possibly, the presence of screw dislocations caused the stacking faults to expand

(a)
(b)
(c)
(d)

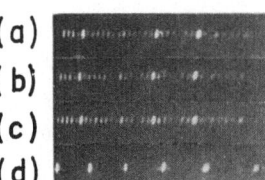

Fig. 34. A succession of $10.l$ reflections on the oscillation photograph of a CdI_2 crystal, (a) before heating, showing reflections of polytype $84R$, (b–d) after 1, 5 and 11 heating runs, showing gradual transformation towards $4H$; other conditions as in Figure 5b.

regularly and thus brought about the observed polytypic transformations, in the manner suggested by Mardix *et al.* for the growth of ZnS polytypes (Section 4.7). A lesser possibility could be that the process was governed by the thermodynamic considerations of the type envisaged in the disorder theory of polytypism (Section 4.3). Only two cases were encountered in which the cell height of the crystal was also found to change upon heating the crystal, including both the conversion of a lower type into a higher type ($12R \rightarrow 32H$) and vice versa ($20H \rightarrow 12R$); the latter has been

Fig. 35.   X-ray photographs of a CdI$_2$ crystal depicting the conversion of (a) polytype $20H$ into (b) a lower polytype $12R$, after heating; conditions as in Figure 34.

depicted in Figure 35. The rarity of such transformations showed that they took place under some exceptional circumstances. The observed quicker transformation of the lower polytypes into $4H$ could be attributed to the increasing departure of the polytype structure from the stable $4H$ structure with increase in the unit cell size, because then it would naturally require a smaller number of layer displacements, and therefore lesser effort, for transforming the lower polytypes into $4H$ than the higher ones.

Fig. 36.   X-ray photographs of a CdI$_2$ crystal showing (a) reflections of the common type $4H$ along with heavy streaking, before heating and (b) appreciable reduction in streaking after heating, conditions as in Figure 34.

Fig. 37.   X-ray photographs of a CdI$_2$ crystal showing (a) $8H$ reflections with moderate streaking, before heating and (b) elimination of streaking, after heating; the polytype has also transformed to $4H$; conditions as in Figure 34.

One example each of the generally observed reduction and elimination of streaking after heating has been presented in Figures 36 and 37, respectively. Since streaking arises from the existence of random stacking faults in the structure, these observations signify that the structure tends to 'heal' by a gradual elimination of the faults. Obviously, the necessary changes in the layer orientation for this purpose are induced by the thermal energy supplied during the process of heating. Presumably, they are brought about by the movement of partial dislocations along the appropriate layers. Most probably, the partials are already present within the original structure, being held up against some obstacles like impurity atoms, other dislocations, etc. If they had to be created afresh, additional random faults would also be introduced in the structure, resulting in increase in streaking, which was rarely observed. That the dislocations were not created afresh was also supported by the observation that any increase in arcing occurred only during the first heating run, as discussed in the next paragraph. The unidentified types, which usually display heavy streaking on their X-ray photographs, invariably tended to transform into definite polytypes upon heating, an illustration of which has been presented in Figure 38. Such a tendency is

(a)
(b)
(c)

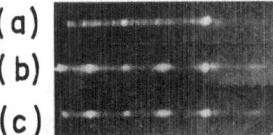

Fig. 38.   X-ray photographs of a CdI$_2$ crystal (a) before heating, showing heavy continuous streaking superposed on the reflections of an unidentified type plus 2H, (b) after first heating run, showing nearly complete transformation of the structure into 4H and discontinuous streaking, concentrated around the main reflections, and (c) after second heating run, showing complete transformation to 4H and superposed light discontinuous streaking; conditions as in Figure 34.

understandable. The polytypes remain unidentified when their reflections are unevenly spaced. Assuming that a polytype is produced by the introduction of stacking faults in the basic 4H or some other structure, the presence of such reflections with superposed heavy streaking signifies the fact that the faults attempt to stabilise into a definite polytypic structure but do not always succeed, so that in the unsuccessful cases most of them remain randomly distributed in the structure. The supply of thermal energy during heating helps the faults to achieve stabilisation, although the structure, which now transforms into a definite polytype, may still retain some random faults. The Figure 38 depicts another common observation that the initially observed continuous streaking becomes discontinuous upon heating the crystal, such that the streak is seen to become more intense around the main reflections than in between them. It follows that the heating has the effect of gradually bringing down the degree of randomness of the faults, such that they increasingly conform to a definite structure.

On about half of the total number of crystals studied, the X-ray photographs revealed development of arcing after the first heat treatment. A typical example has been provided in Figures 39a and b. Since arcing results from the vertical alignment of edge dislocations of the same sign into tilt boundaries (vide Section 5.2), it implies that the heating induces a part of the existing dislocations to move systematically and form tilt boundaries. Any development of arcing was found to occur only after the first heating run. The arc lengths stayed almost constant in the subsequent heating runs, as illustrated in Figure 39. If a polytype originally displayed arcing, the first heat treatment had the effect of enhancing the extent of arcing, thus indicating that more dislocations, earlier held up against obstacles, had moved into the existing boundaries and increased their angles of tilt $(\theta = b/h)$. But in such cases, too, it was observed that the arc lengths did not change in the subsequent heating runs, which confirmed that the heat treatment did not produce fresh dislocations in the crystals, because had it been so, the arc lengths should have increased in each heating run. Besides, it also showed that the initial movement sent the bulk of the dislocations into the tilt boundaries, while the rest of them remained held up against difficult obstacles. The obstacles were not altogether impossible to surmount, because it was found that usually the reflections became diffuse after heating, until after several repeated heating runs they split into a pair of close reflections. An example has been furnished in Figure 40. Obviously, the supplied thermal energy kept on driving the

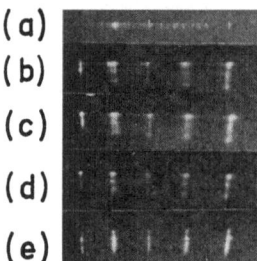

Fig. 39.   X-ray photographs of a CdI₂ crystal (a) before heating, showing reflections of the polytype 42H, (b) after first heating run, showing development of arcing (and transformation of the polytype into 4H) and (c–e) after, seventh and tenth heating runs, respectively, showing practically no change in the extent of arcing but a gradual improvement in the resolution of spots on the arcs; conditions as in Figure 34.

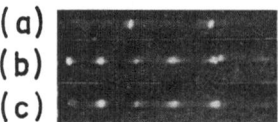

Fig. 40.   X-ray photographs of a CdI₂ crystal (a) before heating, showing reflections of type 2H, (b) after one heating run showing diffuse reflections of type 4H and (c) after six heating runs, showing splitting of each reflection into a pair of close reflections; conditions as in Figure 34.

dislocations. However, the small magnitude of the arc length (Figure 40c) showed that the angle of tilt, θ, of the corresponding tilt boundary was small, signifying, in turn, that the dislocations driven into the boundary had been widely spaced and, therefore, relatively few in number. Also it was observed that the resolution of the spots on an arc steadily improved with each heating run. If the arc looked continuous to begin with, spottiness developed upon it, so that it looked increasingly discontinuous, as illustrated in Figures 39c–e. The improvement in the resolution of spots showed that the alignment of the dislocations in the tilt boundary became increasingly more perfect after each heat treatment, so that the two adjacent crystal blocks, situated on the either side of the boundary, were separated by an increasingly sharp bend.

Tewari and Srivastava [204, 205] have also reported both gradual and quick transformation of CdI₂ polytypes into 4H, after heating for a couple of hours at relatively low (around 250 °C) and relatively high (around 280 °C) temperatures, respectively. They have also interpreted the results in terms of rearrangement of stacking faults, for which they have presented electron microscopic evidence, too [204].

The structural changes consequent upon the heating of PbI_2 crystals have been found to be different from the CdI_2 crystals [77]. The low identity period polytypes, including the most common type $2H$, tend to transform into heavily disordered polytypes of longer periodicities. The arcing of reflections, if not originally present, has also been found to develop after heating. Obviously, unlike the CdI_2 crystals, the effect of heating in the PbI_2 crystals is to induce plenty of additional stacking faults, which are only able to get partially ordered. This is intriguing because the PbI_2 crystals, which are isostructural with the CdI_2 crystals, exhibit far less streaking and negligible arcing at room temperatures and, as discussed earlier (Section 5.3), such behaviour could be qualitatively justified on theoretical grounds by comparing the slip systems of the two compounds. It follows that the slip systems behave quite differently at high temperatures, viz. around 300 °C, at which the heat treatment has been carried out. More theoretical and experimental studies are needed before definite conclusions can be drawn in this regard.

Amongst other substances which are prominently polytypic but do not have layered structures, experiments on heating of ZnS and SiC crystals have also been carried out, which deserve attention. Results of detailed studies in SiC crystals have been recently reported by several different groups of workers [134, 135, 137, 162, 185, 186, 206], some of which were referred to earlier in a different context (Section 4.3). The studies have been mainly confined to the small period polytypes. The results differ from each other, sometimes fairly widely, possibly owing to differences in impurity content [135] and dislocations [186]. Yet they bear some similarity and under normal conditions of growth a rough picture of the effect of heating may be drawn as follows. The heating generally induces strong random disorder in the crystals and gradually brings about a transformation of polytypes, dependent both on time and temperature of heating. The final product of transformation is always $6H$ (a notable exception occurs when high pressures are employed, as described in the next section), although doubts have been expressed if at relatively high temperatures (e.g. above 2200 °C) the observed structural changes result from a genuine solid state transformation or a recrystallization process. The polytype $2H$ is found to be most unstable, with appreciable disorder appearing in it at temperatures as low as 400 °C [186]. Its range of stability extends upto about 1400 °C, near which it transforms into the cubic type $3C$. The ranges of stability of $3C$, $4H$, $15R$ and $6H$ are estimated as upto about 1600 °C, about 1800 °C, between 2000 and 2200 °C and between 2200 and 2600 °C, respectively. The values have been observed to be different for different crystal habits, e.g. plates, whiskers and needles, but again it may be so on account of differences in the impurity and dislocation content.

The results of heating in the ZnS crystals [108, 127, 158, 160, 161] encompass the cubic $3C$ and the hexagonal $2H$ type, as well as the higher polytypes. $3C$ has been reported to undergo a phase transformation into $2H$ and vice versa at temperatures around 1020 °C and 1240 °C, respectively [108]. The higher polytypes and the cubic 3C phase have been observed to form when the 2H crystals are tempered at temperatures between 400 and 1000 °C; the results differ from one worker to another. The higher types have also been observed to form around 800 °C during the cooling period of the 2H crystals [160, 161]. The heating of the crystals generally induces some disorder in the crystals, which strongly increases as the transformation temperature is approached. The reflections become increasingly sharp as the temperature of transformation is gradually exceeded and the crystals heated for a couple of hours. Doping with minute quantities of impurities like Cu, Ga, etc. has been found to lead to a sudden increase of disorder, as also to promotion of the formation of high period types, thus indicating that impurities reduce the stability of 2H [161]. Like the SiC crystals, the ZnS crystals of different habits show variable tendencies towards polytypism and development of disorder upon heating.

Annealing of the polytypic phase 21H of AgI, along with the phases 2H and 3C, has been observed to lead to a loss of the former phase, with the latter two phases gaining in quantity at its cost [119]. The annealing was separately carried out at 200–300 °C for one week and at 125 °C for two weeks, and led to a loss of about 25% in both cases.

6.1.2. *Effect of Pressure*

It has been observed that the application of moderate mechanical stresses to the ZnS 2H crystals in a direction parallel to the (00.1) planes causes both stacking faults and new ordered structures to appear [207, 208]. Prolonged pressure transforms most of the stressed region into 3C, i.e. the phase which is stable at room temperatures [207]. These results can be interpreted similarly to the transformations with temperature. In SiC, Sokhor *et al.* [209] first reported a transformation from α- to β-SiC, by treating α-SiC powders for short intervals of time at 1200–1400 °C and 30–70 kbar pressures; they proposed a direct relationship between the magnitude of applied pressure and degree of conversion. But subsequently Whiteney and Shaffer [210] have failed to reproduce these results. However, they found that the transformations, both $\beta \to \alpha$ and vice versa, did take place in the presence of impurity phases. Boron nitride additions led to $\beta \to \alpha$ transitions while the presence of non-soluble phases like iron led to the reverse transition. Hence they concluded that high pressures alone have little effect, unless secondary phases are also present. A similar conclusion has been reached by Kieffer *et al.* who have observed an interesting anomalous transformation in silicon carbide [211]. Normally, the cubic β-SiC is known to be a low-temperature phase, stable below about 2200 °C, above which only the hexagonal α-phase is supposed to be stable. But by treating powdered SiC in a nitrogen atmosphere of 30–40 atm pressure, at about 2500 °C, the diffraction lines of α-SiC were found to disappear completely, leaving behind only the lines of β-SiC, which implied a $\alpha \to \beta$ conversion. Further, by removing the nitrogen and heating the converted β-SiC under argon atmosphere of 1 atm pressure, the transformation was reversed, yielding α-SiC of the 6H variety. In fact, the transformation could be played back and forth as many times as desired, behaving somewhat like a hysteresis phenomenon. These experiments confirm that the high pressures become effective in inducing phase transformations only in the presence of suitable impurity phases.

The polytype transformations obtained at high pressures in perovskites have already been described in Section 3.4.

6.1.3. *Effect of Growth Conditions*

For studying the effect of growth conditions and for further understanding the role of stacking faults in polytype formation, experiments have been conducted to unfold the history of growth of cadmium iodide crystals, which have an excellent basal cleavage [70, 212]. Each crystal was repeatedly cleaved with the help of an adhesive tape and X-ray photograph of the exposed face was taken after each cleavage, until almost the entire thickness of the crystal had been cleaved off. A minimum thickness of nearly 20 μm could be removed in a single cleavage. The investigations, extending over nearly 50 crystals, have yielded information which confirms as well as supplements the results obtained from the heating experiments and can be similarly interpreted in terms of creation and ordering of stacking faults. In nearly one-fourth of the crystals, the transformation occurred without a change in the cell height, viz. as $nH_1 \rightarrow nH_2 \rightarrow nH_3 \rightarrow \cdots$. In the rest of the crystals the cell height also changed, viz. as $n_1H \rightarrow n_2H \rightarrow n_3H \rightarrow \cdots$. In four crystals, the complete atomic structure of the component polytypes was worked out at each stage of transition, which disclosed that the transformed structure invariably had a close resemblance with the preceding structure, thus implying that the transformation involved little change in the free energy of the crystal. The transformations can be successfully interpreted through the mechanism of screw dislocation-controlled expansion of stacking faults, employed earlier for explaining the generation of ZnS polytypes (Section 4.7), but with the difference that they take place during the growth of the crystals, instead of after the completion of growth [212]. Unlike the case of the ZnS crystals, the Burgers vector strength of the screw dislocation may change during growth due to its dissociation into smaller ones or by adsorption of foreign particles on or near to it. It is also possible that the existing dislocation vanishes altogether, either by ending at another dislocation or by moving out of the crystal, and a new dislocation subsequently originates. An alternative way of explaining the transformations could be through the thermodynamic influence of vibration entropy (Section 4.3), but it seems hardly pertinent for the polytypes of cadmium iodide growing at room temperatures.

The polytypes on the lower faces growing in contact with the flat bottom of the crystallising dish were found to be more disordered than those existing on the upper faces, which further confirmed the existence of widely fluctuating conditions in the initial stages of growth, giving rise to random stacking faults. A polytype transformation was always observed to be preceded by an increase or fresh appearance of disorder in the crystal, implying a fluctuation in the growth conditions. The disorder subsided or reduced when the new polytype was formed, thus indicating an ordering of the faults, possibly aided by a screw dislocation as mentioned before. In fact, when an appreciably disordered polytype was encountered, it was never found to persist through a considerable thickness of the crystal.

Appearance of arcing was frequently observed during cleavage but did not seem to have a bearing on the main observations regarding structural changes, although it did suggest that sometimes the conditions during crystal growth had been favourable for the arrangement of edge dislocations into tilt boundaries.

Wallace [213] has employed X-ray diffraction topography for identifying the syntactically coalesced polytypes and measuring their laminar thicknesses in SiC crystals. While the method is ingenious, it merely achieves polytype identification, without yielding information about other structural changes regarding disorder and arrangement of dislocations. Also, it is considerably limited in resolution, so that its benefit is restricted to relatively small-period polytypes. However, it is specially useful in mapping the distribution of polytype domains on the principal surfaces of a large-sized crystal, which would otherwise require a series of X-ray diffraction photographs. Such distributions are frequently obtained in epitaxial growths.

6.2. CONCLUDING REMARKS

Studies made in the last two decades have shown that polytypism is a widely prevalent phenomenon. The studies hold promise of both theoretical and applied interest in various fields. Since the polytypes of a substance essentially bear the same chemical composition and mainly differ in the value of their cell heights, their structure-dependent properties differ on a more or less known scale. The known polytypic substances do not display the same degree of polytypism, which has been observed to vary within wide limits. Certain prominent polytypic substances like cadmium iodide appear to possess an almost unlimited number of polytypes. The polytypes of a substance have very nearly the same internal energies, since they all have the same nature of nearest-neighbour interaction. This is the basic cause for the observed simultaneous occurrence of several of them together under identical conditions of growth. For the same reason, all substances with layered structure may be expected to be polytypic to a smaller or larger extent, unless some specific factors exist for obstructing the process of polytype formation, which essentially involves various permutations and combinations of the juxtapositions of the constituent layers. To put in a nutshell, polytypism appears to be an inherent property of layered structures.

The chief contribution of the recent investigations has been to bring to light the central role played by the stacking faults in polytype formation. The other factors, namely impurities, pressure, dislocation content, temperature and supersaturation, too, undoubtedly affect the phenomenon (in this connection, the remarkable continuously reversible transformation $3C \rightleftarrows 6H$, observed in SiC (Section 6.1.2) is particularly worth noticing), but they themselves may be acting by wielding their influence on the distribution of the faults. The most promising mechanism available for explaining the periodic arrangement of the faults, and hence for the generation of polytypes, is the screw dislocation-aided expansion of the faults. The existence of large screw dislocations in polytypic substances is substantiated by the observation of growth spirals of good visibility on the crystals of many of them. But since a screw is not always associated with a growth spiral, the best course for verifying its existence will be by examination of the crystals through X-ray diffraction topography, as done for zinc sulphide [179]. However, the influence of thermodynamic factors cannot be overlooked, particularly for the polytypes of the substances like SiC and ZnS, which grow at high temperatures. Even for the CdI_2 crystals grown at room temperatures, the thermodynamic influence has been observed to manifest, but in a different way, viz. all the polytypes invariably tend to acquire the stable $4H$ configuration upon heating, thus implying that thermodynamically all polytypes other than $4H$ represent metastable crystalline states of cadmium iodide. Also, the statistical distribution of the CdI_2 polytypes has been found to conform with Jagodzinski's theory. Thus, while the thermodynamic factors do appear to exercise an overall control on the growth of polytypes, in many substances, or possibly in most substances, the control is weak and has to be aided by the external agency of screw dislocations. Another external factor, viz. the presence of foreign nuclei, is suspected to be operative in the creation of a few polytypes, particularly the ones of high periodicities, in accordance with the epitaxial mechanism of Vand and Hanoka or the original mechanism suggested by Frank.

With the determination of complete crystal structure of a large number of polytypes of the prominent polytypic compounds, a vast amount of precise data has accumulated, which has already found use in several ways, in understanding the formation of polytypes and the observed phase transitions in them. The data is likely to be useful in more ways in the future. Very recently it has been employed for a statistical analysis of Zhdanov sequences in the polytypes of SiC, CdI_2 and ZnS, to yield semi-empirical, semi-theoretical rules regarding the occurrence and distribution of polytypes [214]. A word of comment is due here about the reported existence of more than one structural series, e.g. $(22)_n11$ in cadmium iodide, which have been speculated upon in literature as lending support to Frank's dislocation theory of polytypism. It has been pointed out earlier that the existence of the series fits equally well into any other theory of polytypism [22]. Besides, it is always simpler to work out the atomic structure of the members of the series than the other polytypes, so that the investigator has a natural tendency preferentially to select them for the structure analysis. Thus, the reported structure series most probably owe their existence more to choice than chance.

The commonly observed syntactic coalescence of polytypes clearly indicates that the various polytypes of a substance negligibly differ in their internal energies. This renders the task of accurate estimation of the differences in their energies a very difficult one. Nevertheless, some attempts have been made of late in this direction. Weltner [163] has computed Madelung energies for some small period ZnS polytypes with a view to examine the feasibility of their formation through a cooperative phenomenon (vide Section 4.4.) Similarly, Srinivasan and Parthasarathi [215] have employed semi-empirical potential functions to estimate changes in internal energy arising from interaction at the atomic level and have concluded that these appear to govern the relative proportion of cubic and hexagonal types of packing in the polytypes of SiC, ZnS, and CdI_2. The calculations are presently somewhat crude, but when sufficiently refined they should go to help in a large way in verifying the theoretical ideas concerning the origin of polytypes.

For further understanding the part played by the stacking faults in the formation of polytypes, certain useful suggestions have been made earlier [22]. They relate to estimating the susceptibility of the basic or other polytypic structures to the introduction of stacking faults, estimation of activation energy needed for polytype transformation, estimation of the inherent stability of the basic structure, and estimation of the possible obstacles to the movement of dislocations. It has also been suggested before that the possible influence of the interaction of strain fields of edge dislocations on the periodic arrangement of layers deserves a detailed theoretical investigation [22].

Acknowledgements

The authors are indebted to Dr O. N. Srivastava for substantial help in writing the subsection on stacking faults in Section 3. Also, assistance from Miss Krishna Mazumdar and Mr Prem Chand Jain in the preparation of the manuscript is thankfully acknowledged.

References

1. H. Baumhauer, *Z. Kristallogr.* **50** (1912), 33.
2. H. Baumhauer, *Z. Kristallogr.* **55** (1915) 249.
3. Ajit Ram Verma and P. Krishna, *Polymorphism and Polytypism in Crystals*, Wiley, New York, 1966.
4. G. C. Trigunayat and G. K. Chadha, *Phys. Stat. Sol.(a)* **4** (1971), 9.
5. H. Ungemach, *Bull. Soc. Fr. Mineral. Cristallogr.* **58** (1935), 97.
6. G. S. Zhdanov and Z. V. Minervina, *J. Phys.* **9** (1945), 151.
7. R. W. G. Wyckoff, *Crystal Structures*, vol. 1, Interscience Publishers, U.S.A., 1963.
8. R. C. Evans, *An Introduction to Crystal Chemistry*, Cambridge Univ. Press, U.K., 1966.
9. T. F. W. Barth, *Amer. J. Sci.* **27** (1934), 273.
10. N. W. Thibault, *Amer. Mineral.* **29** (1944), 249; **29** (1944), 327.
11. L. S. Ramsdell, *Amer. Mineral.* **29** (1944), 431; **30** (1945), 519; **32** (1947), 64.
12. L. S. Ramsdell and J. A. Kohn, *Acta Crystallogr.* **4** (1951), 111; **5** (1952), 215.
13. F. C. Frank, *Disc. Faraday Soc.* **5** (1949), 48.
14. W. K. Burton, N. Cabrera, and F. C. Frank, *Phil. Trans. Roy. Soc. (London)* **A243** (1951), 299.
15. Ajit Ram Verma, *Nature* **167** (1951), 939.
16. Ajit Ram Verma, *Phil. Mag.* **42** (1951), 1005.
17. S. Amelinckx, *Nature* **168** (1951), 431.
18. S. Amelinckx, *J. Chim. Phys.* **48** (1951), 475.
19. Ajit Ram Verma, *Crystal Growth and Dislocations*, Butterworths, London, 1953.
20. F. C. Frank, *Phil. Mag.* **42** (1951), 1014.
21. P. Krishna and Ajit Ram Verma, *Phys. Stat. Sol.* **17** (1966), 437.
22. G. C. Trigunayat, *Phys. Stat. Sol. (a)* **4** (1971), 281.
23. Ajit Ram Verma and G. C. Trigunayat, *Solid State Chemistry* (ed. by C. N. R. Rao), Marcel Dekker, U.S.A., 1973, p. 51.
24. P. T. B. Shaffer, *Acta Crystallogr.* **B25** (1969), 477.
25. H. Ott, *Z. Kristallogr.* **61** (1925), 515.
26. G. Hägg, *Ark. Kem. Mineral. Geol.* **16B** (1943), 1.

27. G. S. Zhdanov, *Compt. Rend. Acad. Sci. URSS* **48** (1945) 43.
28. L. Pauling, *Nature of the Chemical Bond*, Cornell University Press, New York, 1945.
29. H. Jagodzinski, *Acta Crystallogr.* **2** (1949), 201.
30. B. B. Zvyagin, *Kristallographia* **6** (1961), 714. English translation in *Soviet Phys. Crystallogr.* **6** (1962), 571.
31. M. Ross, H. Takeda, and D. R. Wones, *Science* **151** (1966), 191.
32. T. Yuasa, T. Tomita and M. Tokonami, *J. Phys. Soc. Japan* **23** (1967), 136.
33. O. Brafman, E. Alexander, and I. T. Steinberger, *Acta Crystallogr.* **22** (1966), 347.
34. M. J. Buerger, *X-Ray Crystallography*, John Wiley, New York, 1953, p. 69.
35. N. F. M. Henry and K. Lonsdale (eds.), *International Tables for X-Ray Crystallography*, Vol. I, Kynoch Press, Birmingham, England, 1952.
36. G. C. Trigunayat, *Bull. Nat. Inst. Scs. India* **14** (1959), 109.
37. R. S. Mitchell, *J. Chem. Phys.* **22** (1954), 1977.
38. G. Honjo, S. Miyake, and T. Tomita, *Acta Crystallogr.* **3** (1950), 396.
39. P. R. van Loan, *Amer. Mineral.* **52** (1967), 946.
40. A. F. Wells, *Structural Inorganic Chemistry*, Clarendon Press, Oxford, 1962, p. 78.
41. W. Kleber and P. Fricke, *Z. Phys. Chem. (Leipzig)* **224** (1963), 353.
42. R. M. Bozorth, *J. Amer. Chem. Soc.* **44** (1922), 2232.
43. R. S. Mitchell, *Z. Kristallogr.* **108** (1956), 296.
44. Z. G. Pinsker, *Acta Physicochim. URSS* **14** (1941), 503.
45. G. K Chadha and G. C. Trigunayat, *Acta Crystallogr.* **23** (1967), 726.
46. G. Lal, G. K. Chadha, and G. C. Trigunayat, *Acta Crystallogr.* **B27** (1971), 2293.
47. R. K. Jain and G. C. Trigunayat, Acta Crystallogr. **A26** (1970), 463.
48. P. C. Jain and G. C. Trigunayat, *Z. Kristallogr.*, in press.
49. G. C. Trigunayat and Ajit Ram Verma, Acta Crystallogr. **15** (1962) 499.
50. V. K. Agarwal, G. K. Chadha, and G. C. Trigunayat, *Z. Kristallogr.* **134** (1971), 161.
51. G. K. Chadha, *Z. Kristallogr.* **139** (1974), 147.
52. O. N. Srivastava, Ph.D. Thesis, Banaras Hindu Univ., Varanasi. 1964.
53. V. K. Agrawal, G. K. Chadha, and G. C. Trigunayat, *Acta Crystallogr.* **B26** (1970), 1911.
54. G. Lal, Ph.D. Thesis, University of Delhi, Delhi, 1972.
55. V. K. Agrawal and G. C. Trigunayat, *Acta Crystallogr.* **24B** (1968), 971.
56. G. K. Chadha, Ph. D. Thesis, University of Delhi, Delhi, 1967.
57. Gyaneshwar, Ph.D. Thesis, Meerut University, Meerut, India, 1973.
58. Gyaneshwar, G. K. Chadha, and G. C. Trigunayat, *Acta Crystallogr.* **B29** (1973), 1791.
59. Gyaneshwar, G. K. Chadha, and G. C. Trigunayat, *Z. Kristallogr.*, in press.
60. G. K. Chadha and G. C. Trigunayat, *Z. Kristallogr.* **126** (1968), 76.
61. R. K. Jain, G. K. Chadha, and G. C. Trigunayat, *Acta Crystallogr.* **B26** (1970), 1785.
62. G. Lal, G. K. Chadha, and G. C. Trigunayat, *Z. Kristallogr.* **134** (1971), 91.
63. Gyaneshwar, G. K. Chadha and G. C. Trigunayat, *Z. Kristallogr.* in press.
64. R. S. Tewari and O. N. Srivastava, *J. Appl. Crystallogr.* **5** (1972), 347.
65. O. N. Srivastava and Ajit Ram Verma, Z. Kristallogr. **117** (1962), 450.
66. O. N. Srivastava and Ajit Ram Verma, *Acta Crystallogr.* **19** (1965), 56.
67. O. N. Srivastava and Ajit Ram Verma, *Acta Crystallogr.* **17** (1964), 260.
68. G. K. Chadha and G. C. Trigunayat, *Acta Crystallogr.* **22** (1967), 573.
69. R. Prasad and O. N. Srivastava, *Z. Kristallogr.* **131** (1970), 376.
70. Gyaneshwar and G. C. Trigunayat, *Phys. Stat. Sol. (a)*, **14** (1972), 191.
71. V. K. Agrawal and G. K. Chadha, *Z. Kristallogr.* **137** (1973), 179.
72. R. Prasad and O. N. Srivastava, *Acta Crystallogr.* **A27** (1971), 259.
73. P. Terpstra and H. G. K. Westenbrink, *Kon. Akad. Wetensch. Proc. Amsterdam* **29** (1926), 431.
74. R. S. Mitchell, *Z. Kristallogr.* **111** (1959), 372.
75. Z. G. Pinsker, L. Tatarinova, and V. Novikova, *Acta Physicochim. URSS* **18** (1943), 378.
76. J. I. Hanoka and V. Vand, *J. Appl. Phys.* **39** (1968), 5288.
77. R. Prasad and O. N. Srivastava, *J. Crystal Growth* **19** (1973), 11.
78. V. K. Agrawal, G. K. Chadha, and G. C. Trigunayat, *Acta Crystallogr.* **A26** (1970), 140.
79. Mahesh Chand and G. C. Trigunayat, *Z. Kristallogr.*, in press.
80. Mahesh Chand and G. C. Trigunayat, *Acta Crystallogr.*, in press.
81. V. K. Agrawal, Ph. D. Thesis, Delhi Univ., Delhi, 1970.

82. S. B. Hendricks and M. E. Jefferson, *Amer. Mineral.* **24** (1939), 729.
83. J. V. Smith and H. S. Yoder, Jr., *Mineral. Mag.* **31** (1956), 209.
84. H. Takeda, *Acta Crystallogr.* **22** (1967), 845.
85. M. Rieder, *Z. Kristallogr.* **132** (1970), 161.
86. G. W. Brindley, B. M. Oughton, and K. Robinson, *Acta Crystallogr.* **3** (1950), 408.
87. B. E. Brown and S. W. Bailey, *Amer. Mineral.* **47** (1962), 819; *ibid.* **48** (1963), 42.
88. R. Steadman and P. M. Nuttall, *Acta Crystallogr.* **16** (1963) 1; *ibid.* **17** (1964), 404.
89. C. Frondel, *Amer. Mineral.* **47** (1962), 781.
90. A. A. Kashaev and V. A. Drits, *Soviet. Phys. Crystallogr.* **15** (1970), 40.
91. Z. G. Pinsker, *Electron Diffraction* (transl. by J. A. Spink and E. Feigl), Butterworths, London, 1953, p. 271.
92. E. A. Monroe, D. B. Sass, and S. H. Cole, *Acta Crystallogr.* **A25** (1969) 578.
93. R. S. Mitchell, *Z. Kristallogr.* **117** (1962), 309.
94. J. A. Kohn and D. W. Eckart, *Acta Crystallogr. Suppl.* **16** (1963), A118.
95. D. R. Hudson, A. F. Wilson, and I. M. Threadgold, *Mineral. Mag.* **36** (1967), 305.
96. M. Deliens and H. Goethals, *Mineral. Mag.* **39** (1973), 152.
97. J. H. Fang and P. D. Robinson, *Amer. Mineral.* **55** (1970), 1534.
98. H. Ungemach, *Bull. Soc. Fr. Mineral. Cristallogr.* **58** (1935), 97.
99. M. C. Bandy, *Amer. Mineral.* **23** (1938), 669.
100. H. Ungemach, *Z. Kristallogr.* **91** (1935), 1.
101. P. J. Traill, *Can. Mineral.* **7** (1963), 524.
102. F. E. Wickman and D. K. Smith, *Amer. Mineral.* **55** (1970), 1843.
103. J. W. Frondel and F. E. Wickman, *Amer. Mineral.* **55** (1970), 1857.
104. B. E. Brown and D. J. Beerntsen, *Acta Crystallogr.* **18** (1965), 31.
105. T. Yuasa, T. Tomita, and M. Tokonami, *J. Phys. Soc. Japan* **23** (1967), 136.
106. Z. Inoue, H. Komatsu, H. Tanaka, and Y. Inomata, *Intern. Conf. on Silicon Carbide*, U.S.A., 1973, Abstract No. 24.
107. Z. Inoue, Y. Inomata, and H. Tanaka, Mineral. *J. Japan* **6** (1972), 486.
108. H. Hartmann, *Kristall und Technik* **1** (1966), 27.
109. S. Kume, E. Kodera, T. Aikami, and J. Kakinoki, *J. Phys. Soc. Japan* **32** (1972), 228.
110. I. Kieflawi and S. Mardix, *Acta Crystallogr.* **B25** (1969), 1195.
111. S. Mardix, I. Kiflawi, and Z. H. Kalman, *Acta Crystallogr.* **B25** (1969), 1586.
112. I. Kiflawi, S. Mardix, and I. T. Steinberger, *Acta Crystallogr.* **B27** (1971), 378.
113. I. Kiflawi, Z. H. Kalman, S. Mardix, and I. T. Steinberger, *Acta Crystallogr.* **B28** (1970), 2110.
114. A. S. Pashinkin, G. N. Tishchenko, I. V. Korneeva and B. N. Ryzhenko, *Kristallografia* **4** (1960), 261.
115. J. B. Goodenough, J. A. Kafalas, and J. M. Longo, *Preparative Methods in Solid State Chemistry* (ed. by P. Hagenmuller), Academic Press, New York, 1972, Chapter 1.
116. E. Tronc and M. Huber, *J. Phys. Chem. Solids* **34** (1973), 2045.
117. R. Haser, M. Pierrot, and C. E. De Broin, C. R. *Acad. Sci. (France)* **51** (1969), 268.
118. L. M. Pant, *Acta Crystallogr.* **B24** (1968), 1205.
119. B. L. Davis and R. L. Petersen, *Crystal Lattice Defects* **1** (1970), 275.
120. R. Siems, P. Delavignette and S. Amelinckx, *Phil. Mag.* **9** (1964), 121.
121. S. Amelinckx and P. Delavignette, *Direct Observation of Imperfections in Crystals*, Interscience Publishers, New York, 1962, p. 295.
122. M. Shiojiri, H. Morikawa, and E. Suito, *J. Appl. Phys. Japan* **6** (1967), 409.
123. R. Prasad and O. N. Srivastava, *J. Phys. D: Applied Phys.* **3** (1970), 91.
124. M. Kronberg, *Acta Meteorol.* **9** (1961), 970.
125. A. J. Forty, *Phil. Mag.* **5** (1960), 787.
126. A. J. Forty, *Phil. Mag.* **6** (1961), 895.
127. W. L. Roth, *Physics and Chemistry of II–VI Compounds*, North-Holland Publishing Company, 1967, p. 119.
128. P. B. Hirsch, R. B. Nicholson, A. Howie, D. W. Pashley, and M. J. Whelan, *Electron Microscopy of Thin Crystals*, Butterworths, London, 1965.
129. S. Amelinckx, *The Direct Observation of Dislocations*, Academic Press, New York, 1964.
130. R. Prasad and O. N. Srivastava, *J. Appl. Phys. Japan* **8** (1969), 810.
131. H. Espig, *Abhandl. Sachs Akad. Wiss. Leipzig, Math.-Phys. Kl.* **38** (1921), 53.

132. D. Lundqvist, *Acta Chem. Scand.* **2** (1948), 177.
133. A. Hayashi, *J. Mineral. Soc. Japan* **4** (1960), 363.
134. W. F. Knippenberg, *Phillips Res. Rep.* **18** (1963), 161.
135. G. A. Bootsma, W. F. Knippenberg, and G. Verspui, *J. Crystal Growth* **8** (1971), 341.
136. H. Jagodzinski and H. Arnold, *Silicon Carbide*, Pergamon Press, 1960, p. 136.
137. H. Jagodzinski, *Soviet Phys. – Crystallogr.* **16** (1972), 1081.
138. L. W. Strock and V. A. Brophy, *Amer. Mineral.* **40** (1955), 94.
139. A. Kremheller, *J. Electrochem. Soc.* **107** (1960), 422.
140. B. J. Skinner and P. B. Barton, Jr., *Amer. Mineral.* **45** (1960), 612.
141. M. Aven and J. A. Parodi, *J. Phys. Chem. Solids* **13** (1960), 56.
142. J. Nickerson, P. Goldberg, and D. H. Baird, *J. Electrochem. Soc.* **110** (1963), 1228.
143: G. Hägg, *Coll. Int. Centre Nat. Rech. Sci.* **10** (1948), 5.
144. J. I. Hanoka, K. Vedam, and H. K. Henisch, *J. Phys. Chem. Solids, Suppl. Crystal Growth*, Pergamon Press, Oxford, 1967, p. 369.
145. G. C. Trigunayat, *Z. Kristallogr.* **122** (1965), 463.
146. Ajit Ram Verma, *Proc. Roy. Soc.* **A240** (1957), 462.
147. A. J. Forty, *Phil. Mag.* **43** (1952), 72.
148. R. S. Mitchell, *Z. Kristallogr.* **108** (1957), 341.
149. G. K. Chadha and G. C. Trigunayat, *J. Phys. Chem. Solids, Suppl. Crystal Growth*, Pergamon Press, Oxford, 1967, p. 313.
150. M. R. Tubbs, *Acta Crystallogr.* **B27** (1971), 857.
151. Krishna Mazumdar and G. C. Trigunayat, to be published.
152. R. S. Mitchell, *Z. Kristallogr.* **109** (1957), 1.
153. P. Krishna and Ajit Ram Verma, *Z. Kristallogr.* **121** (1965), 36.
154. H. E. Buckley, *Z. Electrochem.* **56** (1952), 275.
155. H. E. Buckley, *Proc. Phys. Soc. B* **65** (1952), 578.
156. H. Jagodzinski, *Neue Jahrb. Mineral. Mh.* **3** (1954), 49.
157. H. Jagodzinski, *Acta Crystallogr.* **7** (1954), 300.
158. H. Müller, *Neues Jahrb. Mineral., Abh.* **84** (1952), 43.
159. F. G. Smith, *Amer. Mineral.* **40** (1955), 658.
160. H. Gobrecht, H. Nelkowski, J. W. Baars and G. Brandt, *Int. Lumineszenz Symp., München 1965*, Verlag K. Thiemig, München, 1965, p. 407.
161. J. Baars and G. Brandt, *J. Phys. Chem. Solids* **34** (1973), 905.
162. Y. Inomata and Z. Inoue, *J. Ceram Ass. Japan (Yogyo Kyokai Shi)* **78** (1970), 133 (In Japanese); also several other references of Inomata and his coworkers given in this paper.
163. William Weltner, Jr., *J. Chem. Phys.* **51** (1969), 2469.
164. C. J. Schneer, *Acta Crystallogr.* **8** (1955), 279.
165. W. L. Bragg and E. J. Williams, *Proc. Roy. Soc.* **A145** (1934), 699; **A151** (1935), 540.
166. H. A. Bethe, *Proc. Roy. Soc.* **A150** (1935), 552.
167. K. N. Rai and P. Krishna, *Indian J. Pure Appl. Phys.* **6** (1968), 118.
168. E. Alexander, Z. H. Kalman, S. Mardix, and I. T. Steinberger, *Phil. Mag.* **21** (1970), 1237.
169. L. D. Landau and E. M. Lifschitz, *Lehrbuch der Theoretischen Physik*, V. Akademie, Berlin, 1966.
170. H. Peibst, *Z. Phys. Chem.* **223** (1963), 193.
171. G. S. Zhdanov and Z. V. Minervina, *J. Phys. USSR* **9** (1945), 244.
172. G. S. Zhdanov and Z. V. Minervina, *Acta Physicochim. URSS* **20** (1945), 386.
173. V. Vand and J. I. Hanoka, *Mat. Res. Bull.* **2** (1967), 241.
174. V. Vand, *Phil. Mag.* **42** (1951), 1384.
175. J. I. Hanoka and R. T. Greer, *J. Appl. Phys.* **40** (1969), 3057.
176. V. G. Bhide and Ajit Ram Verma, *Z. Kristallogr.* **111** (1959), 142.
177. S. Mardix and I. T. Steinberger. *Israel J. Chem.* **3** (1966), 243.
178. S. Mardix, Z. H. Kalman, and I. T. Steinberger, Acta Crystallogr. **A24** (1968), 464.
179. S. Mardix, A. R. Lang, and I. Blech, *Phil. Mag.* **24** (1971), 683.
180. F. S. d'Aragona, P. Delavignette and S. Amelinckx, *Phys. Stat. Sol.* **41** (1966), K115.
181. S. Mardix, E. Alexander, O. Brafman, and I. T. Steinberger, *Acta Crystallogr.* **22** (1967), 808.
182. I. T. Steinberger, E. Alexander, Y. Brada, Z. H. Kalman, I. Kiflawi, and S. Mardix, *J. Crystal Growth* **13–14** (1972), 285.

183. S. Mardix and I. T. Steinberger, *J. Appl. Phys.* **41** (1970), 5339.

184. E. Lendvay and P. Kovacs, *Acta Phys. Hung.* **20** (1966), 31.

185. J. P. Golightly and L. J. Beaudin, *Mat. Res. Bull.* **4** (1969), S119.

186. J. A. Powell and H. A. Will, *J. Appl. Phys.* **43** (1972), 1400.

187. B. K. Daniels, *Phil. Mag.* **14** (1966), 487.

188. K. N. Rai, *Phys. Stat. Sol .(a)* **8** (1971), 271.

189. K. N. Rai, O. N. Srivastava, and P. Krishna, *Phil. Mag.* **21** (1970), 1247.

190. J. W. Baars, *II–VI Semiconducting Compounds* W. A. Benjamin, Inc., New York, 1967, p. 631.

191. O. Brafman and I. T. Steinberger, *Phys. Rev.* **143** (1966), 501.

192. W. J. Choyke, D. R. Hamilton, and L. Patrick, *Phys. Rev.* **133** (1964), 1163.

193. G. C. Trigunayat, *Nature* **212** (1966), 808.

194. H. Jagodzinski, *Acta Crystallogr.* **2** (1949) 208; **2** (1949), 298.

195. V. K. Agrawal and G. C. Trigunayat, *Acta Crystallogr.* **A25** (1969), 401.

196. V. K. Agrawal and G. C. Trigunayat, *Acta Crystallogr.* **A25** (1969), 407.

197. V. K. Agrawal and G. C. Trigunayat, *Acta Crystallogr.* **A26** (1970), 426.

198. V. K. Agrawal, *Acta Crystallogr.* **A26** (1970), 567.

199. R. S. Tewari, R. Prasad, and O. N. Srivastava, *Acta Crystallogr.* **A29** (1973), 154.

200. V. K. Agrawal, *Acta Crystallogr.* **A29** (1973), 310.

201. V. K. Agrawal, *Acta Crystallogr.*, in press.

202. Gulzari Lal and G. C. Trigunayat, *J. Crystal Growth* **11** (1971), 177.

203. Gulzari Lal and G. C. Trigunayat, *J. Solid State Chem.* **9** (1974), 132.

204. R. S. Tewari and O. N. Srivastava, *J. Appl. Crystallogr.* **5** (1972), 347.

205. R. S. Tewari and O. N. Srivastava, *Z. Kristallogr.* **L37** (1973), 184.

206. P. Krishna and R. C. Marshall, *J. Crystal Growth* **9** (1971), 319.

207. I. T. Steinberger and S. Mardix, *II–VI Semiconducting Compounds* (ed. by D. G. Thomas), W. A. Benjamin, Inc., New York, 1967, p. 167.

208. G. Shachar, S. Mardix, and I. T. Steinberger, *J. Appl. Phys.* **39** (1968), 2485.

209. M. I. Sokhor, V. G. Kondakov, and L. I. Feldgun, *Soviet Phys. Dokl.* **12** (1968), 749.

210. E. D. Whitney and P. T. B. Shaffer, *High Temps. – High Pressures* **1** (1969), 107.

211. A. R. Kieffer, P. Ettmayer, E. Gugel, and A. Schmidt, *Mat. Res. Bull.* **4** (1969), S153.

212. Gyaneshwar and G. C. Trigunayat, to be published.

213. C. A. Wallace, *Z. Kristallogr.* **126** (1968), 444.

214. R. Srinivasan and V. Parthasarathi, *Z. Kristallogr.*, in press.

215. R. Srinivasan and V. Parthasarathi, *Z. Kristallogr.*, in press.

THERMAL BEHAVIOR OF STACKING FAULTS

SCHWAB S. MAJOR, Jr.

University of Missouri-Kansas City, Kansas City, Mo. 64110, U.S.A.

Abstract. A stacking fault is considered as a local region of strain distinct from the surrounding unstrained lattice. For a stacking fault of area ΔA, if γ is the stacking fault energy, $\gamma \cdot \Delta A = \Delta G > 0$, where ΔG is the change in free energy in the faulted region when the fault is formed. It is shown that the dependence of ΔG on temperature T may be expressed $(1/\Delta G)(\partial \Delta G/\partial T) = -B/kT^2$, where $B > 0$, and k is the Boltzmann constant. Further, for a stacking fault on a close-packed plane in a non-allotropic material, $(1/\gamma)(\partial \gamma/\partial T) < 0$. For an otherwise perfect crystal, a higher strain state exists in the faulted region than in the surrounding unstrained lattice. The lowest energy states of the un-strained lattice do not exist in the faulted region; this factor is dominant in B which includes effects of changes in elastic moduli.

The dependence of the stacking fault energy γ on temperature T, referred to γ_0 at T_0, may be expressed as $\gamma/\gamma_0 = [1 - 2\alpha(T - T_0)] \exp [(9\theta/8) \cdot (T^{-1} - T_0^{-1})]$, where α is the linear thermal expansion coefficient in the fault plane, and θ the Debye temperature of the crystal. The model is applied specifically to silver, in which the stacking fault is pictured as a locally layered structure. Agreement between theory and experiment is very good. The dominant influence on the behavior of γ vs. T results from the lowest vibrational states of the lattice. For a stacking fault surrounded by an un-strained lattice, the lowest states are not accessible to the faulted region. Except at low temperature, the contribution of the lowest states is nearly independent of temperature. The model appears to be valid generally for $T > \theta/2$. It is plausible to expect that the model can be made to predict $(1/\gamma)(\partial \gamma/\partial T)$ for layered structures as graphite.

List of Symbols

ΔA	area of stacking fault (cm²)
a	nearest neighbor separation (cm)
α	lattice linear thermal expansion coefficient
B	lattice potential distortion factor (erg)
β	lattice area thermal expansion coefficient
c_{ij}	lattice elastic coefficient
γ	stacking fault energy (erg cm⁻²)
γ_0	stacking fault energy at 300 K
ΔG	change in free energy when fault is formed (erg)
g	average thermodynamic energy per atom
H_i	Hamiltonian of ith state in ensemble (erg)
$\langle H \rangle$	grand canonical average of all states H_i
$\langle \Delta H \rangle$	grand canonical average of all changes H_i in faulted region
η	enthalpy (erg)
K	fault elastic coefficient
k	Boltzmann constant (erg deg⁻¹)
N_A	number of atoms per unit area in atomic plane
N_f	number of atoms in faulted region
Q	grand partition function
Q_f	grand partition function in faulted region
S	thermodynamic entropy (erg deg⁻¹)
σ	ensemble entropy ($\sigma = S/k$ dimensionless)
T	absolute temperature (deg Kelvin)
τ	ensemble temperature ($\tau = kT$ erg)
θ	Debye characteristic temperature (deg Kelvin)
Θ	ensemble characteristic temperature ($\Theta = k\theta$ erg)

F. Lévy (ed.), Crystallography and Crystal Chemistry of Materials with Layered Structures. 341–357. All Rights Reserved.
Copyright © 1976 by D. Reidel Publishing Company, Dordrecht-Holland.

1. Behavior of Free Energy

1.1. INTRODUCTION

Attempts have been made experimentally to deduce the temperature dependence of stacking fault energy in both bulk [1, 2] and thin [3, 4, 5, 6] specimens of various materials. Interpretation of the experimental results has been difficult and usually has not clarified the behavior in a fundamental way. In bulk specimens stacking faults are not observed directly; consequently the results are statistical over a large number of faults and dependent on the accuracy of stress-strain measurements and the thermal history of the bulk sample. In thin specimens, stacking faults may be observed directly, but the boundary conditions on the fault are known incompletely. Recent work [4] indicates that the geometric behavior of individual faults with varying temperature in silver films is independent of how the fault was formed and of the boundary conditions on the fault. It is reasonable to extend this conclusion to individual faults in bulk specimens. It is then plausible to inquire into the thermal behavior of the stacking fault energy of an individual fault in a bulk specimen even though such a fault cannot be observed directly.

The present investigation attempts to clarify the nature of the temperature dependence of stacking fault energy. This first part applies the methods of statistical thermodynamics to describe the individual stacking fault. The treatment will be restricted to bulk specimens since the boundary conditions in thin specimens are known incompletely. The stacking fault will be pictured intuitively as a region of local strain surrounded by an unstrained lattice. The analysis centers around the free energy in the faulted region and its behavior with changes in temperature.

Section 2 applies the theoretical model to predict the experimentally observed behavior reported for bulk specimens of pure silver. The agreement between the theoretical model and the experimental behavior in silver is very good.

1.2. THERMODYNAMIC BASIS

There have been several attempts to treat stacking faults thermodynamically [7, 8, 9]. In these treatments, the stacking fault energy γ is related to the change in Gibbs free energy ΔG which occurs in the faulted region when a fault is formed over an area ΔA; specifically,

$$\gamma \cdot \Delta A = \Delta G. \tag{1}$$

If a stacking fault exists, then $\gamma > 0$; consequently $\Delta G > 0$. This sign convention agrees with that of other investigators [8].

Since in bulk specimens changes in ΔA cannot be observed, only the explicit functional relation between γ and temperature T can be related directly to experiment. This relation may be expressed

$$(1/\gamma)\,(\partial\gamma/\partial T)_N + (1/\Delta A)\,(\partial\Delta A/\partial T)_N = (1/\Delta G)\,(\partial\Delta G/\partial T)_N. \tag{2}$$

The partial derivatives are evaluated for fixed composition N. The second derivative on the left may be identified as the area thermal expansion coefficient in the fault plane. The evaluation of the derivative on the right is the subject of the subsequent discussion.

The treatment given here differs from those given previously. The earlier treatments were concerned primarily with wide stacking faults when impurities were present; the behavior of ΔA was that observed under those conditions. Temperature effects were obtained incidently. The present paper inquires directly into the behavior of γ as an explicit function of T. Since ΔA is not observable in bulk specimens, its behavior as observed in thin specimens is ignored.

The association of ΔG with the faulted region tacitly implies that the faulted region may be considered distinct from the surrounding crystal lattice. The justification for such an assumption is made in the following section.

1.3. FAULTED REGION

Usually, for stacking faults in a close-packed plane, the fault has been assumed to result in a local second order phase transition from fcc to hcp, or *vice versa* [7, 8]. This approximation requires *ad hoc* the introduction of an entropy term, since for a second order phase transition there is no change in entropy [9]. The change in entropy ΔS which occurs when the fault is formed is given by the functional derivative $(\partial \Delta G/\partial T)_N = -\Delta S$. Experimental values for $(1/\gamma)\,(\partial \gamma/\partial T)$ [1] can be accounted for only if $\Delta S \neq 0$, as the magnitude of $(1/\gamma)\,(\partial \gamma/\partial T)$ is too large to be attributed only to thermal expansion. Therefore, the second order phase transition model is not adequate for a precise thermodynamic analysis of γ.

Another approach has supposed a gradual transition, over a large distance, from unstrained lattice to the fault plane [10]. This description ignores the local effects produced by nonlinear strains in the partial dislocation cores and over the fault plane. While this model appears to be useful in describing the effects of segregation, it does not satisfactorily account for local characteristics of the stacking fault *per se*.

In the present treatment, the nature of the faulted region will be described by considering the fault as a locally strained region surrounded by unstrained lattice. An intuitive picture of such a region is shown in Figure 1. The actual extent of the locally strained region is not known. Quantum mechanical calculations of dislocation cores [11] indicate that the strained region may extend not past the second atomic plane on either side of the fault plane. Accordingly, the faulted region and the surrounding unstrained lattice will be taken as two distinct parts.

If the stacking fault has a width not less than six times the nearest neighbor separation, [12] then the faulted region may be pictured as a locally layered structure. Examples are: deformation faults in graphite, growth faults in silver, and the hcp↔fcc phase change at 690 K in cobalt. However, even when pictured as a locally layered structure, local strains must be part of the picture so that $\Delta S \neq 0$.

When the stacking fault width is less than six times the nearest neighbor separation,

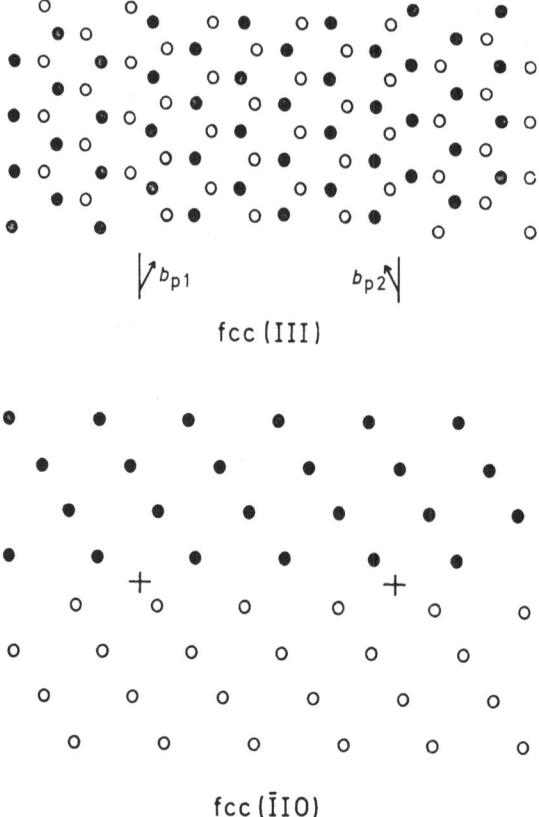

fcc (III)

fcc (Ī10)

Fig. 1. Stacking fault produced by extended screw dislocation in fcc lattice. View normal to (111) plane: Plane of dots $\sqrt{6}\,a/3$ above plane of circles, where a is nearest neighbor separation. Vertical lines below array identify partial dislocation axes. Burgers vectors $\mathbf{b}_{p1}$, $\mathbf{b}_{p2}$ drawn to scale. View normal to (Ī10) plane: At left end, array of dots $a/4$ above plane of page, array of circles $a/4$ below. In center region, both dots and circles in plane of page. At right end, array of dots $a/4$ below plane of page, array of circles $a/4$ above. Crosses between arrays identify partial dislocation axes. It is helpful to hold plane of sketch horizontally at nearly eye level and examine the figure from various angles in this plane. Note that in immediate neighborhood of fault plane the fcc (Ī10) plane resembles an hcp (11Ī0) plane.

the picture of localized layering becomes fuzzy. The localized nonlinear strain model still remains valid.

Stacking faults which may be observed by electron microscopy usually have widths much greater than six times the nearest neighbor separation. Thus, most directly observable stacking faults may be considered as locally layered structures.

Usually, the faulted region is much smaller in extent than the surrounding unstrained lattice. Thus, the surrounding lattice may be taken as an isothermal reservoir. The formation of the fault will be considered an isothermal process. Discussion will be restricted to thermodynamically reversible behavior of the faulted region.

A change in entropy $\Delta S \neq 0$ occurs in the local region when the fault is formed. However, since the faulted region is not isolated, it is not known *a priori* whether ΔS

is positive or negative. Intuitively, it may be argued that the strains along the associated partial dislocation axes are nonlinear. These nonlinear strains and the transition from strained state at the fault plane to unstrained lattice outside the faulted region produce a configuration of disordered phase. On this basis, it is expected that $\Delta S > 0$, in which case ΔG decreases with an increase in temperature. This intuitive prediction is borne out in the subsequent analysis.

Equations (1) and (2) are assumed to apply to the faulted region as described here. The change in Gibbs free energy ΔG is for the entire local region of strain. The value of γ represents the average increase in free energy per unit area of fault when the fault is formed.

1.4. FREE ENERGY OF FAULTED REGION [13]

The purpose of this section is to deduce the general behavior of ΔG in the faulted region when the temperature changes. The argument mainly is analytical using ideas of statistical thermodynamics. The physical significance of the analysis given here is discussed in the section immediately following this one.

An open system in contact with an isothermal reservoir is properly described by the grand canonical ensemble of statistical mechanics. In the subsequent analysis, it is convenient to introduce the ensemble temperature $\tau = kT$ (in units of energy) and the ensemble entropy $\sigma = S/k$ (in dimensionless units). In these expressions, k is the Boltzmann constant, T the absolute temperature, and S the thermodynamic entropy. The temperature derivative transforms as $\partial/\partial\tau = (1/k)(\partial/\partial T)$. All temperature derivatives are understood to be evaluated for fixed composition; the usual subscripts will be dropped.

It will be assumed that the faulted region may be represented by a grand canonical distribution of R states; associated with the ith state is a Hamiltonian H_i [14].* It is not necessary to know the explicit form for H_i in the present treatment. The partition function may be expressed [9]

$$Q = \sum_{i=1}^{R} \exp(-H_i/\tau). \tag{3}$$

When the faulted region is formed isothermally, the change produced in Q is

$$\Delta Q = \sum_{i=1}^{R} \exp[-(H_i + \Delta H_i)/\tau] - Q. \tag{4}$$

The Gibbs free energy for a grand canonical distribution is [9]

$$G = -\tau \ln Q. \tag{5}$$

For an isothermal process, the change in free energy is

$$\Delta G = -\tau \ln[1 + (\Delta Q/Q)]. \tag{6}$$

---

* In general, the Hamiltonian is a function of both configuration and composition.

Equations (4) and (6) may be combined to obtain

$$\exp(-\Delta G/\tau) = (1/Q) \sum_{i=1}^{R} \exp[-(H_i + \Delta H_i)/\tau].$$ (7)

The left side of Equation (7) may be expanded as a Taylor series, and each term $\exp(-\Delta H_i/\tau)$ under the sum may be expanded similarly. The result is

$$\sum_{n=0}^{\infty} (-1)^n (\Delta G)^n /n! \tau^n = \sum_{n=0}^{\infty} (-1)^n \langle (\Delta H)^n \rangle /n! \tau^n.$$ (8)

Since ΔG is produced by an isothermal process, the value τ in Equation (8) is that for the ensemble to be in thermal equilibrium with an isothermal reservoir. Therefore, τ has a unique value; and terms of corresponding order may be equated to obtain

$$(\Delta G)^n = \langle (\Delta H)^n \rangle.$$ (9)

The present discussion is involved only with the first order term,

$$\Delta G = \langle \Delta H \rangle.$$ (10)

The meaning of Equation (10) is obtained by comparing

$$\Delta G = \Delta \eta - \tau \Delta \sigma,$$ (11)

where $\Delta \eta$ is the change in enthalpy and $\Delta \sigma$ the change in entropy produced when the faulted region is formed. The net effect of these changes is $\langle \Delta H \rangle$, which is the grand canonical average of all changes ΔH_i in the Hamiltonian of the faulted region. Since by Equation (1) $\Delta G > 0$, then also $\langle \Delta H \rangle > 0$.

For a reversible change in temperature, the temperature dependence of ΔG may be investigated through Equation (10). The temperature derivative is

$$\partial \Delta G/\partial \tau = -(1/\tau^2)(\langle H \rangle \langle \Delta H \rangle - \langle H \cdot \Delta H \rangle).$$ (12)

On the right side, $\langle H \rangle$ is the grand canonical average of all states H_i in the local region before the fault is formed, and $\langle \Delta H \rangle$ the average over all changes which result from strain when the fault is formed in the same local region. For a crystal lattice $\langle H \rangle > 0$.* It was shown earlier that $\langle \Delta H \rangle > 0$. If $\langle H \cdot \Delta H \rangle > 0$, then the Schwartz inequality [15] may be applied to yield

$$\langle H \rangle \langle \Delta H \rangle - \langle H \cdot \Delta H \rangle \geqslant 0.$$ (13)

If $\langle H \cdot \Delta H \rangle \leqslant 0$, the inequality still holds. Consequently $\partial \Delta G/\partial \tau \leqslant 0$. The equality may be eliminated by appealing to the functional derivative $\partial \Delta G/\partial \tau = -\Delta \sigma$ and the fact that $\Delta \sigma \neq 0$ in the faulted region. Therefore,

$$\partial \Delta G/\partial \tau < 0.$$ (14)

---

* The sign of $\langle H \rangle$ can be justified intuitively by imagining the lattice to consist of individual oscillators. The hamiltonian of each oscillator is positive. Thermodynamic justification for $\langle H \rangle > 0$ may be obtained by following the arguments in Reference [9], pp. 45, 184, 189, 192.

In the derivation of the inequality (14), no specific structure was assumed for either the crystal lattice or the faulted region. Therefore, the inequality applies generally to any local region of strain which exists in thermodynamics equilibrium with a surrounding unstrained lattice.

The functional relations established for ΔG in Equation (10) and $\partial \Delta G/\partial \tau$ in Equation (12), by themselves, are quite general. The result $\partial \Delta G/\partial \tau < 0$ is obtained only after consideration of the crystal lattice and the faulted region. Generally, for a solid, G decreases as τ increases [16]. Therefore, the result $\partial \Delta G/\partial \tau < 0$ means that the value of G in the faulted region decreases more rapidly than the value of G in the surrounding unstrained lattice.

1.5. SIGNIFICANCE OF NONVANISHING TEMPERATURE DEPENDENCE

From Equation (10) and (12) may be obtained

$$(1/\Delta G)(\partial \Delta G/\partial \tau) = -B/\tau^2 . \tag{15a}$$

$$B = \langle H \rangle - \langle H \cdot \Delta H \rangle / \langle \Delta H \rangle > 0 . \tag{15b}$$

To determine the physical significance of B, it is helpful to identify a new ensemble with partition function

$$Q_f = \sum_{i=1}^{R} (\Delta H_i/\tau) \exp(-H_i/\tau) . \tag{16}$$

On this ensemble,

$$\langle H \cdot \Delta H \rangle / \langle \Delta H \rangle = \langle H \rangle_f . \tag{17}$$

The ensemble average $\langle H \rangle_f$ is taken over the same set of single particle states H_i, $i = 1, 2, ..., R$, as that for $\langle H \rangle$. The difference is in the factors $\Delta H_i/\tau$ weighting corresponding states H_i according to changes which result from displacement of atoms from normal lattice positions in the faulted region. Now,

$$B = \langle H \rangle - \langle H \rangle_f > 0 . \tag{18}$$

The quantity $\langle H \rangle$ is a property of the unstrained lattice and $\langle H \rangle_f$ a property of the faulted region. The partition functions Q and Q_f are defined on the same set of single particle states H_i, $i = 1, 2, ..., R$. In Q_f, the weight factors $\Delta H_i/\tau$ change the distribution among states H_i as a result of the elastic limit having been exceeded when slip occurred to form the fault. The difference $\langle H \rangle - \langle H \rangle_f$ may be attributed to distortion in the lattice potential configuration over the fault plane. Figure 2 shows how this distortion might appear in one dimension.

Since the elastic limit of the crystal has been exceeded, the faulted region exists in a strain state different from that of the surrounding unstrained lattice. If the surrounding unstrained region is large in extent, the potential minimum will be higher in the faulted region than in the unstrained lattice. As a result, the lowest level vibrational states which are part of $\langle H \rangle$ do not contribute to $\langle H \rangle_f$; this statement is tantamount to saying that $\Delta H_i = 0$ for the lowest level states in Equation (16).

Between partial dislocations (see Figure 1), in the two atomic planes contiguous to the fault plane, the atoms uniformly occupy a set of interstices alternate to the unfaulted set [17]. As a result of this uniformity, the distortion of the potential configuration probably is not large and may be negligibly small in the region between the partials.* In this region, the difference expressed in Equation (18) occurs essentially in the low level states excluded from $\langle H \rangle_f$.

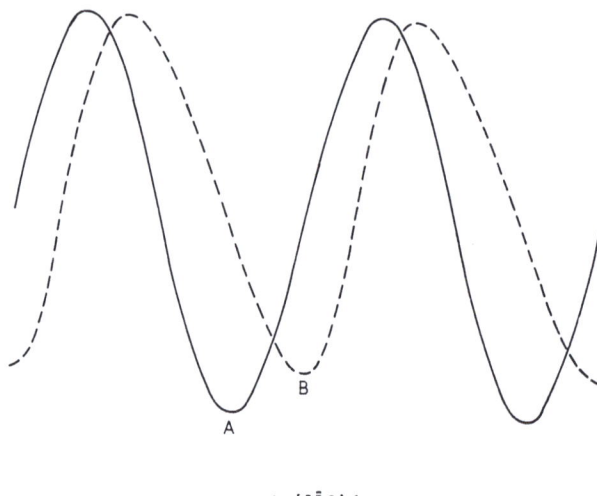

$[1\bar{1}0]$ fcc

Fig. 2.   Lattice potential configuration along line of nearest neighbors in (111) plane fcc. Solid line: configuration along $[1\bar{1}0]$ in unstrained lattice. Minima represent positions of atoms separated by distance a. Dashed line: configuration in plane contiguous to fault plane after fault is formed. Minimum at B in strained state at higher potential level than minimum at A in unstrained state. Displacement from A to B: $a/8$ along $[1\bar{1}0]$ (in plane of page); $a/8\sqrt{3}$ along $[\bar{1}\bar{1}2]$ (perpendicular to plane of page). In atomic plane on other side of fault plane, displacement equal in magnitude but opposite in direction. Cf. Figure 1. Shift from A to B corresponds to shift of dots by amount $(1/2)|\mathbf{b}_p| = a/4\sqrt{3}$ along Burgers vector $\mathbf{b}_{p1}$. Shift of circles equal in magnitude but opposite in direction.

1.6. REMARKS ON TEMPERATURE DEPENDENCE

A formal evaluation of $(1/\gamma)(\partial\gamma/\partial T)$ may be obtained by substituting the right side of Equation (15a) into Equation (2). Thus, since $\tau = kT$,

$$(1/\gamma)(\partial\gamma/\partial T) = -B/kT^2 - \beta, \tag{19}$$

where β is the lattice area thermal expansion coefficient in the fault plane.

The coefficient $B > 0$ represents an average value per atom for all low energy states not accessible to the faulted region (see Section 2). Broadly, B includes the effects of

---

* As the fault narrows so that the partial dislocation cores interact strongly, the distortion of the potential configuration may become large. Generally, if γ is large, the potential well is deep with steep sides; the potential configuration under strain may be distorted greatly. If γ is small, the potential well is shallow; wide stacking faults may occur in which the potential configuration is distorted moderately.

the elastic moduli on $(1/\gamma)\,(\partial\gamma/\partial T)$. Numerical evaluation of B may be possible through the right side of Equation (15b). Present efforts are being directed toward explicit evaluation of $\langle H \rangle$ and $\langle H \rangle_f$.

In general, it is expected that B will be a function of T; consequently the first term on the right of Equation (19) is not readily integrable. It will be found, at least for silver, that an especially simple model permits an approximation for B nearly independent of temperature. This simplification is discussed in Section 2.

For a close-packed plane, $\beta > 0$. Thus, for stacking faults on a close-packed plane in a nonallotropic material, Equation (19) states that the stacking fault energy decreases monotonically as the temperature increases. That is,

$$(1/\gamma)\,(\partial\gamma/\partial T) < 0. \tag{20}$$

The behavior of fault geometry, other than thermal expansion, has been ignored as this behavior cannot be observed in bulk specimens. Such behavior may be observed in thin specimens.

In thin specimens, the boundary conditions on the fault may be quite different than in bulk specimens, [4]* as a result of the proximity of the specimen surfaces. One condition that may exist in thin specimens is that the lattice surrounding the fault may exist in a strained state. In this situation, the lowest vibrational states in the faulted region may not be accessible to the surrounding lattice. The result would be an apparent inversion of the T dependence of γ as indicated by the observed behavior of the fault geometry.

An example of inverted behavior for γ *vs.* T has been observed in cobalt [6]. This inversion can be accounted for as a consequence of lattice strains induced as the cobalt changes from hcp below 690 K to fcc above.

Further investigation is underway to work out a correlation between the observed behavior of fault geometry in thin specimens and the behavior of γ in bulk specimens. The behavior of nonallotropic materials also is being considered.

The investigation reported here was concerned primarily with the T dependence of γ. No attempt was made to calculate specific values for $\gamma(T)$. An attempt to calculate γ has been made by Statz [18] using quantum mechanical techniques. This latter attempt ignored temperature dependence. The middle ground between the two approaches would involve quantum statistical mechanics, but such an approach at the present time appears formidable.

2. Application of Model

2.1. STATEMENT OF PROBLEM

Section 1, presented a model to predict the behavior of stacking fault energy γ with

---

* Large changes in ΔA, as observed for graphite (Reference [5]) and cobalt (Reference [6]) may be attributed to factors other than thermal expansion; e.g. local anisotropy produced by nonlinear strains and changing boundary conditions. See Reference [19], pp. 57–68, 74–75.

varying temperature T. For bulk specimens of pure nonallotropic materials it was determined that

$$(1/\gamma)(\partial\gamma/\partial T) = -B/kT^2 - \beta,\tag{21}$$

where $B>0$ generally is a function of temperature, k the Boltzmann constant, and β the lattice area thermal expansion coefficient in the fault plane. The objective in Section 2 is to apply that model to predict experimentally observed results.

Of the pure metals, silver has been most widely investigated to determine the behavior of stacking fault energy with temperature. For bulk specimens (172 to 873 K), the temperature dependence was deduced from measurements of τ_{III}, the critical resolved shear stress at the onset of stage III hardening [1, 2]. In thin specimens (300 to 740 K), both mechanical [3, 4] and growth [4] faults have been observed by transmission electron microscopy. Attempts were made to relate the temperature dependence of stacking fault energy to observed changes in the radii of extended nodes and in the widths of extended ribbons. The problem in thin specimens is compounded by boundary conditions peculiar to thin specimens [4, 19] and the influence of film structure by techniques of preparation. However, the results of observations on thin specimens of silver appear to correlate with the results deduced for bulk specimens. It has been concluded [1, 2, 3, 4] that in pure silver the stacking fault energy decreases monotonically as the temperature increases.

Since on a close-packed plane $\beta>0$, Equation (21) predicts that $(1/\gamma)(\partial\gamma/\partial T)<0$. Thus, there is general agreement between the model developed in Section 1 and the experimental results for silver.

Other experimental attempts have included observations of deformation faults in graphite platelets (170 to 890 K) [5] and extended nodes in cobalt foils (293 to 983 K) [6]. In graphite, only a qualitative result was obtained; the result was believed not inconsistent with the results for silver. Quantitative results were reported for cobalt. Cobalt is allotropic, changing from hcp below 690 K to fcc above. How the phase change about 690 K may affect the behavior of $(1/\gamma)(\partial\gamma/\partial T)$ in cobalt was indicated in Section 1. For neither graphite nor cobalt are the experimental results as conclusive or as widely correlated as for silver.

Silver is chosen as the material against which to check the model of Section 1. Besides being the only material for which adequate bulk data is available, silver is one of the simplest metals, being non-allotropic and monovalent.

2.2. EXPLICIT TEMPERATURE DEPENDENCE

In this section, the right side of Equation (21) is put in a readily integrable form. The quantities B and β are evaluated explicitly. An expression is derived for γ at T referred to γ_0 at T_0.

In Section 1, it was shown that

$$(1/\Delta G)(\partial\Delta G/\partial T)_N = -B/kT^2,\tag{22}$$

where $\Delta G>0$ is the increase in Gibbs free energy which occurs in the faulted region when the fault is formed. The derivative is evaluated for fixed composition N. The dominant contribution to B is by the lowest energy states which exist in the unstrained lattice but not in the faulted region. In this paper, it will be assumed that B is a function only of the lowest vibrational states of the unstrained lattice.

An especially simple model may be developed if two simplifying assumptions are made: (a) The lattice strain occurs predominantly in the atom planes contiguous to the fault plane. (b) The stacking fault is wide enough that effects of the partial dislocation cores can be neglected. In this simple model, the stacking fault is pictured as a localized layered structure. On these assumptions, all atoms in the same plane may be pictured as having been displaced uniformly. Since the temperature derivative of ΔG is evaluated for fixed composition, $(1/\Delta G)(\partial \Delta G/\partial T) \simeq (1/\Delta g)(\partial \Delta g/\partial T)$, where g is the average thermodynamic potential per atom, and Δg the change in potential when the fault is formed. Thus, the value of the left side of Equation (22) for a sufficiently wide faulted region may be approximated as the average value for a single atom in that region (see Figure 1 for a pictoral illustration of the region described here).

If the lattice distortion [1] is small enough in the faulted region, the dominant contribution to B may be assumed to be the zero point energy for the unstrained lattice.* Except at low temperature, the average zero point energy per atom may be expressed in terms of the Debye characteristic lattice temperature, such that

$$B \simeq (9/8)\, k\theta. \tag{23}$$

Since the variation of Θ with temperature is small, it will be assumed that B essentially is constant independent of temperature. The expression in Equation (23) is particularly applicable to wide stacking faults (on the order of $10^{-3}\ \mu$, or larger) corresponding to relatively small values of γ (less than $\sim 10^2$ erg cm^{-2}).

The lattice area thermal expansion coefficient in the fault plane is

$$\beta = (1/\Delta A)(\partial \Delta A/\partial T)_N, \tag{24}$$

where ΔA is the area of the stacking fault. The derivative is evaluated for fixed composition N. It is convenient to express the right side of Equation (24) in terms of lattice parameters. The simplest relations occur in the close-packed plane, on which stacking faults commonly occur.

In a close-packed plane, there are $2/\sqrt{3}\ a^2$ lattice points per unit area, where a is the nearest neighbor separation. Therefore, if the total number of atoms is N_A in a single plane of area ΔA, then $\Delta A = N_A\,(\sqrt{3}\ a^2/2)$. Consequently,

$$\beta = (2/a)(\partial a/\partial T)_N. \tag{25}$$

The derivative $(1/a)(\partial a/\partial T)_N$ may be identified as the linear expansion coefficient along a line of nearest neighbors.

---

* This assumption was discovered to be plausible when, as a first approximation, B was assumed independent of temperature and Equation (21) was integrated and empirically fit to the experimental data for bulk silver. The fit yielded a value for B very close to the Debye zero point energy.

By means of Equations (23) and (25), Equation (21) may be put in a readily integrable form:

$$(1/\gamma)\,(\partial\gamma/\partial T) = -\,(9/8)\,(\theta/T^2) - (2/a)\,(\partial a/\partial T). \tag{26}$$

The integration is performed with reference to a temperature T_0 to obtain

$$\gamma/\gamma_0 = (a_0/a)^2 \exp\left[(9\theta/8)\,(T^{-1} - T_0^{-1})\right]. \tag{27}$$

The values γ, a are evaluated at T, referred to γ_0, a_0 at T_0.

The temperature dependence of a may be approximated as $a = a_0[1 + \alpha(T - T_0)]$, where α is the linear thermal expansion coefficient at temperature T (referred to T_0). Usually, $\alpha(T - T_0) \ll 1$ so that a very good approximation is $(a_0/a)^2 \simeq [1 - 2\alpha(T - T_0)]$. It is now possible to write Equation (27) as

$$\gamma/\gamma_0 = [1 - 2\alpha\,(T - T_0)]\,\exp\left[(9\theta/8)\,(T^{-1} - T_0^{-1})\right]. \tag{28}$$

Equation (28) is the working equation by which the model of Section 1 will be applied to predict the behavior γ vs. T for pure silver in bulk.

Values for α may be computed directly from the data of Simmons and Balluffi [20]. A very good empirical approximation for silver is $\alpha = \alpha_0[1 + (0.176)\ln(T/T_0)]$, where $\alpha_0 = 1.880 \times 10^{-5}$ (deg K)$^{-1}$ at $T_0 = 300$ K.

The Debye characteristic temperature for silver is $\theta = 226.2$ K [21]. Therefore, $9\theta/8 = 254.5$ K.

2.3. STACKING FAULT ENERGY IN SILVER

Values of γ/γ_0 may now be computed with Equation (28). Table I lists computed values and experimental values of γ/γ_0 taken from the data of Buhler et al. [1]. The computed values determine the curve γ/γ_0 vs. T displayed against the experimental data in Figure 3.

At 295 K, the $\tau_{\rm III}$ model [1] for silver yielded $\gamma_0 = 33 \pm 10$ erg cm^{-2}. The accepted value of γ_0 for silver lies between 20 and 40 erg cm^{-2} [22] some workers [3, 23] believe the lower values to be more accurate. However, the behavior of the ratio γ/γ_0 should be independent of the model and the particular numerical value assigned to γ_0.

The computed curve agrees well with the empirical data. Only at 573 K does the computed value lie outside the experimental uncertainty. The experimenters [1] concluded that the behavior of γ is dominated by that of the shear modulus in the (111) plane. This behavior is monotonically decreasing with increasing temperature [1, 24]. On this basis, it is plausible to attribute the large discrepancies to scatter in the $\tau_{\rm III}$ data.

The behavior γ/γ_0 vs. T is influenced dominantly by the behavior of an elastic coefficient K associated with the fault [1, 4, 5, 6].* The general effects of $(1/K)\,(\partial K/\partial T)$ are included in the coefficient $B \simeq (9/8)k\theta$. Usually, K is computed from elastic

---

* The fault coefficient K behaves linearly with the c_{ij} provided the fault plane is nearly isotropic.

TABLE I

Values of γ/γ_0 for bulk specimens of silver at selected temperatures

T (deg K)	γ^a (erg cm^{-2})	$\gamma/\gamma_0{}^a$ (experiment)	$\gamma/\gamma_0{}^b$ (theory)	$\Delta(\gamma/\gamma_0)^c$ (discrepancy)
100			5.489	
172	57	1.75	1.888	−0.24
200			1.533	
268	46	1.42	1.117	+0.30
295	33	1.015	1.015	0.00
300			1.000	
372	29	0.89	0.846	+0.04
473	25	0.77	0.728	+0.04
500			0.706	
573	33	1.02	0.660	+0.36
673	28	0.86	0.615	+0.24
773	21	0.65	0.583	+0.06
873	18	0.55	0.558	−0.01
1000			0.535	
1200			0.507	
1234			0.503	

[a] Reference 1. Value at 295 K normalized to 1.015. Uncertainty ± 0.31.
[b] Computed from Equation (28); $\theta = 226.2$ K (Reference [20]).
[c] $\Delta(\gamma/\gamma_0) = (\gamma/\gamma_0)_{\text{experimental}} - (\gamma/\gamma_0)_{\text{theoretical}}$.

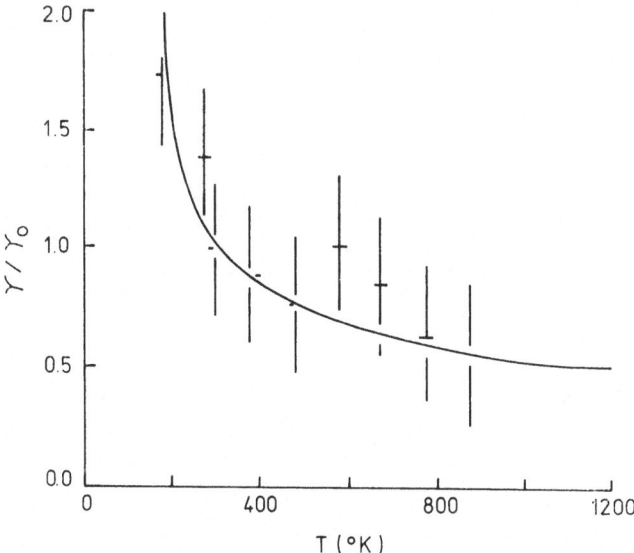

Fig. 3. Stacking fault energy of silver as a function of temperature, displayed as γ/γ_0 vs. T, where γ is the value at T and γ_0 the fiducial value at 300 K. Solid curve computed from Equation (28) with $\theta = 226.2$ K. Individual data points taken from Buhler et al. (Reference [2]).

coefficients c_{ij} for the unstrained lattice. However, the c_{ij} in the faulted region may not have the same value as in the unstrained lattice [27]. It is expected that the temperature dependence of the c_{ij} also may be different in the faulted region than in the unstrained lattice. Such a difference may account for the exponential behavior which exists for γ/γ_0 vs. T rather than linear behavior.

The model is not expected to accurately predict γ/γ_0 vs. T at low temperature as the Debye zero point energy is a high temperature approximation. The actual behavior of γ/γ_0 below approximately $\theta/2$ may be estimated by considering the behavior of the fault elastic coefficient K. The observed behavior of the c_{ij} [26, 28] (unstrained lattice) below 40 K shows that $\partial c_{ij}/\partial T = 0$ as $T \to 0$. Even in the faulted region, it may be expected that $\partial K/\partial T = 0$ as $T \to 0$. The behavior of γ as $T \to 0$ is expected to be similar. Figure 4 displays qualitatively the expected behavior γ/γ_0 vs. T below $\theta/2$.

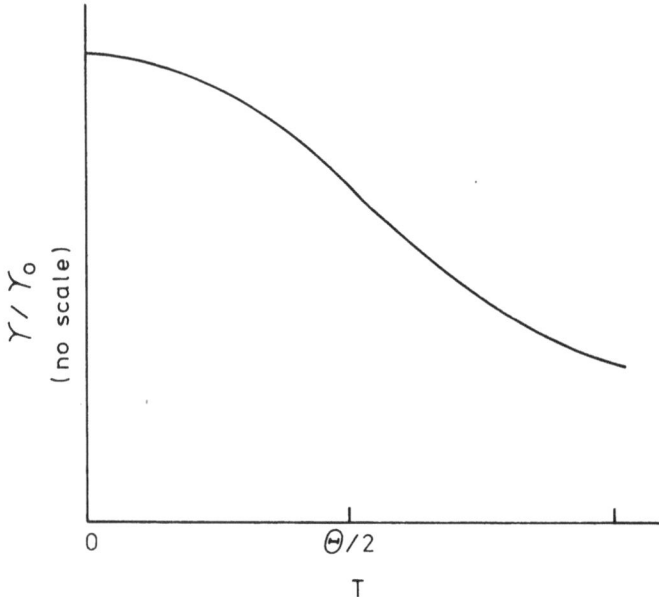

Fig. 4.   Probable behavior of γ/γ_0 vs. T below $T \simeq \theta/2$, as inferred from behavior of elastic coefficients in silver (Reference [13]). No scale is indicated along ordinate since exact values for γ/γ_0 are not known in this temperature range.

An additional estimate on the low temperature behavior of γ/γ_0 may be obtained by computing the zero state value $\gamma(0) = \Delta G(0)/\Delta A(0)$. From Equation (23) is obtained $\Delta G(0) \simeq N_f(9/8) k\theta$, where N_f is the number of atoms in the faulted region. It is plausible to assume that the predominant contribution to $\Delta G(0)$ results from displacements in the two atomic planes adjacent to the fault plane. Thus $N_f/A \simeq 2 N_A'$, where N_A' is the number of atoms per unit area in the atomic plane. For a stacking fault on the (111) plane $\gamma(0) \simeq (3\sqrt{3}/2)\,(k\theta/a^2)$.

For silver, $k = 312 \times 10^{-16}$ erg and $a^2 \simeq 8.20 \times 10^{-16}$ cm^2 at 0 K. These values yield $\gamma(0) \simeq 98.8$ erg cm^{-2}.

At $T=\theta/2$, $\gamma/\gamma_0=4.119$ for silver. If it is assumed that $\gamma(\Theta/2)\simeq\gamma(0)$, then $\gamma_0=24.0$ erg cm^{-2}. This value lies within the accepted range of values at room temperature. This result and the close agreement between the calculated and experimental results above $T=\theta/2$ suggest that the knee of the curve in Figure 4 may break rather sharply below $T=\theta/2$. Thus the model may be assumed to yield accurate results for silver in the temperature range from $T\gtrsim113$ K to the melting point.

2.4. CONCLUSION

A model has been presented which predicts the behavior of stacking fault energy with changes in temperature in bulk specimens. The model was applied to silver as an example of a monovalent, non-allotropic material. The stacking fault was pictured as a localized layered structure in a nonlinear strain state. The quantitative agreement between theory and experiment is very good.

The numerical evaluation of B in Equation (21) was approximated as the Debye zero point energy per atom. Application of the model to other metals, more complicated in electronic structure than silver, is expected to require modification. However, the behavior of γ/γ_0 vs. T has not been investigated adequately in other metals (bulk specimens) so that any modification of the present model would amount to no more than speculation at this point. With this qualification, values for $\gamma a^2/\gamma_0 a_0^2=\Delta G/\Delta G_0=$ $=\exp(9\theta/8)(T^{-1}-T_0^{-1})$ are listed in Table II for platinum, copper, gold, and aluminum. A more precise description ultimately must depend on explicit evaluation of B as given in Section I.

The model, as applied to silver, excludes behavior at low temperature below approximately one-half the Debye temperature. This restriction enters through the Debye zero point energy approximation. The low temperature behavior of γ/γ_0 is believed

TABLE II

Values of $\gamma a^2/\gamma_0 a_0^2 = \Delta G/\Delta G_0$ for bulk specimens of platinum, copper, gold, aluminum at selected temperatures.

T (K)	Platinum (2047) (229)	Copper (1356) (343)	Gold (1336) (162.4)	Aluminum (933) (428)
100	5.573	13.10	3.382	–
200	1.573	1.906	1.356	3.677
300	1.000	1.000	1.000	1.000
500	0.7093	0.5979	0.7842	0.3526
900	0.5640	0.4243	0.6660	0.1752
1300	0.5161	0.3719	0.6260	–
2000	0.4890	–	–	–

Note: Values of $\gamma a^2/\gamma_0 a_0^2 = \Delta G/\Delta G_0$ computed from Equation (27). Numbers in parentheses under name of each metal are, respectively, melting point and Debye temperature (Reference [21]) in deg. Kelvin. Below 300 K, $\Delta G/\Delta G_0 > \gamma/\gamma_0$; above 300 K, $\Delta G/\Delta G_0 < \gamma/\gamma_0$. At $T\simeq\theta/2$ and at melting point, values for $\Delta G/\Delta G_0$ are within 10% of values for γ/γ_0.

to follow that of the elastic coefficients. The values of the coefficients in the faulted region may be different than usually reported for the crystal lattice. Present data are inadequate for more than a crude approximation to the low temperature limit on the validity of the model.

The empirical success of the proposed model in predicting the behavior of γ/γ_0 vs. T in bulk specimens of silver appears to validate the assumptions on which the model was based. Therefore, it is plausible to restate the assumptions here as conclusions: The behavior of $\gamma(T)$ is influenced dominantly by the lowest energy states which exist in the unstrained lattice but not in the faulted region. For the simplest metals, having small values of γ (wide faulted area), this contribution by the lowest energy states is nearly independent of temperature.

To test these conclusions, experimental data is needed for bulk specimens of metals such as platinum, gold, copper, and aluminum. This selection provides a wide range of values for γ and different electronic structures, and the results should help estimate more closely the range of valid limits for the model.

Appendix – Graphite

Data for the thermal behavior of stacking faults in graphite are inadequate for a precise quantitative treatment. Values for the thermal expansion coefficient and the elastic moduli depend on the particular structure. The structure depends on how the graphite was formed. Direct observation of stacking faults in thin single crystals of pyrolytic graphite have indicated that generally $(1/\gamma)\,(\partial\gamma/\partial T)<0$ [Reference 5].

The statistical thermodynamic model developed here was applied successfully by picturing stacking faults as localized layered structures. The results agree with the general behavior of the stacking fault energy in graphite. It is plausible to expect that this model can be made to predict the temperature dependence of stacking fault energy for layered structures such as graphite.

Acknowledgment

The ideas presented here evolved from an experimental problem suggested by Dr J. C. Grosskreutz, Midwest Research Institute. In the early formulative stages, I had several helpful discussions with Dr R. D. Dragsdorf, Kansas State University.

References

1. S. E. Buhler, K. Lucke, and F. W. Rosenbaum, *Phys. Stat. Sol.* **3** (1963), 886.
2. M. Ahlers and P. Haasen, *Z. Metallk.* **53** (1962), 302.
3. P. C. J. Gallagher, *J. Appl. Phys.* **39** (1968), 160.
4. S. S. Major, Jr. and J. C. Grosskreutz, *Japan J. Appl. Phys.* **7** (1968), 574.
5. S. S. Major, Jr. and J. C. Grosskreutz, *J. Appl. Phys.* **37** (1966), 4275.
6. T. Ericsson, *Acta Met.* **14** (1966), 853.
7. H. Suzuki, *J. Phys. Soc. (Japan)* **17** (1962), 322.
8. R. Dewitt and R. E. Howard, *Acta Met.* **13** (1965), 655.

9. L. D. Landau and E. M. Lifshitz, *Statistical Physics*, Addison-Wesley, Reading, 1958, pp. 105–106, 437.
10. T. Ericsson *Acta. Met.* **14** (1966), 1073.
11. V. Gallina, C. P. Galotto and M. Omini, *Phys. Stat. Sol.* **8** (1965), 239.
12. S. S. Major, Jr., Ph. D. Dissertation, Kansas State University, 1967, pp. 154–156.
13. S. S. Major, Jr., Ph. D. Dissertation, Kansas State University, 1967, pp. 79–101.
14. K. Huang, *Statistical Mechanics*, Wiley, New York, 1963, p. 163.
15. R. Courant and D. Hilbert, *Methods of Mathematical Physics*, Interscience, New York, 1953, vol. 1, p. 2.
16. P. M. Morse, *Thermal Physics*, Benjamin, New York, 1965, pp. 119–122.
17. J. Friedel, *Dislocations*, Addison-Wesley, Reading, 1964, pp. 136–139.
18. H. Statz, *Z. Naturforsch.* **17a** (1962), 906.
19. S. S. Major, Jr., Ph. D. Dissertation, Kansas State University, 1967, pp. 57–68.
20. R. O. Simmons and R. W. Balluffi, *Phys. Rev.* **119** (1960), 600.
21. C. Kittel, *Introduction to Solid State Physics*, 3rd edn., Wiley, New York, 1967, p. 180.
22. A. Seeger, *Phil. Mag.* **12** (1965), 1087.
23. M. H. Loretto, L. M. Clarebough, and R. L. Segall, *Phil. Mag.* **10** (1965), 731.
24. J. R. Heighbors and G. A. Alers, *Phys. Rev.* **111** (1958), 707.
25 A. Seeger, and G. Schock, *Acta. Met.* **1** (1952), 519.
26. G. B. Spence, *J. Appl. Phys.* **33** (1962), 729.
27. G. Benedek, and G. F. Nardelli, *Phys. Rev.* **127** (1968), 837.
28. H. B. Callen, *Thermodynamics*, Wiley, New York, 1960, p. 233.

INDEX OF NAMES

Acrivos 89
Afanaseev 81
Agrawal 38
Alami 16
Albers 15
Alexander 311, 315
Al-Hilli 15
Allpress 131
Ambindavato, Madagascar 244
Amelincx 37, 39, 44, 271
Anderson 129, 131, 132, 134, 167
Andersson 130
André 188
Antonova 10
Arnold 305, 310
Arrhens 199
Atterberg 187, 225

Baars 310, 312, 315, 317
Bailey 290
Balchin **1**, 16
Ballaffi 352
Bartam 16
Barz 35
Batisse 235
Baumhauer 269, 305
Beaudin 316
Beevers 150
Benard 25
Benda 53, 69
Berger 238
Bernusset 27
Bertaut 27, 177
Bethe 311
Bettman 152, 157, 158
Bhide 314
Blitz 24
Boehm 191
Bottelberghs 157
Bradley 215
Brafman 275, 317
Bragg 150, 192, 201, 311
Brandt 310, 312, 315
Brindley 208, 216, 218, 243, 256
Brixner 15
Bro 15
Bronger **93**
Brongniart 187
Brophy 305

Brown 201, 185, 290
Brunauer 228
Brunix 27
Buckley 308
Buerger 277
Buhler 352

Caillère **185**, 202, 210, 244, 246, 249
Calvet 216, 224
Camp Berteau 231
Caro 136
Chadha 306, 308, 311
Chan 84
Chapman 230
Chaussidon 185, 211, 215, 216, 239
Ché 208
Cheto 230
Chevenard 248
Chevreton 27, 160, 178, 180
Choyke 317
Chu 12
Clark 189
Clarke 72, 88
Cohen 57
Comes 60, 157
Coward 14
Cruz 257

Dana 189
Daniels 37, 316
De Jong 191
Delafosse 188
Delmaire 168
Demolon 235
Deschamp 157
Deuel 239
De Vries 150
DiSalvo 55, 69, 70
Donetskikh 14
Dragsdorf 356
Duksina 10
Dyal 230

Edwards 11
Ehrenfreund 12
Ehrlich 26
Emmet 228
Engelhardt 217
Epsig 305

Escard 229
Evans 6, 9, 14, 15, 270, 289

Farmer 212, 214, 216
Fehlner 81
Fender 129, 131
Flahaut 173
Fleet 179
Flick 164
Follet 239
Fong 14
Forty 306
Frank 36, 271, 272, 274, 300, *306*, 312, 313, 335
Franzen 26
Frindt 11
Fripiat 185, 208, 214, 229, 237, 238
Fry 191

Gatineau 222, 223
Gibbs 342
Giese 257
Glaeser 224, 233
Gleizes 28
Glick 152
Glinka 188
Gobrecht 310, 315
Golightly 316
Gouy 230
Gowers 14
Greenaway 1, 12
Green-Kelley 224
Grigorieff 200, 207
Grimm 185, 209, 236, 240
Gronvold 26
Grosskreutz 356
Günter 148

Hadding 190
Hägg 274, 305
Hahn 11, 25, 26, 28, 36
Haines 188
Hameda 14
Hanoka 36, 37, 313, 335
Harder 25, 36
Harkins 229
Haüy 254
Hayashi 305
Hazelwood 9, 14
Heine 84
Hendricks 191, 201, 230
Hénin **185**, 202, 227, 246, 249
Hey 217
Hissink 233, 240
Hofmann 224, 236, 263
Hooley 161
Houghton 187
Huang 12

Huber 165
Hughes 63, 90
Huismann 10, 160
Hulliger 7, 21, 160, 173
Hyde 130

Ida 14
Igaki 174, 176
Iglesias 170
Inomata 310

Jagodzinski 36, 274, 305, *308*, 310, 318, 335
Jain 309, 336
Janot 209
Jeannin 25, 27, 28, 165
Jellinek 23, 29, 164, 168, 170, 173
Jelly 185
Jenny 232
Jura 229

Kalikhmann 10
Kieffer 332
Kirkaloy 174
Klemen 224, 263
Knippenberg 305, 310, 317
Kohn 81, 273, 275, 305
Kornfeld 233
Kovacs 316
Krebs 148
Kremheller 305
Krishna 284, 306, 308, 311
Kröger 130
Kummer 150

Landau 312
Landolt(-Bornstein) 1
Langmyhr 26
Laudelout 231
Le Brusq 168
Le Cars 157
Le Châtelier 187, 244
Ledoux 214
Lee 13, 14
Lendvay 316
Les Baux, Provence 190
Liang 63, 90
Lifschitz 312
Lindh 208
Loly 81
Lomer 81
Lundquist 208, 305
Lyon 213

Mackenzie 185, 186
MacKinsky 208
Madelung 335
Magnan 186, 209

Maillard 241
Major **341**
Mantine 240
Mardix 36, 314, 315, 316, 328
Marezio 32
Marschall 201
Mattheiss 81, 84
Mauguin 189, 190, 192, 201, 202
Mazumdar 306, 336
McClellan 213
McEwan 216
McMillan 88
McMurchy 217
McTaggart 28
Méring 209, 222, 223, 239, 256
Mikami 174
Millot 185
Minervina 313
Mitchell 291, 306
Monckton 67, 80
Montet 11
Moore 152
Mooser 21
Morel 240
Morimoto 178, 179
Müller 310
Murray 9
Myers 11

Nakazawa 179
Ness 26
Newnham 218
Nitsche 1, 12, 15, 19
Norrby 26
Norrish 223
Nuttall 290

Oberlin 210
Oftedal 26
Ohashi 174
Orcel 244, 253, 256
Ott 274
Overhauser 81

Parthasarathi 335
Patil 13
Pauling 190, *191*, 198, 200, 238, 274
Pédro 254, 256, 257
Peibst 312, 313
Peierls 55
Peters 152, 157
Pimentel 213
Pinsker 291
Popma 174, 176
Popov 14
Powell 316
Prost 211, 224

Raam 27
Radoslovich 221, 223
Rai 311, 316
Ramsdell *273*, 275, 305
Rautureau 209
Reitmeier 232
Revelli 16
Rice 90
Rimmington 16, 20, 30, 31
Rinne 190
Roborg 189
Ross 275
Roth 150, 155, 157
Rothmund 233
Roy 212
Rüddorf 31, 34, 98
Russell 214, 215, 216

Sabatier 246, 257, 258
Said 13, 14
Sanders 131
Schäfer 16
Schloesing 187
Schneer 311, 312
Schofield 231, 234
Schonberg 11
Schottky 130
Scott 90
Scruby 57, 78
Serratosa 212, 215
Shaffer 332
Shirozu 217
Shockley 300
Siffert 213
Silbernagel 25
Simmons 352
Smeggil 16
Smeltzer 174
Smith 81, 218, 220, 255, 310
Snake Creek, Utah 204
Sobolev 14
Sokhor 332
Srinivasan 335
Srivastava 306, 330, 336
Statz 349
Steadman 290
Steinberg 208
Steinberger 36, 314, 315, 316, 317
Steinfink 170
Strock 305
Strunz 185
Stubican 212
Suntola 14

Tchoubar 256
Tehachapi 230
Teller 228

Tengner 23
Terner 158
Tewary 330
Théry 157
Thibault 305
Thiessen 239
Thompson 11, 38, 53, 54, 64
Tilley **127**, 131
Tracey 35
Tredgold 13
Trigunayat 38, **269**, 306, 308, 309, 311
Tronc 165
Tschermak 189
Tubbs 1, 306
Tuddenham 213

Van Bruggen 174, 176
Vand 36, 37, 313, 335
Van Gool 157
Van Laar 12
Van Landuyt 44
Vedder 215
Verbekt 15
Verma 33, **269**, 271, 284, 306, 308, 314
Vilanova 12

Wadsley 26, 28, 130, 134, 167
Wagner 130

Wallace 334
Way 187, 231
Wells 283
Weltner 311, 312, 335
Wexler 81
White 208, 214
Whitehouse 17, 36
Whiteney 332
Wickman 292
Wiegers 164, 168
Will 316
Williams **51**, 57, 61, 85, 311
Willis 131
Wilson 1, 7, 21, 33, 53, 61, 63, 69, 71, 73, 81,
    86, 90
Winchell 189
Wyckoff 1, 270, 274
Wyoming 230

Yoder 220, 255
Yoffe 1, 7, 21, 33, 53
Youell 243
Young 14, 174, 176

Zeissel 239
Zemlyanov 14
Zhdanov *274*, 313, 335
Zvyagin 275

INDEX OF SUBJECTS

acidity scale 238
adsorption 228
alkalimetals: A
    ALnS₂-compounds 95
allevardite 209, 258, 263
amesite 261
analysis of minerals 202
anauxite 247
antigorite 214, 218, 243, 244, 245, 252, 260
arcing (X-ray diagrams) 38, 282, *319*
attapulgite 229, 260

batavite 262
β-alumina *149*
beidellite 224, 236, 247, 260, 261, 263
berthierine 261
biotite 189, 215, 222, 240, 242, 252, 262, 264, 288
bonding
    covalent 198
    dichalcogenides 7
    ionic 198
    minerals 255
bonds 257
bowlingite 262
bramalite 261
bravaisite 247
bronzes 140, 141
brucite 212
Burgers vector *38*, *294*, 296

cadmium iodide structure *1*, 94, 99, 160, 273, *283*
charge density waves **51**, 55, *60*
chlorite *188–290*
chromocre 261
chrysotile 261
classification of minerals *253*
clay minerals *187*
coloration phenomena 241
cookeite 262
coquimbite 291
cronstedite 261, 289, 290
crystallisation 20

delanouite 252
density 251
desorption 230
diamourite 261

diaspore 190, 209
dichalcogenides **1**, *4*, 22, **51**, *159*
dickite 214, 218, 260, 291
differential thermal analysis *244*
dislocation
    density 319
    dichalcogenides *37*
    edge- 319
    Frank theory 300
    partial 296
    screw- 271, *306*
    Shockley 300
disorder theory *308*, 328
disorder (1-dimensional) 318, 308
distorted structure 77
dombassite 261

electrical properties (chalcogenides) 9, 53
electron energy theory 317
electron microscopy 37, 131, 209, 225, 300, 315, 344
electronic energy
    band structure 9, 82
    band gap 317
epitaxial theory 313
exchange capacity 225, *231*

Fermi surface *81*
ferrite 158, 291
fireclay 261
friedelite 290
fuschite 252, 261

Gapon's equation 231
gel 225, 259
gibbsite 190, 218
glauconite 213, 252, 261, 264
goethite 209
graphite *70*, 270, 292, 343, 350
greenalite 261
grovesite 261
growth
    dichalcogenides **1**, *15*
    polytypes 333

haematite 209, 253
halloysite 191, 209, 230, 236, 246, 248, 257, 261, 264
hectorite 209, 212, 260, 262

heterogenite 291
hexagonal cell 1, 276
homogeneity range 29, 130
hormite 260
hybridisation
    dichalcogenides 7
hydromica 263
hydroxyl 210, *213*

illite 229, 230, 236, 243, 245, 247, 261, 264
impurity 305
imogolite 259
infrared spectroscopy *210*
intercalation compounds 9, 12, *30*, *70*, **93**, *99*,
    107, 161
interstratification *257*, 264
isomorphic series 189

jefferisite 262

kamererite 252
kaolin 188
kaolinite *190–291*
Kohn anomaly 81

magnetic resonance 208
magnetic susceptibility
    dichalcogenides 52
    intercalation compounds 100, 106
manadoite 262
marcasite 22, 94, 176
margarite 222, 288
mica *188–288*
mineralisation 19
minnesotaite 262
molybdenum disulphide *4*
montmorillonite *191–264*
Mössbauer spectroscopy 209
muscovite 190, 207, 216, 221, 222, 240, 249,
    252, 264, 288

nacrite 214, 260, 291
nepouite 252
neutron diffraction 131
nickel arsenide 23, 94, 160
nontronite 210, 236, 247, 252, 261
noumelite 261

octahedral coordination 3
    dichalcogenides 22, *55*
    silicates 192
opal 291
optical properties 241, *252*
orthoantigorite 261

palygorskite 209, 236, 247, 259, 260
paragonite 216, 223

parasite 291
particle size (clay) *187*
Pauling's rules 192
Peierls-like distortion *60*
perovskite 137, 293
phase transformation theory 311
phase transition **51**, 53, 56
phlogopite 189, 204, 215, 217, 222, 262, 288
phyllosilicates **185**
plane (of atoms) 254
polymorph 5, 6, 35
polymorphism 254, *270*
polyphyllite *257*
polytype
    impurity 305
    pressure 332
    notations *273*
polytypism *35*, 165, 254, **269**, 324
pyrite 21, 22, 94, 176
pyrophillite 207, 216, 257, 258, 261
pyrosmalite 289
pyrrohite 177

quartz 246, 270

radius (atomic) *198*
Ramsdell notation 273
reciprocal lattice 278
rectorite 258, 263
rhenium oxide *137*
rheological properties *187*, 209
rhombohedral cell 276
roscoelite 252
rutile 21, 94

saphonite 236
saponite 212, 260, 262
sauconite 252, 262
schallerite 290
sepiolite 209, 229, 236, 247, 259, 260
sericite 261
serpentine 189, 223
sheet (of atoms) *254*, 259
Shockley partials 38, 300, 314, 315
silica-gel 225
size of particles 225
smectite 256, 259, 263, 264
smythite 180
sphalerite 293
stacking fault *37*, **269**, *294*, *318*, *341*
    expansion 314
stevensite 260, 262
stoichiometry 29, **127**
streaking (X-ray diagram) 38, 285, 308, 318
structure element (minerals) 201
sudoite 262
superconductivity 10, 12, 34

surface 225, *226*
synchisite 291
synchroshear 40, 298
syntactic coalescence 36, 281, 285, 288, 291, 292, 313, 334

taffeite 291
talc 189, 209, 255, 257, 258, 262
ternary chalcogenides **93**
tetrahedral coordination 192
thermal behaviour
  silicates 241
  stacking faults 326, *341*
thermal expansion *248*, 351
thermogravimetric analysis *241*
Thompson notation 38
transition metal dichalcogenides **1, 51**, *159*
trigonal prismatic coordination 4, 160
troilite 177

turbostratic sheet 255

vapour transport technique 15–21
vermiculite 206, 230, 236, 250, 260–264
vitreous fusion 245
volchonskoite 252, 261

wurtzite 94, 293

xilotile 259, 260
X-ray 259
  diffraction diagrams 221, 270, *276*, 313
  fluorescence 208
  topography 37, 272, 315

Zhdanov notation 274
zincblende 94, 293
Zvyagin notation 275

INDEX OF FORMULAS

Ag 342, 350
AgAl$_{11}$O$_{17}$ 155
AgI 294, 332
Ag$_6$Mo$_{10}$O$_{33}$ 140
Ag$_2$S 121
Ag$_x$V$_2$O$_5$ 143
Al 355
AlCl$_3$ 247
AlF$_3$ 145
Al(NO$_3$)$_3$ 247
Al$_2$O$_3$ *149*
Au 355

BN 332
BaCu$_4$S$_3$ 121
BaCrO$_3$ 293
BaFeO$_3$ 293
Ba$_2$FeSbO$_6$ 293
BaIrO$_3$ 293
Ba$_2$Me^{II}Fe$_{12}$O$_{22}$ 158
BaMnS$_2$ 122
Ba(OH)$_2$ 233
BaRuO$_3$ 293
BaTaS$_3$ 293
BaTiO$_3$ 293
BiNbO$_4$ 147
BiTe 167

C$_{6000}$Br 70
C$_2$H$_6$ 312
CH$_3$COONH$_4$ 240
CH$_3$(CH$_2$)$_3$COCOONa 294
CaAl$_2$(OH)$_2$(Si$_2$Al$_2$)O$_{10}$ 288
CaCl$_2$ 231
Ca(OH)$_2$ 231
Ca$_2$Nb$_2$O$_7$ 147
CdBr$_2$ 291, 319, 324
CdCl$_2$ 1, 96
CdI$_2$ 1, 36, *270–283*, 294, *305–326*
CdS 293
CdSe 293
CdTe 293
Co 343, 349, 350
CoTe 23
CoTe$_2$ 2, 23
CoMX$_2$ 102
CrS *173*
Cr$_7$S$_8$ 170, *173*
Cr$_5$S$_6$ 170, *173*

Cr$_3$S$_4$ 170
Cr$_2$S$_3$ 170
Cr$_5$S$_8$ 170
Cr$_x$Al$_{2-x}$O$_3$ 129
Cr$_x$MX$_2$ 102
CsAg$_3$S$_2$ 120
CsCdCl$_3$ 293
Cs$_2$Co$_3$S$_4$ 112
CsCrS$_2$ 106
CsFeS$_2$ 111
CsFeSe$_2$ 111
Cs$_2$Mn$_3$S$_4$ 108, 122
Cs$_{0.25}$MoO$_3$ 140
Cs$_{0.6}$MoS$_2$ 107
Cs$_2$Ni$_3$S$_4$ 113
Cs$_2$Pd$_3$S$_4$ 113
Cs$_2$Pd$_3$Se$_4$ 115
Cs$_2$Pt$_3$S$_4$ 116
Cs$_2$Pt$_4$S$_6$ 118
CsTiS$_2$ 172
Cs$_{0.6}$TiS$_2$ 98
Cs$_x$TiS$_2$ 172
Cu 355
CuS 94
CuTe 94, 109
Cu$_x$V$_2$O$_5$ 144

EDTA 240

FeF$_3$ 145
Fe$_{1/3}$NbS$_2$ 12
Fe$_2$O$_3$ 209, 242
Fe$_3$O$_4$ 209
Fe$_{1-x}$O 133, 135
Fe(OH)$_2$ 242
FeS *177*
Fe$_2$S$_2$ *176*
Fe$_7$S$_8$ 170, 174, *178*
Fe$_3$S$_4$ 180
FeS$_2$ *176*
FeS$_x$ 153
FeTi$_2$S$_4$ 172
Fe$_x$TiS$_2$ 171
Fe$_x$MX$_2$ 102

Ga-Al$_2$O$_3$ 155

HCl 247
H$_3$Fe(CN)$_6$ 294

H_4MoO_6 148
H_2MoO_4 148
$H_2Mo_2O_6$ 149
$H_2MoO_4 \cdot H_2O$ 148
$H_{2n-2}Mo_nO_{4n-2}$ 149
H_2SO_4 247
$H_2WO_4 \cdot H_2O$ 148
HfS_2 2, 12, 15, 17, 20, 37, 46, 51
$HfSe_2$ 2, 12, 17
HfS_xSe_{2-x} 16
$HfTe_{2-x}$ 16
HgI_2 112

In_2S_3 37, 46
$IrSe_2$ 22
$IrTe_2$ 2

KAg_4S_3 120
$KAl_2(OH)_2(Si_3Al)O_{10}$ 288
$KCo(CN)_6$ 294
$KCrS_2$ 105
KCu_3S_2 119, 122
KCu_4S_3 119
$KFeS_2$ 111
$KFeSe_2$ 111
$KMg_3(OH)_2(Si_3Al)O_{10}$ 288
$KMnAs$ 122
$KMnP$ 122
$K_2Mn_3S_4$ 108
$K_{0.6}MoS_2$ 107
$K_{0.5}MoS_2$ 107
$K_{0.4}MoS_2$ 34
K_2NiF_4 147
$K_2Ni_3S_4$ 113
$K_2Ni_3Se_4$ 113
$K_2Pb_3IrS_6$ 118
$K_2Pd_3S_4$ 113
$K_2Pb_3IrS_6$ 118
$K_2Pd_3Se_4$ 113
$K_2Pd_3SnS_6$ 118
$K_2[Pt(CN)_4]Br_{0.30} \cdot 3H_2O$ 84
K_2PtS_4 116
$K_2Pt_3SnS_6$ 118
$K_2Pt_4S_6$ 117
$K_2Pt_4Se_6$ 118
$K_{0.8}TiS_2$ 98
$KTiS_2$ 172
K_xTiS_2 172

$LiCl$ 231
$LiOH$ 233
$LiCrS_2$ 99, 104
$Li_{0.8}(NH_3)_{0.8}MoS_2$ 107
$Li_{0.6}TiS_2$ 98
$LiTiS_2$ 99, 104, 172
Li_xTiS_2 172
$Li_xTi_{1.1}S_2$ 35

LiV_2O_5 143
$Li_xV_2O_5$ 143
$LiVS_2$ 99, 101, 104

M_xMoO_3 (M=alkalimetal)   140
$M_xV_2O_5$ (M = Ag, Cu, 3d-transition metal)   141
MTi_2S_4 (M=3d-transition metal)   *169*
$MgAl_2O_4$ 150
$MgCl_2$ 240
MnO 242
MnO_4 242
MnS 109
$MnMX_2$ 102
MoF_3 138, 145
$MoO_{2.9975}$ 139
MoO_3 137, *138*, 149
Mo_nO_{3n-2} 140
$MoO_{3-x}F_x$ 138
$MoO_{3-x}(OH)_x$ 138
MoS_2 2, *4*, *6–13*, *31–52*, 107, 292, 299, 305
$MoSe_2$ 2, 9, 13, 41, 107
$MoTe_2$ 2, 9, 13, 44

$(NH_4)_2CO_3$ 247
NH_4OH 233, 247
$Na—Al—O$ *149*
$NaAl_{11}O_{17}$ 136, *149*
$NaCl$ 96, 231
$NaCrS_2$ 102, 105
$NaCrSe_2$ 106
$NaFeO_2$ 96
$Na_2MgAl_{10}O_{17}$ *157*
$Na_{1.67}Mg_{0.67}Al_{10.33}O_{17}$ 157
$Na_{1.36}Mg_{1.88}Al_{14.96}O_{25}$ 158
$Na_{1.67}Mg_{2.67}Al_{14.33}O_{25}$ 158
$Na_{0.6}MoS_2$ 107
Na_xMoS_2 35
$NaNbO_2F_2$ 147
$NaOH$ 233
$NaTiS_2$ 102, 105, 173
$Na_{0.8}TiS_2$ 98
Na_xTiS_2 172
$Na_{0.9}TiSe_2$ 98
NaV_2O_5 142
$Na_xV_2O_5$ 142
NaV_2O_4F 142
$NaV_2O_{5-x}F_x$ 143
$Na_xV_2O_{5-x}F_x$ 142
$NaVS_2$ 102
$NaVSe_2$ 102
$NaMX_2$ *102*
NbC_{1-x} 133
NbF_4 *145*
Nb_2O_5 136
NbS_2 2, 4, 13, 31, 33, 68
$NbS_2(pyridine)_{1/2}$ 31
NbS_3 9

NbSe$_2$ 2, *9–15*, 32, 35, 52, *67*, 80, 292
NbSe$_2$(pyridine)$_{1/2}$ 32
NbTe$_2$ 2, 9, 15, 44, 52
NiAs 23, 94
NiBr$_2$ 294
NiTe$_2$ 2
Ni$_x$TiS$_2$ 172
Ni$_x$MX$_2$ 102

PbF$_4$ 145
PbFe$_{12}$O$_{19}$ 152, 158
PbI$_2$ 36, *270–286*, 305–331
PbO 94
PdS$_2$ 22
PdSe$_2$ 22
PdTe$_2$ 2, 7, 22
PrO$_{2-x}$ 133
Pt 355
PtS$_2$ 2, 8, 22
PtSe$_2$ 2, 22
PtTe$_2$ 2

Rb$_2$Ag$_4$S$_3$ 121
RbAg$_3$S$_2$ 120
RbCuS$_3$ 119
Rb$_2$Co$_3$S$_4$ 112
RbCrS$_2$ 106
RbFeS$_2$ 111
RbFeSe$_2$ 111
RbMn$_3$S$_4$ 108
Rb$_{0.3}$MoS$_2$ 34
RbNiF$_3$ 293
Rb$_2$Ni$_3$S$_4$ 113
Rb$_2$Pd$_3$S$_4$ 113
Rb$_2$Pd$_3$Se$_4$ 115
Rb$_2$PtS$_2$ 116
Rb$_2$Pt$_3$S$_4$ 116
Rb$_2$Pt$_4$S$_6$ 118
Rb$_2$Pt$_4$Se$_6$ 118
Rb$_2$Pt$_3$SnS$_6$ 118
RbTiS$_2$ 172
Rb$_x$TiS$_2$ 172
Rb$_{0.5}$WS$_2$ 20
ReO$_3$ 33, *137*
ReS$_2$ 22, 110
ReSe$_2$ 9, 22, 44, 110
RhSe$_2$ 22
RhTe$_2$ 2, 22
Rh$_x$MX$_2$ 102

(Si$_{2.8}$Al$_{1.2}$)O$_{10}$(Mg$_{2.8}$Fe$^{3+}_{0.2}$)(OH)$_2$ 241
(Si$_{8-x}$Al$_x$)O$_{20}$Al$_4$(OH)$_4$K$_x$ 207
SiC 269–331
2SiO$_2$, Al$_2$O$_3$, 2H$_2$O 188
Si$_2$O$_5$Al$_2$(OH)$_4$ 207
Si$_4$O$_{10}$Al$_2$(OH)$_2$ 207
SiTe$_2$ 2

SmF$_3$ 145
SmZrF$_7$ 145
SnF$_4$ 137, *144*
SnS$_2$ 2–40, 294
SnSSe 2
SnS$_x$Se$_{2-x}$ 14, 15
SnSe$_2$ 2, 9, *14*, 17, 37
Sn$_x$Zr$_{1-x}$S$_2$ 16
SrMnO$_3$ 293
Sr$_2$TiO$_4$ 147
Sr$_3$Ti$_2$O$_7$ 147
Sr$_4$Ti$_3$O$_{10}$ 147
SrTiO$_3$ 147
Sr$_n$Ti$_n$O$_{3n-1}$ 147
Sr$_{n+1}$Ti$_n$O$_{3n+1}$ 147

TaF$_3$ 145
TaS$_2$ 2–13, *35–84*
TaS$_2$(C$_5$H$_5$N)$_{1/2}$ 31, 89
TaS$_2$(C$_5$H$_5$N)$_{1/4}$ 31
TaS$_2$(EDA)$_{1/4}$ 73, 89
TaS$_x$Se$_{2-x}$ 16
TaSe$_2$ *2–80*, 292
TaTe$_2$ 9, 44
Ta–TiS$_2$ 69
TcS$_2$ 22
TiO$_{0.73}$ 129
TiO$_{1.23}$ 129
Ti$_6$S 15
TiS 15, 24, *161*, 168
Ti$_{1-x}$S 15, 293
Ti$_8$S$_9$ 168
Ti$_4$S$_5$ 168
Ti$_3$S$_4$ 15, 168
Ti$_2$S$_3$ 15, 24, 164
Ti$_5$S$_8$ 164
Ti$_3$S$_5$ 164
TiS$_{1.81}$–TiS$_{1.92}$ 129
TiS$_2$ *2–64*, *159–164*
TiS$_3$ *15*
Ti–S 15, 24, 162
TiS$_2$+alkalimetal *100*
Ti$_{2x}$Se$_{2-x}$ 16
TiS$_{2x}$Te$_{2-x}$ 16
TiSe 26
TiSe$_2$ 2, 12, 15, 17, 26
Ti–Se *64*, 90
TiSe$_x$Te$_{2-x}$ 16
TiTe$_2$ 2, 12, 15, 17, *27*
TiMX$_2$ 102
TlFeS$_2$ 111
TlFeSe$_2$ 111
Tl$_2$Ni$_3$PtS$_6$ 118
Tl$_2$Pd$_3$PtS$_6$ 118
Tl$_2$Pd$_3$SnS$_6$ 118
Tl$_2$ 118
Tl$_2$Pt$_3$SnS$_6$ 118

$Tl_2Pt_3TaS_6$ 118
$Tl_2Pt_3ZrS_6$ 118
TlS_2 2, 119
Tl_xMX_2 102

UO_{2+x} 133, 135

VSe_2 2, 33, 52, *63*, 85
VMX_2 102
V_2O_5 137, *141*
V_3O_7 141
V_4O_9 141
V_6O_{13} 141
VO_2 141
$V_2O_{4.975}F_{0.025}$ 141
$V_2O_{4.987}F_{0.013}$ 141

WO_3 145
WO_{3-x} 133

WS_2 2, 13, 31, 107
WSe_2 2, 9, 13, 15, 107
$W_{1-x}Ta_xSe_2$ 15
WTe_2 15, 44

ZnS 270–332
$ZnSe$ 293
$ZnTe$ 293
ZrF_4 145
ZrS 169
Zr_2S_3 169
ZrS_2 2, 28, 37, 46, 169
Zr—S 28
ZrS_xSe_{2-x} 16
$ZrSe_2$ 2, 8, 13, 15, 17, *28*
Zr—Se *28*
$ZrTe_2$ 2, 15, *28*
Zr—Te *28*
$ZrS_2(+$alkalimetal$)$ *101*

ABBREVIATIONS

A	Alkalimetal	MO	Molecular orbital
CDW	Charge density wave	MX_2	Transition metal dichalcogenide
CS	Crystallographic shear	Oc, octa	Octahedral
em	Electron microscope	PLD	Peierls-like distortion
fcc	Face centered cubic	T	Transition metal
FS	Fermi surface	TMD	Transition metal dichalcogenide
hcp	Hexagonal close-packed	Te	Tetrahedral
Ln	Lanthanoid	Trig pr	Trigonal prismatic
M	Metal	X	Chalcogen: S, Se, Te